SOLUTIONS MANUAL FOR

SOLID MECHANICS IN ENGINEERING

by RAYMOND PARNES

2001 VERSION 1.1

Note :

1. The solutions to a number of problems – "computer-related problems" - require the use of a computer. Since these problems (which appear in the various chapters) may be solved using different algorithms, programs, software, etc., solutions to such problems do not appear in this manual.

2. The solutions herein are believed to be correct. The author will be be grateful to be informed of any errors.

1.1 $\sigma_{AB} = \dfrac{P-30}{A_{AB}}$; $\sigma_{BC} = P/A_{BC}$; $A_{BC}/A_{AB} = 1/4$

a)

$\sigma_{AB} = \sigma_{BC} \Rightarrow (P-30) = P\,(A_{AB}/A_{BC}) = 4P \Rightarrow P = -30/3 = -10\text{kN}$ (comp.)

b) $\sigma_{AB} = -\sigma_{BC} \Rightarrow (P-30) = -P\,(A_{AB}/A_{BC}) = -4P \Rightarrow P = 6\text{kN}$ (tens.)

1.2. a) $\bar{\tau} = P/\pi D t \Rightarrow P = \pi D t \bar{\tau} = \pi(20\times10^{-3})(4\times10^{-3})(90\times10^{6}) = 7200\pi = 22619\ \text{N}$

b) $\bar{\sigma} = P/\pi D^2/4 = \dfrac{(22619)(4)}{\pi(20\times10^{-3})^2} = 72\ \text{MPa}$.

1.3 $P = \sigma A = (120\times10^{6})(2400\times10^{-6}) = 288{,}000\ \text{N} = 288\ \text{kN}$

$\varepsilon = P/AE \Rightarrow P = AE\varepsilon = (2400\times10^{-6})(100\times10^{9})(5\times10^{-4}) = 120{,}000\ \text{N} = 120\ \text{kN}$

1.4 $\Delta L = PL/AE$; $\sigma = P/A \Rightarrow A = P/\sigma = \dfrac{40{,}000}{20\times10^{6}} = 0.002\ \text{m}^2 = 20\ \text{cm}^2$

$D = 2\sqrt{A/\pi} = 5.04\ \text{cm}$

$L = \dfrac{\Delta L}{\varepsilon} = \Delta L\cdot E/\sigma = \dfrac{(2\times10^{-3})(70\times10^{9})}{20\times10^{6}} = 7\ \text{m}.$

1.5

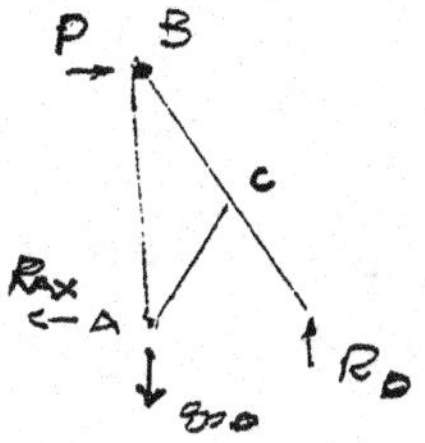

$R_D = (600)(320)/240 = 80\text{kN}$; $R_{Ay} = 80\ \text{kN}$; $R_{Ax} = 60\ \text{kN}\leftarrow$.

$\Sigma F_{Ax} = 0 \Rightarrow (120/200)\,F_{AC} = 600 \Rightarrow F_{AC} = 100\text{kN}$ (tens.)

$\Sigma F_{Ay} = 0 \Rightarrow F_{AB} + 4/5\,(100) - 80 = 0 \Rightarrow F_{AB} = 0$

$A = \pi D^2/4 = 7.068\ \text{cm}^2 = 7.068\times10^{-4}\ \text{m}^2$

$\sigma_{AB} = 0$, $\sigma_{AC} = 100\times10^{3}/(7.068\times10^{-4}) = 141.5\ \text{MPa}$

1.6

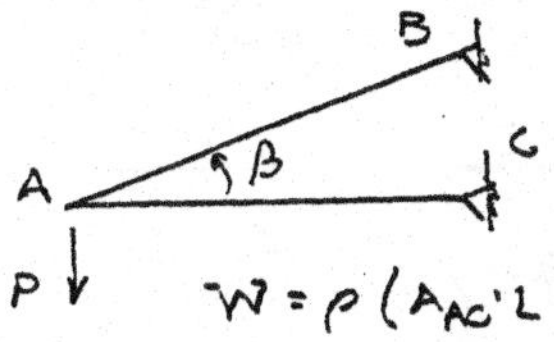

$F_{AB} = P/\sin\beta$, $F_{AC} = P\cos\beta/\sin\beta$.

$A_{AB} = F_{AB}/\sigma$; $A_{AC} = F_{AC}/\sigma$

$W = \rho\,(A_{AC}L + A_{AB}\,L/\cos\beta) = \rho PL/\sigma\left[\dfrac{\cos\beta}{\sin\beta} + \dfrac{1}{\sin\beta\cos\beta}\right] = \rho PL/\sigma\left(\dfrac{1}{\tan\beta} + \dfrac{2}{\sin 2\beta}\right)$

$dW/d\beta = -\dfrac{1}{\sin^2\beta\cos^2\beta}\,(2\cos^2\beta - \sin^2\beta)$; $dW/d\beta = 0 \Rightarrow \sin^2\beta = 2/3 \Rightarrow \beta = 54.74°$

1.7 $P = \tau(\pi D)(a) = (4\times10^{6})(0.3\pi)(.2) = 754\ \text{kN}$

1.8 (a). $\varepsilon(x) = \alpha\delta T = \alpha x\cdot\Delta T_0/L = (2.34\times10^{-4})\,x$; (b) $u(x) = \int_0^x \varepsilon(\xi)\,d\xi = \dfrac{\alpha\Delta T_0}{2L}x^2 = 11.7\times10^{-5}\cdot x^2$

c). $\Delta L = u(2) = 4.68\times10^{-4}\ \text{m} = 0.468\ \text{mm}.$

d) $\bar{\varepsilon} = \Delta L/L = 2.34\times10^{-4}.$

Chapter 2 - Problems

2.1

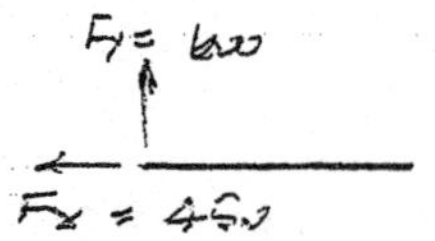

a) $F = 450\,N$, $V = 600\,N$, $M = 1200\,N\text{-}m$.

2.2

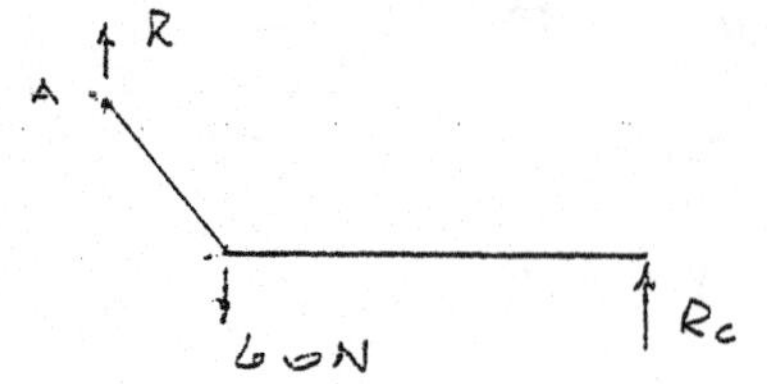

$\sum M_A = 0 \Rightarrow 120 R_c - 40(60) = 0 \Rightarrow R_c = 20\,N$

$\sum F_y = 0 \Rightarrow R_A = 40\,N.$

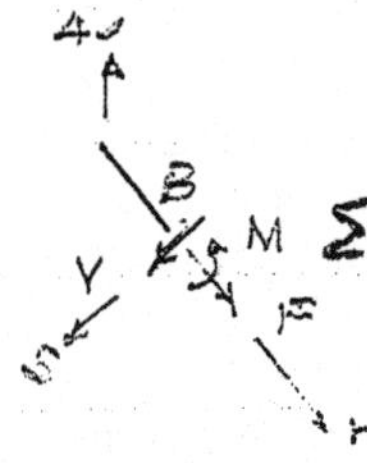

At B: $\sum F_n = 0 \Rightarrow F = 40(\sqrt{2}/2) = 20\sqrt{2}$, $\sum F_s = 0 \Rightarrow V = 20\sqrt{2}$.

$M = 40(10\sqrt{2}) = 400\sqrt{2}\ N\text{-}cm.$

At D: $F = 0$, $V = 20\,N$, $M = 600\ N\text{-}cm$

2.3

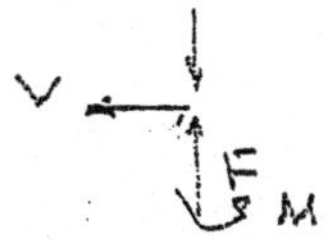

a) **At A.**

$F = 600\,N$, $V = 0$, $M = 7(600) = 4200\ N\text{-}m.$

At C: $V = 600\,N$, $F = 0$, $M = 1200\ N\text{-}m.$

b)

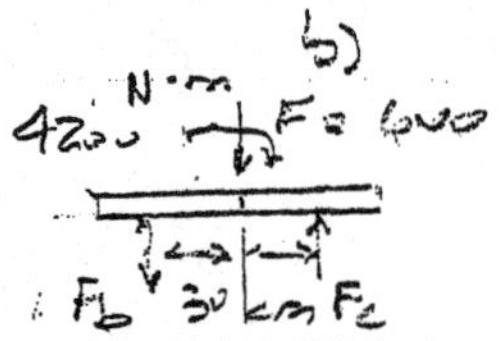

$$\left.\begin{aligned} -F_b + F_c &= 600 \\ .30 F_c + .30 F_b &= 4200 \end{aligned}\right\} \Rightarrow \begin{aligned} F_b &= 6300\,N \ \text{(tens.)} \\ F_c &= 7300\,N \ \text{(comp.)} \end{aligned}$$

2.4 $V = \frac{30}{2}(450) + \frac{30}{2}(150) = 15(600) = 9000\,N$

$M_A = (6750)(3) + (2250)(1) = 22,500\ N\text{-}m.$

2.5 $V = q_0 \int_0^L \sin(\pi x/2L)\,dx = -2q_0L/\pi \cos\frac{\pi x}{2L}\Big|_0^L = 2q_0L/\pi$

$M = q_0 \int_0^L (L-x)\sin(\pi x/2L)\,dx = 2q_0L^2/\pi - q_0\int_0^L x\sin(\pi x/2L)\,dx$

$= 2q_0L^2/\pi - q_0\, 4L^2/\pi^2 = \frac{2q_0L^2}{\pi}(1 - 2/\pi)$

2.6

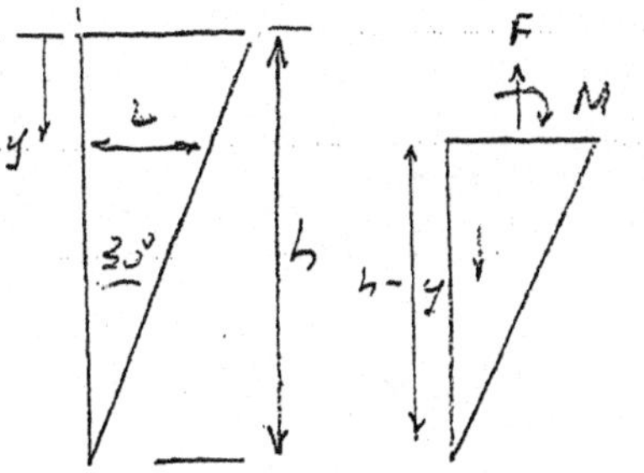

$$b = (h-y)\tan 30^\circ$$

$$F = \rho \frac{tb}{2}(h-y) = \rho \frac{t}{2}(h-y)^2 \tan 30^\circ = \frac{\sqrt{3}}{6}\rho t (h-y)^2$$

$$M = F\cdot(b/2 - b/3) = \frac{Fb}{6} = \rho \frac{t}{12}(h-y)^3 \tan^2 30^\circ = \frac{\rho t (h-y)^3}{36}$$

2.7 $r = y\,8/24 = y/3$, $R = 8$, $h = 24$, $r = yR/h$

$$Vol = \frac{\pi}{3}[R^2 h - r^2 y] = \frac{\pi}{3}[R^2 h - y^3 R^2/h^2]$$

$$= (\pi/3)R^2/h^2\,[h^3 - y^3] \Rightarrow F = \frac{\pi\rho R^2}{h^2}(h^3 - y^3), \quad F(y=0) = \pi R^2 \rho h$$

2.8 b) $F(x) = Ac\int_0^x (1+\alpha\xi^2/L^2)^{-1/2}\,d\xi =$

$$= \frac{AcL}{\alpha^{1/2}}\int_0^{\alpha^{1/2}x/L}(1+y^2)^{-1/2}\,dy \quad [y^2 = \alpha\xi^2/L^2;\ dy = (\alpha^{1/2}/L)\,d\xi]$$

$$= \frac{AcL}{\alpha^{1/2}}\{\ln[\alpha^{1/2}x/L + (1+\alpha x^2/L^2)^{1/2}]\}$$

c) $\alpha \ll 1 \Rightarrow F(x) \cong \frac{AcL}{\alpha^{1/2}}\{\ln[\alpha^{1/2}x/L + 1 + \alpha x^2/2L]\} \cong \frac{AcL}{\alpha^{1/2}}[\alpha^{1/2}x/L + \alpha x^2/2L] \to Acx$

2.9 (a) $V(x) = -40x$, $M = -20x^2$

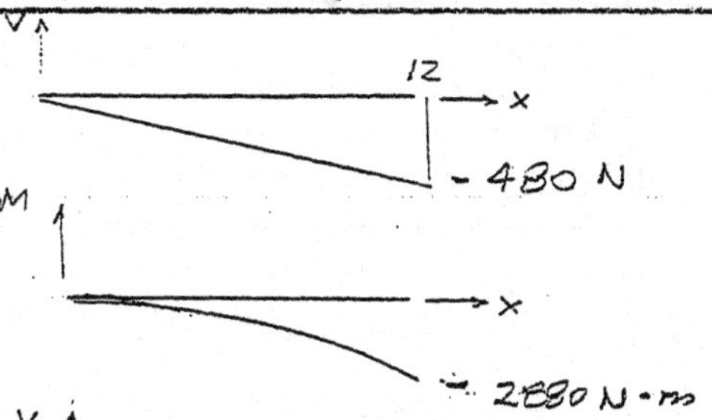

(b) $V = -40x$, $M = -20x^2$; $0 < x < 6$

$V = -240$, $M = -240(x-3)$, $6 < x < 12$

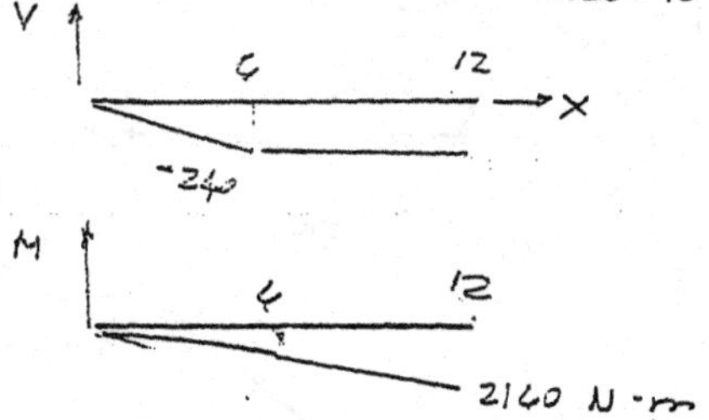

(c). $V = -5x^2/3$, $M = -5x^3/9$

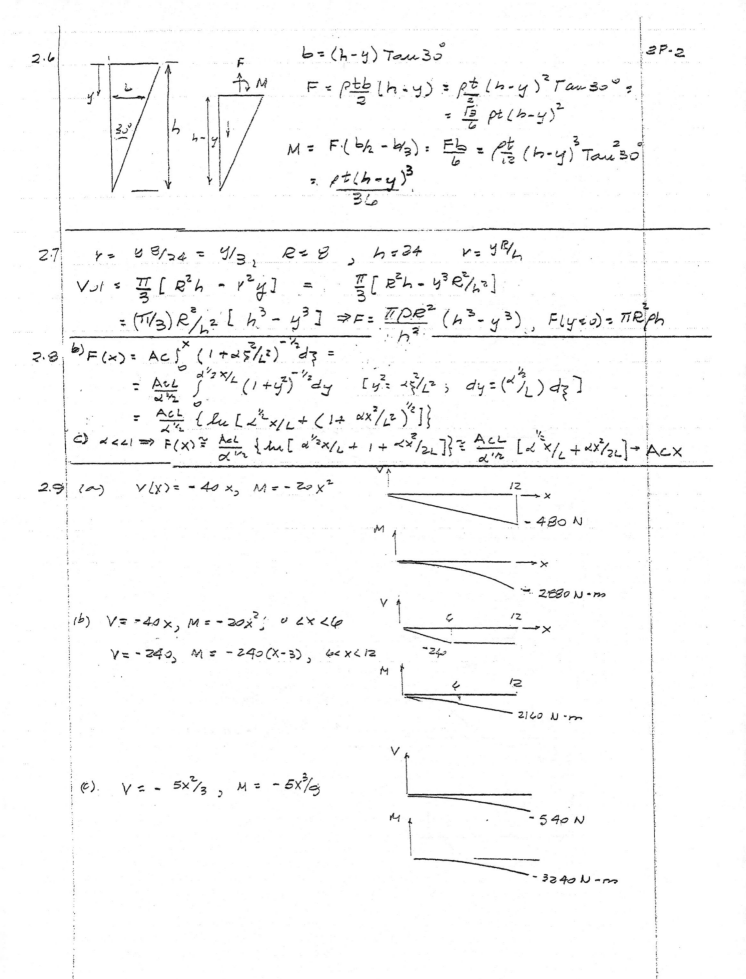

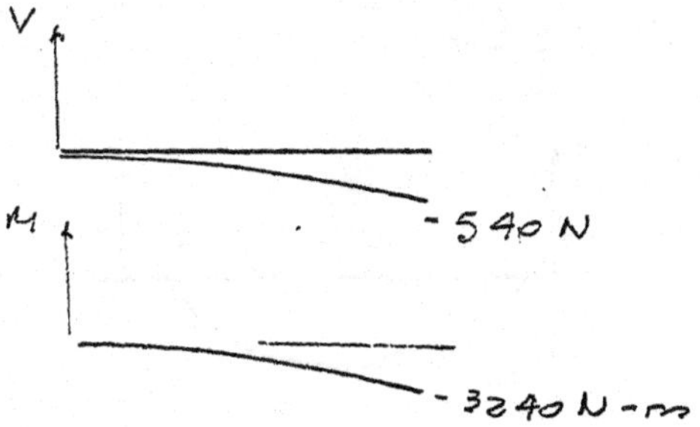

(d) $V = 1200 - 300x$

$M = 1200x - 150x^2$

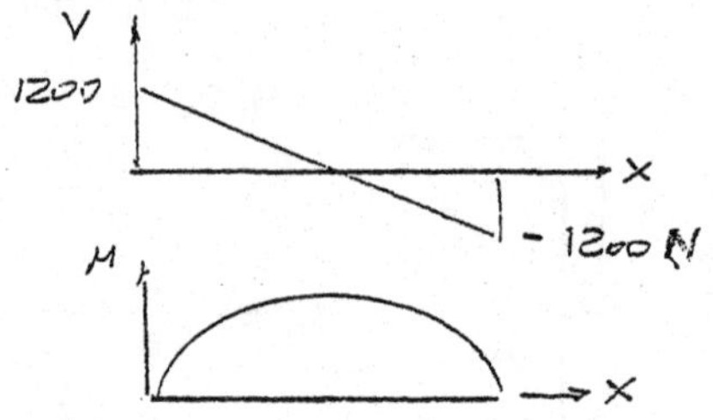

(e)

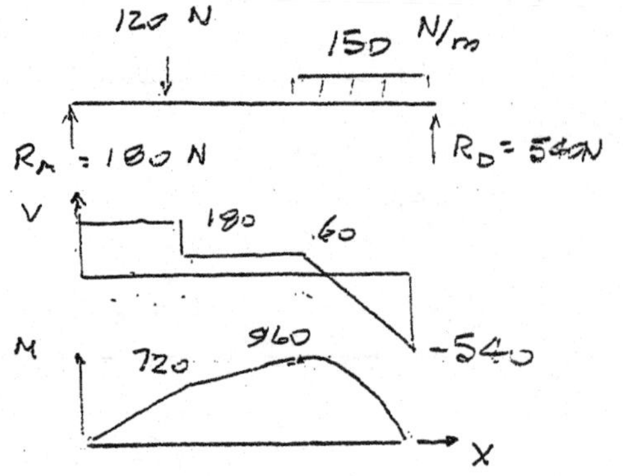

$V = 180,\ M = 180x,\quad 0 < x < 4$

$V = 60,\quad M = 60x + 480\quad 4 \le x \le 8$

$V = -150x + 1260,\ M = -75x^2 + 1260x - 4320$

$8 \le x \le 12$

(f)

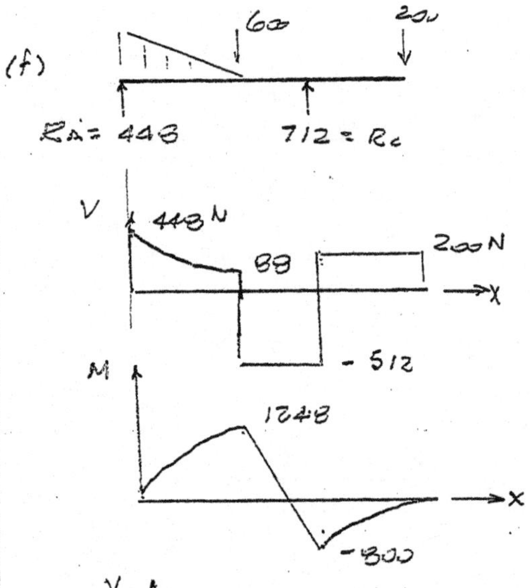

$$\left.\begin{aligned} V &= 10x^2 - 120x + 448 \\ M &= 10x^3/3 - 60x^2 + 448x \end{aligned}\right\}\quad 0 \le x \le 6$$

$V = -512,\ M = -512x + 4320,\quad 6 < x < 10$

$V = 200,\ M = -200(14 - x),\quad 10 < x < 14$

(g)

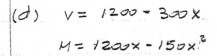

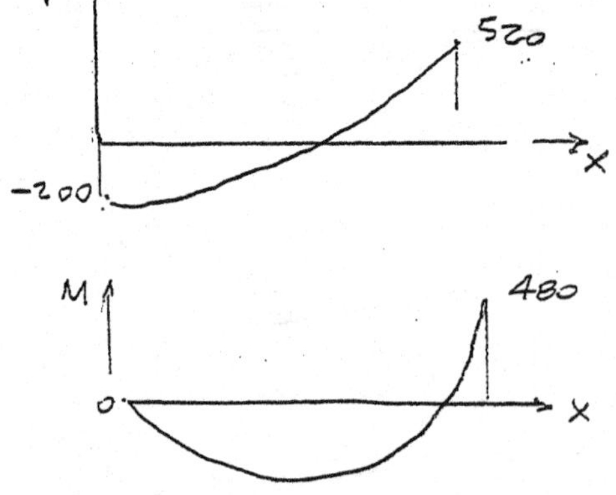

$V = 5x^2 - 200$

$M = 5x^3/3 - 200x = 5x(x^2/3 - 40)$

(h)

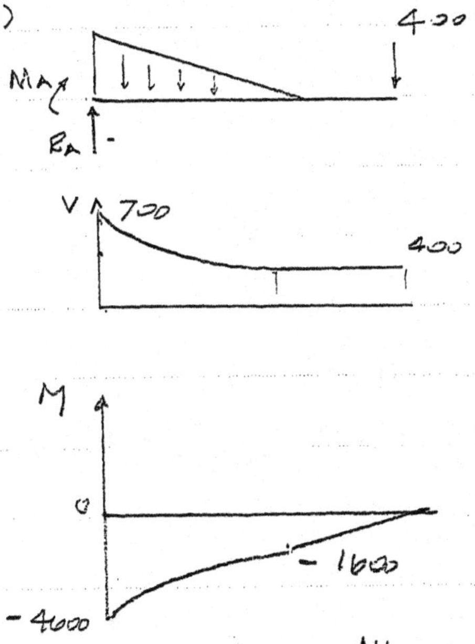

$M_A = -4000 - 600 = -4600$ N-m

$R_A = 700$ N

$V = 25x^2/3 - 100x + 700$

$M = 25x^3/9 - 50x^2 + 700x - 4600$ } $0 \le x \le 6$

$V = 400$, $M = +400(x-10)$, $6 < x \le 10$

(i)

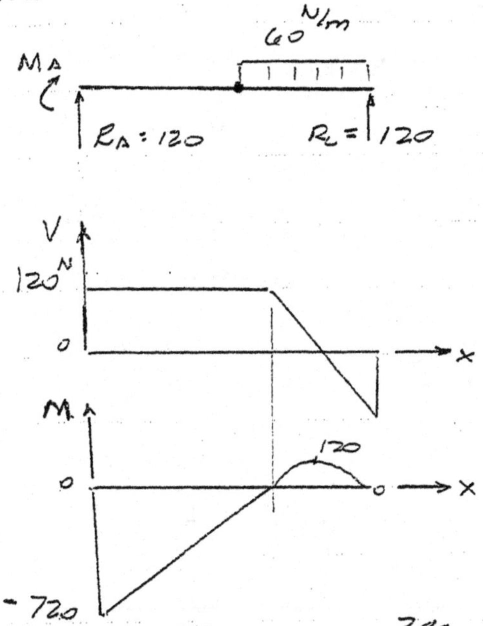

$M_A = -720$ N-m

$V = 120$, $M = 120x - 720$, $0 < x < 6$

$V = -60x + 480$, $M = -30x^2 + 480x - 1800$

(j)

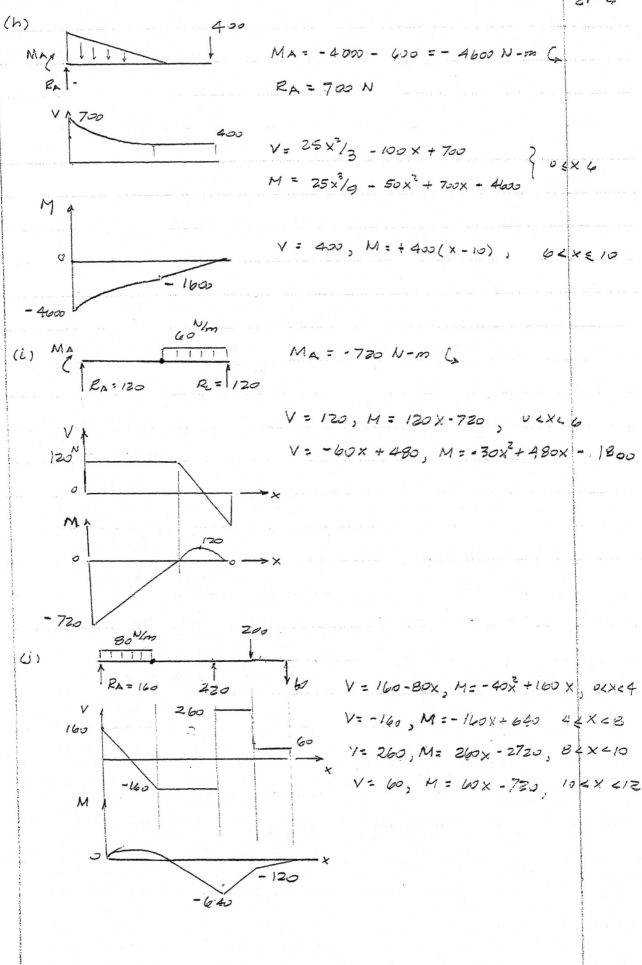

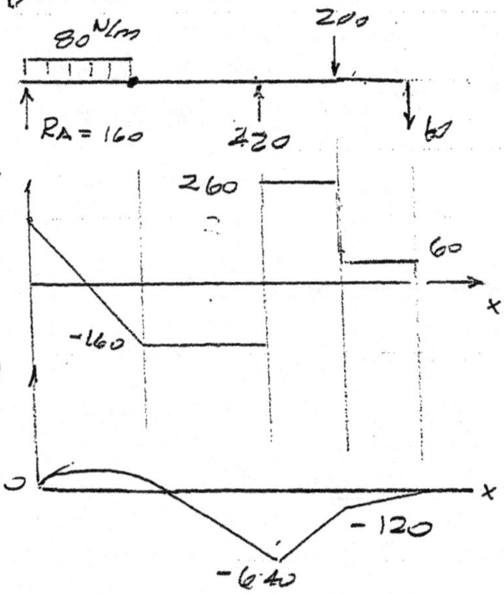

$V = 160 - 80x$, $M = -40x^2 + 160x$, $0 < x < 4$

$V = -160$, $M = -160x + 640$, $4 \le x < 8$

$V = 260$, $M = 260x - 2720$, $8 \le x < 10$

$V = 60$, $M = 60x - 720$, $10 \le x < 12$

2.10

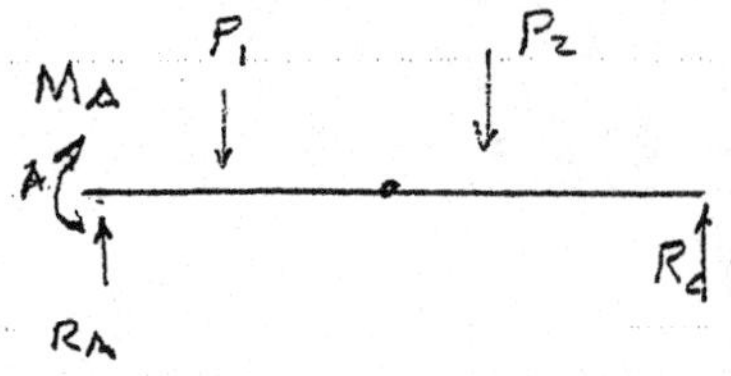

$$R_D = \frac{P_2(a - L/2)}{L/2} = 2P_2(a_2/L - 1/2)$$

$$R_A = P_1 + P_2 - 2P_2(a/L - 1/2)$$

$$= P_1 + 2P_2(1 - a_2/L)$$

$$M_A = -R_A L/2 + P_1(L/2 - a_1)$$

$$= -[P_1 + 2P_2(1 - a_2/L)]L/2 + P_1(L/2 - a_1)$$

$$= -P_1 a_1 - P_2(L - a_2)$$

$$\left.\begin{aligned} V &= P_1 + 2P_2(1 - a_2/L), \\ M &= -P_1 a_1 - P_2(L - a_2) + [P_1 + 2P_2(1 - a_2/L)]x \end{aligned}\right\} \quad 0 < x < a_1$$

$$= P_1(x - a_1) + P_2[2(1 - a_2/L)x - (L - a_2)]$$

$$V = 2P_2(1 - a_2/L)$$

$$M = -P_1 a_1 - P_2(L - a_2) + [P_1 + 2P_2(1 - a_2/L)]x - P_1(x - a_1) \quad a_1 < x < a_2$$

$$= -P_1 a_1 - P_2(L - a_2) + P_1 a_1 + 2P_2(1 - a_2/L)x$$

$$= P_2[-L + a_2 + 2x - 2a_2 x/L]$$

$$\left.\begin{aligned} V &= -2P_2(a_2/L - 1/2) \\ M &= 2P_2(a_2/L - 1/2)(L - x) \end{aligned}\right\} \quad a_2 < x < L$$

2.11

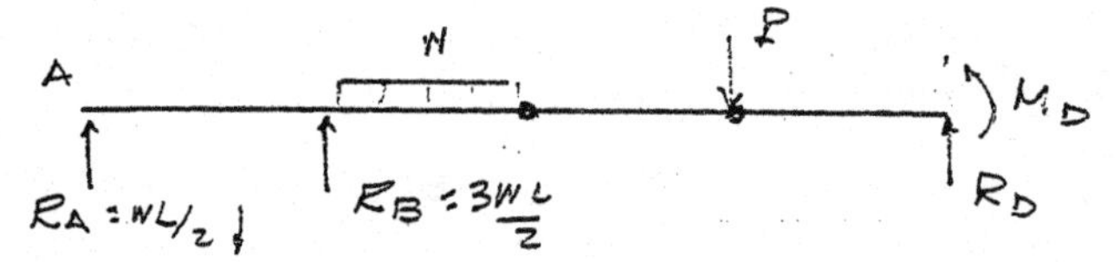

$M_D = PL$

$R_D = P$

$WL = P$

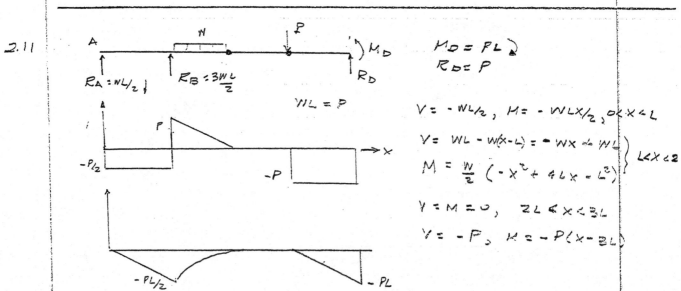

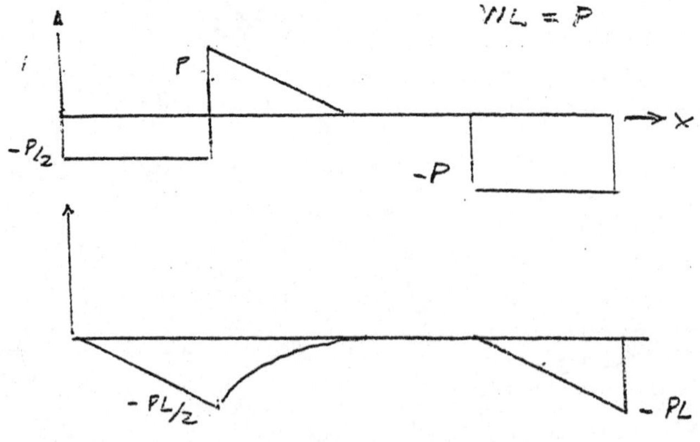

$V = -WL/2$, $M = -WLx/2$, $0 < x < L$

$$\left.\begin{aligned} V &= WL - W(x - L) = -Wx + WL \\ M &= \frac{W}{2}(-x^2 + 4Lx - L^2) \end{aligned}\right\} \quad L < x < 2L$$

$V = M = 0$, $2L < x < 3L$

$V = -P$, $M = -P(x - 3L)$

2.12 a)

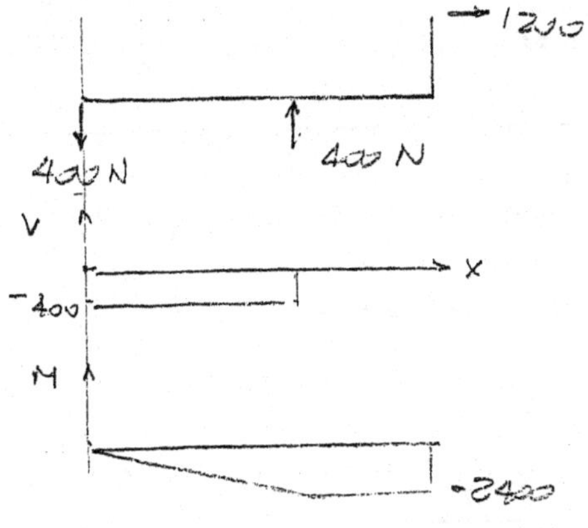

$F = 1200, V = -400, M = -400x, \quad 0 < x < 6$

$F = 1200, V = 0, M = -2400 \text{ N-m} \quad 6 < x < 10$

b)

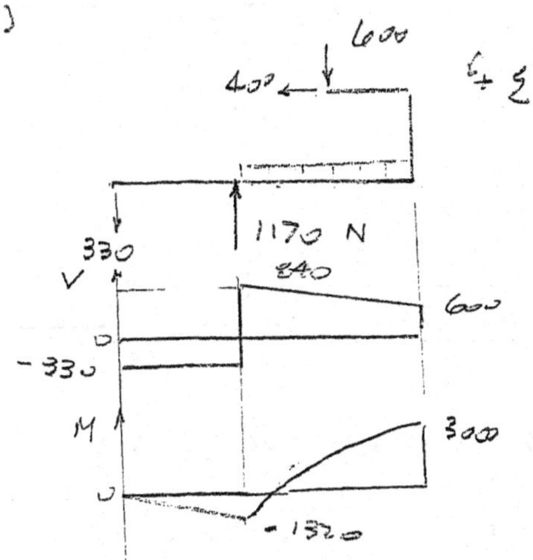

$\circlearrowleft + \sum M_A = 0 \Rightarrow 4R_B - 7(600) - (240)(7) + 400(3) = 0$

$F = -400, V = -330, M = -330x \quad 0 \le x < 4$

$F = -400, V = 1000 - 40x \quad 4 < x < 10$

$M = -20x^2 + 1000x - 5000$

c)

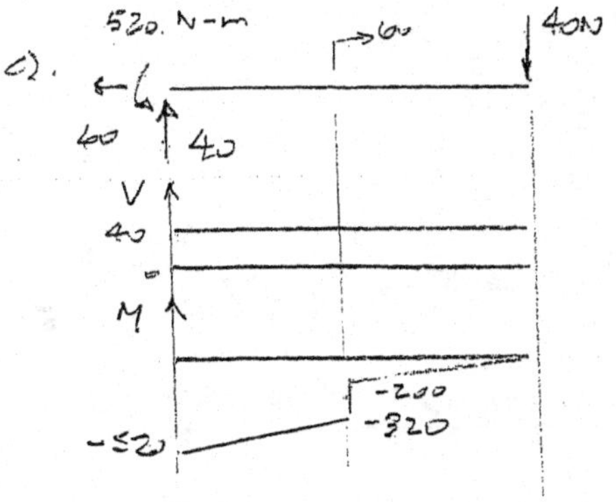

$F = 60 N, V = 40 N, M = -520 + 40x, \quad 0 < x < 5$

$F = 0, \quad V = 40 N, M = -400 + 40x \quad 5 < x < 10$

d)

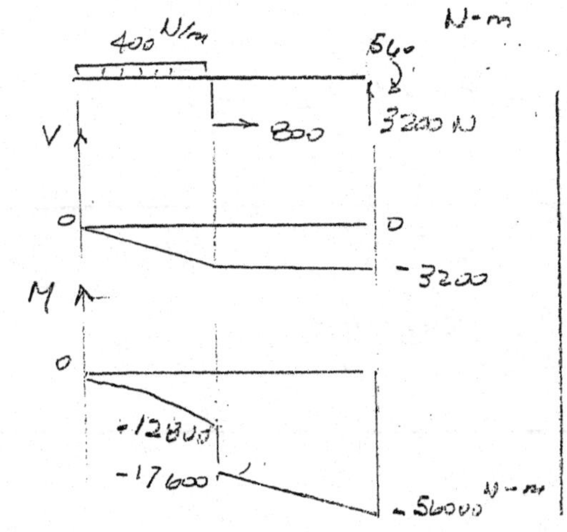

$F = 0, V = -400x, M = -200x^2, \quad 0 < x < 8$

$F = 0, V = -3200, M = -3200x + 8000, \quad 8 < x < 16$

e).

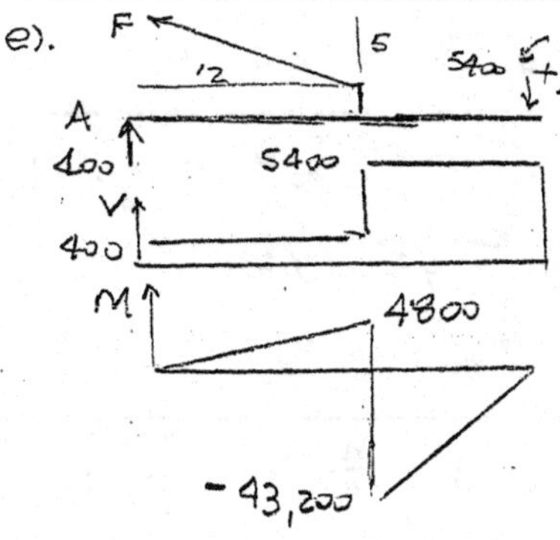

$\circlearrowleft + \sum M_A = \left(\frac{5}{13}F\right)(12) + \left(\frac{12}{13}F\right)4 - 20(5400) = 0$

$F = 13{,}200 N$

$F_x = 12{,}000^N, F_y = 5000^N$

$F = -12000^N, V = 400, M = 400x \Big\} \; x < 8$

$F = 0, V = 5400, M = 5400(x - 20) \Big\} \; 8 < x$

2.13 $V_Y = 0$, $V_z = 20N$, $T = M_x = 600$ N-cm, $M_y = 20(100-x)$ N-cm

---

2.14 AB: $F = P_z$, $V_x = P_x$, $V_y = P_y$; $M_x = -P_y(c-z)$, $M_y = P_x(c-z)$, $T = M_z = 0$

BC: $F = P_y$, $V_x = P_x$, $V_z = P_z$; $M_x = -P_y c + P_z(b-y)$, $T \equiv M_y = P_x c$

$M_z = P_x(b-y)$

CD: $F = P_x$, $V_y = P_y$, $V_z = P_z$, $T = M_x = P_z b - c P_y$, $M_y = c P_x$; $M_z = -b P_x + (a-x) P_y - (a-x) P_z$

2.15

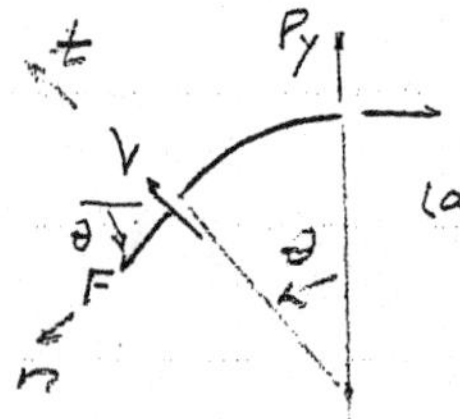

(a) $\Sigma F_n = 0 \Rightarrow F = P_x \cos\theta + P_y \sin\theta$

$\Sigma F_t = 0 \quad V = -P_x \sin\theta + P_y \cos\theta$

$M = R[P_y \sin\theta - P_x(1-\cos\theta)]$

SOLUTION TO 2.16 ON P. 2P-8

(b) $F = V = 0$, $M = M_0$, $T = T_0$

2.17

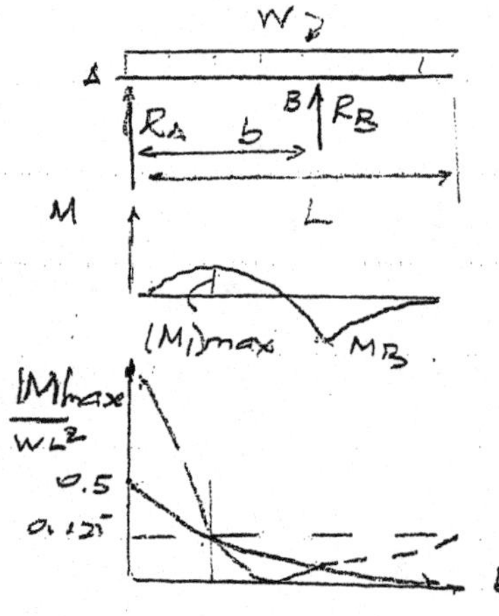

$bR_B = wL^2/2 \Rightarrow R_B = wL^2/2b \Rightarrow R_A = wL - wL^2/2b = wL(1 - L/2b)$

$M_1(x) = wL(1-L/2b)x - wx^2/2$, $0 \le x \le b$; $M_2(x) = -\frac{w}{2}(L-x)^2$, $b \le x \le L$

$M_1(x)|_{max} \Rightarrow \partial M_1/dx = 0 \Rightarrow x = L(1-L/2b)$

$\therefore M_1|_{max} = \frac{wL^2}{2}(1-L/2b)^2$;

$M_{2B} = -\frac{wL^2}{2}(1-b/L)^2$

$|M_1|_{max} = |M_2| \Rightarrow (1-L/2b)^2 = (1-b/L)^2 \Rightarrow b/L = \sqrt{2}/2$ Other Root $b/L = 0.2929$

$\downarrow$

$M_{max} = 0.04289\, wL^2$

---

stress equilibrium eqs. yield:

2.18 (i) $2axyz + 6xyz + 6xyz = 0 \Rightarrow a = -6$

(ii) $3y^2z + bxz^3 + c(12y^2z - 20xz^3) = 0 \Rightarrow c = -1/4$, $b = -5$

(iii) $3yz^2 + 12cyz^2 = 0 \Rightarrow c = -1/4$ ✓

2.19 (i) $\partial\sigma_x/\partial x + \partial\tau_{yx}/\partial y + \partial\tau_{zx}/\partial z = 0$

(ii) $\partial\tau_{xy}/\partial x = 0 \Rightarrow \tau_{xy} = f_1(y,z)$

(iii) $\partial\tau_{xz}/\partial x = 0 \Rightarrow \tau_{xz} = f_2(y,z)$

$\therefore$ (i) $\partial\sigma_x/\partial x = -(f_1(y,z) + f_2(y,z)) \equiv g(y,z)$

$\sigma_x = x \cdot g(y,z) + a$

2.20 $\sigma_Y A = \left(\frac{y}{h}\right) W \Rightarrow \sigma_Y = \frac{W}{A} \cdot \frac{y}{h} = \frac{W}{\pi R^2} \cdot (y/h)$.

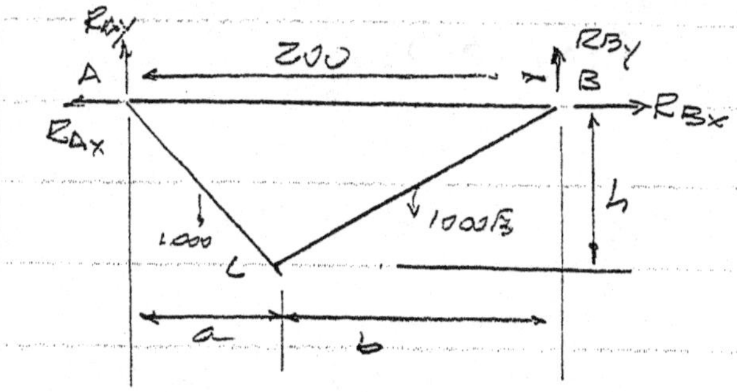

$m = 10$ N/cm

$AC = 100, \; BC = 100\sqrt{3}, \; h = 50\sqrt{3}$

$a = 50, \; b = 150$

$\circlearrowleft + \sum M_A = 0 \Rightarrow 200 R_{By} - 1000(a/2) - (1000\sqrt{3})(a + b/2) = 0$

$R_{By} - 2.5a - 5\sqrt{3}(125) = 0$

$R_{By} = 125(1 + 5\sqrt{3}) = 1207.532$

BC: $\circlearrowleft + \sum M_C = 0 \Rightarrow b R_{By} - h R_{Bx} - (1000\sqrt{3}) b/2 = 0$

$R_{By} = (h/b) R_{Bx} + 500\sqrt{3}$

$R_{Bx} = (R_{By} - 500\sqrt{3})(b/h) = (125 + 125\sqrt{3})(3/\sqrt{3})$

$R_{Bx} = 125(1+\sqrt{3})\sqrt{3} = 125(3 + \sqrt{3})$

$R_{Ax} = R_{Bx}, \quad R_{Ay} = (1000 + 1000\sqrt{3}) - R_{By} = 1000(1+\sqrt{3}) - 125(1 + 5\sqrt{3})$

$R_{Ay} = 875 + 375\sqrt{3} = 125(7 + 3\sqrt{3})$

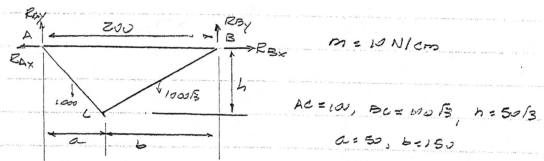

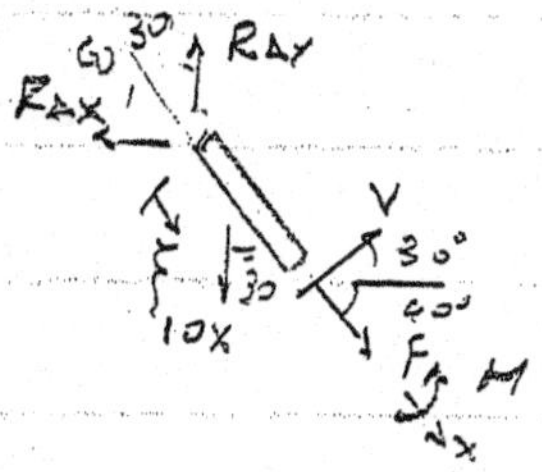

$\sum F_x = F - R_{Ax}\cos 60 - R_{Ay}\cos 30 + 1000x \cos 30 = 0$

$F = R_{Ax}(1/2) + R_{Ay}(\sqrt{3}/2) - 1000\sqrt{3}/2 \cdot x$

$F = 125(3+\sqrt{3})(1/2) + 125(7 + 3\sqrt{3}) - 500\sqrt{3} \cdot x$

$= \frac{1}{2}[375 + 125\sqrt{3} + 875 + 375\sqrt{3}] - 500\sqrt{3}x$

$= \frac{1}{2}(1250 + 500\sqrt{3}) - 500\sqrt{3}x = 625 - 250\sqrt{3}x$

$V = R_{Ax}\sin 60 - R_{Ay}\sin 30 + 10\xi \sin 30$

$= 125(3+\sqrt{3})(\sqrt{3}/2) - (875 + 375\sqrt{3})(1/2) + 5\xi$

$= \frac{1}{2}[375\sqrt{3} + 375 - 875 - 375\sqrt{3}] + 5\xi \quad = \frac{1}{2}[-500] + 5\xi$

$V = 2.50(2\xi - 100)$

$M = R_{Ay}(\xi \sin 30) - R_{Ax} \cdot \xi \cos 30 = (875 + 375\sqrt{3})\xi/2 - (375 + 125\sqrt{3})\sqrt{3}/2 \cdot \xi$
$\quad - (10x)(\frac{x}{2}\sin 30) \qquad - 2.5\xi^2$

$= \frac{\xi}{2}[875 + 375\sqrt{3} - 375\sqrt{3} - 375] - 2.5\xi^2 = \frac{\xi}{2}(500) - 2.5\xi^2$

$M = -2.5\xi(\xi - 100)$

2.21

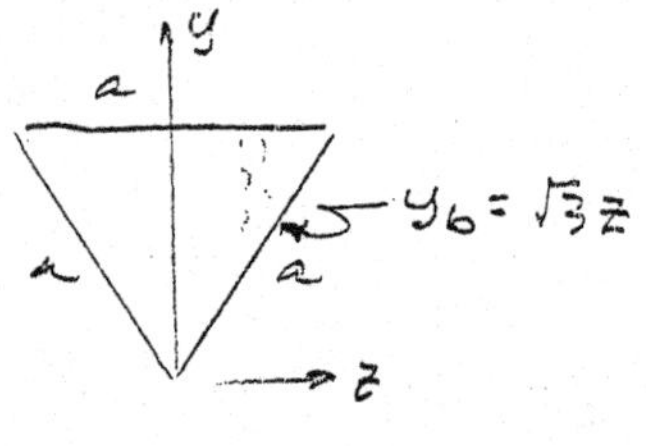

$$F = \iint \sigma_x \, dA = 2\int_{y_b}^{a\sqrt{3}/2} \left(\int_0^{y_b} (Ay + Bz^2)\, dz \right) dy$$

or

$$F = 2 \int_0^{a/2} \int_{y_b}^{a\sqrt{3}/2} (Ay + Bz^2)\, dy\, dz$$

$$F = 2\int_0^{a/2} \left[Ay^2/2 + Bz^2 y \right]_{y_b = \sqrt{3}z}^{a\sqrt{3}/2} dz = 2\int_0^{a/2} \left[3a^2A/8 + Bz^2 a\sqrt{3}/2 - 3Az^2/2 - Bz^3\sqrt{3} \right] dz$$

$$= 2\int_0^{a/2} \left\{ \left[\frac{3a^2}{8} - \frac{3z^2}{2} \right] A + \sqrt{3}\left[az^2/2 - z^3 \right] B \right\} dz$$

$$= 2\left\{ \left(\frac{3}{8}a^2 z - z^3/2 \right) A + \sqrt{3}\left(az^3/6 - z^4/4 \right) B \right\}_0^{a/2}$$

$$= 2\left\{ \left(3a^3/16 - a^3/16 \right) A + \sqrt{3}a^4 \left(1/48 - 1/64 \right) B \right\}$$

$$F = 2\left[a^3A/8 + \sqrt{3}a^4B/192 \right] = a^3A/4 + \sqrt{3}a^4B/96$$

$$M_z = \iint \sigma_x \cdot y\, dA = 2\int_0^{a/2} \int_{y_b}^{a\sqrt{3}/2} (Ay^2 + Bz^2 y)\, dy\, dz$$

$$= 2\int_0^{a/2} \left[Ay^3/6 + Bz^2y^2/2 \right]_{\sqrt{3}z}^{a\sqrt{3}/2} dz = 2\int_0^{a/2} \left[3\sqrt{3}a^3A/48 + Bz^2 \cdot 3a^2/8 - A3\sqrt{3}z^3/6 - \frac{3Bz^4}{2} \right] dz$$

$$= 2\int_0^{a/2} \left\{ \sqrt{3}A\left[a^3/16 - z^3/2 \right] + B\left[3a^2z^2/8 - 3z^4/2 \right] \right\} dz$$

$$= 2\left\{ \sqrt{3}A\left[a^3z/16 - z^4/8 \right] + B\left[a^2z^3/8 - 3z^5/10 \right] \right\}_0^{a/2}$$

$$= 2\left\{ \sqrt{3}A\left[a^4/32 - a^4/128 \right] + B\left[a^5/64 - 3a^5/320 \right] \right\}$$

$$= 2\left\{ 3\sqrt{3}Aa^4/128 + 2Ba^5/320 \right\} = 3\sqrt{3}Aa^4/64 + Ba^5/80$$

$$24a^3A + \sqrt{3}a^4B = 96F \Rightarrow$$

$$3\sqrt{3}a^4A/4 + Ba^5/5 = 16M_z$$

$$|D| = 3a^7 \begin{vmatrix} 8 & \sqrt{3} \\ \sqrt{3}a/4 & a/5 \end{vmatrix} = 3a^8\left(8/5 - 3/4 \right) = 51a^8/20$$

$$A = \frac{1}{D}\begin{vmatrix} 96F & \sqrt{3}a^4 \\ 16M_z & a^5/5 \end{vmatrix} = \frac{16a^4}{D}\begin{vmatrix} 6F & \sqrt{3} \\ M_z & a/5 \end{vmatrix} = \frac{16a^4}{D}\left(6aF/5 - \sqrt{3}M_z \right) = \frac{320}{51a^4}\left(16aF - \sqrt{3}M_z \right)$$

$$B = \frac{3a^3}{D}\begin{vmatrix} 8 & 96F \\ \sqrt{3}a/4 & 16M_z \end{vmatrix} = \frac{48a^3}{D}\begin{vmatrix} 8 & 6F \\ \sqrt{3}a & M_z \end{vmatrix} = \frac{48a^3}{D}\left(8M_z - 6\sqrt{3}aF \right) = \frac{960}{51a^5}\left(8M_z - 6\sqrt{3}aF \right)$$

$$= \frac{1920}{51a^5}\left(4M_z - 3\sqrt{3}Fa \right)$$

F passing through centroid ; $y_c = \frac{a\sqrt{3}}{2} \cdot \frac{2}{3} = a\sqrt{3}/3$.

$$M_{z_c} = M_z - F \cdot a\sqrt{3}/3 = \left(3\sqrt{3}a^4 A/64 + Ba^5/80\right) - \left(a^3 A/4 + \sqrt{3}a^4 B/96\right)\left(\frac{\sqrt{3}a}{3}\right)$$

$$= \left(\frac{3\sqrt{3}}{64} - \frac{\sqrt{3}}{12}\right)a^4 A + \left(\frac{a^5}{80} - \frac{a^5}{96}\right)B$$

$$= \frac{\sqrt{3}}{4}\left(\frac{3}{16} - \frac{1}{3}\right)a^4 A + \frac{1}{16}\left(\frac{1}{5} - \frac{1}{6}\right)a^5 B$$

$$= -\frac{7\sqrt{3}}{192}a^4 A + \frac{a^5}{480}a^5 B$$

2.22 a) $T = \iint \tau_{x\theta}\, r\, dA = 2\pi \int_0^R \tau_{r\theta}(r) r^2 dr = 2\pi k \int_0^R r^3 dr = \frac{\pi k}{2} R^4$

$k = \frac{T}{\pi R^4/2} \Rightarrow \tau_{x\theta} = \frac{2Tr}{\pi R^4}$

b) $T = \tau_0 \iint r\, dA = 2\tau_0 \pi \int_0^R r^2 dr = 2\pi \tau_0 R^3/3 \Rightarrow \tau_0 = 3T/2\pi R^3$

2.23 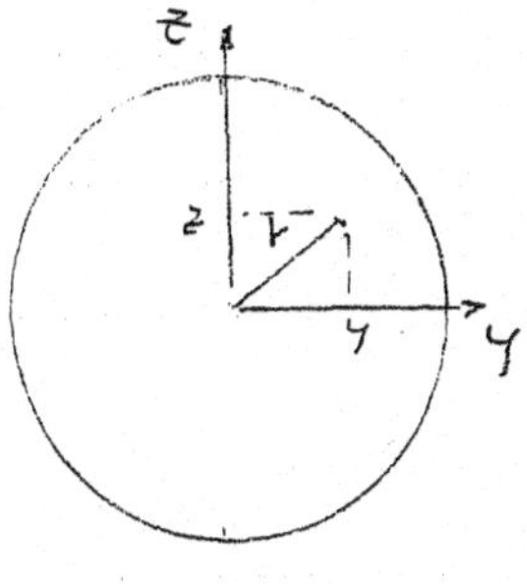

$F = 0$, $V_y = \iint \tau_{xy}\, dA$; $\tau_{xy} = -200\, r\cos\theta$, $\tau_{xz} = 200y$

$y = r\cos\theta$, $z = r\sin\theta$; $V_y = -200 \int_0^{2\pi}\left(\int_0^R r\,dr\right)\cos\theta\, d\theta = 0$

$V_z = 200 \int_0^{2\pi} \left(\int r\,dr\right)\sin\theta\, d\theta = 0$

$M_y = M_z = 0$

$$T = \iint (y\tau_{xz} - z\tau_{xy})\, dA = 200 \iint (y^2 + z^2)\, dA$$

$$= 200 \iint r^2 dA = 200(\pi R^2/2)$$

$$= 100\pi R^2$$

2.24 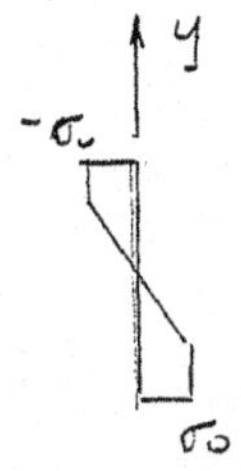

b) $M = \iint_A y\,\sigma_x(y)\, dA = b \int_{-h/2} y\,\sigma_x(y)\, dy$

$$= -\sigma_0 \left\{ \int_{-h/2}^{-c} y\, dy - \int_{-c}^{c} y^2 dy - \int_{c}^{h/2} y\, dy \right\} = \sigma_0\left[bc^2/3 - bh^2/4\right]$$

$$= -\frac{b\sigma_0}{12}(4c^2 - 3h^2)$$

c). $M_{max} \Rightarrow dM/dc = 0 \Rightarrow c = 0$ $\quad M_{max} = bh^2\sigma_0/4$

2.25

$$F = -P = \iint_A (B + cy)\,dA = Bb^2 + cb\int_{-b/2}^{b/2} y\,dA = Bb^2$$

$$B = -P/b^2.$$

$$M_z = Pe = -\iint y\,\sigma_x\,dA = -b\int_{-b/2}^{b/2} (By + cy^2)\,dy = -bBy^2/2\Big|_{-b/2}^{b/2} - c\,b^4/12 = -c\,b^4/12$$

$$c = -12M_z/b^4 = -12Pe/b^4$$

$$\therefore\ \sigma_x = P\left[-12ey/b^4 - 1/b^2\right]$$

$$e = 4b,\quad \sigma_x\Big|_{y=b/2} = -P\left(24b^2/b^4 + 1/b^2\right) = -25P/b^2.$$

2.26

$F = 6000$ N, $M_A = 1{,}000$ N-m.

a)

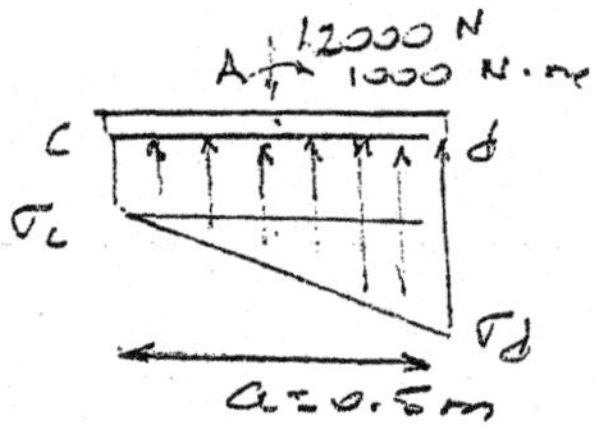

$\sum F_y = 0 \Rightarrow$

$$(.5)^2\left[\frac{\sigma_c + \sigma_d}{2}\right] = 12000 \Rightarrow \sigma_c + \sigma_d = 96{,}000$$

$\sum M_A = 0 \Rightarrow$

$$(0.5)^2\left[\frac{\sigma_d - \sigma_c}{2}\right]\left[\frac{2}{3}a - \frac{a}{2}\right] = 1000 \text{ N-m}$$

$$\left[\frac{\sigma_d - \sigma_c}{8}\right]\left(\frac{.50}{6}\right) = 1000 \Rightarrow \sigma_d - \sigma_c = 96000$$

$$\left.\begin{aligned}\sigma_c + \sigma_d &= 96{,}000\\ -\sigma_c + \sigma_d &= 96{,}000\end{aligned}\right\} \Rightarrow \sigma_d = 96{,}000 \text{ N/m}^2,\quad \sigma_c = 0$$

b)

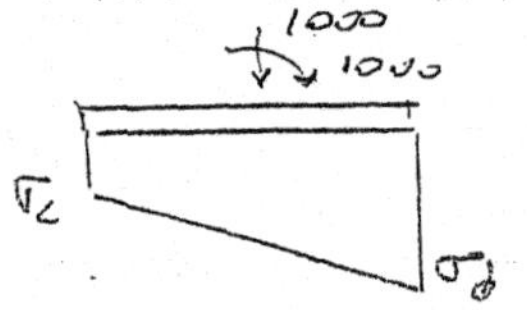

$$\left.\begin{aligned}\sigma_c + \sigma_d &= 8000\\ -\sigma_c + \sigma_d &= 96000\end{aligned}\right\} \Rightarrow \begin{aligned}\sigma_d &= 52{,}000 \text{ N/m}^2, \text{ (comp.)}\\ \sigma_c &= -44{,}000 \text{ N/m}^2 \text{ (tens.)}\end{aligned}$$

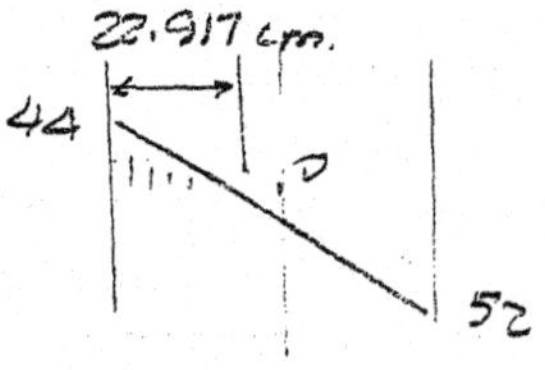

$$M_{tens}\Big|_0 = \frac{(4.4)(22.917)(50)}{2}\left[25 - 22.917\left(\tfrac{1}{3}\right)\right] = 43{,}765 \text{ N-cm}$$

$Fb = M_{tens.} \Rightarrow \quad b = 10 \Rightarrow F = 4377$ N

2.28 $I_{\sigma_1} = \sigma_x + \sigma_y = -4\sigma_x,\quad I_{\sigma_2} = \sigma_x\sigma_y = -5\sigma_x^2;\quad \dfrac{\sigma_x + \sigma_y}{\sigma_x\sigma_y} = \dfrac{-4\sigma_x}{-5\sigma_x^2} = \dfrac{I_{\sigma_1}}{I_{\sigma_2}} = \dfrac{1}{300 \text{ MPa}}$

$$\Rightarrow \sigma_x = 240 \text{ MPa}$$

$$|\tau_{max}| = \left|\frac{\sigma_x - \sigma_y}{2}\right| = 3\sigma_x = 720 \text{ MPa}$$

2.29 $\sigma_n = \sigma_x \cos^2\theta + \sigma_y \sin^2\theta + 2\tau_{xy}\sin\theta\cos\theta$

$\tau_{nt} = -(\sigma_x - \sigma_y)\sin\theta\cos\theta + \tau_{xy}(\cos^2\theta - \sin^2\theta)$

a) $\sigma_n = 3/4(200) + 1/4(400) + 800(\sqrt{3}/4)$

$= 250 + 200\sqrt{3}$

$\sigma_t = 1/4(200) + 3/4(400) - 800(\sqrt{3}/4)$

$= 350 - 200\sqrt{3}$

$\tau_{nt} = 200(\tfrac{1}{2}\cdot\sqrt{3}/2) + 400(3/4 - 1/4)$

$= 50\sqrt{3} + 200$

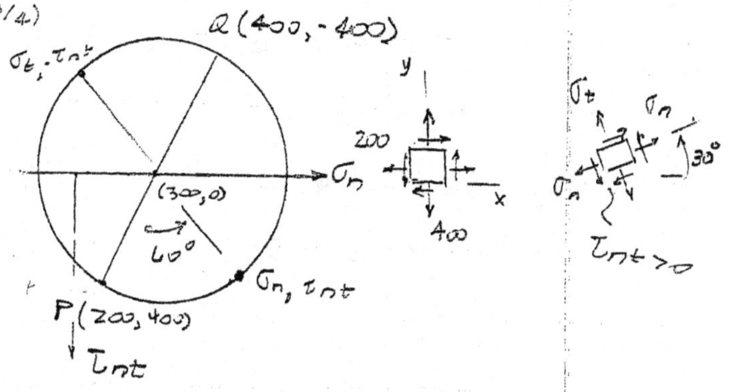

b). $\sigma_n = -400(3/4) + 600(-\sqrt{3}/4)$

$= -300 - 150\sqrt{3}$

$\sigma_t = -400(1/4) + 600(\sqrt{3}/4)$

$= -100 + 150\sqrt{3}$

$\tau_{nt} = 400(-\sqrt{3}/4) + 300(3/4 - 1/4)$

$= 150 - 100\sqrt{3}$

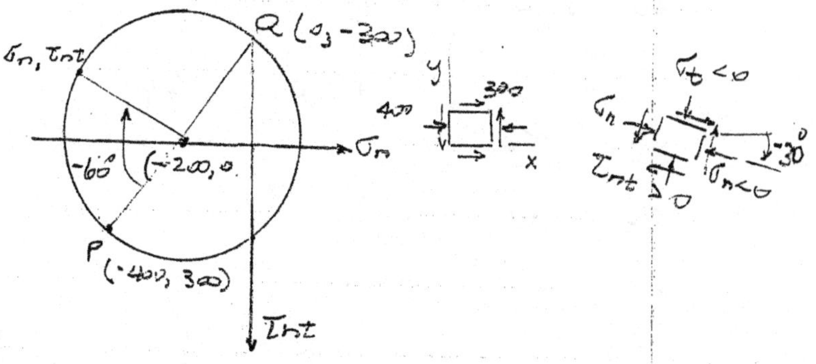

c). $\sigma_n = 300$; $\sigma_t = -300$ MPa, $\tau_{nt} = 0$

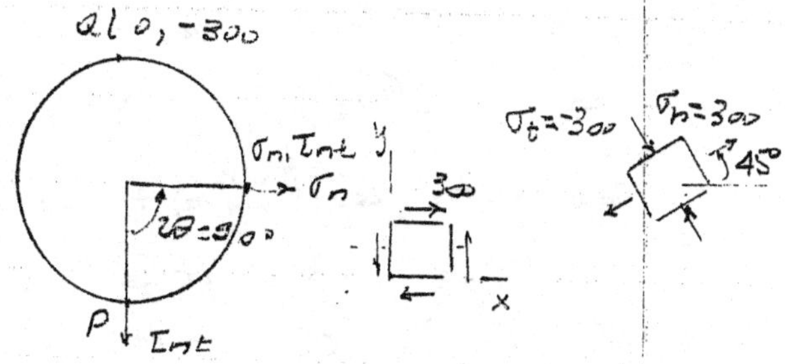

d). $\sigma_n = 1200(1/4) + 800(3/4) - 1600(\sqrt{3}/2)(-1/2)$

$= (900 + 400\sqrt{3})$ kPa

$\sigma_t = 1200(3/4) + 800(1/4) - 1600(-1/2)(-\sqrt{3}/2)$

$= (1100 - 400\sqrt{3})$ kPa

$\tau_{nt} = -400(-1/2)(\sqrt{3}/2) - 800(1/4 - 3/4)$

$= (100\sqrt{3} + 400)$ kPa

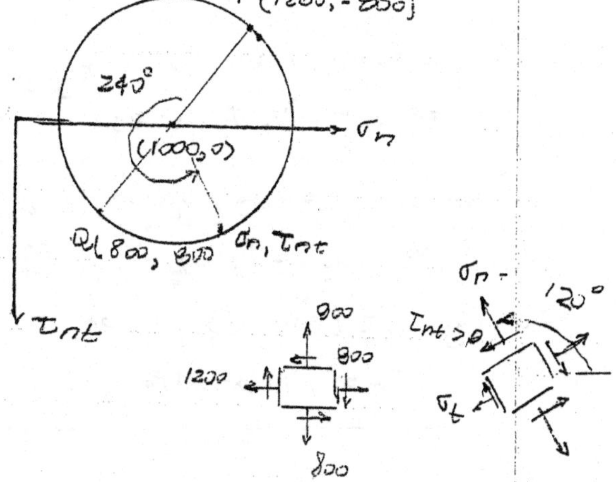

e). With respect to n, t: $\theta_n = -30^\circ$, $\theta_t = 60^\circ$

$$\sigma_x\Big|_{30^\circ} = -100(3/4) - 50(1/4) + 200(\sqrt{3}/2)(-1/2)$$
$$= -87.5 - 50\sqrt{3}$$

$$\sigma_y\Big|_{\theta=60^\circ} = -100(+1/4) - 50(3/4) + 200(\sqrt{3}/2)(1/2)$$
$$= -62.5 + 50\sqrt{3}$$

$$\tau_{xy} = 25(-1/2)(\sqrt{3}/2) + 100(3/4 - 1/4)$$
$$= -6.25\sqrt{3} + 150$$

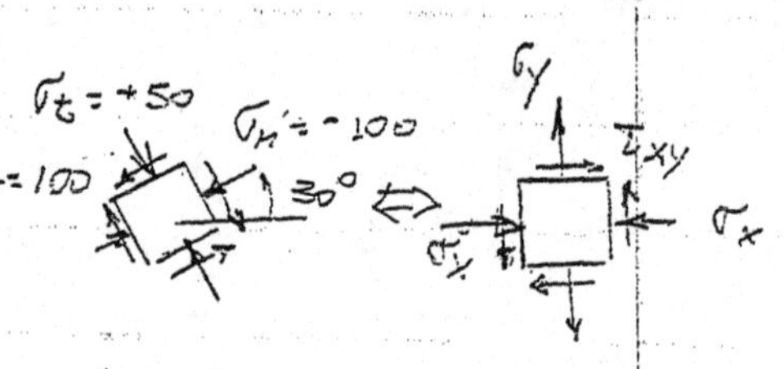

f) $\sigma_n = 200(1/2) + 100(1/2) + 2\tau_{xy}(1/2)$

$$50 = 150 + \tau_{xy} \Rightarrow \tau_{xy} = -100$$

$$\sigma_t = 200(1/2) + 100(1/2) - 200(1/\sqrt{2})(-1/\sqrt{2})$$
$$= 250$$

$$\tau_{nt} = -100(1/2) = -50$$

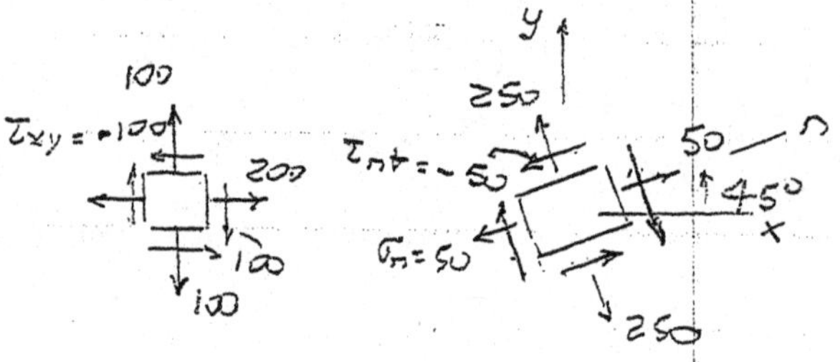

g) $\overset{100}{\sigma_n} = \sigma_x(1/4) + 200(3/4) + 2\tau_{xy}(1/2)(\sqrt{3}/2)$

$$\Rightarrow \sigma_x/4 + \sqrt{3}/2\,\tau_{xy} = -50 \quad (i)$$

$$\overset{0}{\tau_{nt}} = -(\sigma_x - 200)(1/2)(\sqrt{3}/2) + \tau_{xy}(1/4 - 3/4)$$

$$\Rightarrow \tau_{xy} = -\sqrt{3}\sigma_x/2 + 100\sqrt{3} \quad (ii)$$

Solving (i) & (ii): $\sigma_x = 400$ MPa, $\tau_{xy} = -100\sqrt{3}$

$\sigma_t = 500$ MPa

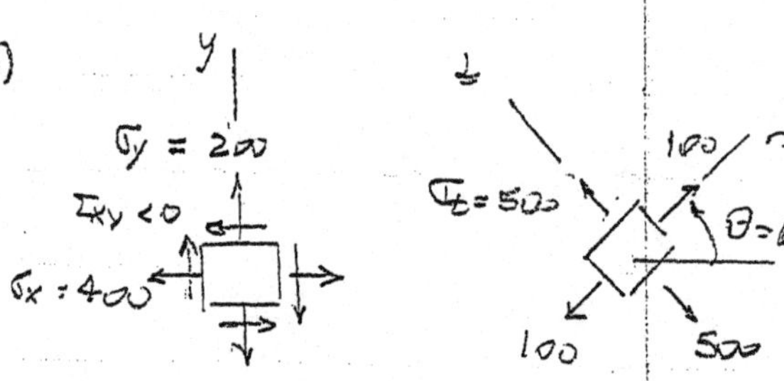

h). $\overset{100}{\sigma_n} = \sigma_x(1/4) - 200(3/4) + 2\tau_{xy}(1/2)(\sqrt{3}/2)$

$$\Rightarrow \sigma_x/4 + \sqrt{3}/2\,\tau_{xy} = 250 \quad (i)$$

$$\overset{0}{\tau_{nt}} = -(\sigma_x + 200)(1/2)(\sqrt{3}/2) + \tau_{xy}(1/4 - 3/4)$$

$$\Rightarrow \tau_{xy} = -\sqrt{3}\sigma_x/2 - 100\sqrt{3} \quad (ii)$$

Solving (i) & (ii): $\sigma_x = -800$, $\tau_{xy} = 300\sqrt{3}$

$\sigma_t = -1100$ MPa

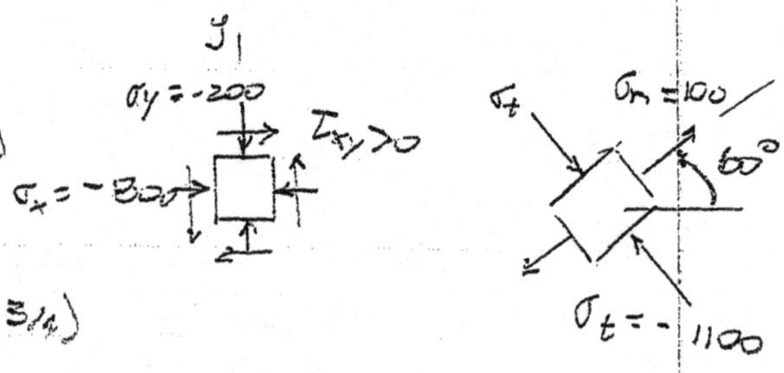

Problem 2.31 on p. 2P-18, Problem 2.30 (Bottom of this page)

2.34 $\tan\theta = z/y$, $\tan\psi = \tau_{xz}/\tau_{xy} = c_2 y / c_1 z = -y/z$ if $c_2 = -c_1$

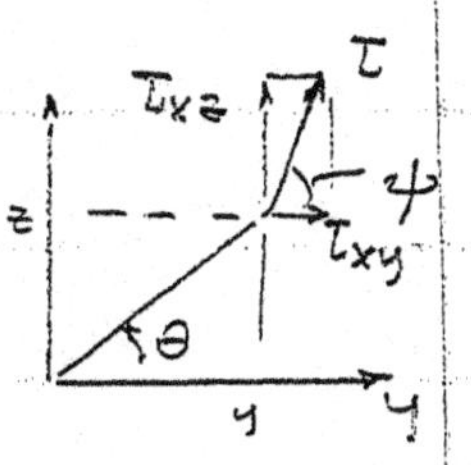

∴ $\tan\theta \tan\psi = -1 \Rightarrow \psi = \theta \pm \pi/2$; i.e. circumf. direct.

2.32

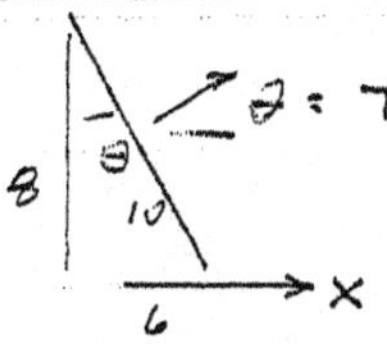

$\theta = \tan^{-1}(a/b) = \tan^{-1}(3/4)$. $\sin\theta = 3/5$ $\cos\theta = 4/5$

$$\tau_{nt} = -(\sigma_x - \sigma_y)\sin\theta\cos\theta + \tau_{xy}(\cos^2\theta - \sin^2\theta)$$

$\tau_{nt} = -(-300 - \sigma_y)(12/25) + 400(7/25) \lessgtr \pm 600$

$-600 < 144 + 12\sigma_y/25 + 16(7) < 600$

$-600 < 256 + 12\sigma_y/25 < 600$

$-856 < 12\sigma_y/25 < 344$

$-\frac{856}{12}\cdot 25 < \sigma_y < \frac{344}{12}\cdot 25 \Rightarrow -1783 < \sigma_y < 716.7$ kPa

---

2.33 $\sigma_n < 400$ kPa. $\sin\theta = \cos\theta = \sqrt{2}/2$

$\sigma_n = \sigma_x\cos^2\theta + \sigma_y\sin^2\theta + 2\tau_{xy}\sin\theta\cos\theta$

$\sigma_n = \frac{1}{2}\sigma_x - 300(1/2) < 400 \Rightarrow \sigma_x = 1100$ kPa.

2.30 (a) $\cos\alpha = \sqrt{2}/2$, $\sin\alpha = \sqrt{2}/2$: $\cos 2\alpha = 0$, $\sin 2\alpha = 1$

$\sigma_n \equiv \sigma_x$, $\sigma_s = \frac{1}{2}\sigma_x + \frac{1}{2}\sigma_y + \tau_{xy}$

$\sigma_t = \sigma_y$

∴ $\tau_{xy} = \sigma_s - \frac{1}{2}(\sigma_x + \sigma_y) = \sigma_s - \frac{1}{2}(\sigma_n + \sigma_t)$

b) $\cos\alpha = 1/2$, $\sin\alpha = \sqrt{3}/2$; $\cos 2\alpha = -1/2$, $\sin 2\alpha = \sqrt{3}/2$

$$\left.\begin{aligned} \sigma_s &= \sigma_x/4 + 3\sigma_y/4 + \sqrt{3}/2\,\tau_{xy} \\ \sigma_t &= \sigma_x/4 + 3\sigma_y/4 - \sqrt{3}/2\,\tau_{xy} \end{aligned}\right\} \Rightarrow \sqrt{3}\tau_{xy} = \sigma_s - \sigma_t \Rightarrow \tau_{xy} = \frac{\sqrt{3}}{3}(\sigma_s - \sigma_t)$$

2-35 $\sigma_{1,2} = \dfrac{\sigma_x+\sigma_y}{2} \pm \left[\left(\dfrac{\sigma_x-\sigma_y}{2}\right)^2+\tau_{xy}^2\right]^{1/2}$, $\tan 2\theta = \dfrac{\tau_{xy}}{\left(\frac{\sigma_x-\sigma_y}{2}\right)}$

(a) $\sigma_{1,2} = 30 + [\,\overline{30}^2 + \overline{40}^2\,]^{1/2} = 30 \pm 50$ (τ_{max})

$\sigma_{1,2} = -20,\ 80$ MPa

$\tan 2\theta = 40/60 = 2/3 \Rightarrow \theta_{1,2} = 16.85^\circ,\ 106.85^\circ$

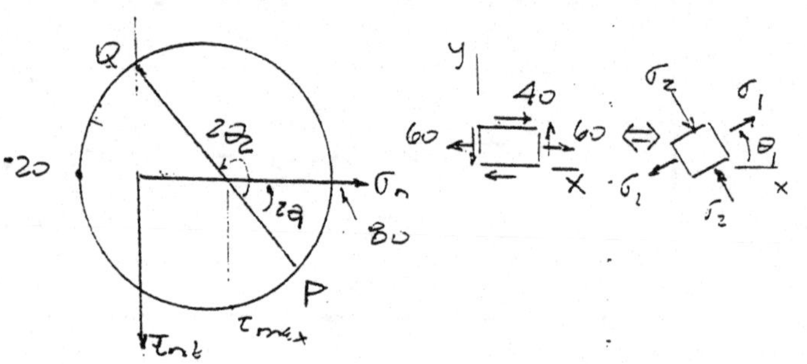

b) $\sigma_{1,2} = \pm[\,\overline{200}^2 + \overline{200}^2\,]^{1/2} = \pm\sqrt{8}\times 10^2$ ($\tau_{max} = 282.8$)

$\sigma_{1,2} = \pm 282.8$ kPa

$\tan 2\theta = \dfrac{-200}{200} = -1 \Rightarrow \theta_{1,2} = -22.5^\circ,\ 67.5^\circ$

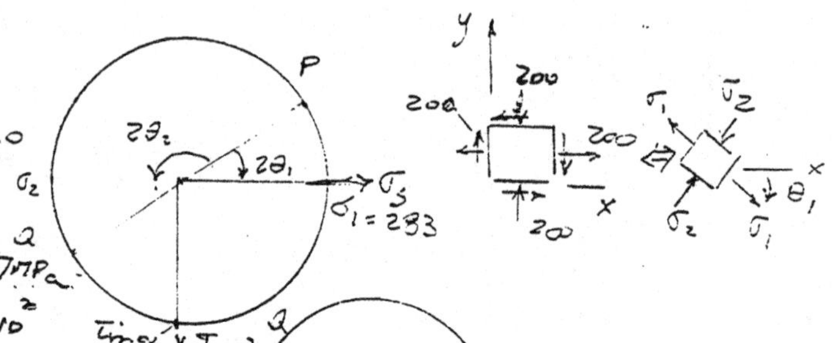

c) $\sigma_{1,2} = 500 \pm [\,\overline{400}^2 + \overline{200}^2\,]^{1/2} = 150 \pm \sqrt{20}\times 10^2$ ($\tau_{max} = 447$ MPa)

$\sigma_{1,2} = 947,\ 52?.8$ MPa $\quad \tau_{max} = 447$ MPa

$\tan 2\theta = \dfrac{200}{400} = 1/2 \Rightarrow \theta_{1,2} = 13.3^\circ\ 103.3^\circ$

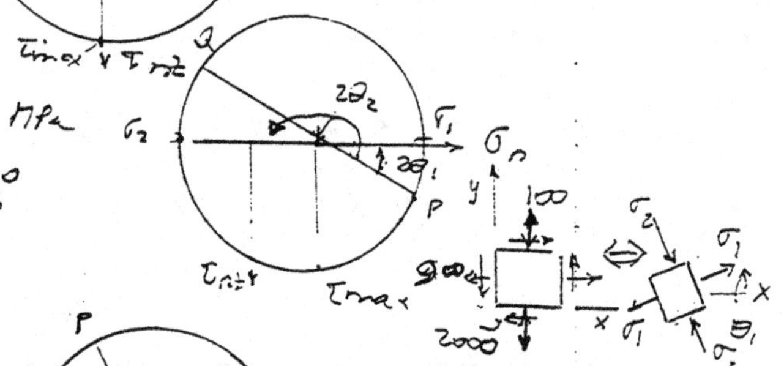

d). $\sigma_{1,2} = 600 \pm [\,\overline{400}^2 + \overline{600}^2\,]^{1/2} = 600 \pm \sqrt{52}\times 10^2$ ($\tau_{max} = 721$)

$\sigma_{1,2} = 1321,\ -121$ kPa

$\tan 2\theta = \dfrac{-600}{-200} = +3 \Rightarrow \theta_{1,2} = -54.2^\circ,\ 35.8^\circ$

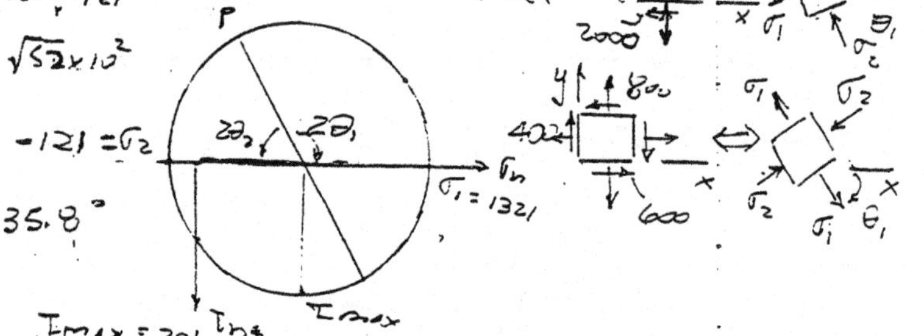

e). $\sigma_{1,2} = -150 \pm [(50)^2 + \overline{200}^2]^{1/2} = -150 + \sqrt{4.25}\times 10^2$ ($\tau_{max} = 206$)

$\sigma_{1,2} = 56.2,\ -356$ MPa

$\tan 2\theta = \dfrac{200}{-50} = -4 \qquad \theta_{1,2} = 52^\circ,\ -38^\circ$

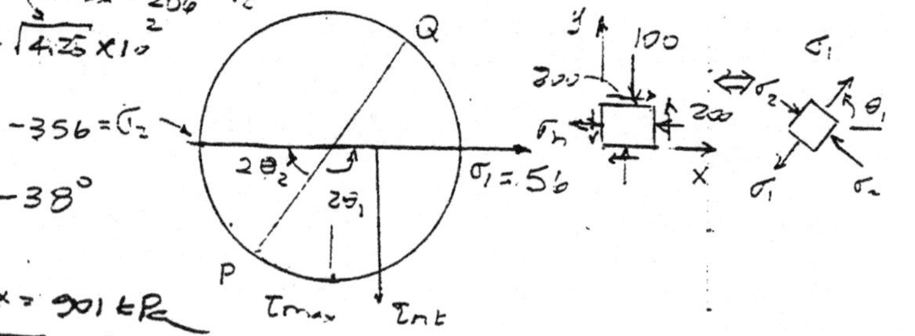

$\tau_{max} = 901$ kPa

f) $\sigma_{1,2} = 1250 \pm [\,\overline{750}^2 + \overline{500}^2\,]^{1/2} = 750 \pm \sqrt{81.25}\times 10^2$

$\sigma_{1,2} = 2151,\ 349$ kPa $\quad \tau_{max} = 901$ kPa

$\tan 2\theta = \dfrac{-500}{750} = -2/3 \Rightarrow \theta_{1,2} = -16.8,\ 73.2$

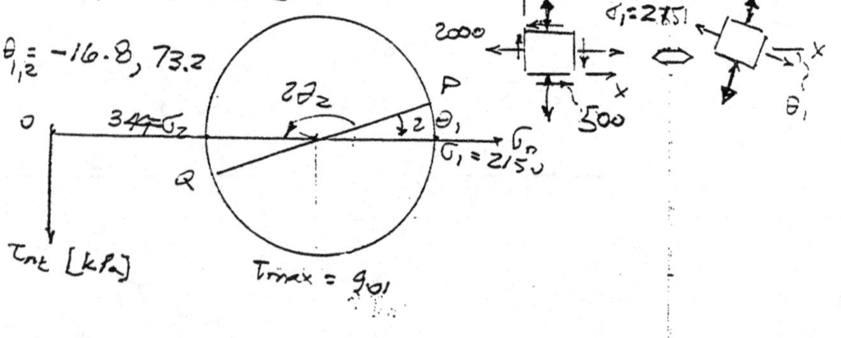

g). $\sigma_{1,2} = -40 \pm [(80)^2 + 20^2]^{1/2} = -40 \pm \sqrt{68 \times 10}$ ($\tau_{max} = 82.5$)

$\sigma_{1,2} = 42.5, -122.5$ MPa

$\tan 2\theta = \frac{-20}{-80} = 0.25 \Rightarrow \theta_{1,2} = -83^\circ, 7^\circ$

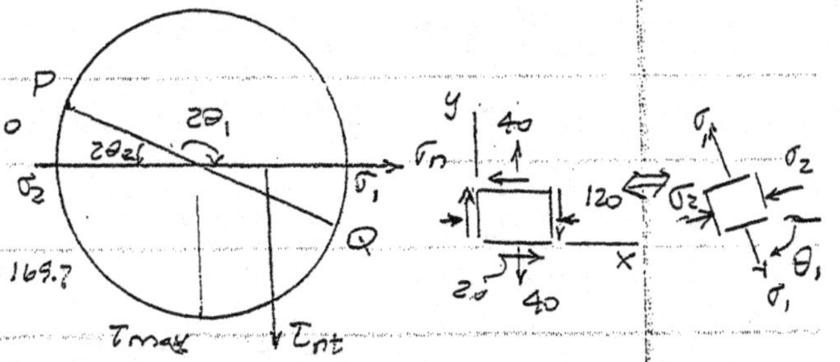

h) $\sigma_{1,2} = 120 \pm [120^2 + 120^2]^{1/2} = 120 \pm 120\sqrt{2}$ ($\tau_{max} = 169.7$)

$\sigma_{1,2} = 290, -49$ MPa

$\tan 2\theta = \frac{120}{240} = 0.5 \Rightarrow \theta_{1,2} = 13.28, 103.3^\circ$ or -76.7

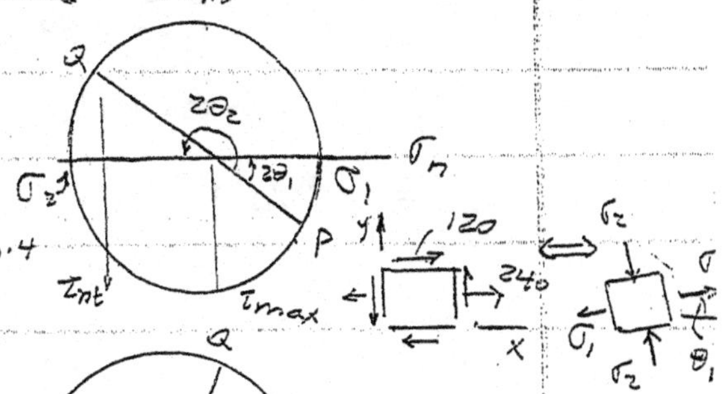

i) $\sigma_{1,2} = -50 \pm [150^2 + 320^2]^{1/2} = -50 \pm \sqrt{12.49 \times 10^2}$ ($\tau_{max} = 353.4$)

$\sigma_{1,2} = 303.4, -403.4$ kPa

$\tan 2\theta = \frac{320}{-150} = -2.133 \Rightarrow \theta_{1,2} = 57.6^\circ, -32.4^\circ$

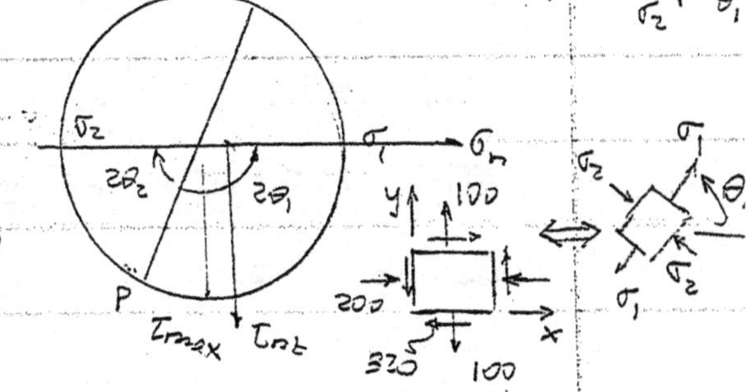

j). $\sigma_{1,2} = 120 \pm [120^2 + 120^2]^{1/2} = 120 + 120\sqrt{2}$ ($\tau_{max} = 169.7$)

$\sigma_{1,2} = 289.7^\circ, -49.7^\circ$ MPa

$\tan 2\theta = \frac{120}{-120} = -1 \Rightarrow \theta_{1,2} = 67.5^\circ, -22.5^\circ$

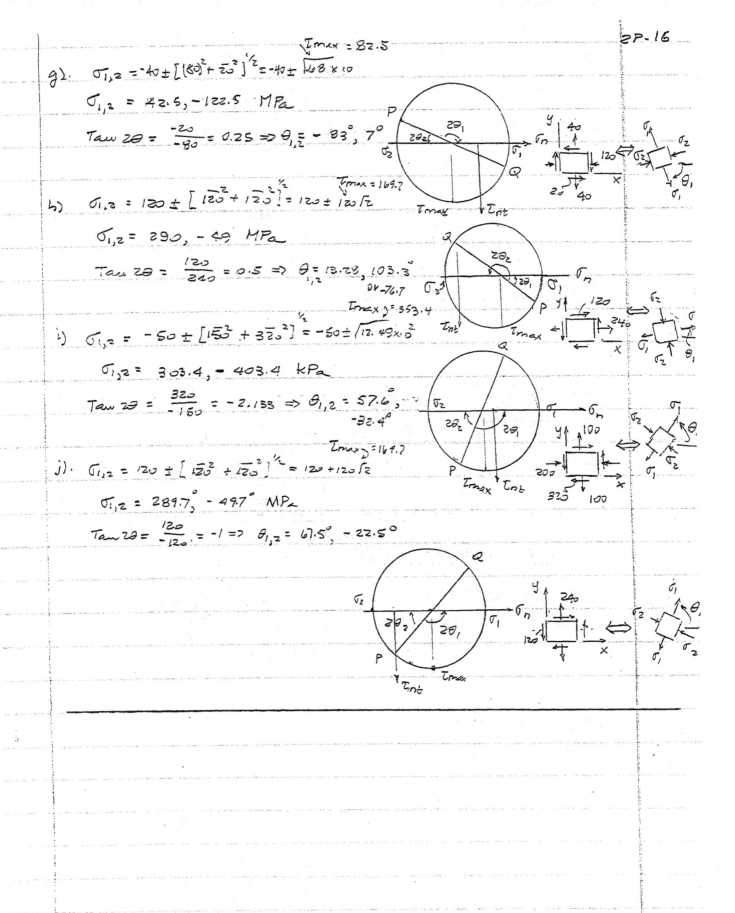

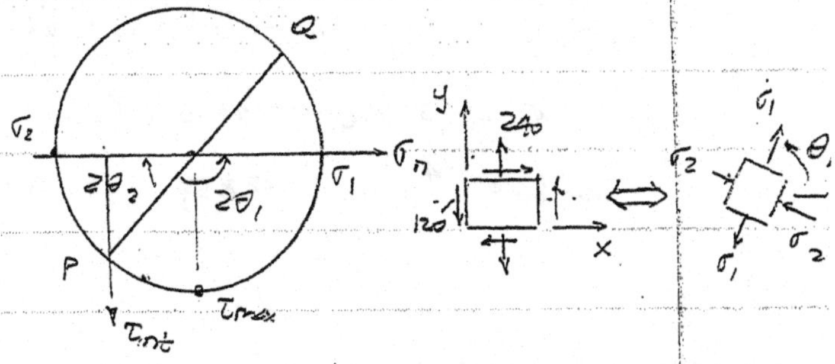

2.36 a) $\sigma_{1,2} = 50 \pm [\overline{30}^2 + \overline{40}^2]^{1/2}$

$\sigma_{1,2} = 100, 0$ MPa

$\tan 2\theta = 40/30 = 1.333, \ \theta_{1,2} = 26.6^\circ, -63.4^\circ$

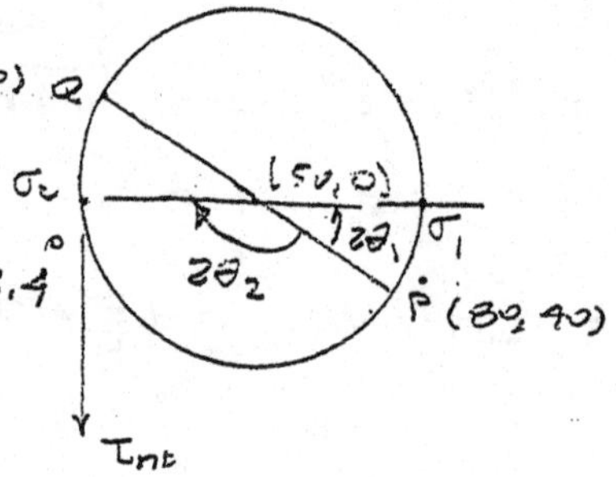

b) $\sigma_{1,2} = -170 \pm [\overline{30}^2 + \overline{40}^2]^{1/2} = -170 \pm 50$

$\sigma_{1,2} = -120, -220$ MPa

$\tan 2\theta = -40/-30 = 1.333; \ \theta_{1,2} = -63.4^\circ, 26.6^\circ$

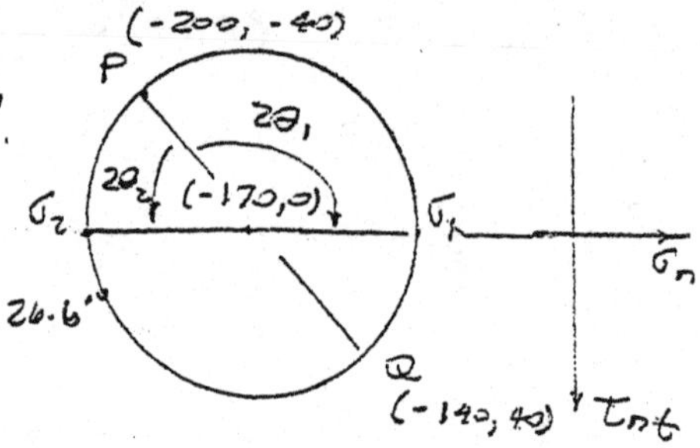

c) $\sigma_{1,2} = 155 \pm [\overline{25}^2 + \overline{60}^2]^{1/2} = 155 \pm \sqrt{422.5 \times 10}$

$\sigma_{1,2} = 155 \pm 65 = 220, 90$ MPa

$\tan 2\theta = 60/-25 = 2.4, \ \theta_{1,2} = 56.3^\circ, -33.7^\circ$

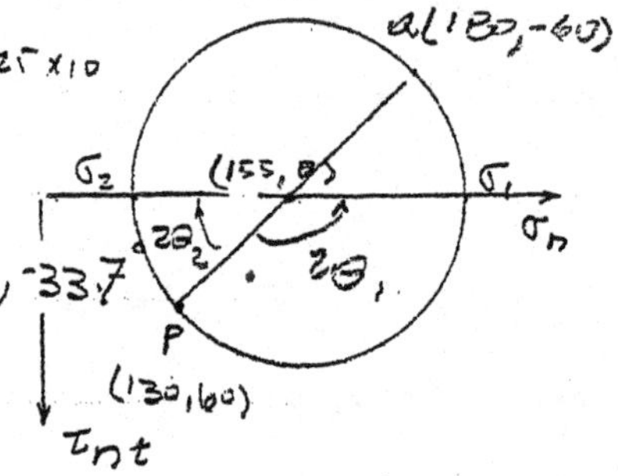

d) $\sigma_{1,2} = 30 \pm [\overline{170}^2 + \overline{40}^2]^{1/2} = 30 \pm \sqrt{305} \times 10$

$\sigma_{1,2} = 204.6, -144.6$ kPa

$\tan 2\theta = -40/170 = -0.235, \ \theta_{1,2} = -6.62, 83.38^\circ$

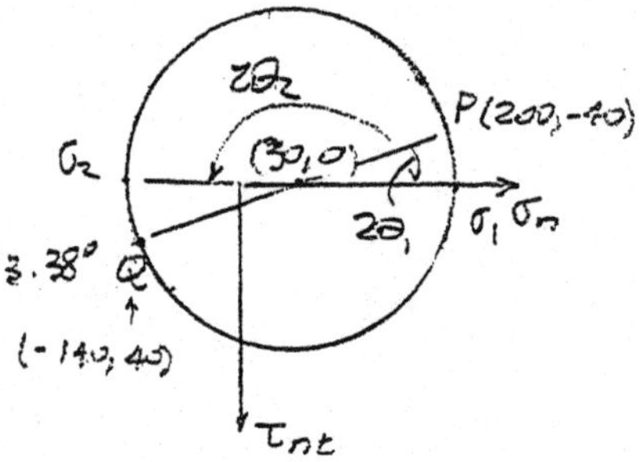

e) $\sigma_{1,2} = 60 \pm [\overline{40}^2 + \overline{30}^2] = 60 \pm 50$

$\sigma_{1,2} = 110, 10$ MPa

$\tan 2\theta = -30/-20 = 1.50, \ \theta_{1,2} = -61.8^\circ, 28.2^\circ$

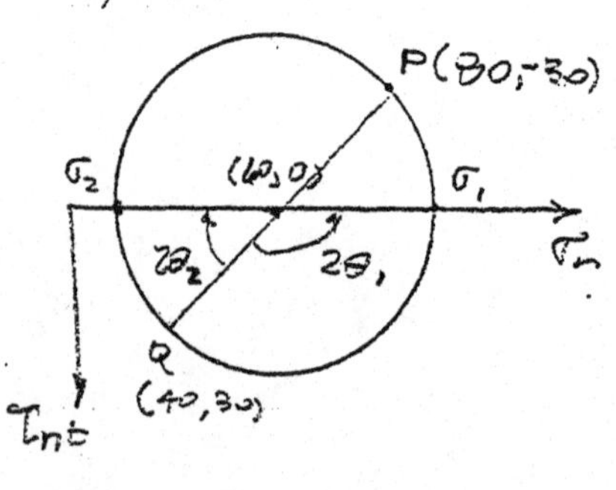

2.31 a). $\sigma_n = 100(\sin^2\theta + \cos^2\theta) + 1000 \sin\theta\cos\theta = 400$

$100 + 500\sin 2\theta = 400$

$\sin 2\theta = 3/5 \Rightarrow \theta = 18.43^\circ$

b). $\sigma_n = 100(\cos^2\theta - \sin^2\theta) + 300\sin\theta\cos\theta = 0$

$100\cos 2\theta + 150\sin 2\theta = 0$

$\Rightarrow \tan 2\theta = -2/3 \Rightarrow \theta = -16.85^\circ$

c). $c(\cos^2\theta - \sin^2\theta) - 2c\sin\theta\cos\theta = 0$

$c\cos 2\theta - c\sin 2\theta = 0$

$\tan 2\theta = 1 \Rightarrow \theta = 22.5^\circ$

2.37 a) $\sigma_n = \sigma_x = 100$; $\theta_s = 45^\circ$, $\theta_t = 90 \Rightarrow \sigma_t = \sigma_y = 20$

$\sigma_s = 100(1/2) + 20(1/2) + 2\tau_{xy}(1/2) = 50$

$\tau_{xy} = -10$ $\quad\quad$ $\tan 2\theta = \frac{-10}{40} = -1/4 \Rightarrow \theta_1 = -7.02^\circ$

$\sigma_{1,2} = 60 \pm [40^2 + 10^2]^{1/2} = 101.2,\ 18.8$ MPa

b). $\sigma_n = \sigma_x = 100$, $\theta_s = 60^\circ$, $(\sin\theta_s = \sqrt{3}/2, \cos\theta_s = 1/2)$, $\theta_t = 120^\circ$ $(\sin\theta_t = \sqrt{3}/2, \cos\theta_t = -1/2)$

$\sigma_s = 100(1/4) + \sigma_y(3/4) + 2\tau_{xy}(1/2)(+\sqrt{3}/2) = -20$

$100 + 3\sigma_y + 2\sqrt{3}\tau_{xy} = -80 \Rightarrow$ $\quad\quad$ $3\sigma_y + 2\sqrt{3}\tau_{xy} = -180$

$\sigma_t = 100(1/4) + \sigma_y(3/4) + 2\tau_{xy}(-\sqrt{3}/4) = 60$

$100 + \sigma_y - 2\sqrt{3}\tau_{xy} = 240 \Rightarrow$ $\quad\quad$ $3\sigma_y - 2\sqrt{3}\tau_{xy} = 140$

$\therefore 6\sigma_y = -40 \Rightarrow \sigma_y = -20/3$, $\tau_{xy} = -80/\sqrt{3} = -80\sqrt{3}/3$

$\sigma_{1,2} = 117.2,\ -23.7$ MPa. $\quad$ $\tan 2\theta = \frac{-80\sqrt{3}/3}{320/3} = -0.433 \Rightarrow \theta_1 = -11.7^\circ$

2.38 (a), (b): $\sigma_x = F/A$, $\tau_{x\theta} = 4FR^2/J = 4FR^2/Ak^2$; $\tan 2\theta = \frac{4R^2/k^2A}{1/2A} = 8R^2/k^2$;

$\theta = \frac{1}{2}\tan^{-1}(8R^2/k^2)$

c). $\sigma_x = 0$, $\tau_{x\theta} = TR/J \Rightarrow \sigma_{1,2} = TR/J$, $\tan 2\theta \to \infty \Rightarrow \theta = \pm 45^\circ$

2.39 a)

$\sigma_1 = -20 \pm [\overline{100}^2 + \tau_{xy}^2]^{1/2} = 220$

$\overline{100}^2 + \tau_{xy}^2 = (240)^2$

$\tau_{xy} = \pm 218.2$ If $\tau_{xy} > 0$ $\tan 2\theta = 218/100 = 2.18 \Rightarrow \theta_{1,2} = 32.7^\circ, -57.3^\circ, \sigma_2 = -260$

$\sigma_2 = -260$ If $\tau_{xy} < 0$ $\tan 2\theta = -2.18 \Rightarrow \theta_{1,2} = -32.7^\circ, 57.3^\circ$

b) $\sigma_1 = 100 \pm [(-20)^2 + \tau_{xy}^2]^{1/2} = 220$

$400 + \tau_{xy}^2 = \overline{120}^2 \Rightarrow \tau_{xy} = \pm 118.3$ (i) $\tau_{xy} = 118.3$; $\tan 2\theta = \frac{118.3}{-20} = -5.92$, $\theta_1 = 49.8^\circ$

(ii) $\tau_{xy} = -118.3$; $\tan 2\theta = -118.3/-20 = 5.92 \Rightarrow \theta_1 = -49.8^\circ, \theta_2 = 40.2^\circ$

$\sigma_2 = -20$

c). $\sigma_1 = \sigma_x/2 + 20 + \left[\left(\frac{\sigma_x - 40}{2}\right)^2 + 900\right]^{1/2} = 80$

$(\sigma_x/2 - 60)^2 = [\,1/4\,(\sigma_x^2 - 80\sigma_x + 1600) + 900]$

$\sigma_x^2/4 - 60\sigma_x + 3600 = \sigma_x^2/4 - 20\sigma_x + 1300$

$40\sigma_x = 2300 \Rightarrow \sigma_x = 57.5$ MPa

$\tan 2\theta = -30/8.75 = -3.43$

$\theta_{1,2} = -36.9^\circ, 53.1^\circ$

$\sigma_2 = 17.5$ MPa

d) $\sigma_2 = \sigma_x/2 + 20 - \left[\left(\frac{\sigma_x - 40}{2}\right)^2 + \overline{30}^2\right]^{1/2} = -80$

$(\sigma_x/2 + 100)^2 = \left[\left(\frac{\sigma_x - 40}{2}\right)^2 + \overline{30}^2\right]^{1/2}$

$\sigma_x^2/4 + 100\sigma_x + \overline{100}^2 = [\,\sigma_x^2/4 - 20\sigma_x + 400 + 900]$

$120\sigma_x = 1300 - 10{,}000 = -8700$

$\sigma_x = -72.5$ MPa ; $\tan 2\theta = \frac{-30}{-56.25} = 0.533$; $\theta_1 = -75.97, \theta_2 = 14^\circ$

$\sigma_1 = 47.5$ MPa

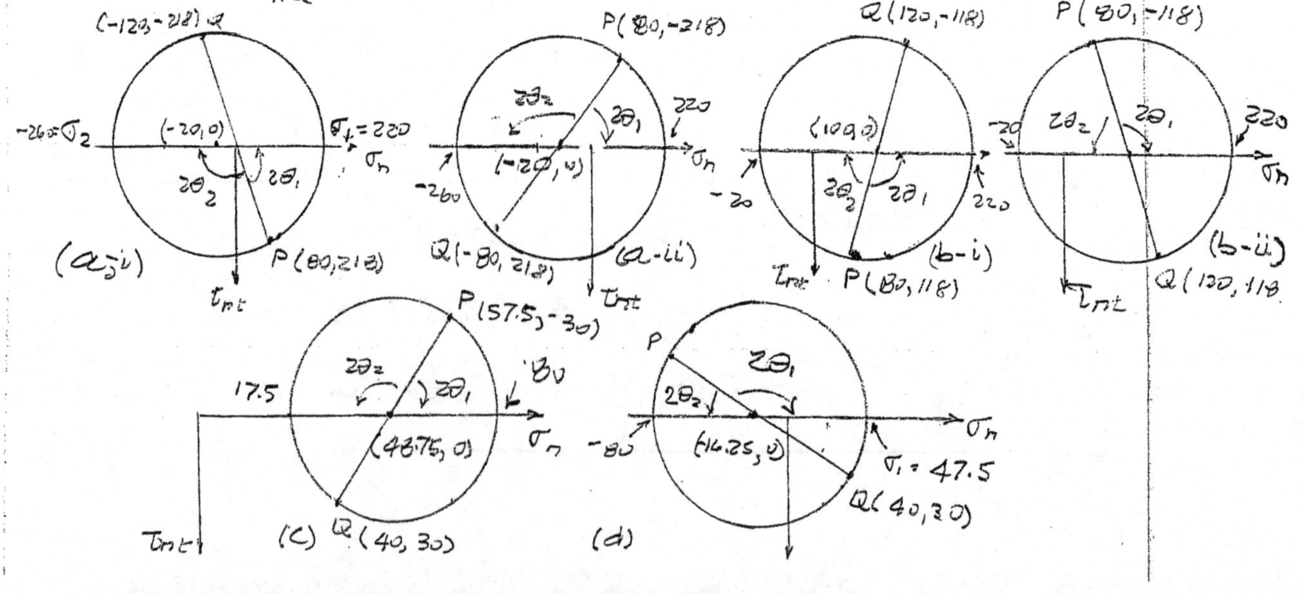

2.40

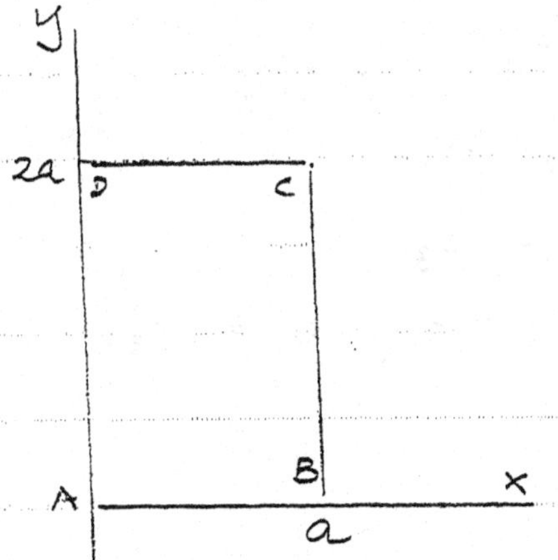

$\sigma_x = c_1 \sin \pi x/2a, \quad \sigma_y = c_2 y^2 \sin \pi x/2a$

$\tau_{xy} = c_3 y \cos \pi x/2a$

$k = \pi/2a$

Eq. eq (i) $c_1 k \cos kx + c_3 \cos kx = 0 \Rightarrow c_3 = -c_1 k$

Eq. eq (ii) $-c_3 k y \sin kx + 2 c_2 y \sin kx = 0$

$c_2 = c_3 k/2 = -c_1 k^2/2.$

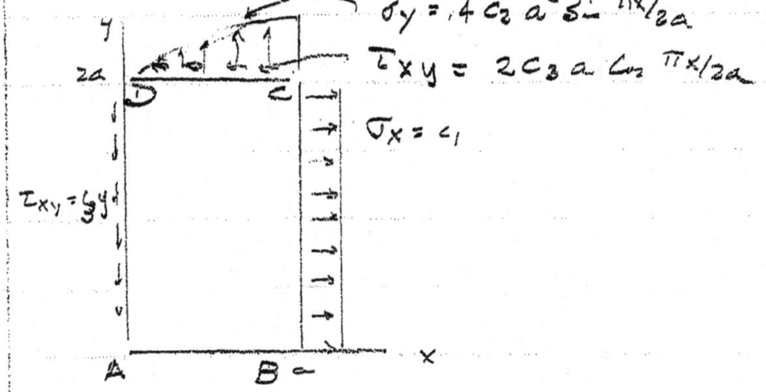

$$\Sigma F_x = \int_B^C \sigma_x(y)\,dy + \int_D^C \tau_{xy}(x)\,dx = c_1(2a) + 2c_3 a \int_0^a \cos kx$$

$$= 2c_1 a + 2c_3 \frac{a}{k} \sin kx\Big|_0^a = 2a\left(c_1 + \frac{c_3}{k} \sin ka\right) = 0 \quad \text{since } c_3 = -c_1 k.$$

$$\Sigma F_y = \int_A^D \tau_{xy}(y)\,dy + \int_D^C \sigma_y(x)\,dx$$

$$= -c_3 \int_0^{2a} y\,dy + 4c_2 a^2 \int_0^a \sin kx\,dx$$

$$= -c_3 y^2/2\Big|_0^{2a} - 4c_2 \frac{a^2}{k} \cos kx\Big|_0^a = -2c_3 a^2 + 4c_2 a^2/k = 2a^2(-c_3 + 2c_2/k)$$

$$= 0 \quad \text{since } c_3 = 2c_2/k.$$

$$\overset{+}{\curvearrowleft}\Sigma M_A = -\sigma_x\Big|_{BC} (2a)\cdot a + \int_D^C \sigma_y(x)\,x\,dx - 2a \int_D^C \tau_{xy}(x)\,dx$$

$$= -2c_1 a^2 + 4a^2 c_2 \int_0^a x \sin kx\,dx - 4a^2 c_3 \int_0^a \cos kx\,dx$$

$$= a^2\left\{-2c_1 + 4c_2\left[-\frac{x}{k}\cos kx\Big|_0^a + \frac{1}{k}\int_0^a \cos kx\,dx\right] - 4\frac{c_3}{k} \sin kx\Big|_0^a\right\}$$

$$= a^2\left\{-2c_1 + 4c_2\left[-\frac{a}{k}\cos ka + \frac{1}{k^2}\sin ka\right] - 4\frac{c_3}{k}\sin ka\right\}$$

$$= a^2\left\{-2c_1 + \frac{4c_2}{k^2} - \frac{4c_3}{k}\right\} = a^2\{-2 - 2 + 4\}c_1 = 0 \checkmark$$

2.41

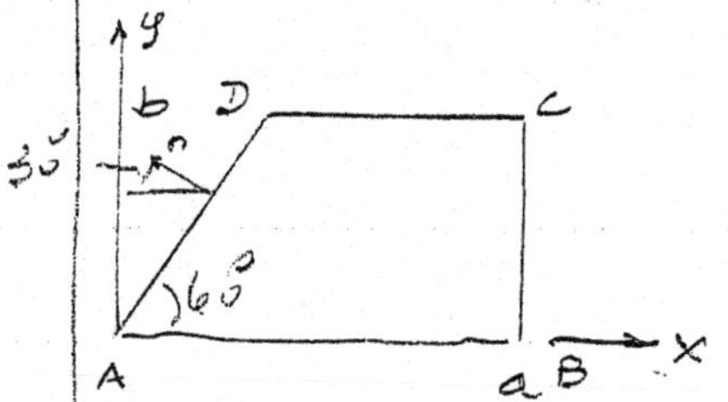

Given: $\sigma_x = c_1 x^2 y$, $\sigma_y = c_2 y^3$

$\tau_{xy} = c_3 x y^2$.

Eq. eq.: $2c_1 xy + 2c_3 xy = 0 \Rightarrow c_3 = -c_1$

$c_3 y^2 + 3c_2 y^2 = 0 \Rightarrow c_2 = -\frac{1}{3} c_3 = +\frac{c_1}{3}$

$\tau_{xy} = -c_1 b^2 x$

$\sigma_y = c_2 b^3 = \frac{c_1}{3} b^3$

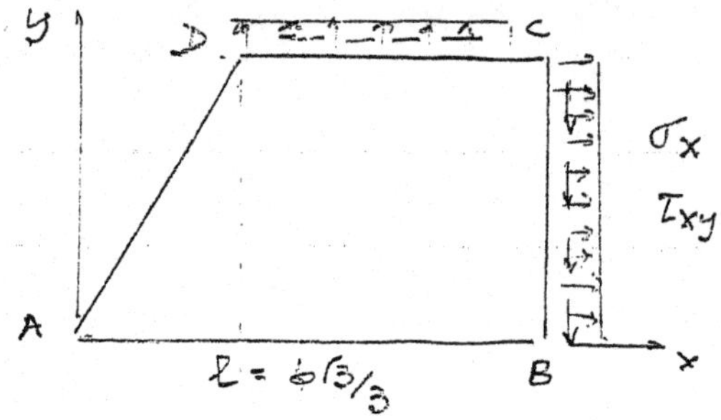

$\sigma_x = c_1 a^2 y$

$\tau_{xy} = +c_3 a y^2 = -c_1 a y^2$

A-B: $X_n = Y_n = 0$

B-C: $X_n = c_1 a^2 y$, $Y_n = -c_1 a y^2$

C-D: $X_n = -c_1 b^2 x$, $Y_n = c_2 b^3 = c_1 b^3/3$

A-D: $l_x = -\sqrt{3}/2$, $l_y = 1/2$, $y = \sqrt{3}x \Rightarrow x = \sqrt{3}y/3$

$$X_n = (c_1 x^2 y)(-\sqrt{3}/2) - (c_1 x y^2)(1/2) = -\frac{c_1}{2}[\sqrt{3} x^2 y + x y^2]$$

$$X_n(y) = -c_1/2\,[\sqrt{3}y^3/3 + \sqrt{3}/3\, y^3] = -c_1\sqrt{3}\,y^3/3$$

$$Y_n = (-\sqrt{3}/2)(-c_1 x y^2) + 1/2\,(\tfrac{c_1}{3}\cdot y^3) = \frac{c_1}{2}[\sqrt{3} x y^2 + \tfrac{1}{3} y^3]$$

$$Y_n(y) = c_1/2\,[y^3 + y^3/3] = 2c_1 y^3/3.$$

A-B: $F_x = F_y = 0$

B-C: $F_x = \int_0^b X_n(y)\,dy = c_1 a^2 b^2/2$, $F_y = -\int_0^b Y_n(y)\,dy = -c_1 a b^3/3$.

C-D: $F_x = \int_\ell^a X_n(x)\,dx = -c_1 b^2 \int_\ell^a x\,dx = -\frac{c_1 b^2}{2} x^2\Big|_\ell^a = -\frac{c_1}{2} b^2 [a^2 - b^2/3]$

$F_y = \int_\ell^a Y_n(x)\,dx = c_1 b^3/3\,[a - b\sqrt{3}/3]$

A-D: $dy/ds = \sqrt{3}/2 \Leftrightarrow ds = \frac{2\sqrt{3}}{3}dy$

$$F_x = \int X_n\,ds = \frac{2\sqrt{3}}{3}\int_0^b X_n(y)\,dy = \frac{2\sqrt{3}}{3}\left(-\frac{c_1\sqrt{3}}{3}\right)\int_0^b y^3\,dy$$

$$= -(2c_1/3)(b^4/4) = -c_1 b^4/6$$

$$F_y = \int Y_n\,ds = \frac{2\sqrt{3}}{3}\int_0^b Y_n(y)\,dy = \frac{4c_1\sqrt{3}}{3}\int_0^b y^3\,dy = c_1\sqrt{3}b^4/3.$$

$$\sum F_x = c_1\{a^2b^2/2 - \frac{b^2}{2}(a^2 - b^2/3) - b^4/6\} = c_1 b^2[a^2/2 - a^2/2 + \frac{b^2}{6} - \frac{b^2}{6}] = 0 \checkmark$$

$$\sum F_y = c_1\{-ab^3/3 + b^3/3(a - b\sqrt{3}/3) + \sqrt{3}b^4/3\}$$
$$= c_1 b^3/3[-a + a - \sqrt{3}b/3 + \sqrt{3}b/3] = 0 \checkmark$$

$$\overset{\curvearrowleft}{+}\sum M_A = -\int_B^C X_n(y)\cdot y\,dy + a\int_B^C Y_n(y)\,dy + \int_C^D x Y_n(x)\,dx - b\int_C^D X_n(x)\,dx$$
$$-\int_A^D y X_n(s)\,ds + \int_A^D x\,Y_n(s)\,ds = 0$$

$ds = (2\sqrt{3}/3)\,dy$ on AD $\qquad x = y/\sqrt{3} = \sqrt{3}y/3$

$\therefore$

$$\sum M_A = -\int_0^b c_1 a^2 y^2\,dy - a^2 c_1\int_0^b y^2\,dy + \frac{c_1 b^3}{3}\int_l^a x\,dx + b^3 c_1\int_l^a x\,dx$$
$$+\frac{2\sqrt{3}}{3}\left\{-\left(-\frac{c_1\sqrt{3}}{3}\right)\int_0^b y^4\,dy + \frac{2c_1}{3}\int_0^b \frac{\sqrt{3}}{3}y^4\,dy\right\} = 0$$

$$\sum M_A = c_1\left\{-a^2\int_0^b y^2\,dy - a^2\int_0^b y^2\,dy + \frac{b^3}{3}\int_l^a x\,dx + b^3\int_l^a x\,dx\right.$$
$$\left. + 2\left[\frac{1}{3}\int_0^b y^4\,dy + \frac{2}{9}\int_0^b y^4\,dy\right]\right\} = 0$$

$$\Sigma M_A = c_1 \left\{ -2a^2 \int_0^b y^2 dy + \frac{4b^3}{3} \int_l^a x\, dx + \frac{10}{9} \int_0^b y^4 dy \right\} = 0$$

$$\therefore -2a^2 y^3/3 \Big|_0^b + (4b^3/3)\, x^2/2 \Big|_l^a + \frac{2}{9} y^5 \Big|_0^b = 0$$

$$-2a^2 b^3/3 + \frac{4b^3}{6}(a^2 - l^2) + \frac{2}{9} b^5 = 0$$

$$-\frac{2}{3} l^2 + \frac{2}{9} b^2 = 0$$

$l = \sqrt{3}b/3$

$$\therefore -\frac{2}{3} \cdot \frac{1}{3} b^2 + \frac{2}{9} b^2 = 0 \quad \checkmark$$

3.1 a) $\Delta L = 1\,cm$, $\bar{\varepsilon} = \Delta L/L = 1/200 = 0.005$

b). $\varepsilon = cx \Rightarrow \Delta L = c\int_0^L x\,dx = cL^2/2 = 1 \Rightarrow c = \frac{2}{L^2} = \frac{2}{4\times 10^4} = 0.5\times 10^{-4}$

$\varepsilon_{max} = (0.5\times 10^{-4})(200) = 10^{-2}$.

c). $\varepsilon(L/2) = cL/2 = 0.5\times 10^{-2}$,

d). $u(x) = cx^2/2 = (0.25\times 10^{-4})x^2$

3.2

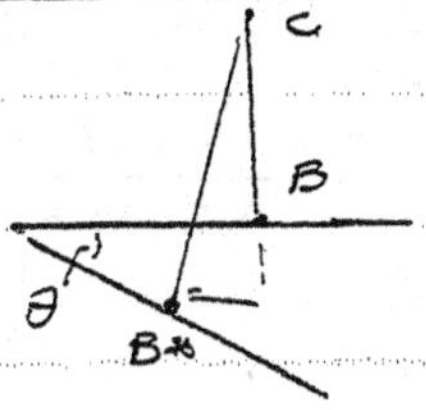

$\Delta_{BC} = [(L - L\cos\theta)^2 + (a + L\sin\theta)^2]^{1/2} - a$

$\bar{\varepsilon}_{BC} = \frac{\Delta_{BC}}{a} = 1/a[\,2L^2(1-\cos\theta) + a^2 + 2aL\sin\theta]^{1/2} - 1$

$|\theta| \ll 1 \Rightarrow \bar{\varepsilon}_{BC} \cong 1/a[\,L^2\theta^2 + a^2 + 2aL\theta]^{1/2} - 1 \cong [1 + 2\theta L/a]^{1/2} - 1 = L\theta/a$

3.3 $\Delta L = \int_0^{200} [0.001 + cy/200]\,dy = 0.60$

$0.001y + cy^2/400\Big]_0^{200} = 0.60 \Rightarrow 0.2 + 100c = 0.6$

$c = 0.4\times 10^{-2} \Rightarrow \varepsilon_A = 0.0014$

3.4

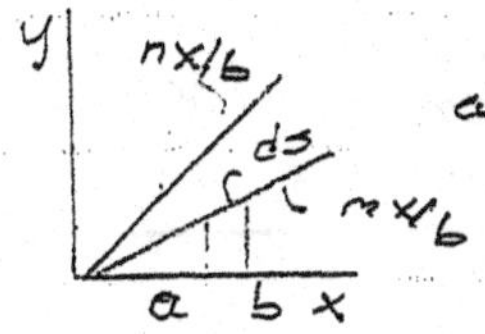

a) $\Delta s = (\Delta x^2 + \Delta y^2]^{1/2} = [1 + (\Delta y/\Delta x)^2]^{1/2}\Delta x$

$ds = [1 + (dy/dx)^2]^{1/2}dx = [1 + m^2/b^2]^{1/2}dx$

$ds^* = [1 + n^2/b^2]^{1/2}dx^*$

$x^*/b = 1/2(x/b)^2 \Rightarrow dx^* = (x/b)dx \Rightarrow ds^* = \frac{1}{b}(1 + n^2/b^2)^{1/2}x\,dx$

$\varepsilon = \frac{\Delta s^* - \Delta s}{\Delta s} = \left[\frac{1 + n^2/b^2}{1 + m^2/b^2}\right]^{1/2}(x/b) - 1$

$L = \int_a^b ds = (1 + m^2/b^2)^{1/2}(b - a)$

$L^* = \int_a^b ds^* = \frac{1}{b}(1 + n^2/b^2)^{1/2}\int_a^b x\,dx = (1 + n^2/b^2)^{1/2}\left(\frac{b^2 - a^2}{2b}\right)$

$\bar{\varepsilon} = \frac{L^* - L}{L} = 1/2\left[\frac{1 + n^2/b^2}{1 + m^2/b^2}\right]\left(\frac{b-a}{b}\right) - 1$

3.5 $\Delta s = [1 + (\Delta y/\Delta x)^2]^{1/2} \Delta x \to ds = [1 + 4a^2 x^2]^{1/2} dx$

$ds^* = [1 + 4b^2 x^{*2}]^{1/2} dx$

$x^* = bx/a \Rightarrow dx^* = (b/a)dx$

$$\varepsilon_n = \frac{ds^* - ds}{ds} = \frac{(1 + 4b^2x^2)^{1/2}}{(1 + 4a^2x^2)^{1/2}} \cdot (b/a) - 1$$

$\varepsilon_n(0) = b/a - 1$, $\varepsilon_n(x = b/a) = \left(\frac{1 + 4b^4/a^2}{1 + 4b^2}\right)^{1/2} (b/a) - 1$

3.6 $$\varepsilon_n - \tilde{\varepsilon}_n = \lim_{\substack{\Delta s \to 0 \\ \Delta s^* \to 0}} \left[\frac{\Delta s^* - \Delta s}{\Delta s} - \frac{\Delta s^* - \Delta s}{\Delta s^*}\right] = \lim \left[\frac{(\Delta s^*)^2 + (\Delta s)^2 - 2\Delta s \cdot \Delta s^*}{\Delta s\, \Delta s^*}\right]$$

$$= \lim \left[\frac{(\Delta s^* - \Delta s)}{\Delta s}\frac{(\Delta s^* - \Delta s)}{\Delta s^*}\right] = \varepsilon_n(P)\,\tilde{\varepsilon}_n(P)$$

3.7 $$\Delta L = \int \varepsilon_\theta(\theta) R\, d\theta = kR \int_0^{2\pi} \cos^2\theta\, d\theta = \frac{kR}{2}\int_0^{2\pi} (1 + \cos 2\theta)\, d\theta = \pi k R$$

3.8 a) $\bar{\varepsilon} = \Delta L / L = 6/40\times10^6 = 0.15\times10^{-6}$

3.9 $A = \pi R^2 \to dA = 2\pi R \cdot dR \Rightarrow \Delta L/L = \Delta R/R \Rightarrow \bar{\varepsilon} = dR/R = dA/2\pi R^2$

$$\bar{\varepsilon} = \frac{0.5\pi}{2\pi(25)^2} = 0.25/(25)^2 = 4\times10^{-4}$$

3.10

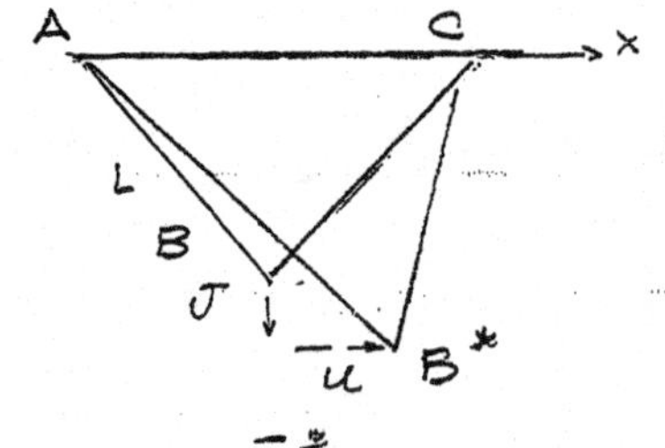

$$(\overline{AB}^*)^2 = (L\sqrt{2}/2 + u)^2 + (L\sqrt{2}/2 + v)^2 = L^2(\sqrt{2}/2 + u/L)^2 + L^2(\sqrt{2}/2 + v/L)^2$$

$$= L^2[1 + \sqrt{2}(u/L + v/L) + u^2/L^2 + v^2/L^2]$$

$$\overline{AB}^* = L[1 + \sqrt{2}(u/L + v/L) + u^2/L^2 + v^2/L^2]^{1/2}$$

$\bar{\varepsilon}_{AB} = \overline{AB}^*/L - 1$; $\bar{\varepsilon}_{AB}\Big|_{\substack{u/L \ll 1 \\ v/L \ll 1}} \cong [1 + \sqrt{2}(u/L + v/L)]^{1/2} - 1 \cong [1 + \frac{\sqrt{2}}{2}(u/L + v/L)] - 1$

$= \frac{\sqrt{2}}{2L}(u + v)$

Similarly:

$$\bar{\varepsilon}_{BC}\Big|_{\substack{u/L \ll 1 \\ v/L \ll 1}} = \frac{\sqrt{2}}{2L}(v - u)$$

3.11

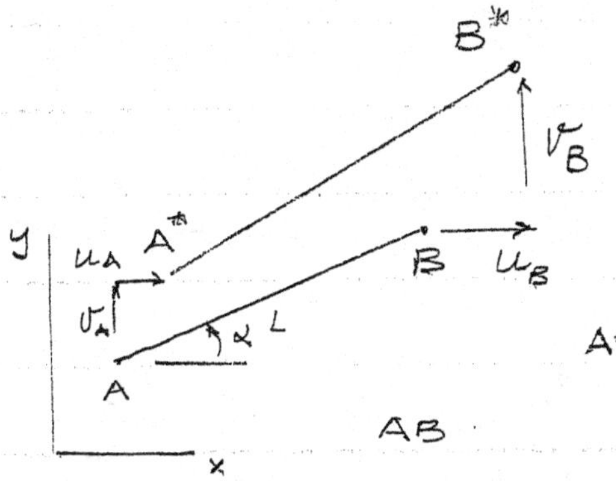

$AB = L \quad , \quad u_A/L \ll 1, \; v_A/L \ll 1 \;$ etc.

$$\overline{A^*B^*} = \left[(L\cos\alpha + u_B - u_A)^2 + (L\sin\alpha + v_B - v_A)^2 \right]^{1/2}$$

$$\overline{A^*B^*} = \left\{ L^2 + 2L(u_B - u_A)\cos\alpha + (u_B - u_A)^2 + 2L(v_B - v_A)\sin\alpha + (v_B - v_A)^2 \right\}^{1/2}$$

$$\cong L\left\{ 1 + \frac{2(u_B - u_A)\cos\alpha + 2(v_B - v_A)\sin\alpha}{L} \right\}^{1/2}$$

$$\bar{\varepsilon}_n = \frac{\overline{A^*B^*}}{L} - 1 \cong \left(\frac{u_B - u_A}{L}\right)\cos\alpha + \left(\frac{v_B - v_A}{L}\right)\sin\alpha$$

3.12

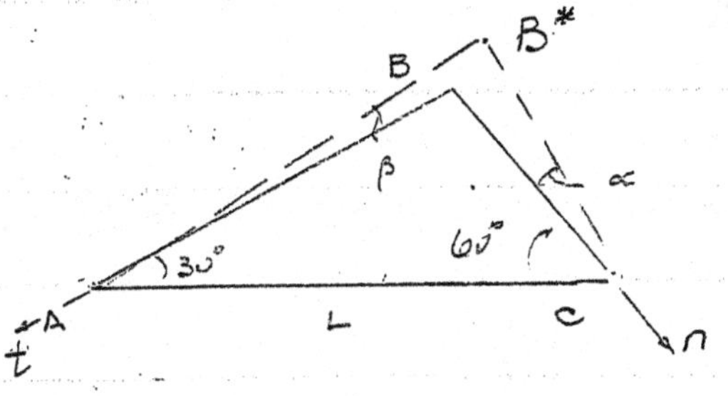

$$\tan\alpha = \frac{(L\cos 30^\circ)\varepsilon_t}{L\sin 30}, \quad \tan\beta = \frac{(L\sin 30^\circ)\varepsilon_n}{L\cos 30^\circ}$$

$$\tan\alpha \approx \alpha = \varepsilon_t \cot 30$$

$$\tan\beta \approx \beta = \varepsilon_n \tan 30^\circ$$

$$\gamma_{nt} = \alpha + \beta = \varepsilon_n \tan 30 + \varepsilon_t \cot 30^\circ = \varepsilon_n\sqrt{3}/3 + \varepsilon_t\sqrt{3} = \sqrt{3}(\varepsilon_t + \varepsilon_n/3) = \sqrt{3}(.007)$$

3.13

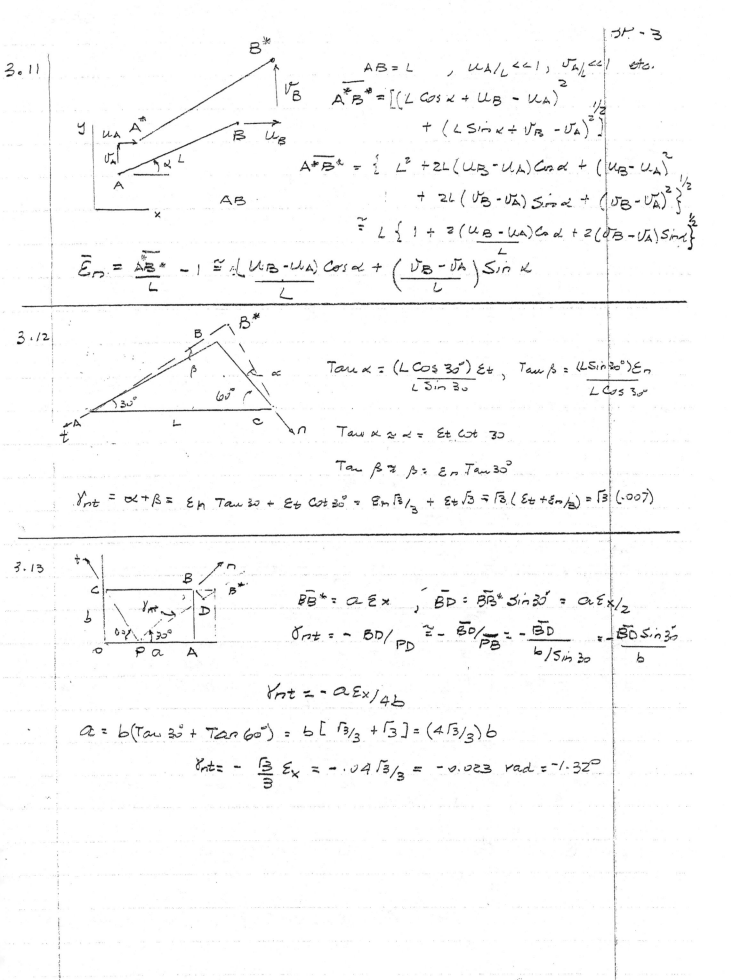

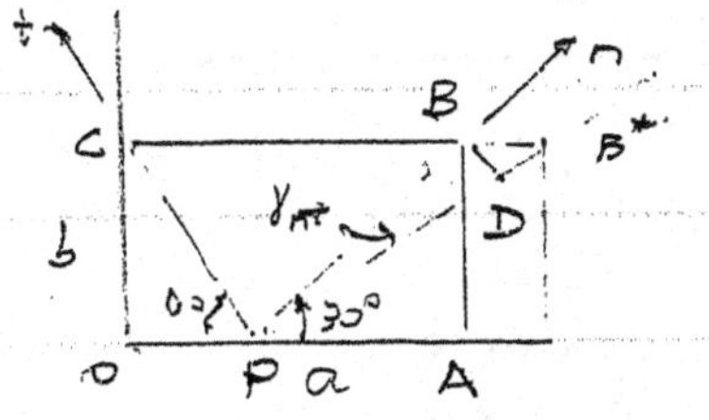

$$\overline{BB^*} = a\varepsilon_x \quad , \quad \overline{BD} = \overline{BB^*}\sin 30^\circ = a\varepsilon_x/2$$

$$\gamma_{nt} = -BD/PD \cong -\overline{BD}/\overline{PB} = -\frac{\overline{BD}}{b/\sin 30} = -\frac{\overline{BD}\sin 30^\circ}{b}$$

$$\gamma_{nt} = -a\varepsilon_x/4b$$

$$a = b(\tan 30^\circ + \tan 60^\circ) = b\left[\sqrt{3}/3 + \sqrt{3}\right] = (4\sqrt{3}/3)b$$

$$\gamma_{nt} = -\frac{\sqrt{3}}{3}\varepsilon_x = -.04\sqrt{3}/3 = -0.023 \text{ rad} = -1.32^\circ$$

3.14

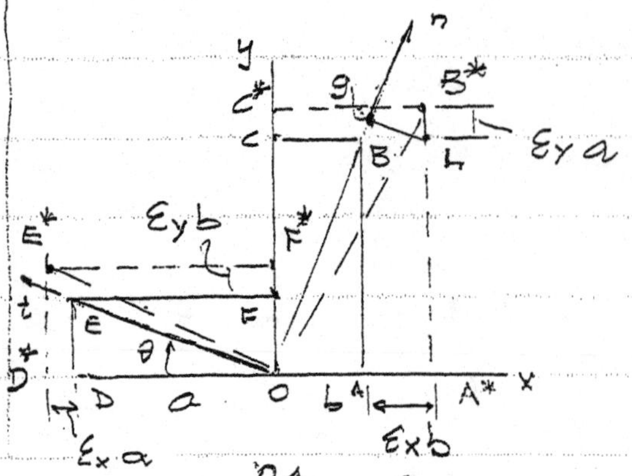

$\gamma_{nt} = \angle EOE^* - \angle BOB^*$, $L = (a^2+b^2)^{1/2}$

$\overline{AA^*} = \overline{Bh} = \varepsilon_x b$; $\overline{CC^*} = \overline{B^*h} = \varepsilon_y a$

$\overline{DD^*} = \overline{Ek} = \varepsilon_x a$; $\overline{FF^*} = \overline{E^*k} = \varepsilon_y b$

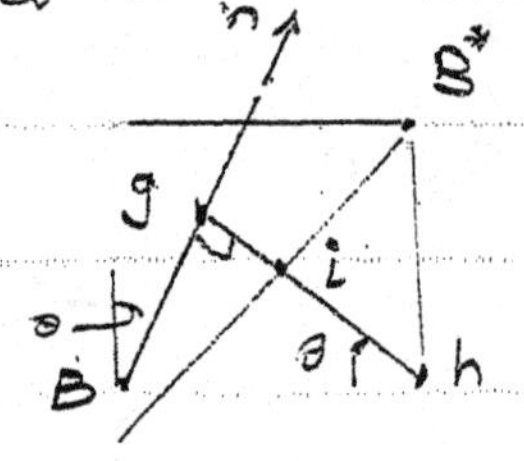

$\overline{gi} = \overline{gh} - \overline{ih} = \varepsilon_x b\cos\theta - \varepsilon_y a\sin\theta$; $\angle BOB^* = \overline{gi}/L$

$\angle BOB^* = \overline{gi}/L$

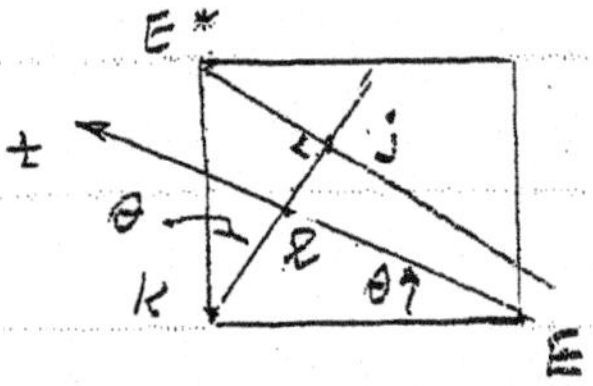

$\overline{jl} = jk - kl = \varepsilon_y b\cos\theta - \varepsilon_x a\sin\theta$; $\angle EOE^* = \overline{jl}/L$

$\gamma_{nt} = \overline{jl}/L - \overline{gi}/L = \varepsilon_y(b/L\cos\theta + a/L\sin\theta) - \varepsilon_x(a/L\sin\theta + b/L\cos\theta)$

$a/L = \cos\theta$, $b/L = \sin\theta$

$\gamma_{nt} = 2(\varepsilon_y - \varepsilon_x)\sin\theta\cos\theta.$

3.15

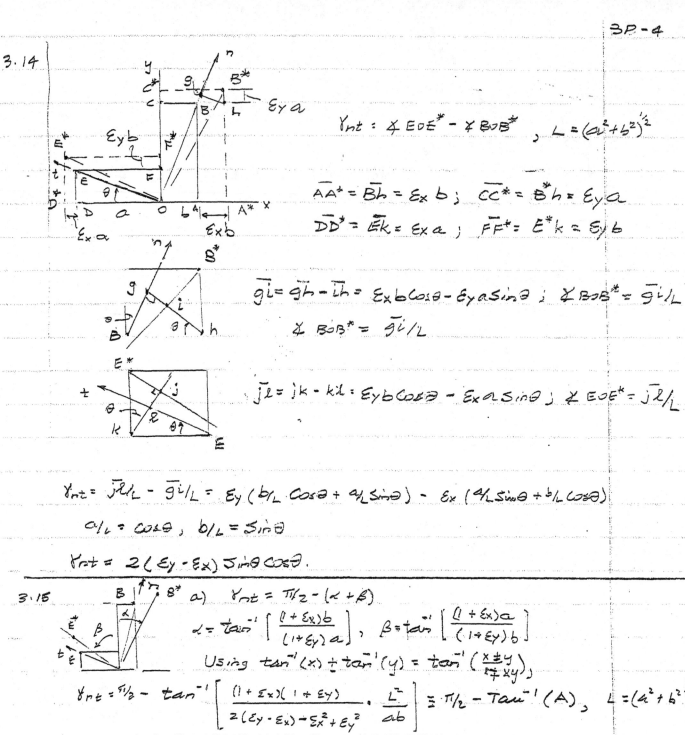

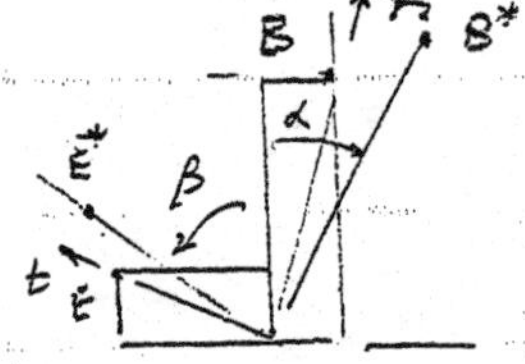

a) $\gamma_{nt} = \pi/2 - (\alpha + \beta)$

$\alpha = \tan^{-1}\left[\dfrac{(1+\varepsilon_x)b}{(1+\varepsilon_y)a}\right]$, $\beta = \tan^{-1}\left[\dfrac{(1+\varepsilon_x)a}{(1+\varepsilon_y)b}\right]$

Using $\tan^{-1}(x) \pm \tan^{-1}(y) = \tan^{-1}\left(\dfrac{x \pm y}{1 \mp xy}\right)$,

$$\gamma_{nt} = \pi/2 - \tan^{-1}\left[\frac{(1+\varepsilon_x)(1+\varepsilon_y)}{2(\varepsilon_y - \varepsilon_x) - \varepsilon_x^2 + \varepsilon_y^2} \cdot \frac{L^2}{ab}\right] \equiv \pi/2 - \mathrm{Tan}^{-1}(A), \quad L = (a^2+b^2)^{1/2}$$

b)

Note: $\pi/2 = \lim_{z\to\infty} \tan^{-1}(z). \Rightarrow \gamma_{nt} = \lim_{z\to\infty} \tan^{-1}(z) - \tan^{-1}(A).$

$$\therefore \gamma_{nt} = \lim_{z\to\infty} \tan^{-1}\left[\frac{z - A}{1 + zA}\right] = \tan^{-1}(1/A) = \tan^{-1}\left[\frac{2(\varepsilon_y - \varepsilon_x) - \varepsilon_x^2 + \varepsilon_y^2}{(1+\varepsilon_x)(1+\varepsilon_y)} \cdot \frac{ab}{L^2}\right]$$

$|\varepsilon_x| \ll 1, |\varepsilon_y| \ll 1 \Rightarrow \gamma_{nt} = \tan^{-1}[2(\varepsilon_y - \varepsilon_x) \cdot ab/L^2] \ll 1$

$\tan^{-1}(x) \sim x$, $(x \ll 1)$

$\gamma_{nt} = 2(\varepsilon_y - \varepsilon_x)\cos\theta\sin\theta$, since $a/L = \cos\theta$, $b/L = \sin\theta$

3.16 a) $\varepsilon_x = \partial u/\partial x = a$; $\varepsilon_y = \partial v/\partial y = c$, $\gamma_{xy} = \partial u/\partial y + \partial v/\partial x = 2b$

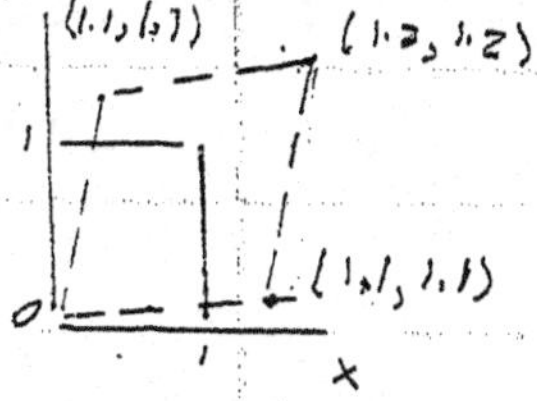

b). $\varepsilon_x = a$, $\varepsilon_y = 0$, $\gamma_{xy} = 0$

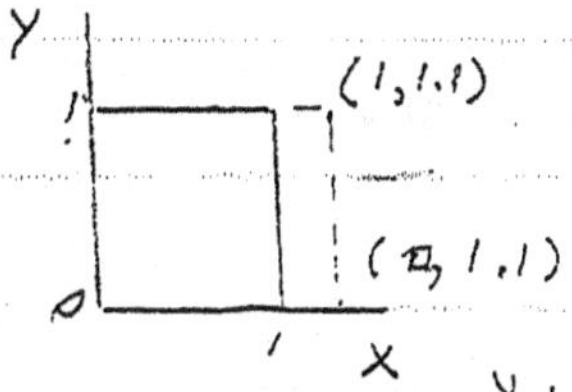

c). $\varepsilon_x = a + by$, $\varepsilon_y = 0$, $\gamma_{xy} = bx - 2a$

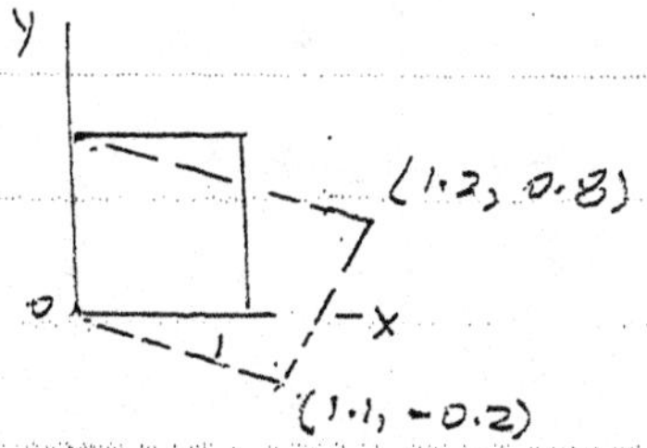

3.17 Solution I:

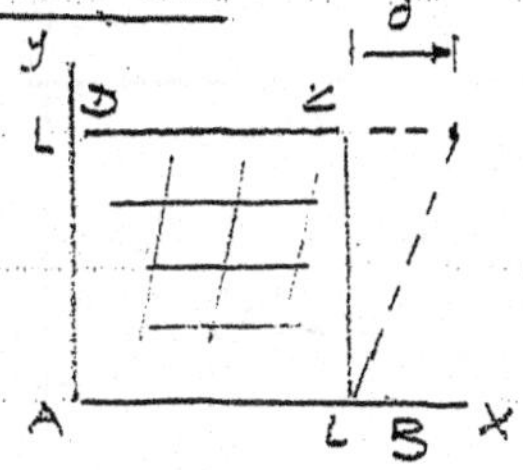

1) $\varepsilon_x(x,L) = \partial u/\partial x = c$ (const) $\Rightarrow u(x,L) = cy + A$ (i)

$\int_0^L \varepsilon_x(x,L)\,dx = cL = \delta \Rightarrow c = \delta/L \Rightarrow \varepsilon_x(x,L) = \delta/L$.

2) $v(x,y) = 0 \Rightarrow \begin{cases} \varepsilon_y = \partial v/\partial y = 0 \\ \gamma_{xy} = \partial u/\partial y + \partial v/\partial x \end{cases}$

Vertical lines remain straight $\Rightarrow \partial u/\partial y = b(x) \Rightarrow u = y\,b(x) + B$, B const.

$u(0,x) = 0 \Rightarrow B = 0 \Rightarrow u(x,y) = y\,b(x)$ (ii)

$\varepsilon_x(x,L) = L\dfrac{db(x)}{dx} = c$ from (i) $\Rightarrow db(x)/dx = c/L = \delta/L^2$.

$b(x) = \delta x/L^2 + D$, (D = const.)

$\therefore\ u = \delta xy/L^2 + Dy \quad u(0,y) = 0 \Rightarrow D = 0 \Rightarrow \gamma_{xy} = \delta x/L^2$.

$\varepsilon_x = \delta y/L^2$

Solution II:

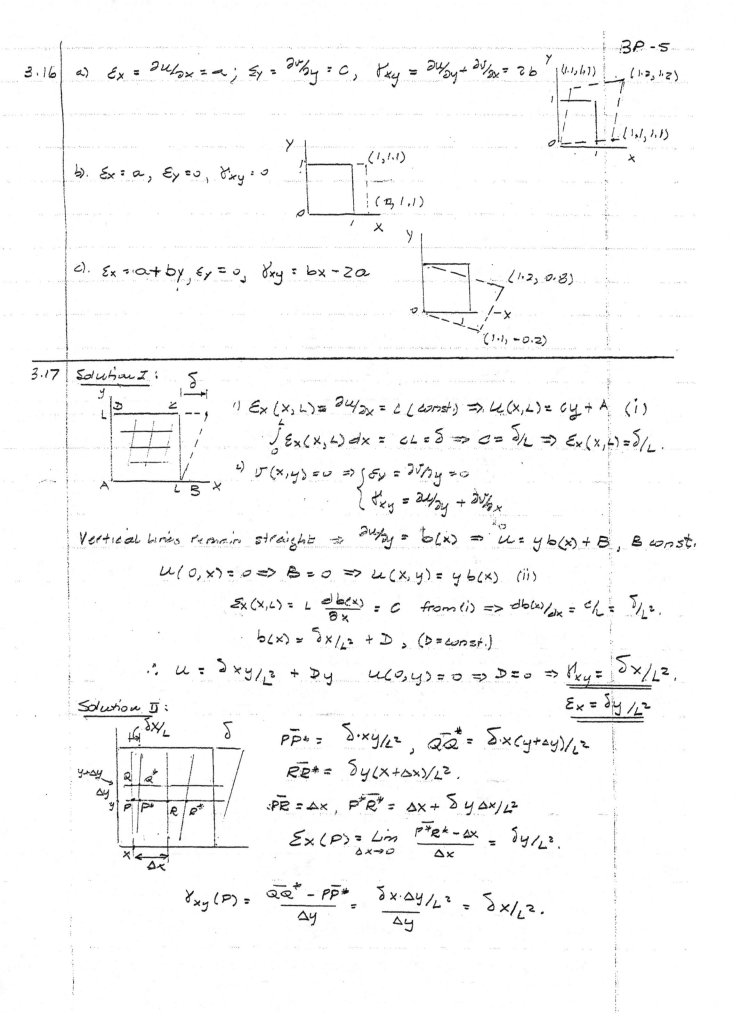

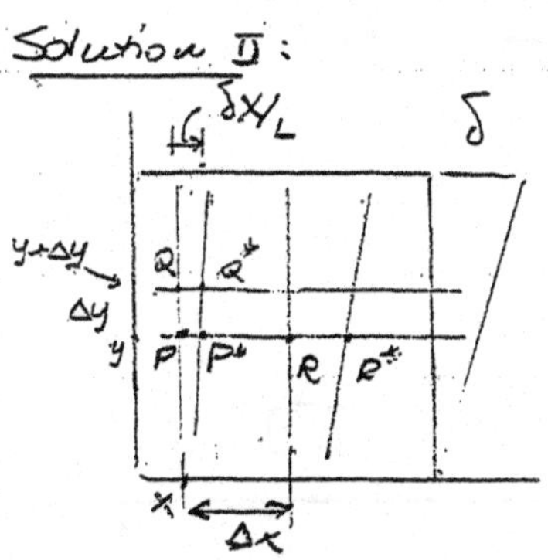

$\overline{PP^*} = \delta\cdot xy/L^2$, $\overline{QQ^*} = \delta\cdot x(y+\Delta y)/L^2$

$\overline{RR^*} = \delta y(x+\Delta x)/L^2$.

$\overline{PR} = \Delta x$, $\overline{P^*R^*} = \Delta x + \delta y\,\Delta x/L^2$

$\varepsilon_x(P) = \lim_{\Delta x\to 0} \dfrac{\overline{P^*R^*} - \Delta x}{\Delta x} = \delta y/L^2$.

$\gamma_{xy}(P) = \dfrac{\overline{QQ^*} - \overline{PP^*}}{\Delta y} = \dfrac{\delta x\cdot\Delta y/L^2}{\Delta y} = \delta x/L^2$.

3.18 $\gamma_{xy} = ax/L + by/H + c(x/L)(y/H)$

$v(x,y) = Ay$ (const.)

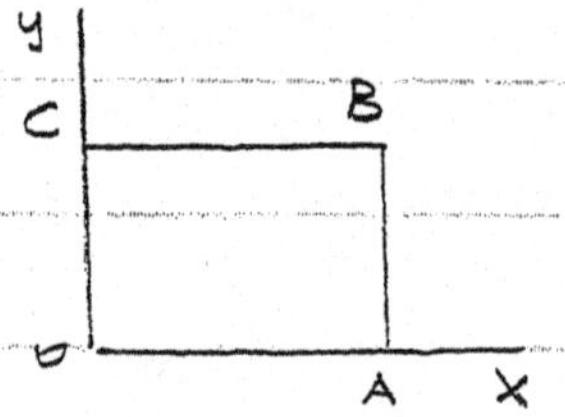

$\gamma_{xy} = \partial u/\partial y + \partial v/\partial x = \partial u/\partial y$ ($\partial v/\partial x \to 0$)

$\therefore u(x,y) = axy/L + by^2/2H + cxy^2/2LH + B(x)$

$\Delta L_{BC} = u_B - u_C = u(L,H) - u(0,L)$

$= (aH + bH/2 + cH/2) - (bH/2) = aH + cH/2.$

$\Delta L_{OA} = u_A - u_O = u(L,0) - u(0,0) = 0$

$\therefore \Delta L_{BC} - \Delta L_{OA} = aH + cH/2 = H(a + c/2)$

3.19 $u = axy/L + A(y)$, $v = bxy/L + B(x)$

$\gamma_{xy} = \left(ax/L + \frac{dA(y)}{dy}\right) + \left(by/L + dB(x)/dx\right) = ax/L + by/L$

$\Rightarrow dA(y)/dy + dB(x)/dx = 0 \Rightarrow dA(y)/dy = -dB(x)/dx = C$

$A(y) = Cy + D, \quad B(x) = -Cx + E$

$\therefore u = axy/L + Cy + D$; $u(0,0) = 0 \Rightarrow D = 0$, $u(0,L) = e \Rightarrow C = e/L$

$v = bxy/L - Cx + E$; $v(0,0) = 0 \Rightarrow E = 0$

$\therefore u(x,y) = (ax + e)(y/L)$, $v(x,y) = (by - e)(x/L)$

3.21 $I_{\varepsilon 2}/I_{\varepsilon 1} = \dfrac{\varepsilon_x\varepsilon_y - \varepsilon_{xy}^2}{\varepsilon_x + \varepsilon_y} = 4\times10^{-3} \Rightarrow \dfrac{-4\varepsilon_x^2 - 4\times10^{-6}}{-3\varepsilon_x} = 4\times10^{-3}$

$4\varepsilon_x^2 - (12\times10^{-3})\varepsilon_x + 4\times10^{-6} = 0$

$\varepsilon_x^2 - (3\times10^{-3})\varepsilon_x + 10^{-6} = 0$

$\varepsilon_x = \tfrac{1}{2}\{(3\times10^{-3}) \pm [9\times10^{-6} - 4\times10^{-6}]^{1/2}\} = \frac{10^{-3}}{2}[3 \pm \sqrt{5}]$

3.22 $R = I_{\varepsilon 2}/I_{\varepsilon 1}$ $\quad -4\varepsilon_x^2 - \varepsilon_{xy}^2 = -3\varepsilon_x R$

$4\varepsilon_x^2 - 3R\varepsilon_x + \varepsilon_{xy}^2 = 0$

$\varepsilon_x = \tfrac{1}{8}\{3R \pm [9R^2 - 16\varepsilon_{xy}^2]^{1/2}\}$

Require $16\varepsilon_{xy}^2 < 9R^2 \Rightarrow |\varepsilon_{xy}| < 3R/4$; $R = 4\times10^{-3} \Rightarrow |\varepsilon_{xy}| < 3\times10^{-3}$

NOTE: SOLUTIONS TO PROBLEMS 3.23 – 3.25, 3.27, 3.28 AND 3.29 ARE IDENTICAL TO CORRESPONDING SOLUTION OF PROBLEMS OF CHAPTER 2 PROVIDED σ AND τ ARE REPLACED BY ε, MPa IS REPLACED BY "μ" (MICRON) AND kPa IS REPLACED BY $10^{-3}/1000$, RESPECTIVELY. [E.G. $\sigma = 4$ MPa $\to \varepsilon = 4\mu$; AND $\sigma = 200$ kPa $\to \varepsilon = 0.2\times10^{-3}$]

3.23 SEE NOTE ABOVE AND SOLUTION TO PROB. 2.29

3.24 SEE NOTE ABOVE AND SOLUTION TO PROB. 2.30

3.25 SEE NOTE ABOVE AND SOLUTION TO PROB. 2.31

3.26 a) $\varepsilon_{1,2} = 30 \pm [100 + \overline{60}^2]^{1/2} = \begin{cases}90.8\\-30.8\end{cases}$

$\tan 2\theta_1 = \frac{60}{-10} = -6$

$\theta_1 = 49.7^\circ, -40.3^\circ$

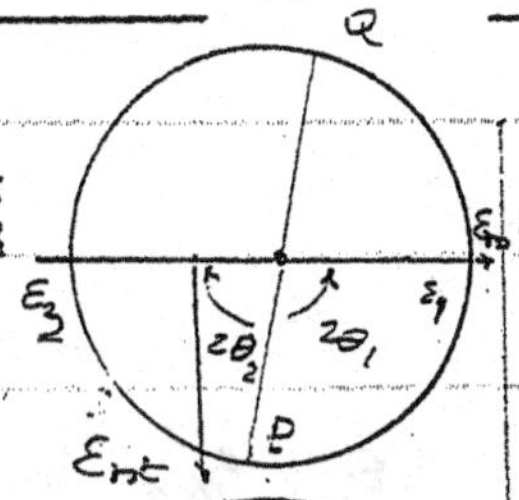

b) $\varepsilon_{1,2} = 30 \pm [100 + \overline{60}^2]^{1/2} = \begin{cases}90.8\\-30.8\end{cases}$

$\tan 2\theta_1 = \frac{-60}{-10} = 6$

$\theta_{1,2} = -49.7^\circ, 40.3^\circ$

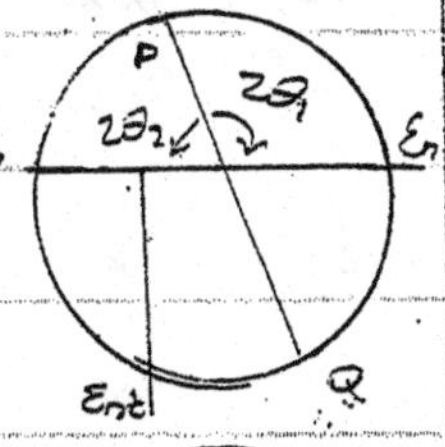

c). $\varepsilon_{1,2} = -10 \pm [900 + \overline{60}^2]^{1/2} = \begin{cases}57.1\\-77.1\end{cases}$

$\tan 2\theta_1 = \frac{60}{30} = 2 \Rightarrow$

$\theta_{1,2} = 31.7^\circ, -58.3^\circ$

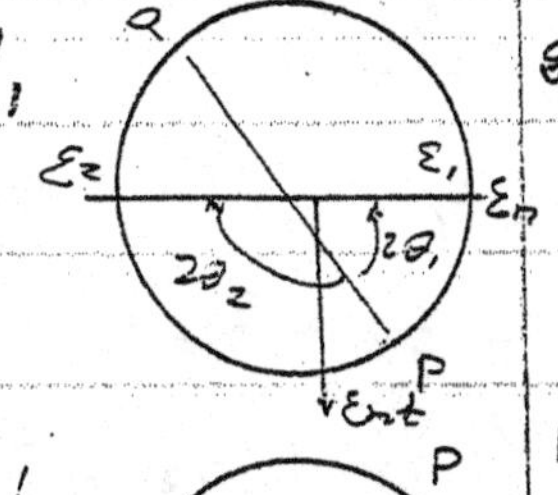

d) $\varepsilon_{1,2} = -10 \pm [900 + \overline{60}^2]^{1/2} = \begin{matrix}57.1\\-77.1\end{matrix}$

$\tan 2\theta_1 = \frac{-60}{30} = -2$

$\theta_1 = -31.7^\circ, 58.3^\circ$

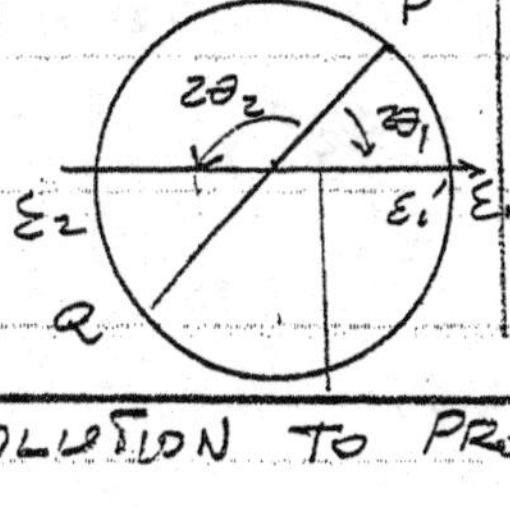

e) $\varepsilon_{1,2} = 10 \pm [900 + \overline{60}^2]^{1/2}$

$= 77.1, -57.1$

$\tan 2\theta = \frac{60}{-30} = -2$

$\theta_1 = 58.3^\circ, -31.7^\circ$

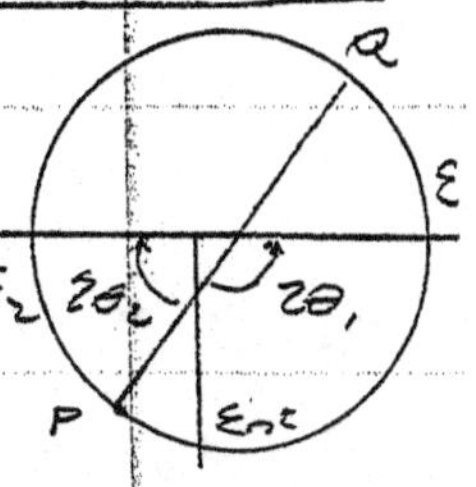

f) $\varepsilon_{1,2} = 10 \pm [900 + \overline{60}^2]^{1/2}$

$= 77.1, -57.1$

$\tan 2\theta = \frac{-60}{-30} = 2$

$\theta_1 = -58.3, 31.7$

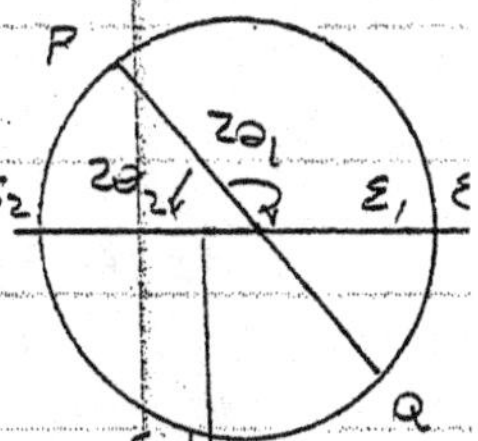

g) $\varepsilon_{1,2} = -30 \pm [100 + \overline{60}^2]^{1/2} =$

$= 30.83, -90.83$

$\tan 2\theta_1 = \frac{60}{10} = 6$

$\theta_1 = 40.3^\circ, -49.7^\circ$

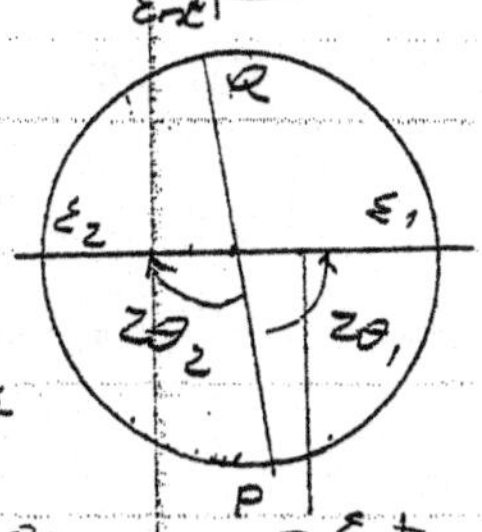

h) $\varepsilon_{1,2} = -30 \pm [100 + \overline{60}^2]^{1/2}$

$= 30.83, -90.83$

$\tan 2\theta_1 = \frac{-60}{10} = -6$

$\theta_1 = -40.3, 49.7$

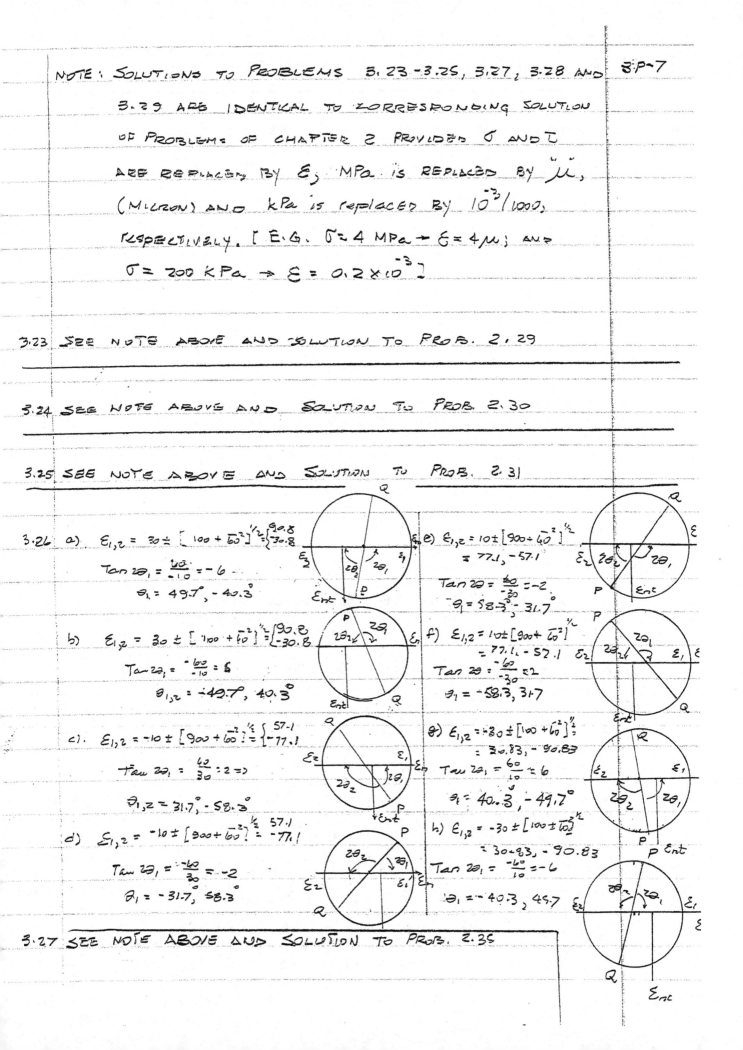

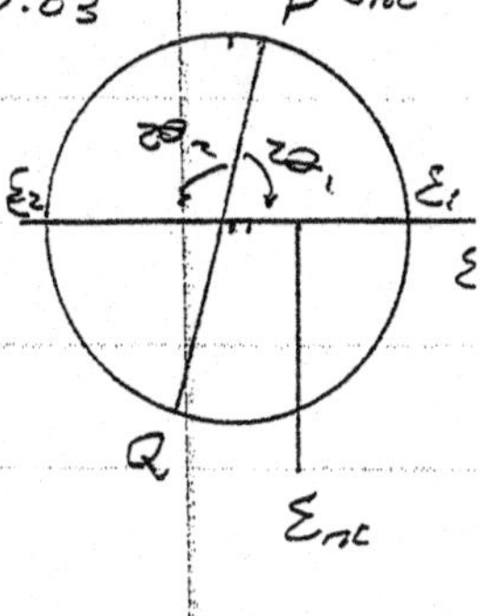

3.27 SEE NOTE ABOVE AND SOLUTION TO PROB. 2.35

3.28 SEE NOTE ABOVE AND SOLUTION TO PROB. 2.37

3.29 SEE NOTE ABOVE AND SOLUTION TO PROB. 2.39

3.30 $\cos^2\theta\,\varepsilon_x + \sin^2\theta\,\varepsilon_y + \sin 2\theta\,\varepsilon_{xy} = \varepsilon_m$

a) $\varepsilon_x = \varepsilon_a = 100\mu$; $\left.\begin{matrix} \tfrac{1}{2}(100\mu) + \tfrac{1}{2}\varepsilon_y + \varepsilon_{xy} = \varepsilon_b = 300\mu \\ \varepsilon_y = \varepsilon_c = -50\mu \end{matrix}\right\} \Rightarrow \varepsilon_y = -50\mu,\ \varepsilon_{xy} = 275\mu$

b). $\varepsilon_x = 200\mu$; $\left.\begin{matrix} \tfrac{1}{2}(200\mu) + \tfrac{1}{2}\varepsilon_y - \varepsilon_{xy} = -600\mu \\ \tfrac{1}{2}(200\mu) + \tfrac{1}{2}\varepsilon_y + \varepsilon_{xy} = 0 \end{matrix}\right\} \Rightarrow \varepsilon_y = -800\mu,\ \varepsilon_{xy} = -500\mu$

c). $\varepsilon_x = -400\mu$; $\left.\begin{matrix} \tfrac{1}{2}(-400) + \tfrac{1}{2}\varepsilon_y - \varepsilon_{xy} = -200\mu \\ \varepsilon_y = 50\mu \end{matrix}\right\} \Rightarrow \varepsilon_y = 50\mu,\ \varepsilon_{xy} = -375\mu$

d). $\varepsilon_x = 100\mu$; $\left.\begin{matrix} \tfrac{3}{4}(100\mu) + \tfrac{1}{4}\varepsilon_y + \tfrac{\sqrt{3}}{2}\varepsilon_{xy} = 200\mu = \varepsilon_b \\ \tfrac{1}{4}(100\mu) + \tfrac{3}{4}\varepsilon_y + \tfrac{\sqrt{3}}{2}\varepsilon_{xy} = 300\mu = \varepsilon_c \end{matrix}\right\} \Rightarrow \varepsilon_y = 300\mu,\ \varepsilon_{xy} = \frac{100\sqrt{3}}{3}\mu$

e). $\varepsilon_x = 0$; $\left.\begin{matrix} +\tfrac{1}{4}\varepsilon_y - \tfrac{\sqrt{3}}{2}\varepsilon_{xy} = -300\mu = \varepsilon_a \\ \tfrac{1}{4}\varepsilon_y + \tfrac{\sqrt{3}}{2}\varepsilon_{xy} = 300\mu = \varepsilon_c \end{matrix}\right\} \varepsilon_y = 0.\ \ \varepsilon_{xy} = 200\sqrt{3}\mu$

f). $\varepsilon_x = -1500\mu$; $\left.\begin{matrix} \tfrac{3}{4}(-1500) + \tfrac{1}{4}\varepsilon_y - \tfrac{\sqrt{3}}{2}\varepsilon_{xy} = -1000 = \varepsilon_b \\ \tfrac{1}{4}(-1500) + \tfrac{3}{4}\varepsilon_y - \tfrac{\sqrt{3}}{2}\varepsilon_{xy} = -500 = \varepsilon_a \end{matrix}\right\} \begin{matrix} \varepsilon_y = -500\mu \\ \varepsilon_{xy} = -\frac{500\sqrt{3}}{3}\mu \end{matrix}$

3.31 As a particular case, let "b"-direction lie on x-axis.

$\varepsilon_x = \varepsilon_b$

$\frac{1}{2}\varepsilon_x + \frac{1}{2}\varepsilon_y - \varepsilon_{xy} = \varepsilon_a \Rightarrow \varepsilon_y - 2\varepsilon_{xy} = 2\varepsilon_a - \varepsilon_b$

$\frac{1}{2}\varepsilon_x + \frac{1}{2}\varepsilon_y + \varepsilon_{xy} = \varepsilon_c \Rightarrow \varepsilon_y + 2\varepsilon_{xy} = 2\varepsilon_c - \varepsilon_b$

$\therefore \varepsilon_y = \frac{1}{2}[2\varepsilon_a + 2\varepsilon_c - 2\varepsilon_b] = \varepsilon_a + \varepsilon_c - \varepsilon_b$

$\varepsilon_{xy} = \frac{1}{2}(\varepsilon_c - \varepsilon_a)$

$\varepsilon_x + \varepsilon_y = \varepsilon_a + \varepsilon_c$

$\varepsilon_x - \varepsilon_y = \varepsilon_b - (\varepsilon_a + \varepsilon_c - \varepsilon_b) = 2\varepsilon_b - (\varepsilon_a + \varepsilon_c)$

$$\varepsilon_{1,2} = \frac{\varepsilon_x + \varepsilon_y}{2} \pm \left[\left(\frac{\varepsilon_x - \varepsilon_y}{2}\right)^2 + \varepsilon_{xy}^2\right]^{1/2} = \frac{\varepsilon_a + \varepsilon_c}{2} \pm \frac{1}{2}\left\{[2\varepsilon_b - (\varepsilon_a + \varepsilon_c)]^2 + (\varepsilon_c - \varepsilon_a)^2\right\}^{1/2}$$

$$= \frac{\varepsilon_a + \varepsilon_c}{2} \pm \frac{1}{2}\left\{4\varepsilon_b^2 - 4\varepsilon_b(\varepsilon_a + \varepsilon_c) + (\varepsilon_a + \varepsilon_c)^2 + (\varepsilon_c - \varepsilon_a)^2\right\}^{1/2}$$

$$= \frac{\varepsilon_a + \varepsilon_c}{2} \pm \frac{1}{2}\left\{4\varepsilon_b^2 - 4\varepsilon_b(\varepsilon_a + \varepsilon_c) + 2(\varepsilon_a^2 + \varepsilon_c^2)\right\}^{1/2}$$

$$\varepsilon_{1,2} = \frac{\varepsilon_a + \varepsilon_c}{2} \pm \frac{1}{2}\left\{4\varepsilon_b^2 + 2\varepsilon_a(\varepsilon_a - 2\varepsilon_b) + 2\varepsilon_c(\varepsilon_c - 2\varepsilon_b)\right\}^{1/2}$$

3.32 $\sin 2\theta(\varepsilon_x - \varepsilon_y) + \cos 2\theta\, \varepsilon_{xy} = \varepsilon_{nt}$

Since ε_x and ε_y have same coefficients, namely $\sin 2\theta$, in 3 inhomogeneous eqs,

$\sin 2\theta_a \varepsilon_x - \sin 2\theta_a \varepsilon_y + \cos 2\theta_a \varepsilon_{xy} = (\varepsilon_{nt})_a$

$\sin 2\theta_b \varepsilon_x - \sin 2\theta_b \varepsilon_y + \cos 2\theta_b \varepsilon_{xy} = (\varepsilon_{nt})_b$

$\sin 2\theta_c \varepsilon_x - \sin 2\theta_c \varepsilon_y + \cos 2\theta_c \varepsilon_{xy} = (\varepsilon_{nt})_c$,

determinant of coeffs, $D = 0$. $\Rightarrow \therefore$ No solution exists for $\varepsilon_x, \varepsilon_y, \varepsilon_{xy}$.

3.33 $I_{\varepsilon_2}/I_{\varepsilon_1} = \varepsilon_1\varepsilon_2/(\varepsilon_1 + \varepsilon_2) = 0.75\times10^{-3} \Rightarrow 3\varepsilon_2^2/4\varepsilon_2 = 0.75\times10^{-3} \Rightarrow \varepsilon_2 = 10^{-3}$

$\therefore \varepsilon_1 = 3\times10^{-3} \Rightarrow I_{\varepsilon_1} = 4\times10^{-3}$

$I_{\varepsilon_2}/I_{\varepsilon_1} = \dfrac{\varepsilon_n\varepsilon_t - \varepsilon_{nt}^2}{4\times10^{-3}} = 0.75\times10^{-3} \Rightarrow \varepsilon_n\varepsilon_t - \varepsilon_{nt}^2 = 3\times10^{-6}$

$\therefore \varepsilon_n(I_{\varepsilon_1} - \varepsilon_n) - \varepsilon_{nt}^2 = 3\times10^{-6} \Rightarrow \varepsilon_n^2 - (4\times10^{-3})\varepsilon_n + \varepsilon_{nt}^2 + 3\times10^{-6} = 0$

$\varepsilon_n = \frac{1}{2}\left\{(4\times10^{-3}) \pm [16\times10^{-6} - 4\varepsilon_{nt}^2 - 12\times10^{-6}]^{1/2}\right\} = \frac{1}{2}\left\{4\times10^{-3} \pm 2[10^{-6} - \varepsilon_{nt}^2]^{1/2}\right\}$.

Require $\varepsilon_{nt}^2 \leq 10^{-6} \Rightarrow |\varepsilon_{nt}| \leq 10^{-3}$

3.34 $\delta A = \iint_A \Delta \, dA = \iint \Delta(x,y)\,dx\,dy = \frac{k}{A}\int_0^b (y-b)^2 dy \int_0^a (x-a)^2 dx$

$\delta A = \frac{k}{9A}\left\{ (y-b)^3\Big|_0^b \cdot (x-a)^3\Big|_0^a \right\} = \frac{-k}{9A}\left\{(-b)^3 \cdot (-a)^3\right\} = \frac{k}{9A}(b^3 \cdot a^3) = ka^2b^2/9$

3.35 $\delta A = \iint \Delta \, dA = \iint (\varepsilon_x + \varepsilon_y)\,dA$; $\varepsilon_x = ay/L^2$, $\varepsilon_y = 2bxy/L^3$.

$\delta A = \frac{1}{L^2}\iint [\, ay + 2bxy/L \,]\,dx\,dy = 1/L^2 \int_0^L [\, ay^2/2\Big|_0^{y=L} + \frac{bxy^2}{L}\Big|_0^{y=L} \,]\,dx$

$= 1/L^2 \int_0^L [\, \frac{a}{2}L^2 + bxL \,]\,dx = 1/L^2 [\, \frac{aL^2}{2}x + bx^2 L/2 \,]_0^L = L/2(a+b)$

3.36 $\varepsilon = \alpha \cdot \delta T$; $\Delta = (\varepsilon_x + \varepsilon_y) = 2\alpha \cdot \delta T(x,y)$

$\delta A = \iint \Delta(x,y)\,dA = 2\alpha\left\{ \delta T_0 \cdot A + \delta T_1 \int_{-a/2}^{a/2}\int_{-b/2}^{b/2} [\cos\left(\frac{\pi x}{a}\right)\cos^2\left(\frac{\pi y}{b}\right)]\,dx\,dy \right.$

$= 2\alpha\{\delta T_0 \cdot A + \delta T_1 [\, \frac{a}{\pi}\sin\left(\frac{\pi x}{a}\right)]_{-a/2}^{a/2} \cdot \frac{1}{2}\int_{-b/2}^{b/2}(1 + \cos\frac{2\pi y}{b})\,dy$

$= 2\alpha\{\delta T_0 \cdot A + \delta T_1 [\, 2a/\pi \cdot b/2]$

$\delta A = 2\alpha ab(\delta T_0 + \delta T_1/\pi)$.

3.37

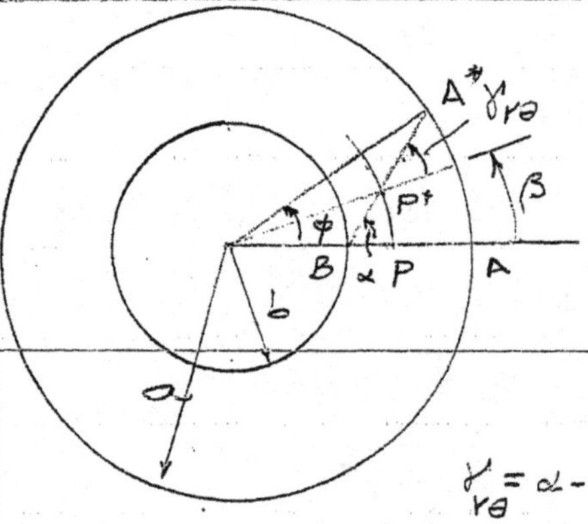

$\gamma_{r\theta} = \alpha - \beta$; $\widehat{AA^*} = (a-b)\alpha \Leftarrow 1^{st}$ order approx ($\widehat{BA^*} \doteq \frac{a-b}{a}$)

$\widehat{AA^*} = a\varphi$

$\therefore \alpha = \frac{a\varphi}{a-b}$; $\widehat{PP^*} = (r-b)\alpha$; 1st order approx

$\widehat{PP^*} = \beta r \Rightarrow \beta = \frac{(r-b)\alpha}{r}$

$\gamma_{r\theta} = \alpha - \beta = [1 - \frac{(r-b)}{r}]\alpha = \frac{b}{r}\alpha = \frac{ab\varphi}{r(a-b)}$

3.38

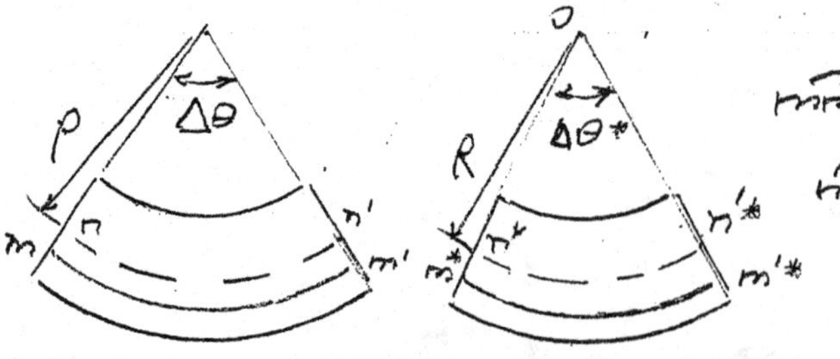

$\widehat{mm'} = (\rho+\eta)\Delta\theta$, $\widehat{m^*m^{*\prime}} = (R+\eta)\Delta\theta^*$

$\widehat{nn'} = \rho\Delta\theta = \widehat{n^*n'^*} = R\Delta\theta^* \Rightarrow \Delta\theta^* = (\rho/R)\Delta\theta$

$\varepsilon_\theta = \frac{\widehat{m^*m^{*\prime}} - \widehat{mm'}}{\widehat{mm'}} = \frac{(R+\eta)\Delta\theta^* - (\rho+\eta)\Delta\theta}{(\rho+\eta)\Delta\theta}$

$\therefore \varepsilon_\theta = \frac{(R+\eta)\rho/R - (\rho+\eta)}{\rho+\eta} = \frac{\eta/R(\rho - R)}{\rho+\eta} = \frac{\eta}{R}\,\frac{(1 - R/\rho)}{1+\eta/\rho}$

3.39

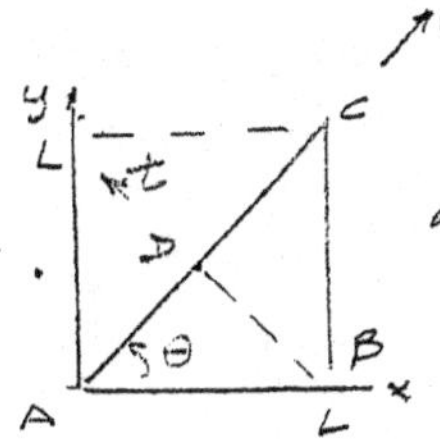

a) $\varepsilon_y(x,L) = 2bL^2y \qquad \Delta_{BC} = \int_0^L \varepsilon_y(x,L)\,dy = 2bL^2\int_0^L y\,dy = bL^4$

(b) $\cos\theta = \sin\theta = \sqrt{2}/2$

$\varepsilon_n = ay/2 + bx^2y + (ax + 2bxy^2)$

On AC: $y = x \Rightarrow \varepsilon_n(x) = 3ax/2 + 3bx^3$

c) $\Delta_{AC} = \int_A^C \varepsilon_n(x)\,ds;\quad ds = \sqrt{2}\,dx \Rightarrow \Delta_{AC} = \sqrt{2}\int_0^L \varepsilon(x)\,dx = 3\sqrt{2}\int_0^L (ax/2 + bx^3)\,dx$

$\Delta_{AC} = \frac{3\sqrt{2}}{4}(a + bL^2)L^2.$

d). $\varepsilon_{nt}(D) = \varepsilon_{nt}(L/2, L/2) = -(aL/2 + bL^3/4) = L/4(2a + bL^2) = -0.0375$

$\gamma_{nt} = -0.075 \text{ rad} = -4.3° \Rightarrow \angle BDC = 85.7°$

3.40

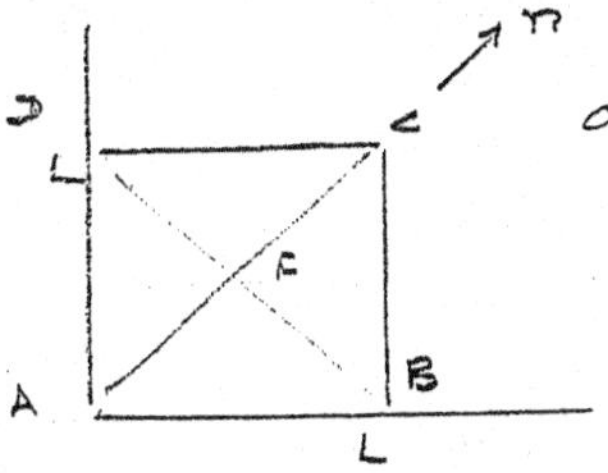

a) $\varepsilon_x = ay^2/L^2,\ \varepsilon_y = ax^2/L^2,\ \varepsilon_{xy} = \frac{2axy}{L^2}$

$\varepsilon_n = a/L^2(y^2 + x^2 + 2xy) = \frac{a}{L^2}(x+y)^2$

b) $\varepsilon_n(x) = 4ax^2/L^2.$

$\Delta_{AC} = \int_A^C \varepsilon_n(x)\,ds = \sqrt{2}\int_0^L \varepsilon_n(x)\,dx = 4\sqrt{2}a/L^2\int_0^L x^2\,dx = 4\sqrt{2}aL/3$

c) $F: (L/2, L/2) \Rightarrow \varepsilon_x = a/4,\ \varepsilon_y = a/4,\ \varepsilon_{xy} = a/2.$

$\varepsilon_{1,2} = a/4 \pm [a/2] = 3a/4, -a/4$ d). Yes.

3.41 $\varepsilon_x = a,\ \varepsilon_y = b,\ \varepsilon_{xy} = (2b+c)/2$

(a). $\varepsilon_{1,2} = \frac{a+b}{2} \pm \left[\left(\frac{a-b}{2}\right)^2 + \frac{1}{4}(2b+c)^2\right]^{1/2} = \frac{a+b}{2} \pm \frac{1}{2}\overbrace{\left[(a-b)^2 + (2b+c)^2\right]^{1/2}}^{\sqrt{R}}$

(b). $\tan 2\theta = \frac{(2b+c)/2}{(a-b)/2} = \frac{2b+c}{a-b}$

(c) $\partial\varepsilon_1/\partial a = 1/2 + 1/4\,R^{-1/2}[2(a-b)] = 0 \Rightarrow R^{1/2} = -(a-b)$ (i)

$\partial\varepsilon_1/\partial b = 1/2 + 1/4\,R^{-1/2}[-2(a-b) + 4(2b+c)] = 0 \Rightarrow R^{1/2} = a - 5b - 2c$ (ii)

$\partial\varepsilon_1/\partial c = 1/4\,R^{-1/2}[2(2b+c)] = 0 \Rightarrow 2b + c = 0 \Rightarrow c = -2b$ (iii)

(ii)&(iii) $\Rightarrow R^{1/2} = a - b$ (iv); (i)&(iv) $\Rightarrow a = b, \Rightarrow a = -c/2.$

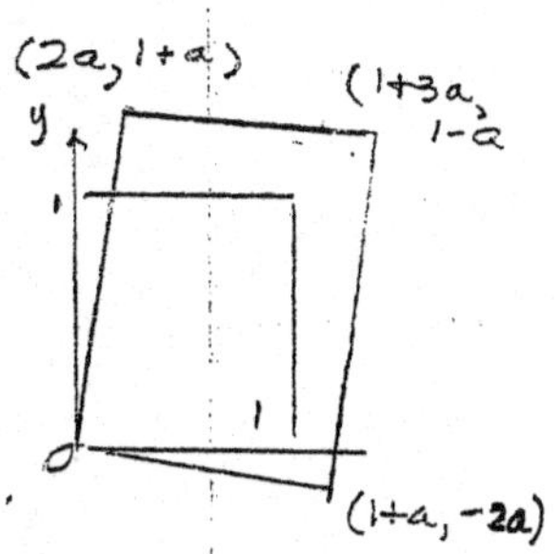

$\tan 2\theta = 0/0$ (not determined). $\Rightarrow$ All directions are principal.

d). $\varepsilon_1 = \varepsilon_2 = a$

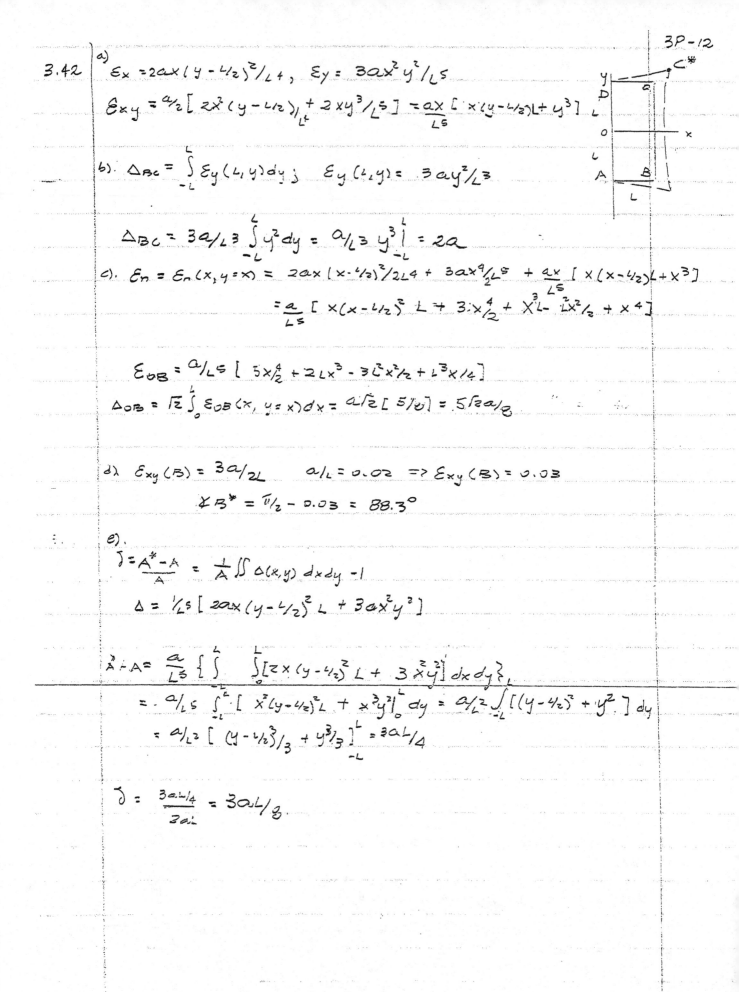

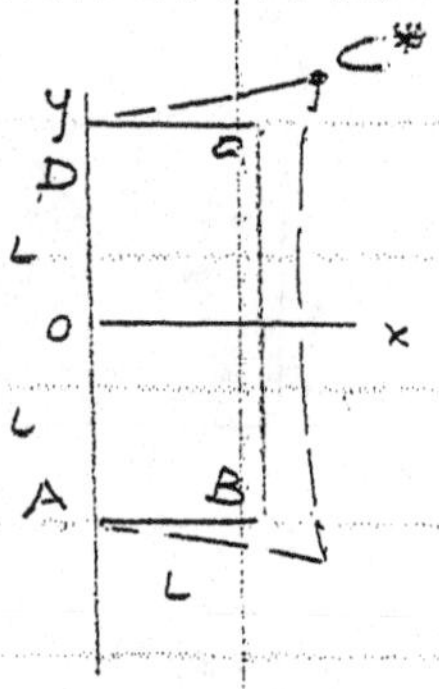

3.42 a) $\varepsilon_x = 2ax(y-L/2)^2/L^4$, $\varepsilon_y = 3ax^2y^2/L^5$

$\varepsilon_{xy} = a/2\,[2x^2(y-L/2)/L^4 + 2xy^3/L^5] = \frac{ax}{L^5}[x(y-L/2)L + y^3]$ L

b). $\Delta_{BC} = \int_{-L}^{L} \varepsilon_y(L,y)\,dy$; $\varepsilon_y(L,y) = 3ay^2/L^3$

$\Delta_{BC} = 3a/L^3 \int_{-L}^{L} y^2 dy = a/L^3\, y^3 \Big|_{-L}^{L} = 2a$

c). $\varepsilon_n = \varepsilon_n(x, y=x) = 2ax(x-L/2)^2/2L^4 + 3ax^4/2L^5 + \frac{ax}{L^5}[x(x-L/2)L + x^3]$

$= \frac{a}{L^5}[x(x-L/2)^2 L + 3x^4/2 + x^3L - L^2x^2/2 + x^4]$

$\varepsilon_{OB} = a/L^5\,[5x^4/2 + 2Lx^3 - 3L^2x^2/2 + L^3x/4]$

$\Delta_{OB} = \sqrt{2}\int_0^L \varepsilon_{OB}(x, y=x)\,dx = a\sqrt{2}\,[5/8] = 5\sqrt{2}a/8$

d). $\varepsilon_{xy}(B) = 3a/2L$ $\quad a/L = 0.02 \Rightarrow \varepsilon_{xy}(B) = 0.03$

$\angle B^* = \pi/2 - 0.03 = 88.3°$

e).

$\delta = \frac{A^* - A}{A} = \frac{1}{A}\iint \Delta(x,y)\,dx\,dy - 1$

$\Delta = 1/L^5\,[2ax(y-L/2)^2 L + 3ax^2y^2]$

$A^* - A = \frac{a}{L^5}\left\{\int_{-L}^{L}\int_0^L [2x(y-L/2)^2 L + 3x^2y^2]\,dx\,dy\right\}$

$= a/L^5 \int_{-L}^{L} [x^2(y-L/2)^2 L + x^3y^2|_0^L\,dy = a/L^2 \int_{-L}^{L} [(y-L/2)^2 + y^2]\,dy$

$= a/L^2\,[(y-L/2)^3/3 + y^3/3]_{-L}^{L} = 3aL/4$

$\delta = \frac{3aL/4}{2aL} = 3aL/8.$

3.43

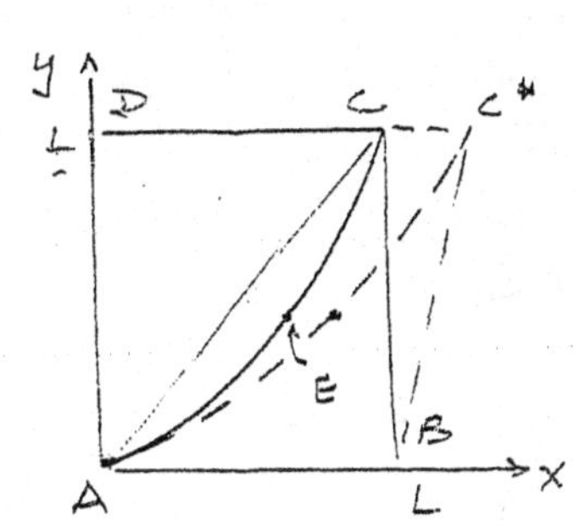

$\varepsilon_x = ky/L^2$, $\varepsilon_y = 0$, $\varepsilon_{xy} = \frac{1}{2}\frac{\partial u}{\partial y} = kx/2L^2$

(a) $\varepsilon_n = ky/2L^2 + kx/2L^2 = \frac{k}{2L^2}(x+y)$

On AC: $ds = \sqrt{2}\,dx$

$$\Delta_{AC}^{L} = \int \varepsilon_n(x, y=x)\,ds = \sqrt{2}k/L^2 \int_0^L x\,dx = \sqrt{2}k/2 = 0.7071\,k$$

(b). On AEC: $y = x^2/L \Rightarrow \Delta s = (\Delta x^2 + \Delta y^2)^{1/2} \Rightarrow ds = [1 + (dy/dx)^2]^{1/2} dx$

$$ds = (1 + 4x^2/L^2)^{1/2}\,dx.$$

$$\varepsilon_n(x, y = x^2/L) = \frac{k}{2L^2}\,x(1 + x/L)$$

$$\Delta_{AEC} = \frac{k}{2L^2}\int_0^L x(1+x/L)(1 + 4x^2/L^2)^{1/2}\,dx$$

By integration:

$$\Delta_{AEC} = \frac{k}{128}\left\{16(5\sqrt{5}-1)/3 + 18\sqrt{5} + \ln\left(\frac{1}{2+\sqrt{5}}\right)\right\} = 0.72735\,k$$

c).

$$\bar{\varepsilon}_{AC} = \frac{\Delta L_{AC}}{\sqrt{2}L} = k/2L$$

$$L_{AEC}: \int_0^L (1 + 4x^2/L^2)^{1/2}\,dx = L/4\,[2\sqrt{5} + \ln(2+\sqrt{5})] = 1.49L > \sqrt{2}L$$

$$\bar{\varepsilon}_{AEC} = \frac{\Delta_{AEC}}{L_{AEC}} = 0.72735k/1.49L = 0.4918\,k/L$$

3.44 a) $\varepsilon_x = \partial u/\partial x = c \Rightarrow u(x,y) = cx + A(y)$; $u(0,y) = 0 \Rightarrow A(y) = 0$

$\varepsilon_{xy} = \frac{1}{2}\partial v/\partial x = ax/L + by/L \Rightarrow v(x,y) = ax^2/2L + bxy/L + B(y)$; $v(0,y) = 0 \Rightarrow B(y) = 0$

$u(x,y) = cx$; $v(x,y) = ax^2/2L + bxy/L$

b). $\varepsilon_x = c$; $\varepsilon_y = bx/L$, $\varepsilon_{xz} = ax/L + by/L$. $\Rightarrow \varepsilon_{1,2} = \frac{1}{2L}\{(cL+bx) \pm [(cL-bx)^2 + 4(ax+by)^2]^{\frac{1}{2}}\}$

c). $\varepsilon_{xy} = 0 \Rightarrow y = -ax/b$; $a = 2b \Rightarrow y = -2x$

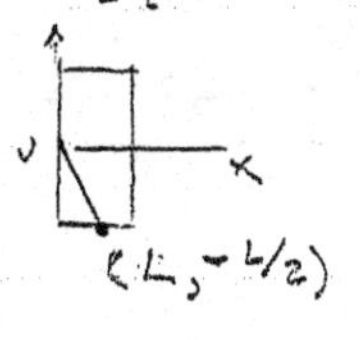

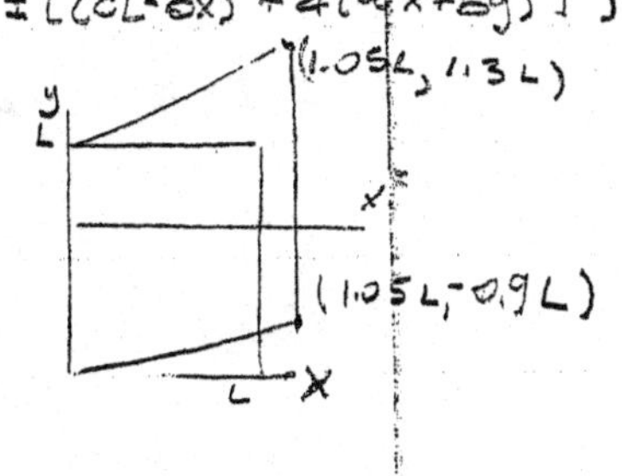

d). $\varepsilon_1 = 2b = c$; $\varepsilon_2 = bx/L = cx/2L$

$|\varepsilon_2|_{max} = b/2 = c/4$

3.45

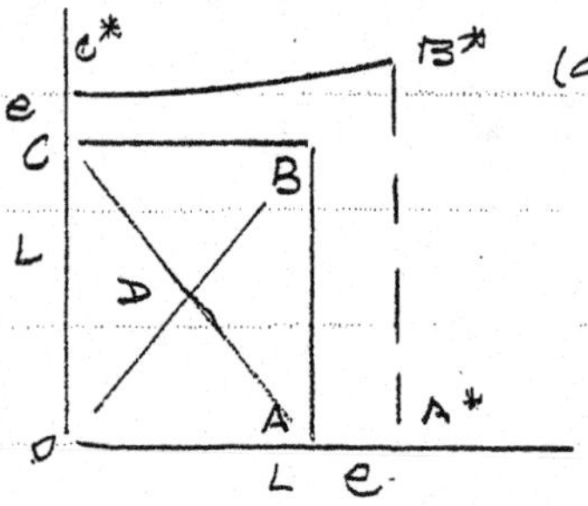

(a) (i) $u(0,y)=0$, $v(x,0)=0$

(ii) $u(L,0)=e$, $v(0,L)=e$

(iii) $\varepsilon_x(x,0)=C_1$, $\varepsilon_y(0,y)=C_2$.

(iv) $v(L,L)=2e$

$$u=u(x),\quad \varepsilon_{xy}=axy/2L^3$$

(iii) $\partial u(x)/\partial x=C_1 \Rightarrow u(x)=C_1x+A$; $\overset{(i)}{u}(x=0)=0 \Rightarrow A=0 \Rightarrow u(x)=C_1x$

(ii) $u(x=L)=e \Rightarrow C_1=e/L \Rightarrow u(x)=ex/L$

$u=u(x) \Rightarrow \varepsilon_{xy}=\partial v/\partial x=axy/L^3 \Rightarrow v(x,y)=ax^2y/2L^3+B(y)$

(iii) $\Rightarrow \left[ax^2/2L^3+dB(y)/dy\right]_{x=0}=C_2 \Rightarrow B(y)=C_2y+D$

$\therefore v(x,y)=ax^2y/2L^3+C_2y+D$; (i) $\Rightarrow v(x,0)=D=0$

(ii) $v(0,L)=C_2L=e \Rightarrow C_2=e/L$.

(iv) $v(L,L)=a/2+e=2e \Rightarrow a=2e$

$\therefore u(x,y)=ex/L$; $v(x,L)=e\left[x^2y/L^3+y/L\right]$

(b) $\varepsilon_x=e/L$, $\varepsilon_y=e(x^2/L^3+1/L)=e/L^3(x^2+L^2)$; $\varepsilon_{xy}=exy/L^3$

$$\varepsilon_n\Big|_{OB}=(e/L)(1/2)+\frac{e}{2L^3}(x^2+L^2)+\left.exy/L^3\right|_{y=x}=\frac{e}{2L^3}\left[L^2+(x^2+L^2)+2x^2\right]$$

$$\varepsilon_n\Big|_{OB}=\frac{e}{2L^3}(2L^2+3x^2).\qquad \Delta_{OB}=\int\varepsilon_n(x)ds=\sqrt{2}\int_0^L\varepsilon_n(x)dx=\frac{\sqrt{2}e}{2L^3}\int_0^L(2L^2+3x^2)dx$$

$$\Delta_{OB}=\frac{3\sqrt{2}}{2}e$$

$$\bar{\varepsilon}=\frac{\Delta_{OB}}{\sqrt{2}L}=3e/2$$

c). $\varepsilon_n(x=L/2)=11e/8L$

d). $\varepsilon_{nt}=-\left(\dfrac{\varepsilon_x-\varepsilon_y}{2}\right)\sin 2\theta+\varepsilon_{xy}\cos 2\theta=-\left(\dfrac{\varepsilon_x-\varepsilon_y}{2}\right)$

$\varepsilon_{nt}\Big|_D=\varepsilon_{nt}(x=L/2)=-\tfrac{1}{2}\left[e/L-5e/4L\right]=e/8L$

$\measuredangle EDB=\pi/2-\gamma_{nt_D}=\pi/2-e/4L=\pi/2-1/40=88.57^\circ$

3.46

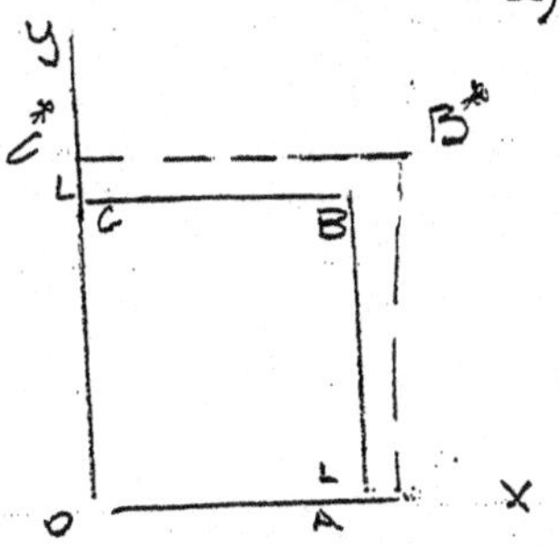

a)

(i) $u(0,y)=0$, $v(x,0)=0$

(ii) $v(0,L)=e$, $u(L,0)=e$

(iii) $\partial u/\partial y\big|_{y=0}=c_1$, $\partial v/\partial x\big|_{y=0}=c_2$; (iv) $u(L,L)=2e$

(v) $v=v(y)$; (vi) $\partial u/\partial y+\partial v/\partial x=axy^2/L^4$.

b) (iii) & (v) $\Rightarrow$ $\partial v(y)/\partial y=c_2 \Rightarrow v(y)=c_2 y+A(x)$

(i) $v(y=0)=A(x)=0 \Rightarrow v(y)=c_2 y$

(ii) $v(L)=c_2 L=e \Rightarrow c_2=e/L$ $\qquad v(y)=ey/L$

(vi) $\partial u/\partial y=axy^2/L^4 \Rightarrow u(x,y)=axy^3/3L^4+B(x)$

(iii) $\left.\dfrac{\partial u}{\partial x}\right|_{y=0}=\dfrac{dB(x)}{dx}=c_1 \Rightarrow B(x)=c_1x+D$

(i) $u(0,y)=D=0 \Rightarrow u(x,y)=axy^3/3L^4+c_1x$

(ii) $u(L,0)=c_1L=e \Rightarrow c_1=e/L$.

(iv) $u(L,L)=a/3+e=2e \Rightarrow a=3e \Rightarrow u(x,y)=e\left(xy^3/L^4+x/L\right)$.

c). $(L^*_{OB})^2=(L+e)^2+(L+2e)^2=2L^2+6Le+2e^2=2L^2(1+3e/L\cdots)$

$L^*_{OB}=\sqrt{2}L(1+3e/L)^{1/2}\doteq\sqrt{2}L(1+3e/2L\cdots)$; $\Delta_{OB}\doteq 3\sqrt{2}e/2$

$\bar{\varepsilon}=\dfrac{L^*_{OB}-L_{OB}}{L_{OB}}=\dfrac{L^*_{OB}}{\sqrt{2}L}-1=3e/2L$.

d). $\varepsilon_x=e(y^3/L^4+1/L)=e/L(x^3/L^3+1)$; $\varepsilon_y=e/L$, $\varepsilon_{xy}=3exy^2/2L^4$

$\varepsilon_n\big|_{OB}=\left[e/2L\left(y^3/L^3+1\right)+e/2L+3exy^2/2L^4\right]_{y=x}=e/L\left[1+2x^3/L^3\right]$

$\Delta_{OB}=\int_0^L \varepsilon_n(x)\,ds=e\sqrt{2}/L\int_0^L(1+2x^3/L^3)\,dx=\dfrac{e\sqrt{2}}{L}\left(x+x^4/2L^3\right)\Big|_0^L=3\sqrt{2}\,e/2$.

(e). $\Delta_D=\int_0^{L/2}\varepsilon_n(x)\,ds=e\sqrt{2}/L\left[x+x^4/2L^3\right]_0^{L/2}=(17\sqrt{2}/32)\,e=0.75130e$ in direct OC.

$|\omega_D|=(u_D^2+v_D^2)^{1/2}=\sqrt{145}/16=0.7526\,e$

(f) $\delta A=\int(\varepsilon_x+\varepsilon_y)\,dA=e/L\int_0^L\left(\int_0^L\left(\dfrac{x^3}{L^3}+2\right)dx\right)dy=9Le/4$

4.1 $\varepsilon_x = 1/E\,(\sigma_x - \nu\sigma_y) = 29/E$, $\varepsilon_y = 1/E\,(\sigma_y - \nu\sigma_x) = -36/E$

$\varepsilon_{xy} = \frac{\tau_{xy}}{2G} = \frac{\tau_{xy}(1+\nu)}{E} = 52/E$

$$\varepsilon_{1,2} = \frac{-3.5}{E} \pm \frac{1}{E}\left[\left(\frac{65}{2}\right)^2 + \overline{52}^2\right]^{1/2} = \frac{1}{E}[3.5 \pm 61.3] = 57.8/E,\ -64.8/E$$

$$\varepsilon_1 = 0.289\mu,\quad \varepsilon_2 = -0.324\mu$$

4.2 STRAIN: $\tan 2\theta = \frac{2\varepsilon_{xy}}{\varepsilon_x - \varepsilon_y} = \frac{\tau_{xy}/G}{\frac{1}{E}[(\sigma_x - \nu\sigma_y) - (\sigma_y - \nu\sigma_x)]} = \frac{\tau_{xy}/G}{\frac{1}{E}[(1+\nu)(\sigma_x - \sigma_y)]}$

$$= \frac{\tau_{xy}}{\frac{G(1+\nu)}{E}(\sigma_x - \sigma_y)} = \frac{\tau_{xy}}{\left(\frac{\sigma_x - \sigma_y}{2}\right)}$$ STRESS PRINC. AXIS

4.3 $\left.\begin{array}{l}\sigma_x - \nu\sigma_y = E\varepsilon_x \\ -\nu\sigma_x + \sigma_y = E\varepsilon_y\end{array}\right\} \Rightarrow \left.\begin{array}{l}\nu\sigma_x - \nu^2\sigma_y = \nu E\varepsilon_x \\ -\nu\sigma_x + \sigma_y = E\varepsilon_y\end{array}\right\} \Rightarrow (1-\nu^2)\sigma_y = E(\varepsilon_y + \nu\varepsilon_x)$

or $\sigma_y = \frac{E(\varepsilon_y + \nu\varepsilon_x)}{(1-\nu^2)}$

Similarly: $\sigma_x = \frac{E}{1-\nu^2}(\varepsilon_x + \nu\varepsilon_y)$

$\therefore \varepsilon_z = -\nu/E\,(\sigma_x + \sigma_y) = -\frac{\nu}{1-\nu^2}[(\varepsilon_x + \nu\varepsilon_y) + (\varepsilon_y + \nu\varepsilon_x)] = -\frac{\nu(1+\nu)}{1-\nu^2}(\varepsilon_x + \varepsilon_y)$

$$\varepsilon_z = -\frac{\nu(\varepsilon_x + \varepsilon_y)}{1-\nu}$$

4.4 $\varepsilon_z = 1/E[\sigma_z - \nu(\sigma_x + \sigma_y)] = 0 \Rightarrow \sigma_z = \nu(\sigma_x + \sigma_y)$.

$\varepsilon_x = 1/E[\sigma_x - \nu(\sigma_y + \sigma_z)] = 1/E[\sigma_x - \nu\sigma_y - \nu^2(\sigma_x + \sigma_y)]$

$= 1/E[(1-\nu^2)\sigma_x - \nu(1+\nu)\sigma_y] = \frac{1-\nu^2}{E}\left[\sigma_x - \frac{\nu}{1+\nu}\sigma_y\right] = \frac{1}{E^*}[\sigma_x - \nu^*\sigma_y]$

Similarly: $\varepsilon_y = 1/E^*[\sigma_y - \nu^*\sigma_x]$.

4.5 $\sigma_x = \frac{E}{1-\nu^2}\,69\,[\mu]$ $\sigma_y = \frac{E}{1-\nu^2}\,48\,[\mu]$ $\tau_{xy} = 30\,G[\mu] = \frac{30E}{2(1+\nu)} = \frac{15\,E(1-\nu)}{1-\nu^2}[\mu]$

$$\sigma_{1,2} = \frac{E}{1-\nu^2}\left\{\frac{117}{2} \pm \left[\left(\frac{21}{2}\right)^2 + \overline{15}^2(1-\nu)^2\right]^{1/2}\right\} = \frac{E}{1-\nu^2}[58.5 \pm (220.5)^{1/2}]\ [\mu]$$

$= \frac{E}{1-\nu^2}[58.5 \pm 14.85]\,\mu$; $E/(1-\nu^2) = 76.9$ GPa

$\sigma_1 = 56.4$ MPa, $\sigma_2 = 33.6$ MPa

4.6

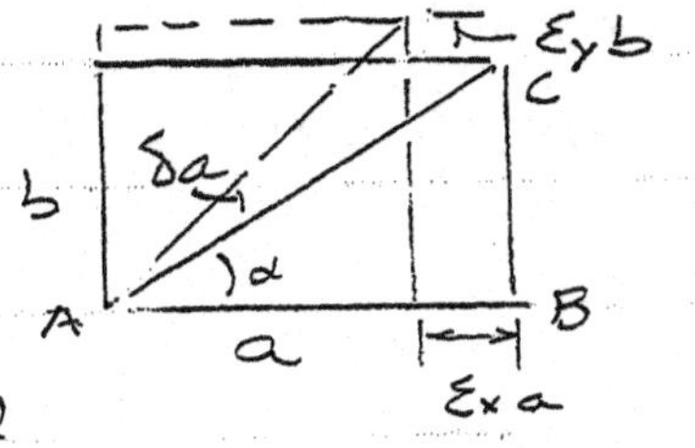

$$\varepsilon_x = \sigma_0/E\,,\quad \varepsilon_y = \nu\sigma_0/E$$

a)

$$\Delta = \tan(\alpha+\delta\alpha) - \tan\alpha = \frac{(1+\nu\sigma_0/E)}{(1-\sigma_0/E)}\frac{b}{a} - b/a = b/a\left(\frac{1+\nu\sigma_0/E}{1-\sigma_0/E} - 1\right)$$

b).

$$\Delta = b/a\{(1+\sigma_0/E+\cdots)(1+\nu\sigma_0/E) - 1\} = b/a\{1+\sigma_0/E+\nu\sigma_0/E\cdots - 1\}$$

$$= b/a\{(1+\nu)\sigma_0/E\}.$$

c). $\Delta = 1.5\,\sigma_0 b/Ea$

d). (i) $\Delta = \frac{1}{2}\left[\frac{1+(0.25)(10^{-3})}{1-10^{-3}} - 1\right] = 6.256255\times10^{-4}$

(ii) $\alpha+\delta\alpha = \tan^{-1}(\Delta+\tan\alpha) \Rightarrow \delta\alpha = \tan^{-1}(\Delta+\tan\alpha) - \alpha = \tan^{-1}(\Delta+0.5) - \tan^{-1}(.5)$

$\delta\alpha = 0.0286693^\circ$

e). $\alpha = \tan^{-1}(b/a)$, $\alpha+\delta\alpha = \tan^{-1}\left[\frac{(1+\nu\varepsilon_x)b}{(1-\varepsilon_x)a}\right] = \tan^{-1}\left[\frac{1+\nu\sigma_0/E}{1-\sigma_0/E}(b/a)\right].$

$\therefore\ \delta\alpha = \tan^{-1}\left[\left(\frac{1+\nu\sigma_0/E}{1-\sigma_0/E}\right)(b/a)\right] - \tan^{-1}(b/a)$

$$\tan^{-1}x - \tan^{-1}y = \tan^{-1}\left(\frac{x-y}{1+xy}\right)$$

$$\delta\alpha = \tan^{-1}\left[\frac{\left[\frac{1+\nu\sigma_0/E}{1-\sigma_0/E} - 1\right]b/a}{1+\left(\frac{1+\nu\sigma_0/E}{1-\sigma_0/E}\right)(b/a)^2}\right] = \tan^{-1}\left[\frac{(1+\nu)(\sigma_0/E)(b/a)}{1-\sigma_0/E+(1+\nu\sigma_0/E)(b/a)^2}\right]$$

$$\delta\alpha = \tan^{-1}(5.00375\times10^{-3}) = 0.028669^\circ$$

Solving for 3 unknowns with 3 eqs:

4.7

$$\left.\begin{aligned} \sigma_x - \nu\sigma_y - \nu\sigma_z &= E\varepsilon_x \\ -\nu\sigma_x + \sigma_y - \nu\sigma_z &= E\varepsilon_y \\ -\nu\sigma_x - \nu\sigma_y + \sigma_z &= E\varepsilon_z \end{aligned}\right\} \quad D = \begin{vmatrix} 1 & -\nu & -\nu \\ -\nu & 1 & -\nu \\ -\nu & -\nu & 1 \end{vmatrix} \begin{aligned} &(1-\nu^2)+\nu \\ &= (1-\nu^2) + \nu(-\nu-\nu^2) - \nu(\nu^2-\nu) \end{aligned}$$

$$\Rightarrow D = (1+\nu)^2(1-2\nu).$$

$$\sigma_x = \frac{1}{D}\begin{vmatrix} E\varepsilon_x & -\nu & -\nu \\ E\varepsilon_y & 1 & -\nu \\ E\varepsilon_z & -\nu & 1 \end{vmatrix} = \frac{E}{D}\left[\varepsilon_x(1-\nu^2) - \varepsilon_y(-\nu-\nu^2) + \varepsilon_z(\nu^2+\nu)\right]$$

$$\Rightarrow \sigma_x = \frac{E\left[\nu(\varepsilon_x+\varepsilon_y+\varepsilon_z) + (1-2\nu)\varepsilon_x\right]}{(1+\nu)(1-2\nu)} = \frac{\nu E}{(1+\nu)(1-2\nu)}\cdot\Delta + \frac{E}{1+\nu}\varepsilon_x$$

$\sigma_x = \lambda\Delta + 2G\varepsilon_x$ Similarly for σ_y, σ_z.

4.8 $K = \dfrac{E}{3(1-2\nu)}$ FROM $\lambda = \dfrac{G(E-2G)}{3G-E} \rightarrow (3G-E)\lambda = GE - 2G^2 \rightarrow E = \dfrac{(3\lambda+2G)G}{G+\lambda}$

$\therefore$

$K = \dfrac{E}{3(1-2\nu)} = \dfrac{(3\lambda+2G)}{3(1-2\nu)(1+\lambda/G)}$ From $\lambda = \dfrac{E\nu}{(1-2\nu)(1+\nu)} = \dfrac{2G\nu}{1-2\nu} \Rightarrow \lambda/G = \dfrac{2\nu}{1-2\nu}$

$\therefore K = \dfrac{3\lambda+2G}{3(1-2\nu)\left[1+\frac{2\nu}{1-2\nu}\right]} = \dfrac{3\lambda+2G}{3}$

4.9 $\varepsilon_r = 1/E\left[\sigma_r - \nu(\sigma_x+\sigma_\theta)\right] = -P/E\,[1-2\nu] \Rightarrow \delta D = \varepsilon_r D = -\dfrac{P(1-2\nu)}{E}\cdot D$

(a) $\delta D = -\dfrac{(10\times10^4)(0.20)(0.20)}{1.5\times10^9} = -2.67\times10^{-4}\ m = 0.267$ mm.

(b) $\delta L = 0.667$ mm (c) $\Delta = -3P(1-2\nu)/E = -4\times10^{-3}$

$\Delta V = \Delta\cdot\pi D^2 h/4 = 6.28\times10^{-5}\ m^3 = 62.7\ cm^3$

4.10 b) $P_p = 27200\ N \Rightarrow \sigma_p = P_p/A = \dfrac{31080}{\pi(1.5\times10^{-2})^2/4} = 176$ MPa; $\varepsilon_p = 0.02/5 = 0.004$

$E = \sigma_p/\varepsilon_p = 44$ GPa.

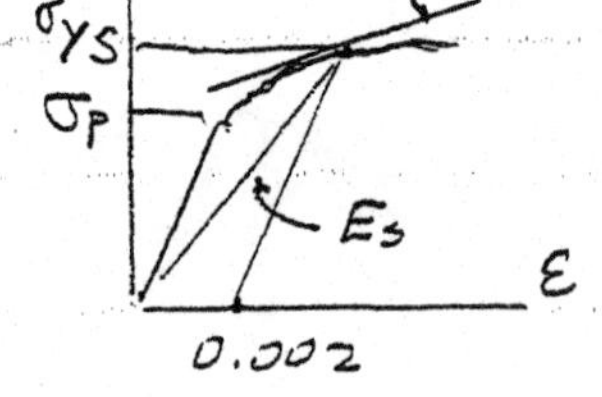

4.11 $\varepsilon_{true} = \ln(1+\varepsilon_{nom}) = \varepsilon_{nom} - \tfrac{1}{2}(\varepsilon_{nom})^2 \cdots$

$\therefore \varepsilon_{true} > \varepsilon_{nom}$ for $|\varepsilon_{nom}| \ll 1$

4.12 $\varepsilon = \sigma/c_1 + c_2(\sigma/c_3)^3 \equiv \sigma/c_1 + \alpha\sigma^3, \quad \alpha = c_2/c_3^3.$

(a)

(i) $d\varepsilon/d\sigma = 1/c_1 + 3\alpha\sigma^2 \Rightarrow E_{t0} = d\sigma/d\varepsilon = c_1 = 1.50\times10^{11} = 150\text{ GPa}$

(ii)

$$E_t = \left(1/c_1 + 3\alpha\sigma^2\right)^{-1}\Big|_{\sigma=175\text{ MPa}} \qquad \alpha = c_2/c_3^3 = \frac{2\times10^2}{(250\times10^6)^3} = 1.28\times10^{-23}$$

$$E_t = \left[1/1.5\times10^{11} + 3(1.28\times10^{-23})(175\times10^6)^2\right]^{-1} = 127.5\text{ GPa}$$

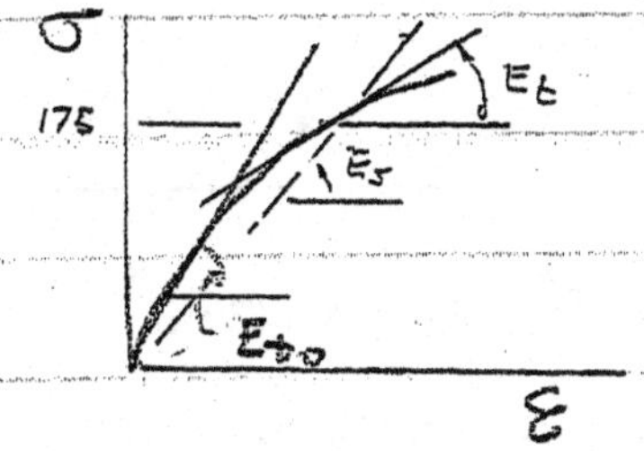

$$\varepsilon_{ys} = \sigma_{ys}/E_s = \left[\sigma_{ys}/c_1 + \alpha\sigma_{ys}^3\right].$$

$$E_s = \frac{\sigma_{ys}}{\sigma_{ys}/c_1 + \alpha\sigma_{ys}^3} = \frac{1}{1/c_1 + \alpha\sigma_{ys}^2} = \frac{c_1}{1+\alpha c_1\sigma_{ys}^2}$$

$$E_s = 141.7\text{ GPa}$$

(b) $$MT = \int_0^{\sigma_{ys}}\left[\varepsilon_{ys} - \varepsilon(\sigma)\right]d\sigma = \varepsilon_{ys}\cdot\sigma_{ys} - \int_0^{\sigma_{ys}}\varepsilon(\sigma)\,d\sigma$$

$$= \sigma_{ys}^2/c_1 + \alpha\sigma_{ys}^4 - \int_0^{\sigma_{ys}}\left[\sigma/c_1 + \alpha\sigma^3\right]d\sigma = \sigma_{ys}^2/c_1 + \alpha\sigma_{ys}^4 - \left[\sigma_{ys}^2/2c_1 + \alpha\sigma_{ys}^4/4\right]$$

$$= \sigma_{ys}^2\left[1/2c_1 + 3\alpha\sigma_{ys}^2/4\right] = (\sigma_{ys}^2/4)\left[2/c_1 + 3\alpha\sigma_{ys}^2\right]$$

$$MT = 111\times10^3\ \text{N}\cdot\text{m/m}^3.$$

4.13

$\varepsilon = \sigma/c_1 + \alpha\sigma^n \ ; \ \alpha = c_2/c_3^n > 0, \ c_1 > 0$

$d\varepsilon/d\sigma = 1/c_1 + n\alpha\sigma^{n-1} = 1/E_t \ , \quad E_s = \sigma/\varepsilon = \dfrac{\sigma}{\sigma/c_1 + \alpha\sigma^n}$

$$E_s/E_t = \frac{\dfrac{\sigma}{\sigma/c_1 + \alpha\sigma^n}}{\left(\dfrac{1}{1/c_1 + n\alpha\sigma^{n-1}}\right)} = \frac{1 + n\alpha c_1\sigma^{n-1}}{1 + \alpha c_1\sigma^{n-1}} = 1 + \frac{(n-1)\alpha c_1\sigma^{n-1}}{1+\alpha c_1\sigma^{n-1}} > 1$$

if $\alpha > 0, \ c_1 > 0, \ n > 1$

6.1 $A = P/\sigma_{all} = \dfrac{12\times10^3}{120\times10^6} = 10^{-4}\,m^2 = 1\,cm^2$

$A = PL/\Delta E = \dfrac{(12\times10^3)(.5)}{(200\times10^{-6})(200\times10^9)} = 1.5\times10^{-4}\,m^2 = 1.5\,cm^2 \Rightarrow d = 13.8\,mm$

6.2 $F_{AB} = 4P$ (comp.), $F_{BC} = P$ (tens.)

$\Delta_C = 0 \Rightarrow 8PL/A_{AB}E_{AB} = P/A_{BC}E_{BC} \Rightarrow A_{AB}/A_{BC} = 8\,E_{BC}/E_{AB}$

6.3 $\delta_c = PL/A(1/E_a + 1/E) \Rightarrow 1/E = \dfrac{A\delta_c}{PL} - 1/E_a = (2.333\times10^{-8})\delta_c - (1.4285\times10^{-11})$

$\delta_c = 1\,mm \Rightarrow 1/E = 9.047\times10^{-12} \Rightarrow E \approx 110.5\,GPa$

$\delta_c = 0.9\,mm \Rightarrow 1/E = 6.71\times10^{-12} \Rightarrow E = 143\,GPa$

$110\,GPa < E < 143\,GPa \rightarrow$ Brass.

6.4 $F(x) = kA\int_a^x \dfrac{d\xi}{\xi^2} = kA(1/a - 1/x)$; $\varepsilon(x) = F/AE = k/E(1/a - 1/x)$

$\Delta L = k/E \int_a^{L+a} \varepsilon(x)\,dx = k/E\,[L/a - \ln(L+a) + \ln a]$

$\Delta L = \dfrac{k}{E}[L/a - \ln(1 + L/a)]$

6.5 $F(x) = A\int_x^a f(\xi)\,d\xi = cA\int_x^a e^{\xi/a}\,d\xi = cAa\,e^{\xi/a}\Big|_x^a = cAa(e - e^{x/a})$

$\Delta L = \int_0^L \varepsilon(x)\,dx = \dfrac{ca}{E}\int_0^a (e - e^{x/a})\,dx = ca/E\,(xe - ae^{x/a})\Big|_0^a = ca^2/E$

6.6 $\varepsilon = \varepsilon_0 \sinh(\sigma/\sigma_0)$ $\quad \sigma = 8000/60 = 133.3\,MPa.$

$\Delta L = \varepsilon L = 0.006\sinh(133.3/100)(0.4) = 0.0042\,m \doteq 4.2\,mm.$

6.7

$\sigma = P/A$

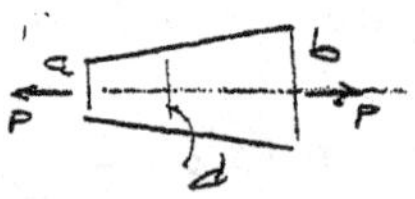

$d = a + (b-a)x/L.$

$\varepsilon = \dfrac{P}{tE[a + (b-a)x/L]} \Rightarrow \Delta L = P/Et\int_0^L \dfrac{dx}{a + (b-a)x/L} = PL/Et(b-a)\left\{\ln\left[a + \left(\tfrac{b-a}{L}\right)x\right]\right\}_0^L$

$\Delta L = \dfrac{PL\ln(b/a)}{Et(b-a)}$

6.8 (a) $\delta = \frac{2F_{BC}h}{E_sA}$, $F = Pa/L \Rightarrow \frac{2Pah}{E_sAL} = \delta \Rightarrow a/L = \frac{E_sA\delta}{2Ph}$

(b)

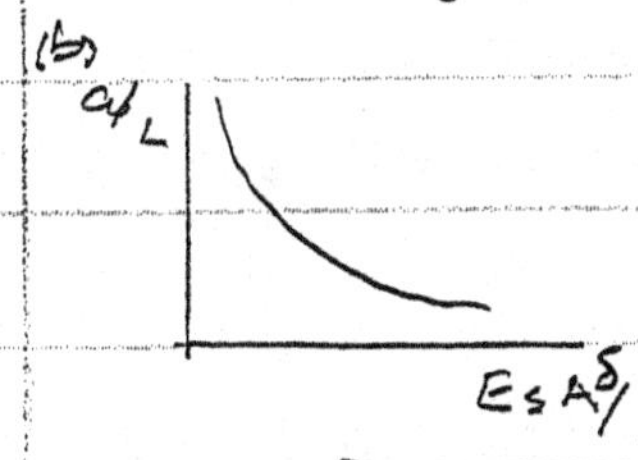

$$a/L = \frac{(200\times10^9)(15\times10^{-6})(2\times10^{-3})}{2(900)(2)} = 1.67$$

6.9

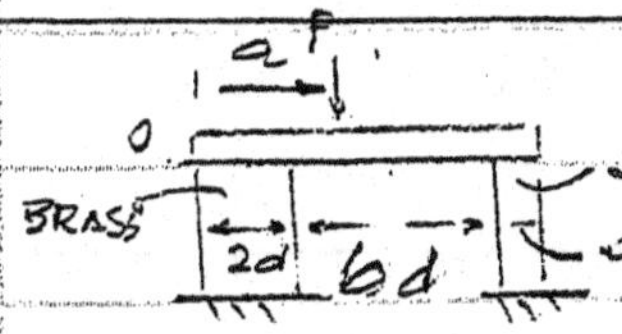

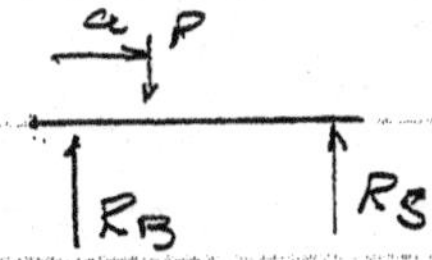

$\varepsilon_s = \varepsilon_b \Rightarrow \frac{R_s}{E_sA_s} = \frac{R_B}{E_bA_b}$

$$R_B = 4(120/200)R_s = 2.4R_s$$

$R_s + R_B = P \Rightarrow 3.4R_s = P$, $R_B = 2.4P/3.4 = 0.7059P$

$\Sigma M_0 = 0 \Rightarrow Pa = R_B d + 8.5R_s d \Rightarrow Pa = P\left(\frac{2.4 + 8.5}{3.4}\right)d = 3.20Pd$

$$a = 3.20\ d$$

6.10

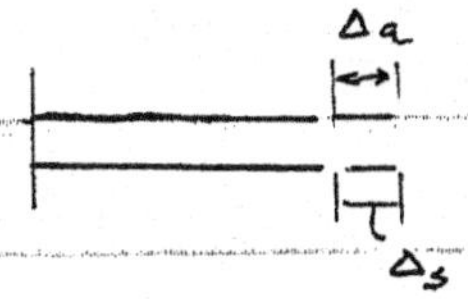

$\Delta_A = \Delta_s \Rightarrow \frac{P_A}{E_aA_a} = \frac{P_s}{E_sA_s} \Rightarrow P_s = P_a\left(\frac{A_sE_s}{A_aE_a}\right)$

$$P = P_a + P_s = P_a(1 + A_sE_s/A_aE_a)$$

$A_s = \pi d^2/4 = 625\pi/4$, $A_a = \pi/4(40^2 - 25^2) = 975\pi/4$

$P = P_a\left[1 + \left(\frac{625}{975}\right)\left(\frac{200}{70}\right)\right] = 2.8315P_a \Rightarrow P_a = P/1.8315 = 120k/1.8315$

$P_a = 65.52^k N \Rightarrow \sigma_a = P_a/A_a = 85.6\ N/mm^2$

$\sigma_A = 85.6\ MPa$; $\sigma_s = 54.4^k/490.8 = 110.8\ MPa$

$$\Delta_A = \Delta_s = \frac{(65.52\times10^3)(250)}{(70\times10^9)(\frac{975\pi}{4}\times10^{-6})} = 0.00031\ m = 0.31\ mm.$$

6.11 $F_{BD} = 18/(3/5) = 30\ kN$ (tens.); $F_{BC} = (30)(4/5) = 24\ kN$ (comp.)

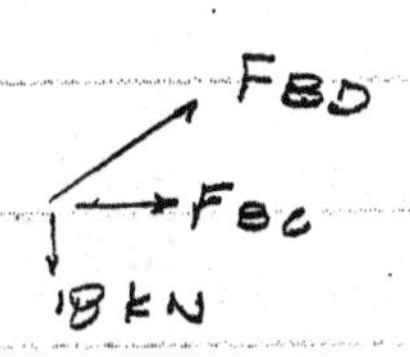

$$\Delta_{BD} = \frac{F_{BD}L_{BD}}{A_{BD}E_{BD}} = \frac{(30k)(5)}{(1.2\times10^{-4})(200\times10^9)} = 0.00625\ m = 6.25mm$$

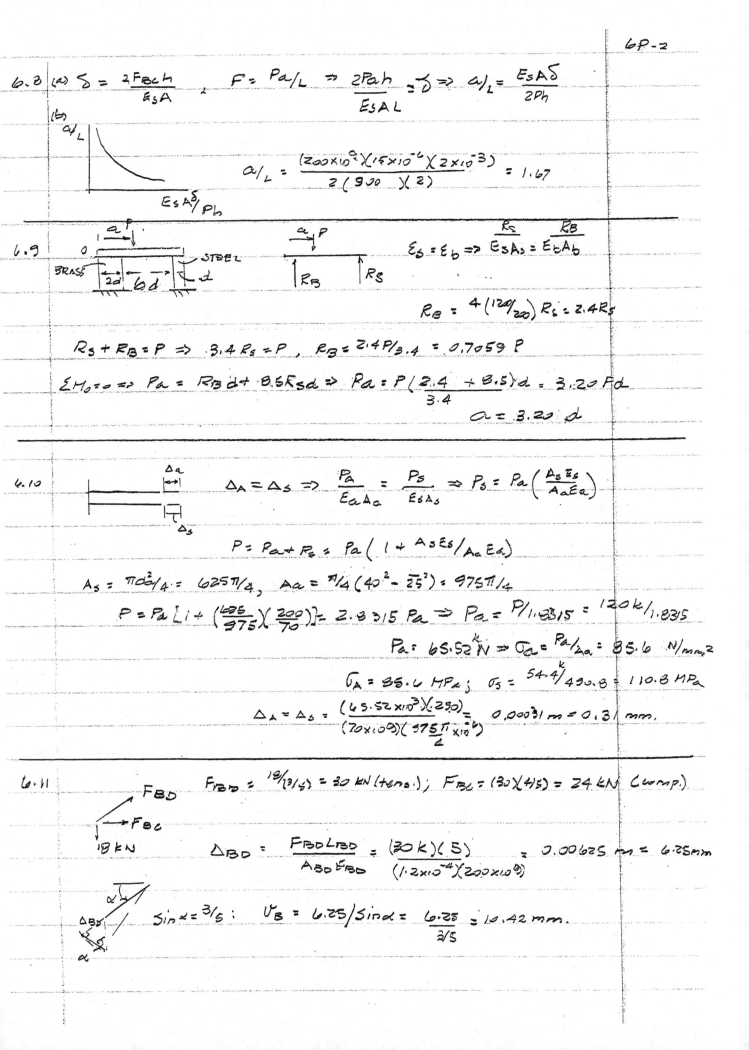

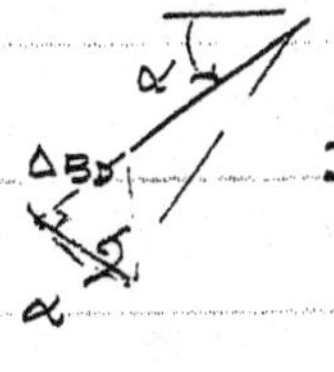

$\sin\alpha = 3/5$: $v_B = 6.25/\sin\alpha = \frac{6.25}{3/5} = 10.42\ mm.$

6.12 $\Delta_{BD} = 6.25$ mm. (from Prob. 6.11)

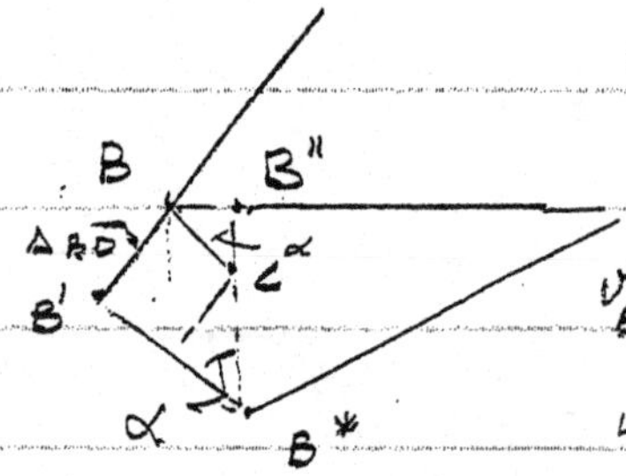

$$\overline{BB''} = \frac{F_{BC}L_{BC}}{E_a A_a} = \frac{(24k)(4)}{(70\times10^9)(3\times10^{-4})} = .00457 = 4.57\text{mm.}$$

$$v_B = \overline{BB^*} = \frac{BB''}{\tan\alpha} + \frac{BB'}{\sin\alpha} = \frac{4.57}{3/4} + \frac{6.25}{3/5} = 16.51 \text{ mm.}$$

$u_B = 4.57$ mm.

6.13

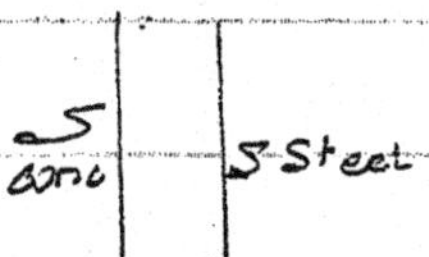

$$P = P_c + P_s \;;\; \Delta_c = \Delta_s \Rightarrow \frac{P_cL}{A_cE_c} = \frac{P_sL}{nA_sE_s}$$

$$P_s = \frac{nA_s}{A_c}\cdot\frac{E_s}{E_c}P_c = \frac{10(2)}{580}\cdot\frac{200}{20}P_c = 0.345P_c \qquad A_c = 600-20 = 580$$

$1.345P_c = P \rightarrow P_c = 500k/1.345 = 371.8k$, $P_s = 128.2k$

$$\sigma_s = P_s/A_s = 128.2\times10^3/20 = 6410 \text{ N/cm}^2 = 64.1 \text{ MPa}$$

$$\sigma_c = P_c/A_c = 641.0 \text{ N/cm}^2 = 6.41\text{MPa}$$

$$\Delta = \left(\frac{\sigma_s}{E}\right)L = \frac{64.1\times10^6}{200\times10^9}(3) = 9.6\times10^{-4} \text{ m} = 0.96 \text{ mm.}$$

6.14

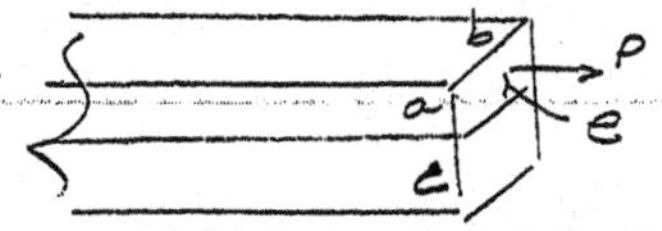

$$\varepsilon_1 = \varepsilon_2 \Rightarrow \sigma_1/E_1 = \sigma_2/E_2 \Rightarrow \sigma_1 = (E_1/E_2)\sigma_2$$

$$P = b[\sigma_1 a + \sigma_2 c] = (b\sigma_2/E_2)[aE_1 + cE_2]$$

$$Pe = (\sigma_1 ab)(a/2) - (\sigma_2 bc)(c/2)$$

$$Pe = b/2[\sigma_1 a^2 - \sigma_2 c^2] = \frac{b\sigma_2}{2E_2}[a^2E_1 - c^2E_2]$$

$$\therefore\; e = \frac{1/2\,[a^2E_1 - c^2E_2]}{aE_1 + cE_2}$$

6.15

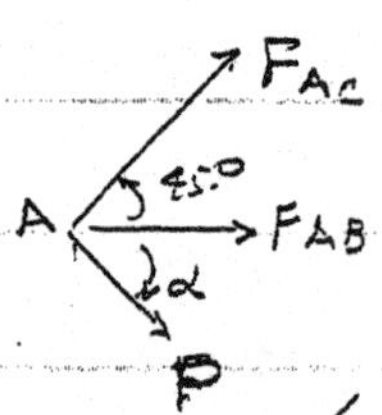

$$F_{AC}\sqrt{2}/2 = P\sin\alpha \Rightarrow F_{AC} = \sqrt{2}P\sin\alpha\;;\; F_{AB} = -P(1+\cos\alpha)$$

$$\Delta_{AC} = F_{AC}L/A_sE_s = \frac{\sqrt{2}PL\sin\alpha}{A_sE_s}\;;\; \Delta_{AB} = -\frac{PL(1+\cos\alpha)\sqrt{2}/2}{A_oE_o}$$

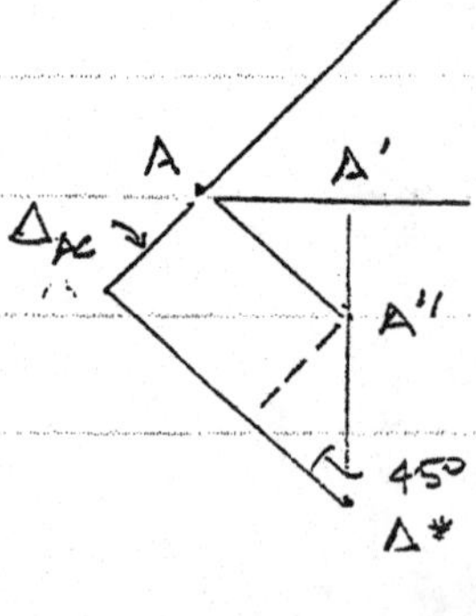

$$u_{A/L} = \Delta_{AB} = -\frac{PL\sqrt{2}(1+\cos\alpha)}{2A_oE_o}\;;\; v_A = \overline{A'A''} + \overline{A''A^*} = \overline{AA'} + \frac{\Delta_{AC}}{\sqrt{2}/2}$$

$$v_{A/L} = \frac{P\sqrt{2}(1+\cos\alpha)}{2A_oE_o} + \frac{2P\sin\alpha}{A_sE_s} =$$

6.16 $P = \sigma_1 A_1 + \sigma_2 A_2$; $\varepsilon_1 = \varepsilon_2 \Rightarrow \sigma_1 = E_1\sigma_2/E_2 \Rightarrow P = \sigma_2(E_1 A_1/E_2 + A_2)$

$$= \frac{\sigma_2}{E_2}(E_1 A_1 + E_2 A_2)$$

$A_1 = \pi R^2/2$, $A_2 = 2R^2$

$$P = (\sigma_2/E_2)[(\pi R^2/2)E_1 + 2R^2 E_2] = \frac{\sigma_2 R^2}{2E_2}(\pi E_1 + 4E_2)$$

$$Pe = \sigma_1\left(\frac{4R}{3\pi}\right)(\pi R^2/2) - \sigma_2(2R^2)(R/2) = R^3(2/3\,\sigma_1 - \sigma_2)$$

$$Pe = R^3\sigma_2\left[\tfrac{2}{3}E_1/E_2 - 1\right] = \frac{R^3\sigma_2}{3E_2}(2E_1 - 3E_2)$$

∴

(a) $e = \frac{2R}{3}\left(\frac{2E_1 - 3E_2}{\pi E_1 + 4E_2}\right)$ (b) $\sigma_2 = \frac{2E_2 P}{R^2}(\pi E_1 + 4E_2)^{-1}$

$$\Delta L = \sigma_2 L/E_2 = \frac{2PL}{R^2}(\pi E_1 + 4E_2)^{-1}$$

6.17

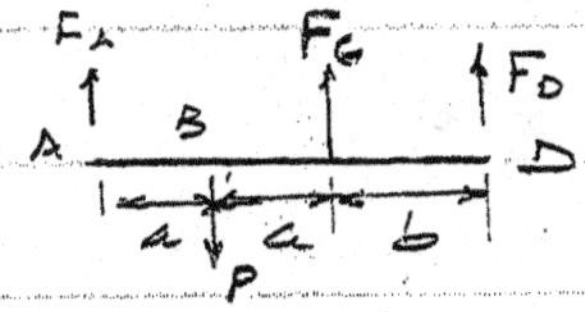

$\Sigma F_y = 0 \Rightarrow F_A + F_C + F_D = P$

$\Sigma M_A = 0 \Rightarrow 2aF_C + (2a+b)F_D = Pa$

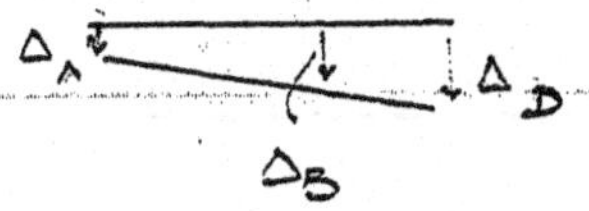

$\Delta_A = \Delta_D + \frac{(2a+b)}{b}(\Delta_C - \Delta_D) \Rightarrow b\Delta_A = (2a+b)\Delta_C - 2a\Delta_D$

$\Delta_A = F_A h/AE$, $\Delta_C = F_C h/AE$, $\Delta_D = F_D h/AE$

$$bF_A - (2a+b)F_C + 2aF_D = 0$$

3 equations in 3 unknowns:

$$\begin{bmatrix} 1 & 1 & 1 \\ 0 & 2a & 2a+b \\ b & -(2a+b) & 2a \end{bmatrix}\begin{Bmatrix} F_A \\ F_C \\ F_D \end{Bmatrix} = \begin{Bmatrix} P \\ Pa \\ 0 \end{Bmatrix}$$

Solution: $F_A = (4 + 3b/a + b^2/a^2)P/D$, $F_C = (2 + b/a + b^2/a^2)P/D$

$F_D = 2P/D$

where $D = 2(4 + 2b/a + b^2/a^2)$

$b = 3a$: $F_A = 11P/19$, $F_C = 7P/19$, $F_D = P/19$

$$A = F_A/\sigma = \frac{11\times 19.0\times 10^3}{19\times 100\times 10^{-6}} = 1.1\times 10^{-4}\ \text{m}^2 = 1.1\ \text{cm}^2.$$

6.18 a) By symmetry: $F_{AD} = F_{CD}$ $\quad \Sigma F_y = 0 \Rightarrow 2F_{CD}\cos\beta + F_{BD} = P$

By symmetry: $\Delta_B \equiv v_B \downarrow$

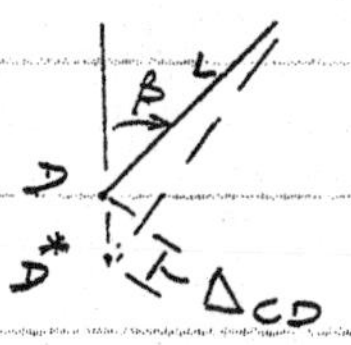

$\Delta_{CD} = \dfrac{F_{CD}L}{AE}$; $v_B = \Delta_{CD}/\cos\beta = \Delta_{CD}L/h = \overline{DD}^*$

$\overline{DD}^* = \dfrac{F_{CD}L}{AE\cos\beta}$

$\overline{DD}^* = F_{BD}h/AE$

$\therefore F_{BD}h/AE = \dfrac{F_{CD}L}{AE\cos\beta} \Rightarrow F_{BD} = \dfrac{F_{CD}L}{h\cos\beta} = \dfrac{F_{CD}}{\cos^2\beta}$

$F_{CD}(2\cos\beta + \dfrac{1}{\cos^2\beta}) = P \Rightarrow F_{CD} = \dfrac{P\cos^2\beta}{1+2\cos^3\beta}$; $F_{BD} = \dfrac{P}{1+2\cos^3\beta}$

b).

$F_{BD} > F_{CD}$; $a = 3m$, $h = 4m \Rightarrow \cos\beta = 4/5$ $\quad P = 50000N$

$F_{BD} = 125P/253 = \dfrac{125 \times 50000}{253} = 24700N$

$A = F_{BD}/\sigma = \dfrac{24700}{120\times10^6} = 205.9 \times 10^{-6}\ m^2 = 2.06\ cm^2$.

6.19 a)

$\overline{DD}^* = F_{CD}L/A_sE\cos\beta$; $\overline{DD}^* = F_{BD}h/A_aE$

$\therefore F_{BD}h/A_aE_a = F_{CD}L/A_sE_s\cos\beta \Rightarrow F_{BD} = \dfrac{F_{CD}L}{h\cos\beta} \cdot R = \dfrac{F_{CD}R}{\cos^2\beta}$; $R = \dfrac{E_aA_a}{E_sA_s}$

$2F_{CD}\cos\beta + F_{BD} = P \Rightarrow F_{CD}(2\cos\beta + \dfrac{R}{\cos^2\beta}) = P$

$\therefore F_{CD} = \dfrac{P\cos^2\beta}{R+2\cos^3\beta}$, $F_{BD} = \dfrac{PR}{R+2\cos^3\beta}$.

$\sigma_s = F_{CD}/A_s = 120\ MPa$, $\sigma_a = F_{BD}/A_a = 90\ MPa$

$\sigma_s/\sigma_a = \dfrac{F_{CD}/A_s}{F_{BD}/A_a} = \dfrac{\cos^2\beta(A_a/A_s)}{E_aA_a/E_sA_s} = \cos^2\beta\,(E_s/E_a) = 120/90 = 4/3$

$\cos^2\beta = (4/3)(70/200) = 7/15 \Rightarrow \cos\beta = 0.4667$

$\beta = 62.2°$

b). $P = 2(\sigma_sA_s\cos\beta + \sigma_aA_a) = 2[(120)(2)(.4667) + 90(4)]10^2 = 94.4\ kN$

6.20 a) $a \le b$ From Prob. 6-7

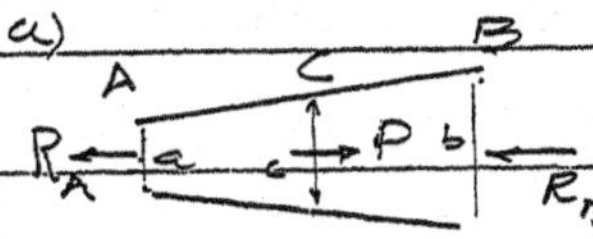

$R_A + R_B = P$

$c = (a+b)/2$

$\Delta_{AB} = \frac{PL}{Et}\,\frac{\ln b - \ln a}{b-a}$

AC:

$$\Delta L_{AC} = \frac{R_A L}{2Et}\left[\frac{\ln\left(\frac{a+b}{2}\right) - \ln a}{\left(\frac{a+b}{2}\right) - a}\right] = \frac{R_A L}{Et}\left(\frac{\ln(a+b) - \ln 2 - \ln a}{b-a}\right) \text{ extension}$$

$$BC:\ \Delta_{BC} = \frac{R_B L}{2Et}\left[\frac{\ln b - \ln\left(\frac{b+a}{2}\right)}{b - (a+b)/2}\right] = \frac{R_B L}{Et}\left(\frac{\ln b - \ln(a+b) + \ln 2}{b-a}\right) \text{ Contraction}$$

$$\Delta L_{AC} = \Delta_{BC} \Rightarrow R_A = R_B\left(\frac{\ln b - \ln(a+b) + \ln 2}{\ln(a+b) - \ln a - \ln 2}\right)$$

$$R_A + R_B = P \Rightarrow R_B = \frac{\ln\left(\frac{a+b}{2a}\right)}{\ln(b/a)}\cdot P$$

b). $b \to a$: $R_B = \frac{\ln 1}{\ln 1}\cdot P = \frac{0}{0}\cdot P$

$\lim_{b\to a} R_B = P/2$ by de l'Hopital rule.

6.21

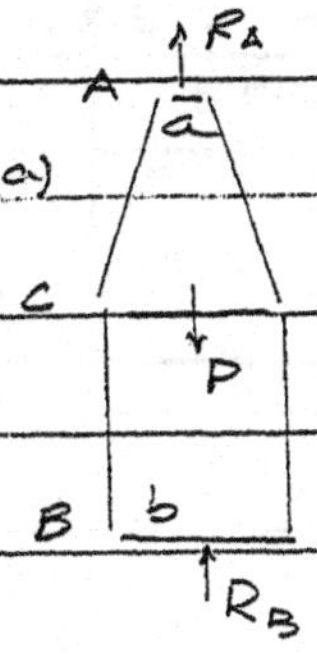

a) $\Delta_{BC} = R_B L/2Ebt$ From Prob. 6.7: $\Delta_{AC} = \frac{R_A L}{2Et}\left(\frac{\ln b - \ln a}{b-a}\right)$

$$\Delta_{AC} = \Delta_{BC} \Rightarrow R_B = bR_A\left(\frac{\ln b - \ln a}{b-a}\right)$$

$$R_A + R_B = P \Rightarrow R_A = \frac{P(b-a)}{b-a+b(\ln b - \ln a)} = \frac{P}{1 + \frac{\ln(b/a)}{1-a/b}}$$

b). $b = a$: $R_A = \frac{P}{1+0/0}$! $\lim_{b\to a}\frac{\ln(b/a)}{1-a/b} = \lim_{b\to a} b/a = 1$

$\therefore R_A = P/2.$

NOTE: $R_B = \frac{P}{1 + \frac{1-a/b}{\ln(b/a)}}$

6.22

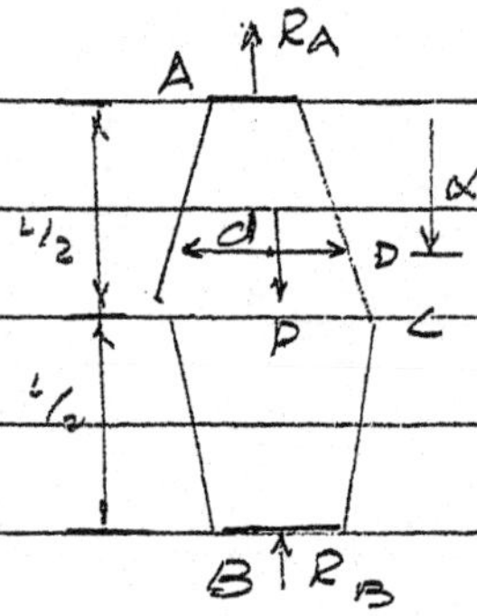

$d = a + 2\alpha(b-a)$; $b - d = (1-2\alpha)(b-a)$

αL $(0 < \alpha < 1/2)$

From Prob. 6.7:

$$\Delta_{AD} = \frac{R_A \alpha L}{Et}\left[\frac{\ln d - \ln a}{d-a}\right] = \frac{R_A L}{2Et}\left\{\frac{\ln[(1-2\alpha)a + 2\alpha b] - \ln a}{b-a}\right\}$$

$$\Delta_{CD} = \frac{R_B(1/2-\alpha)L}{Et}\left[\frac{\ln b - \ln d}{b-d}\right] = \frac{R_B L}{2Et}\left\{\frac{\ln b - \ln[(1-2\alpha)a + 2\alpha b]}{b-a}\right\}$$

$$\Delta_{BC} = \frac{R_B L}{2Et}\,\frac{\ln b - \ln a}{b-a}$$

$\Delta_{AD} = \Delta_{BC} + \Delta_{CD}$

$$R_A\{\ln[(1-2\alpha)a + 2\alpha b] - \ln a\} = R_B\{2\ln b - \ln a - \ln[(1-2\alpha)a + 2\alpha b]\}$$

$R_A + R_B = P$

$$\Rightarrow \quad R_B = P\left\{\frac{\ln[(1-2\alpha)a + 2\alpha b] - \ln a}{2[\ln b - \ln a]}\right\}$$

$$R_A = P\left\{\frac{2\ln b - \ln a - \ln[(1-2\alpha)a + 2\alpha b]}{2[\ln b - \ln a]}\right\}$$

Problem 6.23 on page P6-8

6.24

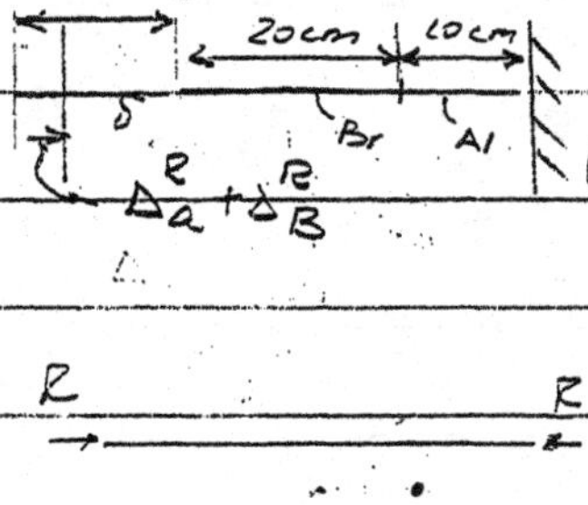

$(\Delta_a^T - \Delta_a^R) + (\Delta_B^T - \Delta_B^R) = \delta$

$\Delta_a^e + \Delta_B^e = (\Delta_a^T + \Delta_B^T) - \delta$

$$R\left[\frac{L_a}{A_a E_a} + \frac{L_B}{A_B E_B}\right] = L(\alpha_a + \alpha_B)\Delta T - \delta$$

$$R = \left(\frac{L_a L_B E_a E_b}{L_a A_B E_B + L_B A_a E_a}\right)\left[(\alpha_a L_a + \alpha_B L_B)\Delta T - \delta\right]$$

$$R = \left[\frac{(0.1)(.2)(70\times10^9)(105\times10^9)}{(0.1)(18\times10^{-4})(105\times10^9) + (0.2)(18\times10^{-6})(70\times10^9)}\right]\left\{[(23.6)(.1) + (18)(.2)]\times10^{-6}(150) - 6\times10^{-4}\right\}$$

$R = 171.5\times10^9$ N $\quad \sigma_a = \frac{171.5\times10^9}{1800} = 95.2$ N/mm^2 $= 95.2$ MPa

$\sigma_b = 114.3$ MPa.

$\Delta L_{Br} = (18)(.2)\times10^{-6}\cdot 150 - (95.2)(.2)/105\times10^9) = .000359$ m $= 0.359$ mm.

$\Delta L_a = \delta - 0.359 = 0.241$ mm.

6.23

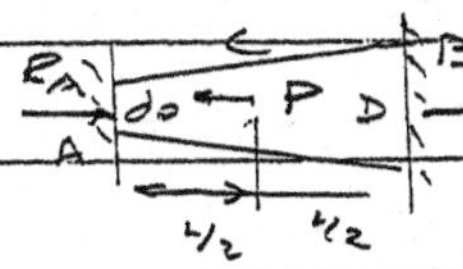

$\Delta_{AC} = \Delta_{BC}$ $\qquad R_A + R_B = P$

$d(x) = 1/L\,[(L-x)d_0 + xD]$

$A = \pi d^2/4 = \frac{\pi}{4L^2}[(L-x)d_0 + D]^2$

$A(x) = \frac{\pi}{4L^2}[\alpha x^2 + \beta x + \gamma]$, $\alpha = (D-d)^2$, $\beta = 2L d_0 (D-d_0)$

$\gamma = L^2 d_0^2$

$$\Delta_{AC} = \frac{R_A}{E}\int_0^{L/2}\frac{dx}{A} = \frac{4R_AL^2}{\pi E}\int_0^{L/2}\frac{dx}{\alpha x^2+\beta x+\gamma}.$$

Note: $\delta \equiv 4\alpha\gamma - \beta^2 = 0$

$$\therefore \int_0 \frac{dx}{\alpha x^2+\beta x+\gamma} = -\frac{2}{\beta+2\alpha x}$$ [Gradshteyn & Ryzhik, (No. 2.172) if $\delta = 0$]

$$\Delta_{AC} = \frac{4R_AL^2}{\pi E}\left[-\frac{2}{\beta+2\alpha x}\right]_0^{L/2} = -\frac{8R_AL^2}{\pi E}\left[\frac{1}{\beta+\alpha L}-\frac{1}{\beta}\right] = \frac{8R_AL^3}{\pi E}\frac{\alpha}{\beta(\beta+\alpha L)}$$

$\beta(\beta+\alpha L/2) = \beta L(D-d_0)(D+d_0)$.

$$\Delta_{AC} = \frac{8R_AL^2}{\pi E}\frac{(D-d_0)}{\beta(D+d_0)} = \frac{4R_AL}{\pi E\, d_0(D+d_0)}.$$

$$\Delta_{BC} = -\frac{8R_BL^2}{\pi E}\left[\frac{1}{\beta+2\alpha x}\right]_{L/2}^{L} = -\frac{8R_BL^2}{\pi E}\left[\frac{1}{\beta+2\alpha L}-\frac{1}{\beta+\alpha L}\right]$$

$$= +\frac{8R_BL^3}{\pi E}\left\{\frac{\alpha}{(\beta+\alpha L)(\beta+2\alpha L)}\right\}$$

Note: $(\beta+\alpha L)(\beta+2\alpha L) = 2DL^2(D-d_0)(D+d_0)$

$$\therefore \Delta_{BC} = \frac{4R_BL}{\pi E D(D+d_0)}$$

Note: For $D = d_0 \Rightarrow \Delta_{AC} = \Delta_{BC} = \frac{RL/2}{\pi E D^2/4} = RL/AE$

$\Delta_{AC} = \Delta_{BC} \Rightarrow R_A/d_0 = R_B/D \Rightarrow R_B = \frac{D}{d_0}R_A.$

$R_A + R_B = P \Rightarrow (1 + D/d_0)R_A = P \Rightarrow R_A = \left(\frac{d_0}{D+d_0}\right)P$, $\quad R_B = \left(\frac{D}{D+d_0}\right)P$

6.25

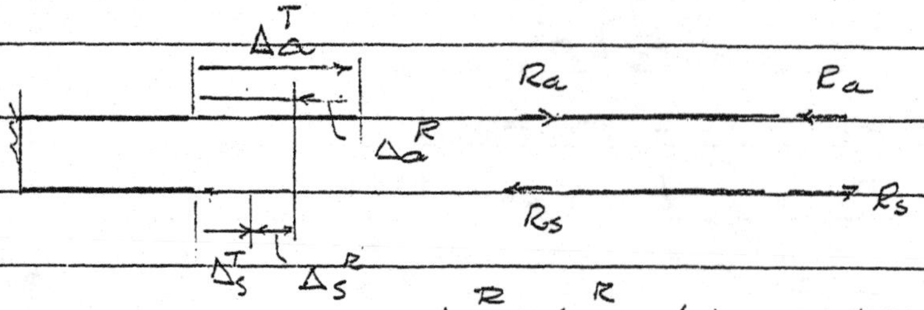

$R_s = R_a$

$$\Delta_a^R + \Delta_s^R = (\alpha_a - \alpha_s)\Delta T \cdot L$$

$$LR\left[\frac{1}{A_aE_a} + \frac{1}{E_sA_s}\right] = (\alpha_a - \alpha_s)\,\Delta T \cdot L$$

$$R\left[\frac{1}{(1.5)(70)\times10^5} + \frac{1}{(0.6)(200)\times10^5}\right] = (23.6 - 11.7)\times10^{-6}(50)$$

$R = 3{,}332\,N$ $\quad \sigma_s = 555.3\,MPa;\ \sigma_a = 222.1\,MPa$

$$\Delta L = \alpha_s L \Delta T + RL/A_sE_s = L\left[\alpha_s\Delta T + R/A_sE_s\right] = 0.0172\,cm = 0.173\,mm$$

6.26 $\quad RL\left[\frac{1}{A_tE_t} + \frac{1}{A_mE_m}\right] = (\alpha_m - \alpha_t)\Delta T L \qquad E_t = \sigma_{oy}/\varepsilon_{oy} = 805\times10^6/0.007 = 115\,GPa,\ E_m = 180\,GPa$

$$R_T = R_m \Rightarrow R_T|_y = (805\times10^6)(0.6\times10^{-4}) = 48{,}300\,N$$
$$R_m|_y = (360\times10^6)(1.5\times10^{-4}) = 54{,}000\,N \qquad R_{Ty} < R_{my}$$

∴ TITANIUM YIELDS FIRST

$$\Delta T = \frac{R\left[1/A_tE_t + 1/A_mE_m\right]}{\alpha_m - \alpha_t} = \frac{(48{,}300)\left[1/(0.6)(115) + 1/(1.5)(180)\right]10^{-5}}{(13.9 - 9.5)10^{-6}} = 1997\,°C.$$

6.27 $\quad \Delta_{cu}^T - \Delta_{cu}^R = \Delta_s^T + \Delta_s^R + \delta$

$$\therefore\ \Delta_c^R + \Delta_s^R = (\Delta_{cu}^T - \Delta_s^T) - \delta$$

$R_c = 2R_s = R$

$$hR\left[1/A_cE_c + 1/2A_sE_s\right] = (\alpha_c - \alpha_s)\cdot h\,\Delta T - \delta$$

$$R = \left(\frac{2A_cA_sE_cE_s}{2A_sE_s + A_cE_c}\right)\cdot\left[(\alpha_a - \alpha_s)\Delta T - \delta/h\right]$$

$$R = \frac{2(10\times10^{-4})(5\times10^{-4})(120)(200)10^{18}}{\left[2(5\times10^{-4})(200) + (10\times10^{-4})(120)\right]10^9}\cdot\left[(16.9 - 11.7)\times10^{-6}(60) - 0.5/2000\right]$$

$$= \frac{2(10)(5)(120)(200)\times10^{10}}{2(5)(200) + (10)(120)\times10^5}\cdot\left[(5.2\times10^{-6})(60) - 0.5/2000\right]$$

$R = 4650.\,N \qquad \sigma_s = R/A_s = 9.3\,MPa,\ \sigma_c = 4.65\,MPa$

6.28 Assume all rods are in tension:

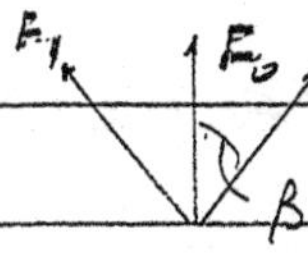

$F_0 + 2F_1 \cos\beta = 0$

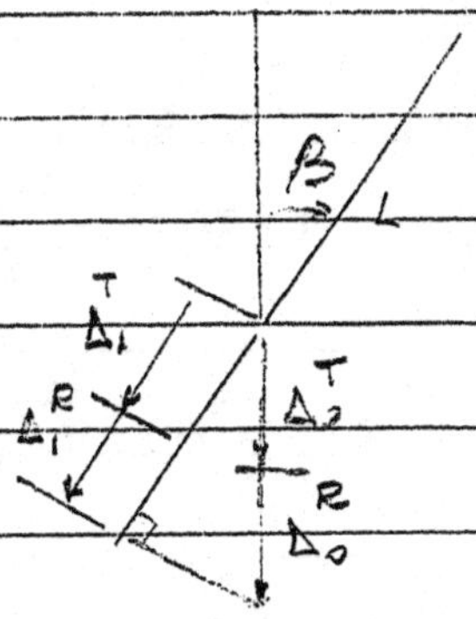

$\Delta_1 = \Delta_1^T + \Delta_1^R , \quad \Delta_0 = \Delta_0^T + \Delta_0^R$

$\dfrac{\Delta_1^T + \Delta_1^R}{\cos\beta} = \Delta_0^T + \Delta_0^R \Rightarrow \Delta_0^R - \Delta_1^R/\cos\beta = \Delta_1^T/\cos\beta - \Delta_0^T$

$\Delta_0^R = F_0 h/AE , \quad \Delta_1^R = F_1 L/AE$

$\therefore \; -\dfrac{2F_1 \cos\beta \, h}{AE} - \dfrac{F_1 L}{AE\cos\beta} = \alpha(L/\cos\beta - h)$

$-\dfrac{F_1 L}{AE}\left(1/\cos\beta + 2\cos\beta \dfrac{h}{L}\right) = \alpha L\left(1/\cos\beta - h/L\right)\Delta T \qquad h/L = \cos\beta$

$-\dfrac{F_1 L/AE}{\cos\beta}\left(1 + 2\cos^2\beta\right) = \dfrac{\alpha L}{\cos\beta}\left(1 - \cos^2\beta\right) \Rightarrow F_1 = -AE\alpha\Delta T \dfrac{1-\cos^2\beta}{1+2\cos^2\beta} = -AE\alpha\Delta T \dfrac{\sin^2\beta}{1+2\cos^2\beta} = F_{AD}$

$F_0 = 2AE\alpha\Delta T \dfrac{(1-\cos^2\beta)\cos\beta}{1+2\cos^2\beta}$

(b). $\beta = 64.4^\circ$ [see p. 6-11]

6-29

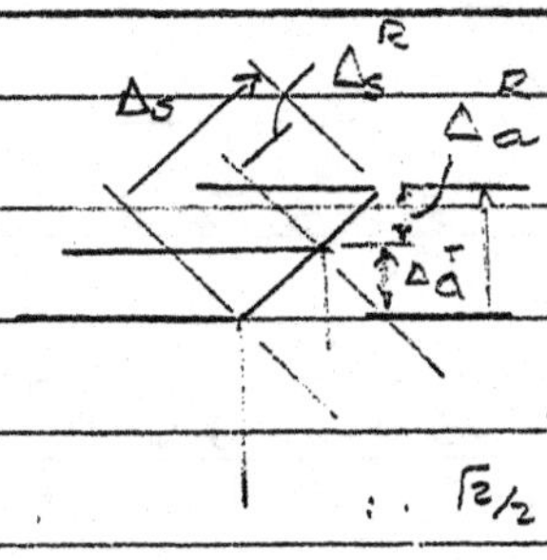

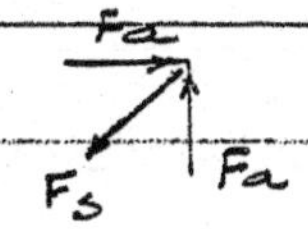

$2F_a \cos 45^\circ = F_s$

$F_s = 2\frac{\sqrt{2}}{2}F_a = \sqrt{2}F_a$

$\Delta_s = \Delta_s^T + \Delta_s^R , \quad \Delta_a = \Delta_a^T - \Delta_a^R \Rightarrow \sqrt{2}\Delta_s/2 = \Delta_a$

$\therefore \; \sqrt{2}/2\left[\Delta_s^T + \Delta_s^R\right] = \Delta_a^T - \Delta_a^R$

$\sqrt{2}/2\,\Delta_s^R + \Delta_a^R = \Delta_a^T - \sqrt{2}/2\,\Delta_s^T$

$\Rightarrow \sqrt{2}/2\, F_s L_s/A_s E_s + F_a L_a/A_a E_a = (\alpha_a L_a - \sqrt{2}/2\,\alpha_s L_s)\Delta T$

$L_s = \sqrt{2}L_a , \quad E_s/E_a = 200/70 , \quad A_s/A_a = 1/20 \Rightarrow A_s E_s/A_a E_a = 1/7$

$\therefore \; \sigma_a = F_a/A_a = \left(\dfrac{\alpha_a - \alpha_s}{7\sqrt{2}+1}\right)\Delta T \cdot E_a = \dfrac{(11.9\times10^{-6})}{10.899}(40)(70\times10^{9}) = 3.06 \text{ MPa}$

$\sigma_s = F_s/A_s = \sqrt{2}F_a/(A_a/20) = 86.5 \text{ MPa}$

6.28 CONTINUED

$$F_o = C(1-\cos^2\beta)\cos\beta(1+2\cos^2\beta)^{-1}, \quad C = 2AE\alpha\Delta T.$$

$$\frac{dF_o}{d\beta} = 0 \Rightarrow 2\cos^2\beta\sin\beta(1+2\cos^2\beta)^{-1} - (1-\cos^2\beta)\sin\beta(1+2\cos^2\beta)^{-1}$$
$$+ (1-\cos^2\beta)\cos\beta(1+2\cos^2\beta)^{-2}(4\cos\beta)\sin\beta = 0$$

$$\therefore 2\cos^2\beta\sin\beta(1+2\cos^2\beta) - (1-\cos^2\beta)\sin\beta(1+2\cos^2\beta) + 4(1-\cos^2\beta)\cos^2\beta\sin\beta = 0$$

$$\sin\beta\{[2\cos^2\beta - (1-\cos^2\beta)](1+2\cos^2\beta) + 4(1-\cos^2\beta)\cos^2\beta\} = 0$$

$$\sin\beta\{(3\cos^2\beta - 1)(1+2\cos^2\beta) + 4(1-\cos^2\beta)\cos^2\beta\} = 0$$

$$\sin\beta\{6\cos^4\beta + \cos^2\beta - 1 + 4\cos^2\beta - 4\cos^4\beta\} = 0$$
$$\sin\beta\{2\cos^4\beta + 5\cos^2\beta - 1\} = 0$$

Let $x = \cos^2\beta \Rightarrow 2x^2 + 5x - 1 = 0 \Rightarrow x = \frac{1}{4}[-5 \pm (33)^{1/2}]$

$$\cos^2\beta = 0.1814 \Rightarrow \cos\beta = 0.4314 \Rightarrow \beta = 64.44^\circ$$

F_o max or min?

$$\frac{d^2F_o}{d\beta^2} = \cos\beta(2\cos^4\beta + 5\cos^2\beta - 1) + \sin\beta(-8\cos^3\beta\sin\beta - 10\cos\beta\sin\beta)$$

$$= \cos\beta[2\cos^4\beta + 5\cos^2\beta - 1 - 8\cos^2\beta\sin^2\beta - 10\sin^2\beta]$$

$$\left.d^2F_o/d\beta^2\right|_{\beta=64.4^\circ} = -4.03 < 0 \Rightarrow F_o \text{ is a max. at } \beta = 64.4^\circ.$$

$$F_o|_{max} = 0.51175\, AE\alpha\Delta T$$

6.30

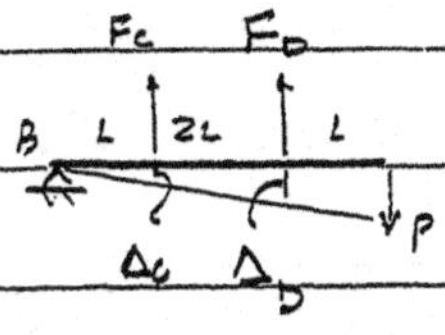

$(\Sigma M)_B = 0 \Rightarrow F_C + 3F_D = 4P \Rightarrow P = \frac{1}{4}(F_C + 3F_D)$

$\Delta_D = 3\Delta_C \Rightarrow$ D yields first.

$F_D = \sigma_0 A,\quad F_C = \sigma_0 A/3$

(a) $P_y = \frac{1}{4}(\frac{1}{3} + 3)\sigma_0 A = 5\sigma_0 A/6$, (b) $\Delta_F = \frac{4}{3}\Delta_D = \frac{4}{3}\sigma_0 h/E$

(c) $F_C = F_D = \sigma_0 A \Rightarrow P_{ult} = \sigma_0 A$

(d) $\Delta_F|_{ult} = 4\Delta_B = 4\sigma_0 h/E.$

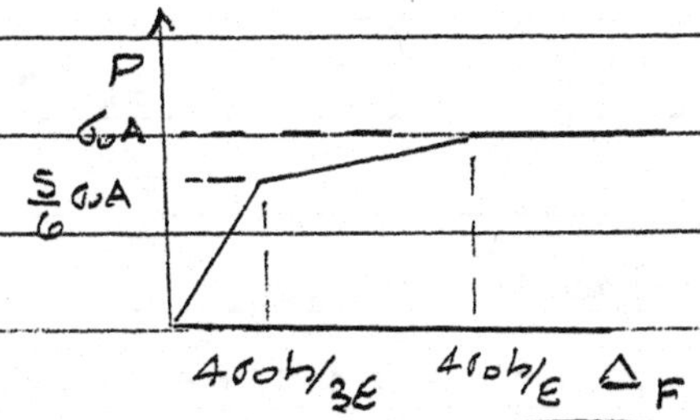

6.31

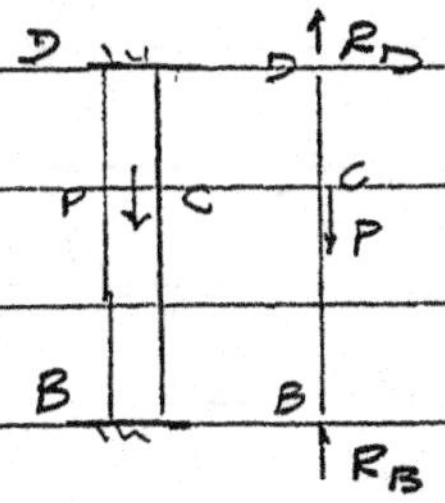

$\Delta_{CD} = \Delta_{BC} \Rightarrow 2R_B L/3 = R_D L/3 \Rightarrow R_D = 2R_B$

$\therefore$ CD yields first : $R_D = \sigma_0 A$; $R_D + R_B = P$

$P_y = 3R_D/2 = 3\sigma_0 A/2$; $\delta_C = \Delta_{CD} = \sigma_0 L/3E$

$P_{ult} = R_D + R_B = 2\sigma_0 A$; $\delta_C = 2\sigma_0 L/3E.$

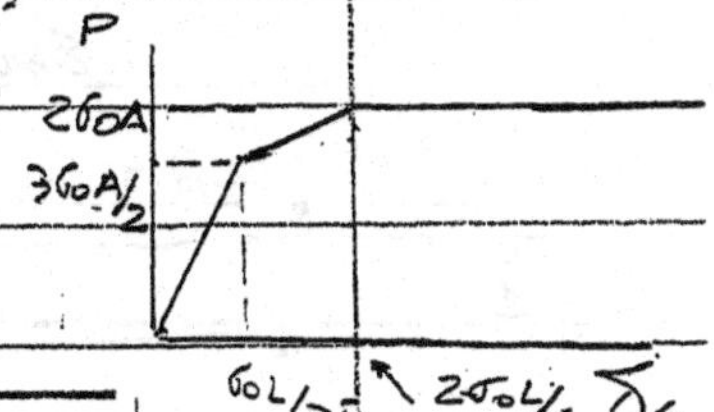

6.32

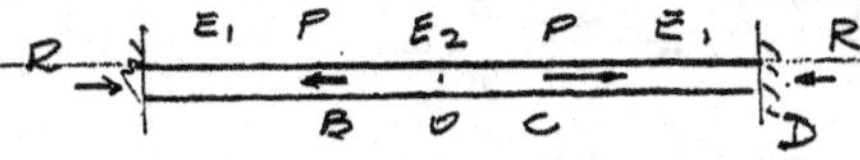

$R = P \qquad E_2/E_1 = 1/4$

$\Delta_{OC} = \Delta_{CD} \Rightarrow \varepsilon_{OC} = 2\varepsilon_{CD}$

$\varepsilon_{1y} = 3\sigma_0/E_1,\ \varepsilon_{2y} = \sigma_0/E_2 = 4\sigma_0/E_1 \Rightarrow 6\sigma_0/E_1 > 4\sigma_0/E_1 \Rightarrow$ BC yields first

At first yield: $F_{BC} = \sigma_0 A,\ F_{CD} = \varepsilon_{CD} E_1 A = (\varepsilon_{OC}/2)E_1 = \frac{1}{2}(\sigma_0/E_2)E_1 = 2\sigma_0 A.$

$2\sigma_0 A \leftarrow$ [$\rightarrow P_y$] $\leftarrow \sigma_0 A \Rightarrow P_y = 3\sigma_0 A$, $\delta_C = \varepsilon_{2y}L/2 = 2\sigma_0 L/E_1$

$P_{ult} = (3\sigma_0 + \sigma_0)A = 4\sigma_0 A$, $\delta_C = L\varepsilon_{CD} = L(3\sigma_0/E_1) = 3\sigma_0 L/E_1$

c)

$P - P_p = k(\delta - \delta_p)$, $k = \frac{3\sigma_0 A}{2\sigma_0 L/E_1} = \frac{3AE_1}{2L}$

$P_p = 0 \Rightarrow \delta_p = \delta - P/k = 6\sigma_0 L/E_1 - 4\sigma_0 A/(3AE_1/2L) = 10\sigma_0 L/3E_1$

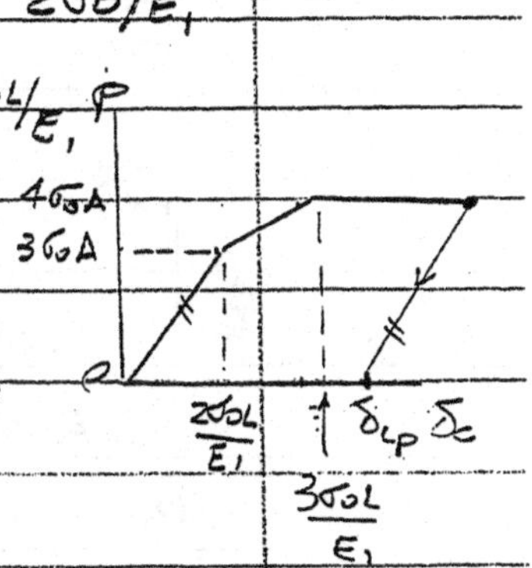

6.33

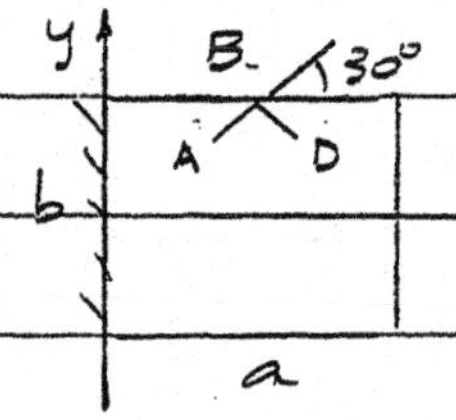

a) Symmetry with respect to the x-axis

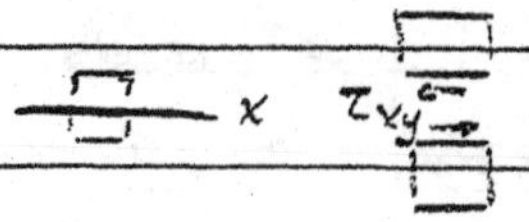

$\Leftarrow$ impossible $\Rightarrow \therefore \tau_{xy}(y=0) = 0$

c). $\frac{\partial \sigma_x}{\partial x} + \frac{\partial \tau_{yx}}{\partial x} + c e^{x/a} = 0 \Rightarrow \partial\sigma_x/\partial x = -c e^{x/a}$

$\frac{\partial \tau_{xy}}{\partial x} + \frac{\partial \sigma_y}{\partial y} = 0 \Rightarrow \frac{\partial \sigma_y}{\partial y} = 0 \Rightarrow \sigma_y = \sigma_y(x)$

But $\sigma_y(x, y = \pm b/2) = 0 \Rightarrow \therefore \sigma_y = 0$.

$\sigma_x = -ac\, e^{x/a} + A(y)$

$\sigma_x(a, y) = -ace + A(y) = 0 \Rightarrow A(y) = ace \Rightarrow \sigma_x = ac(e - e^{x/a})$

d). $\theta\big|_{B-A} = -150°$.

$\varepsilon_{nt} = -\left(\frac{\varepsilon_x - \varepsilon_y}{2}\right)\sin 2\theta + \varepsilon_{xy}\cos 2\theta$

$\varepsilon_x = \sigma_x/E$, $\varepsilon_y = -\nu\sigma_x/E$, $\varepsilon_{xy} = 0$; $\sigma_x = ac(e - \sqrt{e})$

$\varepsilon_{nt} = -\frac{1}{2E}(1+\nu)\sigma_x \sin(-300°) = -(1/2E)(\sqrt{3}/2)(1+\nu)(e - \sqrt{e})ac$

$\varepsilon_{nt}\big|_{a/2} = -\frac{\sqrt{3}ac}{4E}(1+\nu)(e - \sqrt{e})$

e). $u_B = \int_0^{a/2} \varepsilon_x(x)\,dx = \frac{1}{E}\int_0^{a/2} \sigma_x(x)\,dx = \frac{a^2c}{2E}[2 + e - 2\sqrt{e}]$; $v_B = -\frac{\nu\sigma_x b}{2E} = -\nu\frac{abc}{2E}(e - \sqrt{e})$

6.34

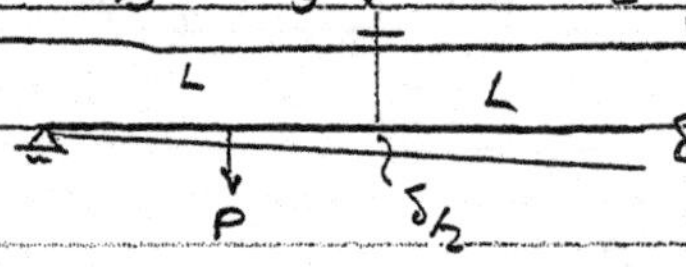

$W = (P/2)\left(\frac{a}{2L}\delta\right) = Pa\delta/4L$

$U = \frac{E\varepsilon^2}{2}Ah = \frac{EAh}{2}\left(\delta/2h\right)^2 = \frac{EA\delta^2}{8h}$

$W = U \Rightarrow Pa\delta/4L = \frac{EA\delta^2}{8h} \Rightarrow a/L = \frac{EA\delta}{2Ph}$

6.35

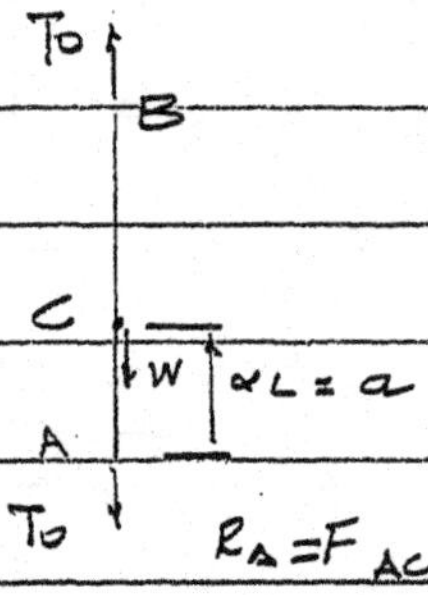

Due to W alone: $R_A^0 + R_B^0 = W$

$\Delta_{AC} = \Delta_{BC} \Rightarrow R_A^0(\alpha L)/AE = R_B^0(1-\alpha)L/AE$

$R_A^0 = R_B^0(1-\alpha)/\alpha$

$\Rightarrow R_B^0 = \alpha W$, $R_A^0 = (1-\alpha)W$

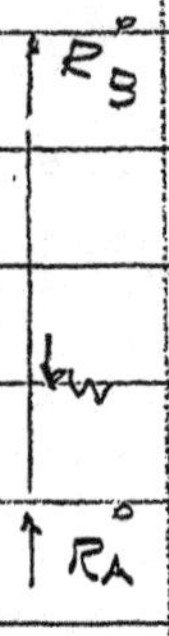

$R_A = F_{AC} = T_0 - R_A^0 = T_0 - (1-\alpha)W = W[(\alpha - 1) + T_0/W]$

$R_B = F_{BC} = T_0 + R_B^0 = T_0 + \alpha W = W[\alpha + T_0/W]$

$R_A = W(\alpha - 1 + \gamma)$, $\gamma \equiv T_0/W > 0$

$R_B = W(\alpha + \gamma)$

6.36 a) Let F_r be friction force.

$F_r = (\pi D) f(y) \Delta y \Rightarrow \pi D k y^2 dy$

$P = \int_0^L dF = \pi D k \int_0^L y^2 dy = \pi D k/3 \; y^3 \Big|_0^L = \pi D k L^3/3$

$k = 3P/\pi D L^3$

$dF = (3P/L^3)\, y^2 dy$

$F(y) = \frac{3P}{L^3} \int_0^y \xi^2 d\xi = \frac{P}{L^3} y^3.$

$\sigma = F/A = \frac{4P}{\pi D^2 L^3} y^3.$

b)

$\Delta L = \frac{1}{E} \int_0^L \sigma(y)\, dy = \frac{4P}{\pi D^2 L^3 E} \int_0^L y^3 dy = \frac{PL}{\pi D^2 E} = PL/4AE.$

6.37

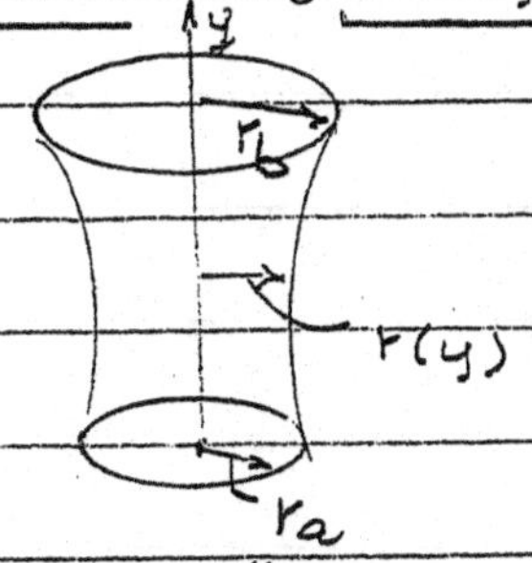

$r(y) = r_a + (r_b - r_a)(y/L)^2$

Let $\beta = r_b/r_a - 1 \Rightarrow r(y) = r_a[1 + \beta(y/L)^2]$.

$A(y) = \pi r^2(y)$

$F(y) = \rho \int_0^y A(\eta) d\eta = \rho \pi r_a^2 \int_0^y [1 + \beta \eta^2/L^2]^2 d\eta = \rho \pi r_a^2 \int_0^y [1 + 2\beta(\eta/L)^2 + \beta^2(\eta/L)^4] d\eta$

$= \rho \pi r_a^2 \{ y + 2\beta y^3/3L^2 + \beta^2 y^5/5L^4 \}$

$\sigma(y) = F(y)/A(y) = \rho \left\{ \frac{y + 2\beta y^3/3L^2 + \beta^2 y^5/5L^4}{1 + 2\beta y^2/L^2 + \beta^2 y^4/L^4} \right\}$

Let $\xi = y/L$: $\frac{\sigma(\xi)}{\rho L} = \frac{\xi + 2\beta\xi^3/3 + \beta^2\xi^5/5}{1 + 2\beta\xi^2 + \beta^2\xi^4}.$

$\varepsilon(y) = \sigma(y)/E \Rightarrow \Delta L = \int_0^L \varepsilon(y)\, dy = L \int_0^1 \varepsilon(\xi) d\xi.$

$\Delta L = \rho L^2/E \cdot \int_0^1 \left\{ \frac{\xi + 2\beta\xi^3/3 + \beta^2\xi^5/5}{1 + 2\beta\xi^2 + \beta^2\xi^4} \right\} d\xi$

$\frac{\Delta L}{\rho L^2/E} = \int_0^1 \frac{\xi(1 + 2\beta\xi^2/3 + \beta^2\xi^4/5)}{1 + 2\beta\xi^2 + \beta^2\xi^4} d\xi.$

Solution valid for small taper, i.e. for

$\frac{r_b - r_a}{L} = r_a\beta/L \ll 1 \Rightarrow -0.1 \le \beta \le 0.1$ for example

6.38 a) Equilib. of cylinder for $d \le r \le D$:

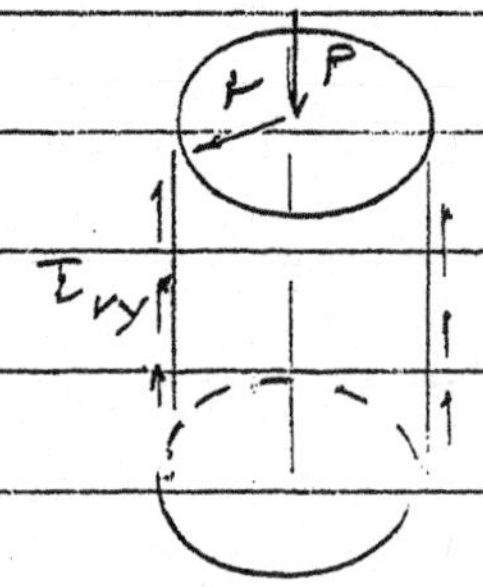

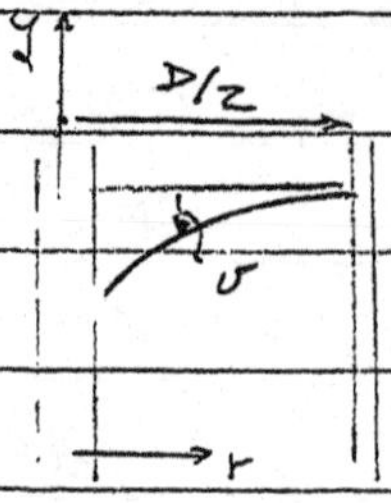

$$P = 2\pi r L \tau_{ry}$$

$$\tau_{ry} = \frac{P}{2\pi L} \cdot \frac{1}{r} \qquad \tau = 2G\varepsilon_{ry}$$

$$\varepsilon_{ry} = -\frac{dv(r)}{dr} = -\frac{P}{4\pi G L} \cdot \frac{1}{r}$$

$$\int_{d/2}^{D/2} dv = -\frac{P}{4\pi G L} \int_{d/2}^{D/2} \frac{1}{r}\, dr = -\frac{P}{4\pi G L} \ln r \Big|_{d/2}^{D/2} = -\frac{P}{4\pi G L} [\ln D/2 - \ln d/2]$$

$$v(D/2) = 0 \Rightarrow \delta = v(r = d/2) = \frac{P}{4\pi G L} \ln (D/d)$$

b). $\delta = v_{200} = \dfrac{10^5 \ln(12.5)}{4\pi(6\times10^9)(0.3)} = 0.0112\, m = 1.12\, cm$

c). $F(y) = (\pi d) \cdot y\, \tau_{ry}(d) = \pi d \cdot \dfrac{P}{\pi L d} \cdot y = \dfrac{P}{L} \cdot y$

$$\sigma_y = \frac{F(y)}{\pi d^2/4} = \frac{4P}{\pi d^2}(y/L)$$

(d)

$$\Delta L = \frac{4P}{\pi d^2 E} \frac{1}{L} \int_0^L y\, dy = \frac{2PL}{\pi d^2 E}$$

(e) $\Delta L = \dfrac{2(10^5)(0.3)}{\pi(.4\times10^{-2})^2(200\times10^9)} = 5.97\times10^{-5}\, m$

$\Delta L = 0.06\, mm$

6.39 a). $\Delta^T = \Delta^R \Rightarrow \varepsilon^T = \varepsilon^R \Rightarrow \alpha \Delta T_E = \sigma_0/E \Rightarrow \Delta T_E = \sigma_0/\alpha E$

b).

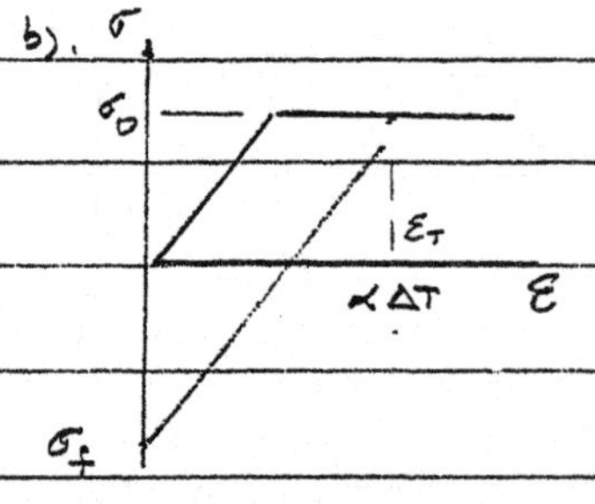

$$\varepsilon_T = \alpha \Delta T \qquad \frac{\sigma_0 - \sigma^f}{\varepsilon_T - \varepsilon^f} = E \Rightarrow \sigma_0 - \sigma^f = E(\varepsilon_T - \varepsilon_f)$$

where σ^f, ε^f are final values.

$$\varepsilon_f = 0 \Rightarrow \sigma^f = \sigma_0 - E\varepsilon_T = \sigma_0 - E\alpha \cdot \Delta T \quad \text{(tension)}$$

c). $\Delta T_E = \dfrac{250\times10^6}{(11.7\times10^{-6})(200\times10^9)} = 106.84\,^\circ C$; $\sigma^f = 109.6\ MPa$

6.40 a) $\varepsilon_f = \varepsilon_m \Rightarrow \sigma_f/E_f = \sigma_m/E_m \Rightarrow \sigma_f/\sigma_m = E_f/E_m$

b). $P = n A_f \sigma_f + A_m \sigma_m$

c). $\bar{\sigma} = P/A = (n A_f \sigma_f + A_m \sigma_m) = v_f \sigma_f + v_m \sigma_m$

d). $E_{eff} \cdot \varepsilon = \bar{\sigma} \Rightarrow E_{eff}\,\varepsilon = v_f E_f \varepsilon + v_m E_m \varepsilon = (v_f E_f + v_m E_m)\,\varepsilon$

$\therefore\ E_{eff} = v_f E_f + v_m E_m.$

e). $E_{eff} = (0.2)(300) + 0.8(2.4) = 61.9$ GPa

$\bar{\sigma} = P/A = (6000 \times 10^3)/(4 \times 10^{-4}) = 15$ MPa

$\varepsilon = \bar{\sigma}/E_{eff} = 2.42 \times 10^{-4}$ $\Delta L = \varepsilon L = 0.97$ mm.

$\sigma_f = E_f \varepsilon = 72.3$ MPa, $\sigma_m = E_m \varepsilon = 0.58$ MPa

---

6.41 $E_{eff} = v_f E_f + v_m E_m = (0.45)(72) + (0.55)(2.4) = 33.72$ GPa

$\bar{\sigma} = 15$ MPa

$\varepsilon = \bar{\sigma}/E_{eff} = 4.45 \times 10^{-4}$ $\Delta L = \varepsilon L = 1.78$ mm

$\sigma_f = E_f \varepsilon = 32.0$ MPa, $\sigma_m = E_m \varepsilon = 1.07$ MPa

7.1 $\tau = TR/J$, $J = 6.434\times10^{-9}\ m^4$; $\tau = (30)(8\times10^{-3})/J = 37.3$ MPa

$\varphi = TL/JG \Rightarrow G = TL/J\varphi$

$3.5° < \varphi < 7.2° \Rightarrow 6.1087\times10^{-2}$ rad $< \varphi < 0.10472$ rad

$3.5° < \varphi \Rightarrow TL/JG > 3.5° \Rightarrow G < TL/6.108\times10^{-2} J = 91.6$ GPa

$TL/JG < 7.2° \Rightarrow G > TL/(0.104)J = 53.4$ GPa; $47.5 < G < 91.6$ GPa STEEL & BRONZE

7.2 $\tau = TR_o/J \Rightarrow J = TR_o/\tau$; $\pi/2(R_o^4 - R_i^4) = TR_o/\tau \Rightarrow R_i = \left[R_o^4 - \frac{2TR_o}{\pi\tau}\right]^{1/4} = 2.2065\times10^{-2}$ m $= 22.065$ mm

$t = 7.935$ mm.

7.3 $\tau_s = \frac{Td/2}{J_s} = \frac{Td/2}{\pi d^4/32} = \frac{16T}{\pi d^3}$ $\tau_h = \frac{TD/2}{\frac{\pi}{32}(D^4 - D_i^4)}$

$R_o = R_i + D/20 \Rightarrow D = D_i + D/10 \Rightarrow D_i = D(1 - 1/10) = 9D/10$

$\therefore \tau_h = \frac{T16D}{\pi D^4[1-(9/10)^4]} = \frac{16T}{\pi D^3[1-(9/10)^4]}$

$\tau_s = \tau_h \Rightarrow \frac{1}{d^3} = \frac{1}{D^3[1-(9/10)^4]} \Rightarrow D = \frac{d}{[1-(9/10)^4]^{1/3}} = \frac{d}{0.7006} = 1.427d$

7.4 $\gamma = \theta r \Rightarrow \tau = c\sqrt{\gamma/2} = \frac{c}{\sqrt{2}}\sqrt{\gamma} = \frac{c\sqrt{\theta}}{\sqrt{2}} r^{1/2}$

$T = \iint_A r\tau(r)\,dA = 2\pi \frac{c\sqrt{\theta}}{\sqrt{2}} \int_0^R r^{5/2}\,dr = \frac{2\pi\sqrt{\theta}}{\sqrt{2}} \frac{R^{7/2}}{7/2} = \frac{2\pi\sqrt{2}\sqrt{\theta}\,R^{7/2}}{7}$

$\theta = \frac{49T^2}{8\pi^2 c^2 R^7}$

7.5 Conclusion

7.6 $\tau = Tr/J$; $T_{in} = 2\pi\int_0^{R_i} \tau(r) r^2\,dr = \frac{2\pi T}{J}\int_0^{D/4} r^3\,dr = \frac{2\pi T D^4/256\cdot4}{\pi D^4/32} = 1/16$

Inner: 6.25 %, outer 93.75 %

7.7 $J = \frac{TR}{\tau} \Rightarrow \pi R^3/2 = T/\tau \Rightarrow R = (2T/\pi\tau)^{1/3} = 4.57\times10^{-2}$ m $= 4.57$ m $\Rightarrow d = 9.14$ cm.

$\varphi = TL/GJ \Rightarrow J = TL/G\varphi \Rightarrow \pi R^4/2 = TL/G\varphi \Rightarrow R = (2TL/\pi G\varphi)^{1/4}$

$\therefore R = \left[\frac{2(6000)(2)}{\pi(26\times10^9)(2.618\times10^{-2})}\right]^{1/4} = 5.788\times10^{-2}$ m $= 5.788$ cm

$d = 11.58$ cm.

7.8 a) $T = \tau_{BC} J_{BC} / R_{BC}$ $\quad J_{BC} = \pi(25)^4/2 = 6.136\times10^5\ mm^4 = 6.136\times10^{-7}\ m^4$

$T = (40\times10^6)(6.136\times10^{-7})/25\times10^{-3} = 981.7\ N\text{-}m$

b. $J_{AB} = T R_{AB}/\tau_{AB} \Rightarrow \pi R^3/2 = T/\tau_{AB} \Rightarrow R_{AB} = (2T/\pi\tau_{AB})^{1/3} = 18.4\ mm$

$D_{AB} = 36.8\ mm.$

7.9 $T = 15\ kN\text{-}m$

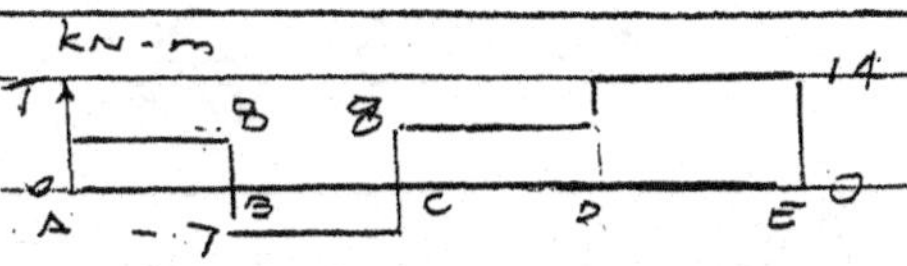

$G = 76\ GPa$

$J = \pi D^4/32 = 4.02\times10^{-6}\ m^4$

a) $\tau_{DE} = TR/J = 139.3\ MPa$; b) $\varphi_{D/B} = \varphi_{C/B} + \varphi_{D/C} = \frac{L}{GJ}[T_{BC} + T_{CD}]$

$\varphi_{D/B} = \frac{L}{GJ}(-7+8)\times10^3 = L/GJ = 4.09\times10^{-3}\ rad = 0.234°$

c. $\varphi_{E/A} = \varphi_{B/A} + \varphi_{C/B} + \varphi_{D/C} + \varphi_{E/D}$

$= \frac{L}{GJ}(8-7+8+14)\times10^3 = (23\times10^3)(1.25)/GJ = 9.407\times10^{-2}\ rad = 5.39°$

7.10

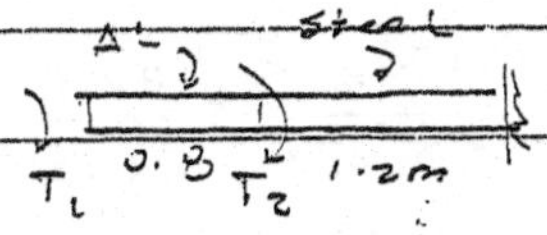

$J = TR/\tau \Rightarrow \pi R^4/2 = TR/\tau \Rightarrow R = (2T/\pi\tau)^{1/3}$

STEEL: $R = [2(1600)/\pi\cdot 80\times10^6]^{1/3} = 2.34\times10^{-2}\ m = 2.34\ cm$

Al: $R = [2(1000)/\pi\cdot 50\times10^6]^{1/3} = 1.96\times10^{-2}\ m = 1.96\ cm$

$\varphi_A = \frac{1}{J}\left(\frac{L_a T_a}{G_a} + \frac{L_s T_s}{G_s}\right) = \frac{1}{J}\left(\frac{(0.8)(600)}{26\times10^9} + \frac{(1.2)(1600)}{76\times10^9}\right) = 4.372\times10^{-8}/J < 2.5°$

$2.5° = 4.363\times10^{-2}\ rad.$

$\frac{\pi R^4}{2} = J > \frac{4.372\times10^{-8}}{4.363\times10^{-2}} = 1.002\times10^{-6} \Rightarrow R > \left[\frac{2(1.002\times10^{-6})}{\pi}\right]^{1/4} = 2.826\times10^{-2}\ m$

$D = 5.65\ cm.$

7.11 $\varphi_a = \frac{T}{G}\int_0^L \frac{dx}{J(x)}$ $\quad r = r_0(1 + x/L) \Rightarrow J = \frac{\pi r_0^4}{2}(1 + x/L)^4$

$\varphi_a = \frac{2T}{G\pi r_0^4}\int_0^L (1 + x/L)^{-4}dx = -\frac{2TL}{G\pi r_0^4}\cdot\frac{1}{3}(1 + x/L)^{-3} = \frac{7TL}{12\pi G r_0^4}$

7.12 ON AB: $T_{AB} = \frac{200}{80}(1000) = 2500\ N\text{-}m.$ $\quad \varphi_B = T_{AB}L_{AB}/GJ$

C-D: $T = 1000\ N\text{-}m \Rightarrow R = (2T/\pi\tau)^{1/3} = 2.197\times10^{-2}\ m \Rightarrow D = 4.395\ cm.$

$\varphi_C = 2.5\varphi_B \Rightarrow \varphi_D = \varphi_C + \varphi_{D/C} = \frac{1}{GJ}[2.5 T_{AB}L_{AB} + T_{CD}L_{CD}] < 1.5°$

$\varphi_D < 1.5° = 2.618\times10^{-2}\ rad.$

$\frac{[(10000)(.4) + 1000(.6)]}{GJ} < 2.618\times10^{-2} \Rightarrow J > \frac{4.6\times10^3}{(2.618\times10^{-2})G} \Rightarrow J > 2.312\times10^{-6}$

$R \geq [2(2.3126\times10^{-6})/\pi]^{1/4} = 3.483\times10^{-2} \Rightarrow d = 6.97\ cm.$

7.13 $\tau = TR/J$, $P = 2\pi N T \Rightarrow T = P/2\pi N$ $\quad P = 100\,kW = (6000 \times 10^3)\ N\text{-}m/min$

$$\tau = PR/2\pi N J = \frac{P}{\pi^2 R^3 N} = \frac{6000 \times 10^3}{\pi^2 (20 \times 10^{-3})^3 (800)} = 95 \times 10^6\ N/m^2 = 95\,MPa$$

7.14 $\tau = TR/J$, $P = 2\pi f T$; $P = 400\ kW = (400 \times 10^3)\ N\text{-}m/sec.$

$$\tau = PR/2\pi f J \Rightarrow J = \pi R^4/2 = \frac{PR}{2\pi f \tau} \Rightarrow R = \left(\frac{P}{\pi^2 f \tau}\right)^{1/3}$$

$$R = \left[\frac{400 \times 10^3}{\pi^2 (50)(90 \times 10^6)}\right]^{1/3} = 2.08 \times 10^{-2}\ m = 2.08\ cm$$

$$D = 4.16\ cm.$$

7.15 $\tau = TR/J$ $\quad P = 2\pi N T \Rightarrow \tau = (P/2\pi N)(R/J) = \frac{P}{\pi^2 R^3 N}$, $P = 600\ kW = 36{,}000\,k\frac{N\text{-}m}{min}$

$$N = \frac{P}{\pi^2 R^3 \tau} = \frac{(36 \times 10^6)}{\pi^2 (2 \times 10^{-2})^3 (80 \times 10^6)} = 5699\ rpm$$

7.16 $\sigma_x = -\alpha \Delta T E$, $\tau = Td/2J = 16 T_0/\pi d^3$.

$$\sigma_{1,2} = -\alpha\Delta T E/2 \pm \left[(\alpha \Delta T E/2)^2 + (16 T_0/\pi d^3)^2\right]^{1/2}; \quad Tan\, 2\theta = \frac{16 T_0/\pi d^3}{\alpha \Delta T E/2} = \frac{32\, T_0}{\pi(\alpha \Delta T) E d^3}$$

$$\sigma_x = (-11.7 \times 10^{-6})(50)(200 \times 10^9) = -117\ MPa; \quad \tau = \frac{16(500)}{\pi(3 \times 10^{-3})^3} = 94.314\ MPa$$

$$\sigma_{1,2} = -117/2 \pm [(-117/2)^2 + 94.314^2]^{1/2} = -58.5 \pm 150.2 \Rightarrow \sigma_{1,2} = 150.3;\ -208.8\ MPa$$

$$Tan\, 2\theta = -94.314/58.5 = -1.6122; \quad \tau > 0 \Rightarrow 2\theta_1 = 121.81^\circ$$

$$\theta_1 = 60.9^\circ, \quad \theta_2 = 150.9^\circ.$$

7.17 

$$J = \pi/2\,(R_o^4 - R_i^4) = \frac{\pi R_o^4}{2}\left(\frac{15}{16}\right) = \frac{15 \pi R_o^4}{32}$$

$$A = \pi(R_o^2 - R_i^2) = 3\pi R_o^2/4.$$

$T = PR_o/4$

$$\sigma_x = P/A = 4P/3\pi R_o^2; \quad \tau = TR_o/J = \frac{PR_o^2}{4J} = \frac{8PR_o^2}{15\pi R_o^4} = \frac{8P}{15\pi R_o^2} \text{ at outer surface}$$

$$\tau = TR_i/J = 4P/15\pi R_o^2 \text{ at inner surface.}$$

At outer surface: $r = R_o \Rightarrow \sigma_{1,2} = \frac{P}{\pi R_o^2}\{2/3 \pm [(2/3)^2 + (8/15)^2]^{1/2}\} = P/\pi R_o^2\,(0.6667 \pm 0.8537)$

$$\sigma_{1,2} = (1.5204,\ -0.1871)\,P/\pi R_o^2; \quad Tan\,2\theta = \frac{8/15}{2/3} = 4/5 \Rightarrow \theta_1 = 19.3^\circ$$

At inner surface: $r = R_i \Rightarrow$ $\quad \sigma_{1,2} = (1.3844,\ -0.0514)\,P/\pi R_o^2$; $Tan\,2\theta = \frac{4/15}{2/3} = 2/5 \Rightarrow \theta_1 = 10.9^\circ$

7.18

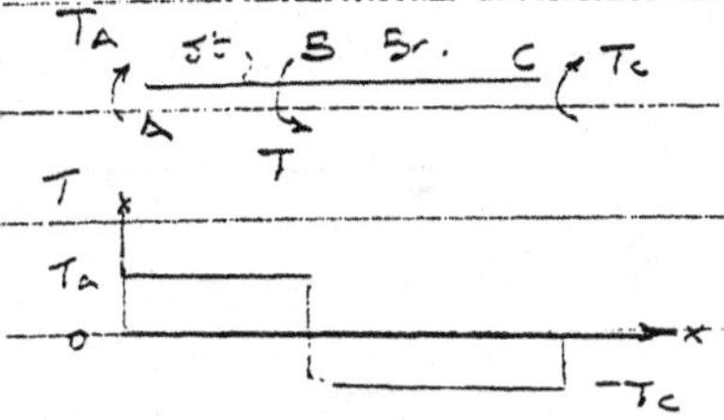

$T_A + T_c = T$

$T_{AB} = T_A, \quad T_{BC} = -T_c$

$\varphi_{B/A} = \varphi_{B/C} \Rightarrow \dfrac{T_A L_{AB}}{J G_S} = \dfrac{T_c L_{BC}}{J G_B}$

a) $T_c = \dfrac{G_B}{G_S}\dfrac{L_{AB}}{L_{BC}} T_A = \left(\dfrac{45}{76}\right)\left(\dfrac{1.5}{2.5}\right) T_A = 0.355 T_A \Rightarrow 1.355 T_A = T$

$T_A = T/1.355 = 0.7379 T, \quad T_c = 0.2621 T$

$J = \pi D^4/32 = 4.0212\times10^{-6}\ m^4; \quad T_A = \tau_{St} J/R = \dfrac{(125\times10^6)(4.0212\times10^{-6})}{40\times10^{-3}} = 1.2566\times10^4$ N-m

$T = 1.7030\times10^4$ N-m $= 17{,}030$ N-m

$T_c = \tau_{Br} J/R = 4.0212\times10^3 \Rightarrow \underline{T = 15{,}342 \text{ N-m}}$

b) $\varphi_B = T_A L_{AB}/J G_S = \dfrac{(0.7379\times15{,}342)(1.5)}{(4.0212\times10^{-6})(76\times10^9)} = 5.5564\times10^{-2}$ rad $= 3.18°$

7.19 a)

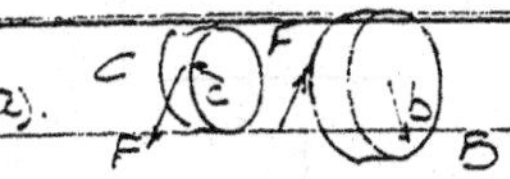

$T_{C/B}/c = T_{B/C}/b \Rightarrow T_{B/C} = (b/c) T_{C/B} = 2.5 T_{C/B}$; $T_D = T_{C/B} \Rightarrow T_{B/C} = 2.5 T_D$

b) $\varphi_{C/D} = 2.5\varphi_{B/A}$ Shaft AB: $T_A + 2.5 T_D = T$

$2.5\, T_A L_{AB}/G J_{AB} = T_D L_{CD}/G J_{CD} \Rightarrow T_D = 1.5 (J_{CD}/J_{AB}) T_A$

$J_{CD}/J_{AB} = (45/60)^4 = (3/4)^4 = 0.31641 \Rightarrow T_D = 2.5(3/5)(0.31641) T_A = 0.4746 T_A$

$\therefore 2.1865 T_A = T \rightarrow T_A = 0.4574 T; \quad T_D = 0.2171 T$

$T = 4000$ N-m: $\tau_{AB} = 2T_A/\pi R^3 = 16 T_A/\pi D_{AB}^3 = 16(0.4574)(5\times10^3)/\pi(60\times10^{-3})^3$

$\tau_{AB} = 53.92$ MPa

$\tau_{CD} = 16 T_D/\pi D_{CD}^3 = 16(0.2171)(5\times10^3)/\pi(45\times10^{-3})^3 = 60.68$ MPa

---

7.20

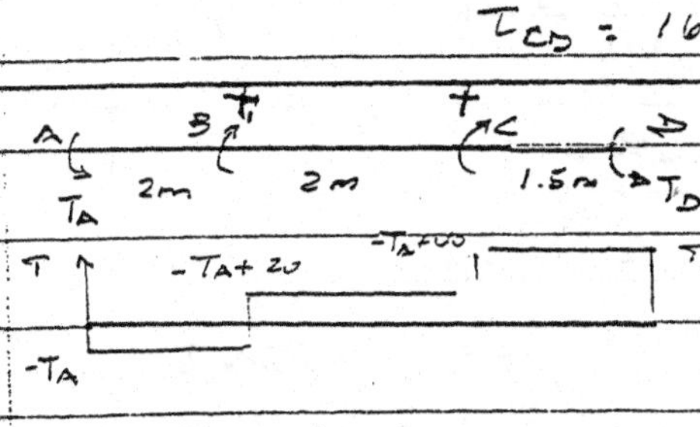

$T_A + T_D = 40$ kN-m.

$\varphi_{D/A} = 0 \Rightarrow \varphi_{B/A} + \varphi_{C/B} + \varphi_{D/C} = 0$

$-\dfrac{2T_A}{G_S J} + \dfrac{2(-T_A+20)}{G_S J} + \dfrac{1.5 T_D}{G_B J} = 0$ $\quad J = 9.818\times10^{-6}\ m^4$

$\dfrac{-4T_A + 40}{G_S} + \dfrac{1.5 T_D}{G_B} = 0 \Rightarrow 1.5 T_D = (G_B/G_S)(4T_A - 40)$

$T_A = 40 - T_D \Rightarrow 1.5 T_D = (G_B/G_S)(120 - 4T_D)$

$G_B/G_S = 45/76 = 0.5921 \quad \therefore 1.5 T_D = 71{,}053 - 2.3684 T_D \Rightarrow T_D = 18{,}367$ N-m.

$T_A = 21{,}633$ N-m

$\tau_S = \dfrac{T_A(50\times10^{-3})}{J}$

$\tau_S = 110.2$ MPa, $\tau_b = 93.5$ MPa $\quad \varphi_B = 2T_A/G_S J = 2(21{,}633)/(76\times10^9)J = 0.0580$ rad

$\varphi_B = 3.322°$

7.21

$t(x) = t_0(1 + x^2/L^2)$

a) $\varphi_{B/A} = 0$ $\quad T(x) = T_A - t_0\int_0^x (1+\xi^2/L^2)\,d\xi = T_A - t_0\left[\xi + \xi^3/3L^2\right]_0^x$

$$T(x) = T_A - t_0(x + x^3/3L^2)$$

$$\varphi_{B/A} = \frac{1}{GJ}\int_0^L T(x)\,dx = \frac{1}{GJ}\left[T_A L - t_0\int_0^L (x + x^3/3L^2)\,dx\right]$$

$$= \frac{1}{GJ}\left\{T_A L - t_0\left[(x^2/2 + x^4/12L^2)\right]_0^L\right\}$$

$$= \frac{1}{GJ}\left[T_A L - t_0(7L^2/12)\right]$$

$\varphi_{B/A} = 0 \Rightarrow T_A = 7t_0L/12$

$T_B = T(L) = T_A - 4t_0L/3 = -9t_0L/12 = -3t_0L/4$

b) $T(x) = t_0\left[7L/12 - (x + x^3/3L^2)\right]$

$$\varphi_B = \frac{t_0}{GJ}\int_0^{L/2}\left[7L/12 - (x + x^3/3L^2)\right]dx = \frac{t_0}{GJ}\left\{7Lx/12 - (x^2/2 + x^4/12L^2)\right\}_0^{L/2}$$

$$\varphi_B = 3t_0L^2/32GJ = \frac{3t_0L^2}{32\frac{\pi d^4}{32}\cdot G} = 3t_0L^2/\pi G d^4$$

7.22 Initially $(\tau_s)_i = TR/J$, $J_s = \pi d^4/32 = 4.0212\times10^{-6}\ m^4$, $(\tau_s)_i = \frac{(10\times10^3)(40\times10^{-3})}{(4.0212\times10^{-6})} = 99.47\,MPa$

$$\varphi_i = TL/GJ_s = \frac{(10\times10^3)(0.8)}{(76\times10^9)(4.0212\times10^{-6})} = 2.6177\times10^{-2}\ rad.$$

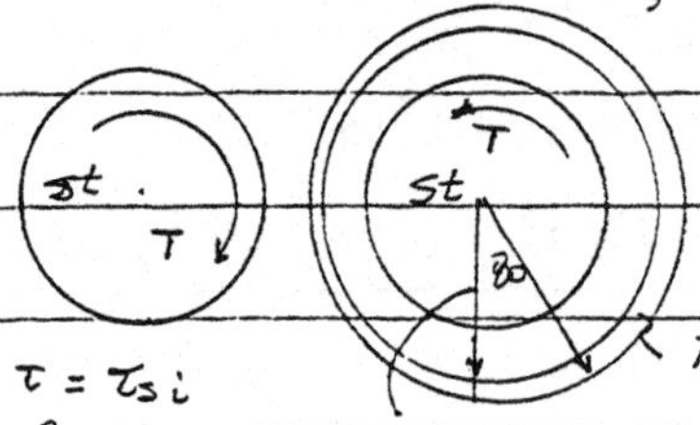

$\tau = \tau_{si}$

$\varphi = \varphi_i$

$\theta_s = \theta_b \Rightarrow \frac{T_s}{G_sJ_s} = \frac{T_B}{GJ_B} \Rightarrow T_s = \left(\frac{G_sJ_s}{G_BJ_B}\right)T_B$

$T_s + T_B = T \Rightarrow \left[1 + \frac{G_sJ_s}{G_BJ_B}\right]T_B = T$

$J_B = \frac{\pi}{2}(80^4 - 70^4)\times10^{-12} = 2.6625\times10^{-5}\ m^4$

$\therefore 1.2733\,T_B = T \Rightarrow T_B = 0.78536\,T = 7854\ N\text{-}m$

$T_s = 2146\ N\text{-}m.$

a). $\tau_B = T_BR/J_B = \frac{(2146)(80\times10^{-3})}{J_B} = 6.45\ MPa$; $\tau_{s2} = T_sR/J_s = (2146)(40)/J_s = 21.35\,MPa$

$\tau_s = \tau_{si} - \tau_{s2} = 78.12\ MPa.$

$$\varphi_s = -T_sL/G_sJ_s = -\frac{(2146)(0.8)}{(76\times10^9)(4.02\times10^{-6})} = 5.6175\times10^{-3}\ rad.$$

$\varphi = (2.6177\times10^{-2}) - (0.56175\times10^{-2}) = 2.056\times10^{-2}\ rad$

$\varphi = 1.18°$

7.23

a). $T_E = \tau_0 J/R$, $J = \pi R^4/2 \Rightarrow T_E = \pi \tau_0 R^3/2 = \frac{\pi}{2}(145\times10^6)(40\times10^{-3})^3 = 14{,}577$ N-m

b). $T = 18$ kN-m

$$T = 2\pi\tau_0\left\{ \frac{1}{b}\int_0^b r^3 dr + \int_b^R r^2 dr \right\} = 2\pi\tau_0\left\{ b^3/4 + r^3/3\Big|_b^R \right\} = 2\pi\tau_0\left[b^3/4 + \tfrac{1}{3}(R^3 - b^3)\right]$$

$$= 2\pi\tau_0\left[R^3/3 - b^3/12 \right] = \frac{\pi\tau_0}{6}(4R^3 - b^3)$$

$$b^3 = -\left(\frac{6T}{\pi\tau_0} - 4R^3\right) \Rightarrow b = 2^{1/3}\left[2R^3 - 3T/\pi\tau_0\right]^{1/3}$$

$$T = 18\text{kN-m} \Rightarrow b = 2^{1/2}\left[2(40\times10^{-3})^3 - 3(18\times10^3)/\pi(145\times10^6)\right]^{1/3}$$

$$b = 0.0299 \text{ m} = 29.9 \text{ mm.}$$

$$\gamma_0 = b\theta \Rightarrow \theta = \gamma_0/b = \frac{\tau_0}{G\cdot b} = \frac{145\times10^6}{(76\times10^9)(2.99\times10^{-2})} = 0.0638 \text{ rad/m} = 3.66^\circ/\text{m}$$

7.24

a) $T_E = \tau_0 J/R_0$; $J = \pi/2(R_0^4 - R_i^4) \Rightarrow T_E = \frac{\pi\tau_0}{2R_0}(R_0^4 - R_i^4)$.

b). $T = T_E \Rightarrow \theta = \gamma_0/R_0 = \frac{\tau_0}{GR_0} = \frac{2T_E}{\pi G}\cdot\frac{1}{(R_0^4 - R_i^4)}$

c). $T_P = 2\pi\tau_0\int_{R_i}^{R_0} r^2 dr = \frac{2\pi\tau_0}{3}(R_0^3 - R_i^3)$.

d). $T \to T_P \Rightarrow \theta = \gamma_0/R_i = \frac{\tau_0}{GR_i} = \frac{\tau_0}{GR_0}\cdot\left(\frac{R_0}{R_i}\right) = \frac{2T_E(R_0/R_i)}{\pi G(R_0^4 - R_i^4)}$

7.25

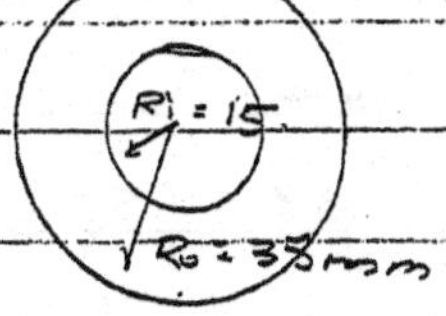

(a). $T_Y = \frac{\tau_0 J}{R_0}$, $J = \pi/2(R_0^4 - R_i^4) = 2.278\times10^{-6}$ m^4

$$T_Y = \frac{(180\times10^6)(2.278\times10^{-6})}{35\times10^{-3}} = 11{,}714 \text{ N-m}$$

b). $\theta = \gamma/R_0 = \tau_0/R_0G = 0.068$ rad $= 3.88^\circ/\text{m} \Rightarrow \varphi = \theta L = 4.65^\circ$

c). $T = 2\pi\left[\frac{\tau_0}{b}\int_0^b r^3 dr + \tau_0\int_b^{R_0} r^2 dr\right] = 2\pi\tau_0\left[b^3/4 + \tfrac{1}{3}(R_0^3 - b^3)\right]$

$T = 2\pi\tau_0(R_0^3/3 - b^3/12)$; $b = 30\times10^{-3}$ m $\Rightarrow T_Y = 13{,}619$ N-m

d). $T_P = 2\pi\tau_0\int_{R_i}^{R_0} r^2 dr = 2\pi\tau_0/3\,(R_0^3 - R_i^3) = 14{,}841$ N-m

e) $T \to T_P \Rightarrow \gamma(R_i) \to \tau_0/G \Rightarrow \theta = \tau_0/GR_i = 0.158$ rad/m

$\theta = 9.05^\circ/\text{m}$

7.26

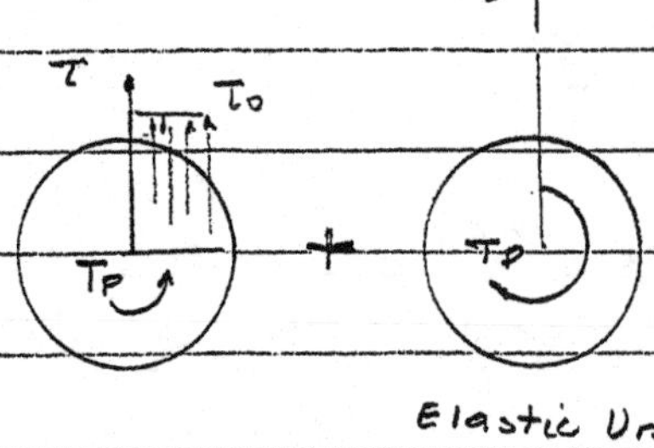

$T_p = 2\pi\tau_o \int_0^R r^2 dr = 2\pi\tau_o R^3/3$

Elastic Unloading: $T = T_p \Rightarrow \tau = T_p r/J$

$\tau_{unload} = -T_p r/J = \left(\dfrac{2\pi\tau_o R^3/3}{\pi R^4/2}\right) r = -\dfrac{4}{3}\tau_o r/R$

Elastic Unloading.

$\tau = \tau_o - \tau_{unload} = \tau_o\left(1 - \frac{4}{3}\frac{r}{R}\right) \quad \tau = 0 \Rightarrow r = 3R/4$

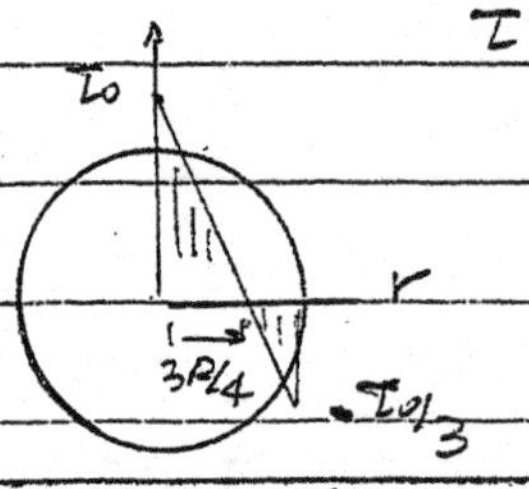

7.27

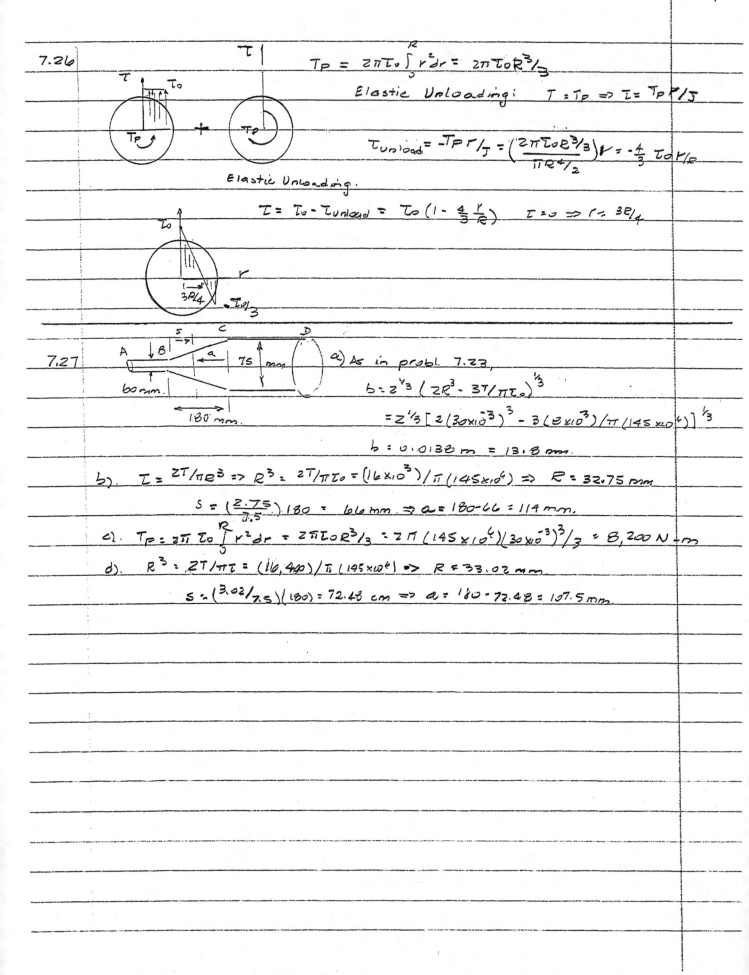

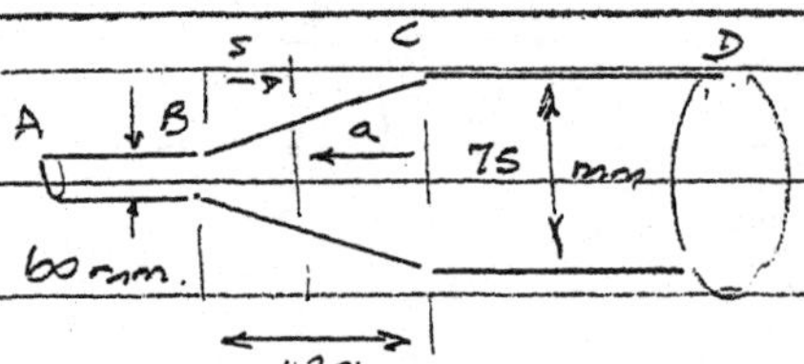

a) As in probl. 7.23,

$b = 2^{1/3}\left(2R^3 - 3T/\pi\tau_o\right)^{1/3}$

$= 2^{1/3}\left[2(30\times10^{-3})^3 - 3(8\times10^3)/\pi(145\times10^6)\right]^{1/3}$

$b = 0.0138\ m = 13.8\ mm.$

b). $\tau = 2T/\pi R^3 \Rightarrow R^3 = 2T/\pi\tau_o = (16\times10^3)/\pi(145\times10^6) \Rightarrow R = 32.75\ mm$

$s = \left(\frac{2.75}{7.5}\right)180 = 66\ mm. \Rightarrow a = 180 - 66 = 114\ mm.$

c). $T_p = 2\pi\tau_o\int_0^R r^2 dr = 2\pi\tau_o R^3/3 = 2\pi(145\times10^6)(30\times10^{-3})^3/3 = 8{,}200\ N\text{-}m$

d). $R^3 = 2T/\pi\tau = (16{,}400)/\pi(145\times10^6) \Rightarrow R = 33.02\ mm$

$s = (3.02/7.5)(180) = 72.48\ cm \Rightarrow a = 180 - 72.48 = 107.5\ mm$

7.28 $T > T_E$: $\tau = r\tau_0/b$, $r < b$; $r > b$: $cG\gamma = (1+c)\tau - \tau_0 \Rightarrow \tau = \dfrac{\tau_0 + cG\gamma}{1+c}$

$\gamma = r\theta \Rightarrow \gamma_0 = \tau_0/G = b\theta \Rightarrow \gamma = (\tau_0/bG)r \Rightarrow cG\gamma = c\tau_0 r/b$

$\therefore r > b \Rightarrow \tau = \dfrac{\tau_0}{1+c}[1 + cr/b]$ $\quad c = 0$ ideal plastic

$$T = \iint \tau r\,dA = 2\pi\tau_0\left\{\frac{1}{b}\int_0^b r^3 dr + \frac{1}{1+c}\int_b^R (1 + cr/b) r^2 dr\right\}$$

$$= 2\pi\tau_0\left\{\frac{1}{4b} r^4\Big|_0^b + \frac{1}{1+c}\left[r^3/3\Big|_b^R + cr^4/4b\Big|_b^R\right]\right\}$$

$$= 2\pi\tau_0\left\{b^3/4 + \frac{1}{1+c}\left[\tfrac{1}{3}(R^3 - b^3) + c/4\,(R^4/b - b^3)\right]\right\}$$

$$= 2\pi\tau_0\left\{b^3/4 + \frac{1}{1+c}\left[R^3(1/3 + cR/4b) - (1/3 + c/4)b^3\right]\right\}$$

$$= 2\pi\tau_0\left\{b^3/4 + \frac{1}{12(1+c)}\left[4R^3 + 3cR^4/b - (4+3c)b^3\right]\right\}$$

$$= \frac{2\pi\tau_0}{12(1+c)}\left\{3(1+c)b^3 + 4R^3 + 3cR^4/b - (4+3c)b^3\right\}$$

$$= \frac{2\pi\tau_0}{12(1+c)}\left\{R^3(4 + 3cR/b) - b^3\right\}$$

$b = \tau_0/G\theta$

$$T = \frac{\pi\tau_0}{6(1+c)}\left\{R^3\left(4 + \frac{3cRG\theta}{\tau_0}\right) - \frac{\tau_0^3}{(G\theta)^3}\right\}, \quad \tau_0 = \frac{2T_E}{\pi R^3}$$

$\therefore$

$$T = \frac{T_E}{3(1+c)}\left[4 + 3c\left(\frac{GR}{\tau_0}\cdot\theta\right) - \left(\frac{\tau_0}{GR\theta}\right)^3\right]$$

$$T/T_E = \frac{1}{3(1+c)}\left[4 + 3c\beta - 1/\beta^3\right]$$

Solve for $\beta \equiv GR\theta/\tau_0$ as function of T/T_E for values of c: $c = 0, 1, 2$.

Note:

$$\beta = GR\theta/\tau_0 \geq \frac{GR}{\tau_0}\cdot\frac{T_E}{GJ} = \frac{T_E R}{J}\cdot\frac{1}{\tau_0} = 1 \Rightarrow \beta \geq 1$$

7.29

$$\theta_a = \theta_b \Rightarrow \frac{T_a}{J_a G_a} = \frac{T_b}{J_b G_b} \; ; \quad T_a + T_b = T$$

$$T_a = \left(\frac{J_a G_a}{J_b G_b}\right) T_b \Rightarrow \left(1 + \frac{J_a G_a}{J_b G_b}\right) T_b = T \Rightarrow T_b = \left(\frac{G_b J_b}{G_a J_a + G_b J_b}\right) T$$

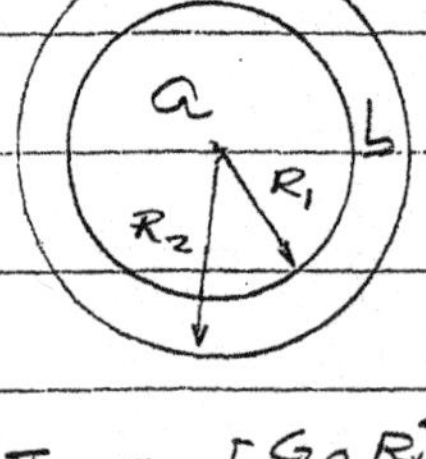

$$T_a = \left(\frac{G_a J_a}{G_a J_a + G_b J_b}\right) T$$

$$J_a = \pi R_1^4 / 2 \qquad T_b = \left(\frac{G_b J_b}{G_a J_a + G_b J_b}\right) T$$

$$J_b = \pi/2 \,(R_2^4 - R_1^4)$$

$$T_a = \left[\frac{G_a R_1^4}{G_a R_1^4 + G_b (R_2^4 - R_1^4)}\right] T \Rightarrow \theta = \frac{2}{\pi}\left[\frac{T}{G_a R_1^4 + G_b (R_2^4 - R_1^4)}\right]$$

$$\theta = \frac{2T}{\pi}\left[(G_a - G_b) R_1^4 + G_b R_2^4\right]^{-1}$$

7.30

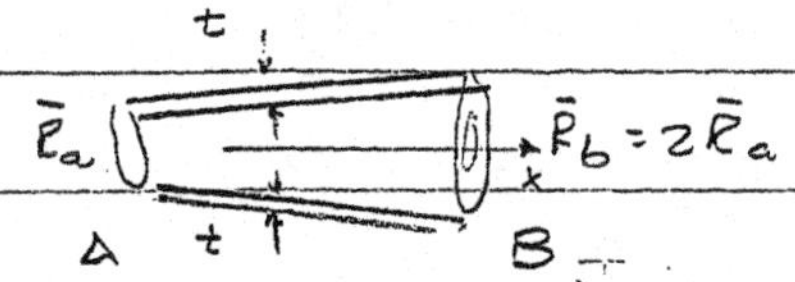

$$\bar{R}(x) = \bar{R}_a \,(1 + x/L)$$

$$\varphi_{B|A} = \frac{T}{G}\int_0^L \frac{dx}{J(x)} \qquad J = \pi/2\left[(\bar{R} + t/2)^4 - (\bar{R} - t/2)^4\right] = \frac{\pi \bar{R}^4}{2}\left[(1 + t/2\bar{R})^4 - (1 - t/2\bar{R})^4\right]$$

$$J(x) = \frac{\pi \bar{R}^4}{2}\left[4t/\bar{R} + t^3/\bar{R}^3\right] = \frac{\pi \bar{R}^3 t}{2}\left(4 + t^2/\bar{R}^2\right)$$

$$J(x) = 2\pi \bar{R}^3 t \left(1 + t^2/4\bar{R}^2\right).$$

$$t/\bar{R} \ll 1 \Rightarrow J(x) = 2\pi \bar{R}^3 t = 2\pi \bar{R}_a^3 t \,(1 + x/L)^3$$

$$\varphi_{B|A} = \frac{T}{2\pi G \bar{R}_a^3 t}\int_0^L (1 + x/L)^{-3} dx = \frac{T}{2\pi G \bar{R}_a^3 t}\left[-\tfrac{L}{2}(1 + x/L)^{-2}\right]_0^L$$

$$= -\frac{TL}{4\pi G \bar{R}_a^3 t}\left(\frac{1}{4} - 1\right) = \frac{3TL}{16\pi G \bar{R}_a^3 t}$$

7.31

Shaft A: $G = G_0 (r/R)$; $\gamma = \theta r \Rightarrow \tau(r) = \theta G(r) \cdot r$

a) $$T = 2\pi \int_0^R r^2 \tau(r)\, dr = 2\pi G_0 \theta / R \int_0^R r^4 dr = 2\pi G_0 \theta / R \cdot R^5/5 = 2\pi G_0 \theta R^4/5$$

$$\theta_A = 5T / 2\pi G_0 R^4.$$

Shaft B: $G = G_0 (1 - r/R) \Rightarrow \tau(r) = \theta G_0 (1 - r/R) r$

$$T = 2\pi \theta G_0 \int_0^R (r^3 - r^4/R)\, dr = 2\pi \theta G_0 \left[R^4/4 - R^4/5\right] = 2\pi \theta G_0 R^4/20$$

$$\theta_B = 10 T / \pi G_0 R^4$$

b). $\theta_A / \theta_B = 1/4$

c). Fibres near outer surface (have larger stresses) to resist applied torque ⇒ Shaft A is stiffer than B.

7.32 $\tau_{x\theta} = TR/J$; $J = \pi/2\,(R_o^4 - R_i^4) = \pi/2\,(160^4 - 55^4)\times 10^{-12} = 5.984\times 10^{-6}\ m^4$

$\tau_{x\theta} = \pm(10{,}027)\,T$ MPa $\quad (T: [N\text{-}m])$

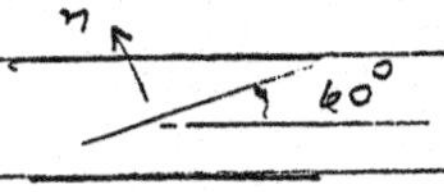

$\theta = 150^\circ \quad \sigma_n = 2\sin\theta\cos\theta\,\tau_{x\theta} = 2(1/2)(-\sqrt{3}/2) = -\sqrt{3}/2\,\tau_{x\theta}$

$\tau_{nt} = \tau_{xy}(\cos^2\theta - \sin^2\theta) = \tau_{xy}(3/4 - 1/4) = \tau_{xy}/2$

$|\sigma_n| = \sqrt{3}/2\,(10{,}027)\,T < 100\ MPa \Rightarrow T < 11{,}516\ N\text{-}m$

$|\tau_{nt}| = 1/2\,(10{,}027)\,T < 50\ MPa \Rightarrow T < 9{,}973\ N\text{-}m$

7.33 a) $\gamma = \theta r \Rightarrow \gamma^n = (\theta r)^n \Rightarrow \tau = k\theta^n r^n$

$$T = 2\pi \int_0^R r^2 \tau(r)\,dr = 2\pi k\theta^n \int_0^R r^{n+2}\,dr = 2\pi k\theta^n \frac{R^{n+3}}{n+3}$$

b)

$$k\theta^n = \frac{(n+3)T}{2\pi R^{n+3}} \Rightarrow \tau = \left(\frac{(n+3)r^n}{2\pi R^{n+3}}\right)T$$

7.34 a) $T(x) = t_o x$; $\gamma = \theta r \Rightarrow \tau = k\gamma^2 = k\theta^2 r^2$

$$T = 2\pi \int_0^R r^2\,\tau(r)\,dr = 2\pi k\theta^2 \int_0^R r^4\,dr = 2\pi k\theta^2 R^5/5$$

$$k\theta^2 = 5T/2\pi R^5 = 5t_o x/2\pi R^5 \Rightarrow \tau_{x\theta} = \frac{5t_o x r^2}{2\pi R^5}$$

b). $\theta(x) = (5t_o/2\pi k R^5)^{1/2}\,x^{1/2}$.

$$\varphi_A = \int_0^L \theta(x)\,dx = \left(\frac{5t_o}{2\pi k R^5}\right)^{1/2} \frac{L^{3/2}}{3/2} = 2/3\left(\frac{5t_o L^3}{2\pi k R^5}\right)^{1/2}.$$

7.35 a) $\gamma = \theta r \Rightarrow \tau_a = G_a\theta r,\ \tau_b = G_b\theta r \qquad (\theta_a = \theta_b = \theta)$

$$\frac{(\tau_a)_{max}}{(\tau_b)_{max}} = \frac{G_a R_1}{G_b R_2}$$

b) $\theta_a = \theta_b \Rightarrow \dfrac{T_a}{G_a J_a} = \dfrac{T_b}{G_b J_b} \qquad J_a = \pi R_1^4/2,\ J_b = \pi/2\,(R_2^4 - R_1^4) = \pi/2\,[R_2^4 - R_2^4/4]$

$J_b = (3\pi/8)R_2^4 = 3\pi R_1^4/2$

$T_a = (G_a/G_b)(J_a/J_b)T_b = 2T_b$; $T_a + T_b = T \Rightarrow T_b = T/3,\ T_a = 2T/3$.

c). $\theta = T_a/G_a J_a = \dfrac{2T/3}{G_a\pi R_1^4/2} = 4T/G_a\pi R_1^4$.

d).

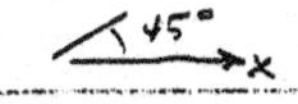

$$\varepsilon_n = \varepsilon_{xy}\sin 2\theta = \varepsilon_{xy} = \gamma/2 = \frac{\theta R_2}{2} = \frac{2TR_2}{G_a\pi R_1^4} = \frac{8T}{G_a\pi R_2^3}$$

$$\sigma_n = E_s\varepsilon_n = \frac{8TE_s}{G_a\pi R_2^3} = \frac{2\sqrt{2}\,E_s T}{\pi G_a}$$

7.36

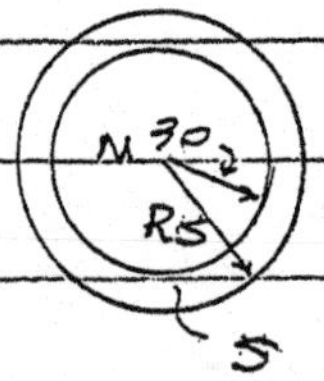

$$\theta_M^{(1)} = \frac{T}{G_M J_M} \qquad \theta = \theta_M^{(2)} = \theta_S = 0.4\,\theta_M^{(1)}$$

$$\theta = \frac{T_m^{(2)}}{G_M J_M} = \frac{T_S}{G_S J_S} \Rightarrow T_S = \left(\frac{G_S J_S}{G_M J_m}\right) T_m^{(2)}$$

$$T_S + T_m^{(2)} = T \Rightarrow \left(1 + \frac{G_S J_S}{G_m J_m}\right) T_m^{(2)} = T$$

$$T_m^{(2)} = \left(\frac{G_m J_m}{G_S J_S + G_m J_m}\right) T \;; \quad \theta_m^{(2)} = (0.4)\,\theta_m^{(1)} \Rightarrow \frac{T}{G_S J_S + G_m J_m} = 0.4\,\frac{T}{G_m J_m}$$

$$2.5\, G_m J_m = G_S J_S + G_m J_m \Rightarrow 1.5\, G_m J_m = G_S J_S$$

$$J_S = 1.5\left(G_m/G_S\right) J_m = 1.5\,(65/86)\, J_m = 1.134\, \pi R_m^2/2$$

$$J_S = 0.567\, \pi R_m^2$$

$$J_S = \pi/2\,(R_S^4 - R_m^2) = 0.567\, \pi R_m^2 \Rightarrow R_S^4 = 2.134\, R_M^4$$

$$R_S = (2.134)^{1/4} R_M = 1.209\, R_M = 36.258 \text{ mm}$$

$$t = 6.26 \text{ mm.}$$

7.37

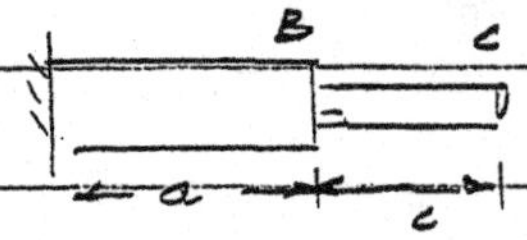

$$T \;\; T_E = \tau_0 J_{BC}/R = \pi \tau_0 R^3/2$$

b). $$\varphi_C = \frac{T}{G}\left[\frac{a}{J_{AB}} + \frac{c}{J_{BC}}\right) = \frac{2T}{\pi}\left(\frac{a}{16R^4} + \frac{c}{R^4}\right) = \frac{T}{8\pi R^4}(a + 16c)$$

c) $$T_b = \tau_0 J_{AB}/2R = \frac{\tau_0}{2R} \cdot 8\pi R^4 = 4\pi \tau_0 R^3 \;; \quad T_E/T_b = 1/8$$

d). $$T_{BC} = 4\pi \tau_0 R^3 \qquad T_{BC} = 2\pi \int_0^R r\,\tau(r)\,dr = 2\pi\left[1/b \int_0^b \tau_0 r^3 dr + \int_b^R \tau_0 r^2 dr\right]$$

e) $$T_b = 0.15\, T_b = T_{BC}$$

$$\therefore T_{BC} = 2\pi\tau_0\left[b^3/4 + 1/3\,(R^3 - b^3)\right] = \frac{\pi \tau_0}{6}(4R^3 - b^3) = 0.15\,(4\pi\tau_0 R^3) =$$

$$4R^3 - b^3 = 3.6R^3 \Rightarrow b = (0.4)^{1/3} R = 0.737R$$

7.38 $\gamma = \theta r \Rightarrow \theta = \gamma_s/r_s = \gamma_a/r_a \Rightarrow \gamma_s = (r_s/r_a)\gamma_a = (15/22)\gamma_a = 0.682\,\gamma_a$

Steel yields first ; $\gamma_s = 0.002$,

$\theta = \theta_s = \theta_a = \gamma_s/r_s = 0.002/(15\times10^{-3}) = 0.1333$ rad/m.

a) $T = T_s + T_a = (G_sJ_s + G_aJ_a)\theta$

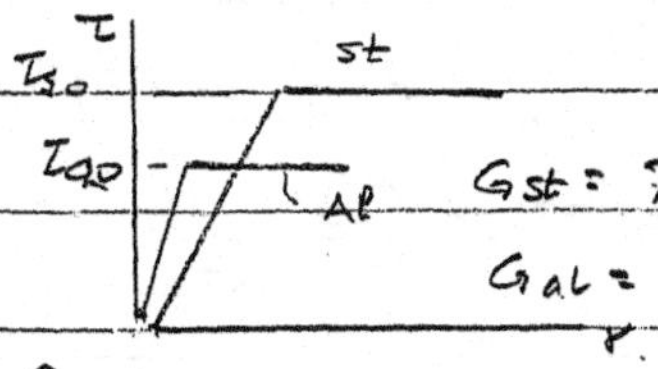

$G_{st} = 75$ GPa

$G_{al} = 25$ GPa

$J_s = \pi/2\,(15)^4\times10^{-12} = 7.952\times10^{-8}$ m^4

$J_a = \pi/2\,(30^4 - 22^4)\times10^{-12} = 9.04377\times10^{-7}$ m^4

$T = [75(7.952\times10^{-8}) + 25(9.04377\times10^{-7})]\times10^9\cdot\theta = 28{,}574\,\theta$

$T_E = (28{,}574)(0.1333) = 3810$ N-m

b) $T_p = 2\pi\left[\tau_{s0}\int_0^{R_s} r^2 dr + \tau_{a0}\int_{R_{ai}}^{R_{ao}} r^2 dr\right]$

$= \frac{2\pi}{3}\left[\tau_{s0}R_s^3 + \tau_{a0}(R_{ao}^3 - R_{ai}^3)\right] = 9.622$ N-m

---

7.39 $T_E = \tau_0 J/R = \pi\tau_0 R^3/2 = 235.6$ N-m ; $\varphi = L\theta = \frac{L\gamma}{R} = \frac{\tau_0 L}{GR} = 0.2961$ rad $= 16.96°$

$\therefore T > T_E$ is required for $\varphi = 25°$.

$\tau_0 = G\gamma = G(\theta b) = G\varphi b/L \Rightarrow b = \tau_0 L/G\varphi$

($25° = 0.43633$ rad) $\therefore b = (150\times10^6)(1.5)/G\varphi = 6.785\times10^{-3}$ m $= 6.79$ mm

$T = 2\pi\left[\frac{1}{b}\int_0^b r^3\tau_0(r)\,dr + \tau_0\int_b^R r^2 dr\right] = 2\pi\tau_0\left[b^3/4 + \tfrac{1}{3}(R^3 - b^3)\right]$

$= \frac{\pi\tau_0}{6}(4R^3 - b^3)$

$R = 10$ mm, $b = 6.79$ mm $\Rightarrow T = 289.6$ N·m

7.40 $\frac{(T_C)_B}{60} = \frac{(T_B)_C}{90} \Rightarrow (T_B)_C = 1.5(T_C)_B$ (See Problem 7.19)

$(T_C)_B = T$, $(T_B)_C = 1.5T$

a). $\tau = 1.5TR/J \Rightarrow T = \tau_{all}J/1.5R$; $J = \pi d^4/32 = \pi/32\,(40\times10^{-3})^4 = 2.5133\times10^{-7}$ m^4

$T = (90\times10^6)J/30\times10^{-3} = 754$ N-m

b)

$\varphi_F = \varphi_{B/A} + \varphi_{F/C} = \frac{T}{GJ}[(1.5)(0.5) + (1)(0.9)] = 1.65T/GJ$

$\varphi_F = (1.65)(754)/(76\times10^9)J = 6.513\times10^{-2}$ rad $= 3.73°$

7.41 $\tau_s = TR/J = 2T_s/\pi R^3$, $\tau_h = \frac{2T_h R_o}{\pi(R_o^4 - R_i^4)} = \frac{2T_h}{\pi R_o^3(1-\beta^4)}$, $\beta = R_i/R_o$

$\pi R^2 = \pi(R_o^2 - R_i^2) = \pi R_o^2(1-\beta^2) \Rightarrow R^2 = R_o^2(1-\beta^2)$

$\tau_s = 2T_s/\pi R_o^3(1-\beta^2)^{3/2}$

$\tau_s = \tau_h \Rightarrow \frac{T_s}{(1-\beta^2)^{3/2}} = \frac{T_h}{1-\beta^4} \Rightarrow T_s/T_h = \frac{(1-\beta^2)^{3/2}}{1-\beta^4} = \frac{(1-\beta^2)^{1/2}}{1+\beta^2}$

$\theta_s = T_s/GJ_s = T_h/GJ_h \Rightarrow \frac{T_s}{T_h} = J_s/J_h = \frac{R_o^4(1-\beta^2)^2}{R_o^4(1-\beta^4)} = \frac{(1-\beta^2)}{1+\beta^2}$

7.42

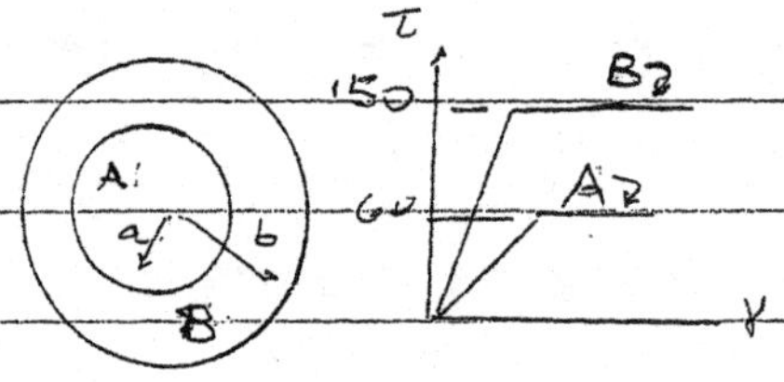

$G_A = 25\,GPa$, $G_B = 80\,GPa$

$\gamma_{YA} = \tau_{0A}/G_A = \frac{60\times10^6}{25\times10^9} = 2.4\times10^{-3}$

$\gamma_{YB} = \tau_{0B}/G_B = \frac{150\times10^6}{80\times10^9} = 1.875\times10^{-3}$

$\gamma = r\theta \Rightarrow \theta_{AY} = \gamma_{YA}/a = \frac{2.4\times10^{-3}}{30\times10^{-3}} = 0.08\,rad/m$; $\theta_{BY} = \gamma_{YB}/G_B = 0.0469\,rad/m$

$\theta_{BY} < \theta_{AY} \Rightarrow$ Material B yields first. $\Rightarrow \theta_y = 0.0469\,rad/m$

$\tau_A = G_A r\theta_y$, $\tau_B = G_B r\theta_y$

$T_y = \iint_{A_A} r\tau_A dA + \iint_{A_B} r\tau_B dA = \theta[G_A \iint_{A_A} r^2 dA + G_B \iint_B r^2 dA] = \theta_y(G_A J_A + G_B J_B)$

$J_A = \pi a^4/2 = 40.5\pi\times10^{-8}\,m^4$; $J_B = \pi/2(b^4 - a^4) = 87.5\pi\times10^{-8}$

$T_y = \pi\theta_y[(25\times10^9)(40.5) + (80\times10^9)(87.5)]\times10^{-8} = 3{,}756\pi = 11{,}799\,N\text{-}m$

b). $\theta_y = 0.0468\,rad/m$ =

c). $T_0 = \int_{A_A} \tau_{AY}\, r\, dA + \iint_{A_B} \tau_{YB}\, r\, dA = 2\pi[(60\times10^6)\int_0^a r^2 dA + (150\times10^6)\int_a^b r^2 dA]$

$a = 30\,mm$, $b = 40\,mm$.

$T_0 = 14{,}860\,N\text{-}m$

7.43 Transmissible torque of coupling: $T_A = (4 A_b \tau_b) D = 4(150\times10^{-6})(60\times10^{6}) D/_2$

$T_A = 1.8\times10^{4} D = 3.6\times10^{3}$ N-m

$J = \pi d^4/_{32} = 3.1063\times10^{-6}$ m^4 $\qquad \tau_{AB} = TR/_J = (1.8\times10^{4})(0.2)R/_J = 43.5$ MPa

$\tau_{AB} = \tau_{BC}$

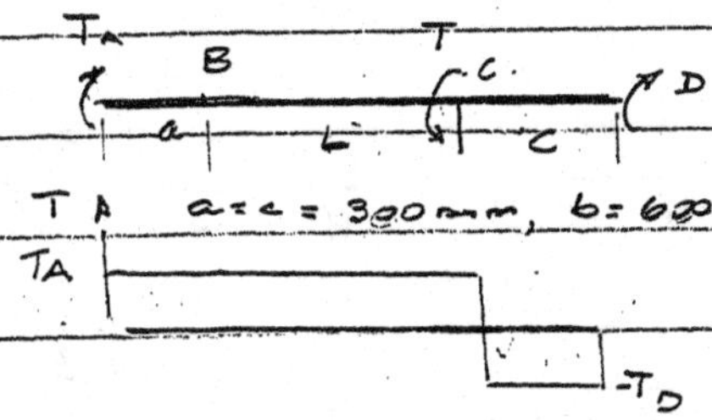

$\varphi_{C/A} = \varphi_{C/D}$

a = c = 300 mm, b = 600 mm.

$\frac{T_A}{J}\left[\frac{a}{G_a} + \frac{b}{G_s}\right] = \frac{T_D c}{G_s J}$

$T_A = \left(\frac{G_a c}{G_s a + G_a b}\right) T_D = 0.20313\ T_D$

$T_D = 4.9231\ T_A = 17{,}7231$ N·m

$T = T_A + T_D = 21{,}323$ N-m

$\tau_{CD} = T_D R/_J = 214$ MPa

$\varphi_B = T_A c/_{G_a J} = (3600)(0.3)/_{G_a J} = 1.337\times10^{-2}$ rad $= 0.766^\circ$

7.44 $\varphi = TL/_{GJ} \Rightarrow T = \varphi GJ/_L$, $\tau = TR/_J = \left(\frac{GR}{L}\right)\varphi = \frac{(76\times10^{9})(0.15)}{2600}(6\pi)$

$\tau = 82.6$ MPa

7.45 $\tau_s = 16T/_{\pi d^3}$, $\tau_h = \frac{TD/2}{\pi/32(D^4 - D_i^4)} = \frac{16TD}{\pi[D^4 - D^4(1-2k)^4]} = \frac{16T}{\pi D^3[1-(1-2k)^4]}$

$\tau_s = \tau_h \Rightarrow D/d = [1-(1-2k)^4]^{-3}$.

7.46 a). $\theta_s = \theta_h \Rightarrow J_s = J_h \Rightarrow d^4 = D^4 - d^4 \Rightarrow D/d = 2^{1/4} =$

b) $\tau_{max}\big|_{AB,CH} = TR/_J = 16T/_{\pi d^3}$, $\tau_{max}\big|_{EF} = \frac{16DT}{\pi(D^4 - d^4)}$

$\tau_{max}\big|_{AB} = \tau_{max}\big|_{EF} \Rightarrow \frac{1}{d^3} = \frac{D}{D^4 - d^4} \Rightarrow D^4 - d^4 = d^3 D$

$(D/d)^4 - D/d - 1 = 0$

7-47 $R_i = kR_o$, $k<1$; $T = \beta P R_o$

$\sigma_x = P/A$, $A = \pi(R_o^2 - R_i^2) = \pi R_o^2 (1-k^2) \Rightarrow \sigma_x = \frac{P}{\pi R_o^2}\frac{1}{(1-k^2)}$

$\tau = TR_o/J$, $J = \frac{\pi}{2}(R_o^4 - R_i^4) = \pi R_o^4/2\,(1-k^4)$

$$T = \beta P R_o \Rightarrow \tau = \frac{2P\beta}{\pi R_o^2}\frac{1}{(1-k^4)}$$

$$\sigma_{1,2} = \sigma_x/2 \pm [(\sigma_x/2)^2 + \tau^2]^{1/2} = \frac{P}{\pi R_o^2}\left\{\frac{1}{2} \pm \left[\frac{1}{4(1-k^2)^2} + \frac{4\beta^2}{(1-k^4)^2}\right]^{1/2}\right\}$$

$$= \frac{P}{\pi R_o^2(1-k^2)}\left\{\frac{1}{2} \pm \left[\frac{1}{4} + \frac{4\beta^2}{(1+k^2)^2}\right]^{1/2}\right\} = \frac{P}{2\pi R_o^2(1-k^2)}\left\{1 \pm \left[1 + \frac{16\beta^2}{(1+k^2)^2}\right]^{1/2}\right\}$$

$$\sigma_{1,2} = \frac{P}{2\pi R_o^2(1-k^2)}\left\{1 \pm \frac{1}{(1+k^2)}\left[(1+k^2)^2 + 16\beta^2\right]^{1/2}\right\}$$

$$\tan 2\theta = \frac{\left(\frac{2\beta}{1-k^4}\right)}{\frac{1}{2(1-k^2)}} = \frac{4\beta(1-k^2)}{1-k^4} \qquad \theta_1 = \tfrac{1}{2}\tan^{-1}\left(\frac{4\beta(1-k^2)}{1-k^4}\right)$$

8.1

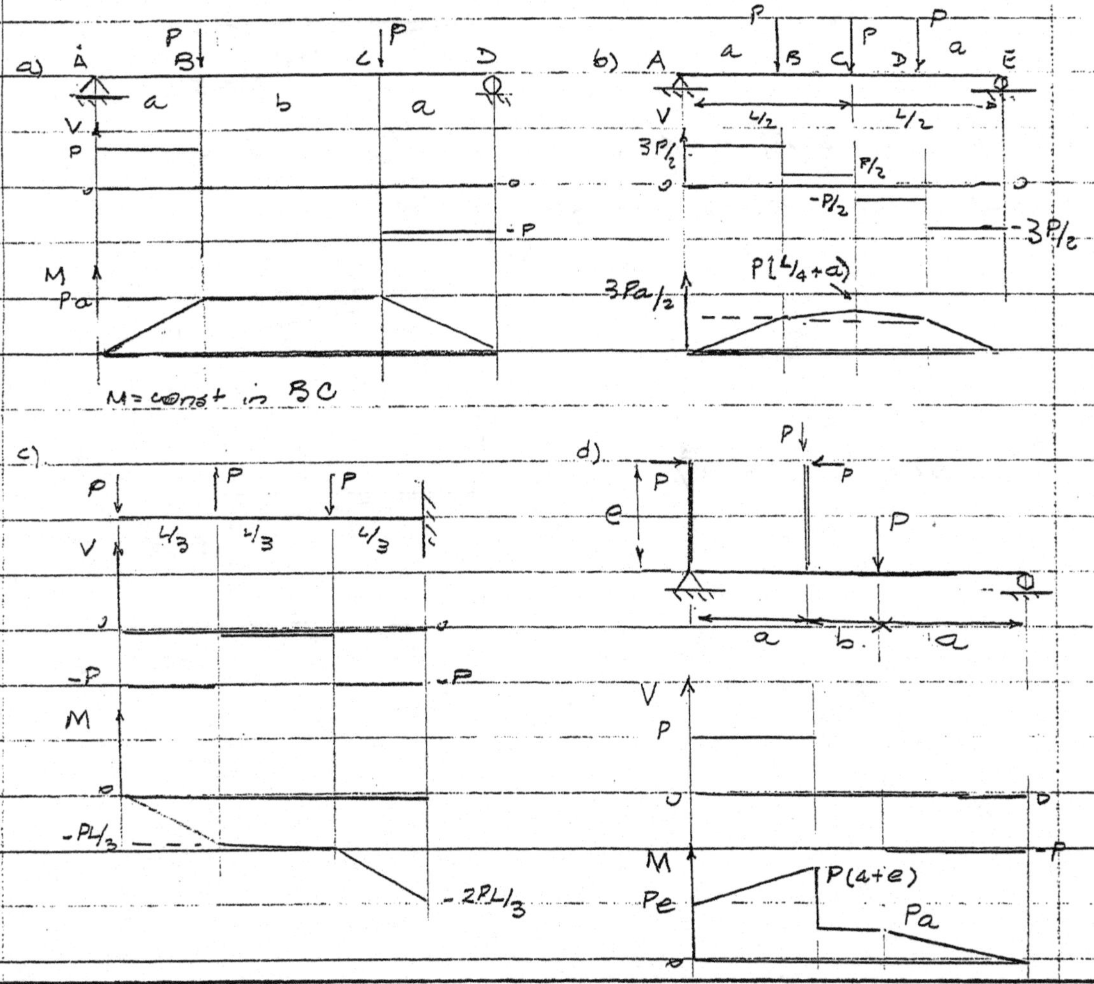

8.2

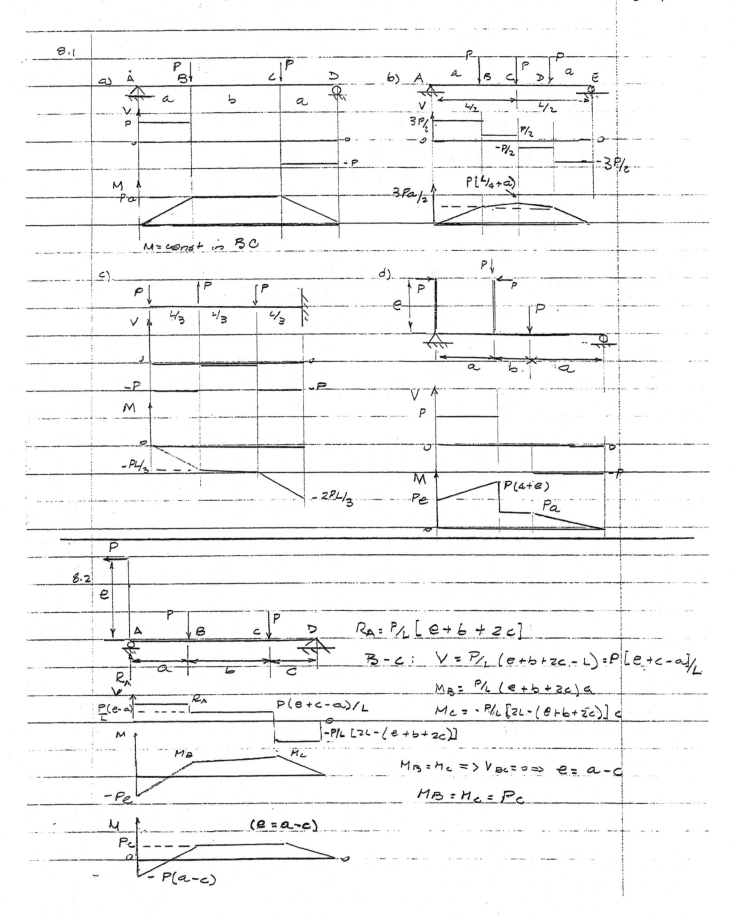

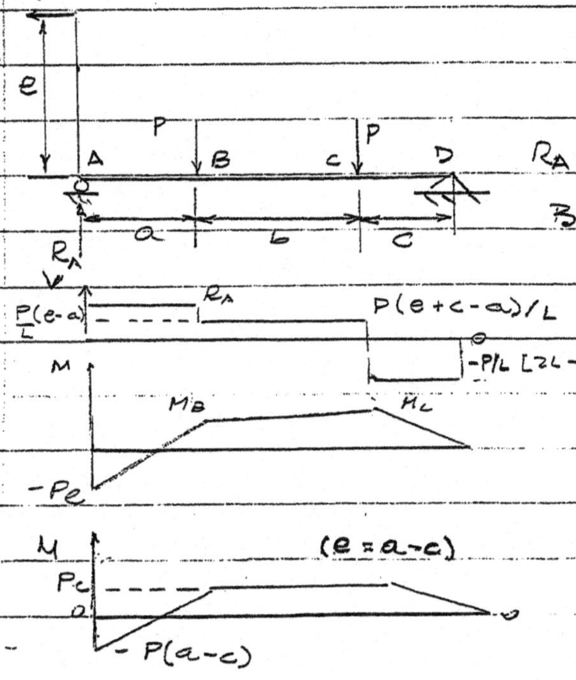

$R_A = \frac{P}{L}[e+b+2c]$

B - C: $V = \frac{P}{L}(e+b+2c-L) = P[e+c-a]/L$

$M_B = \frac{P}{L}(e+b+2c)a$

$M_C = \frac{P}{L}[2L-(e+b+2c)]c$

$M_B = M_C \Rightarrow V_{BC} = 0 \Rightarrow e = a-c$

$M_B = M_C = Pc$

8.3 (a) $V(x) = -40x$, $M = -20x^2$

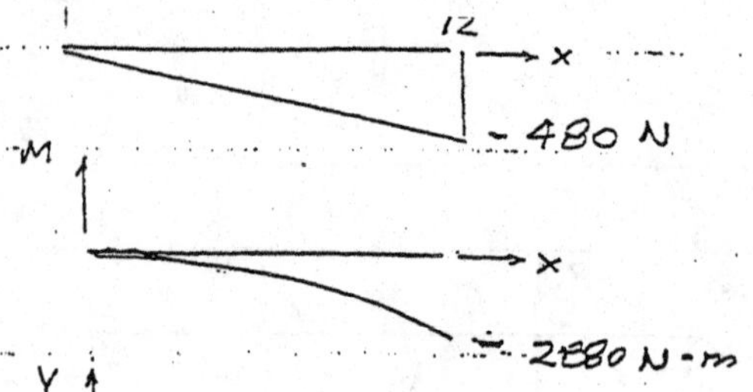

(b) $V = -40x$, $M = -20x^2$; $0 < x < 6$

$V = -240$, $M = -240(x-3)$, $6 < x < 12$

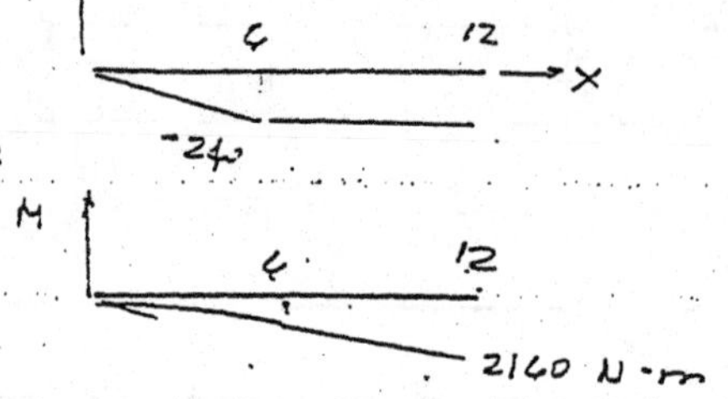

(c) $V = -5x^2/3$, $M = -5x^3/9$

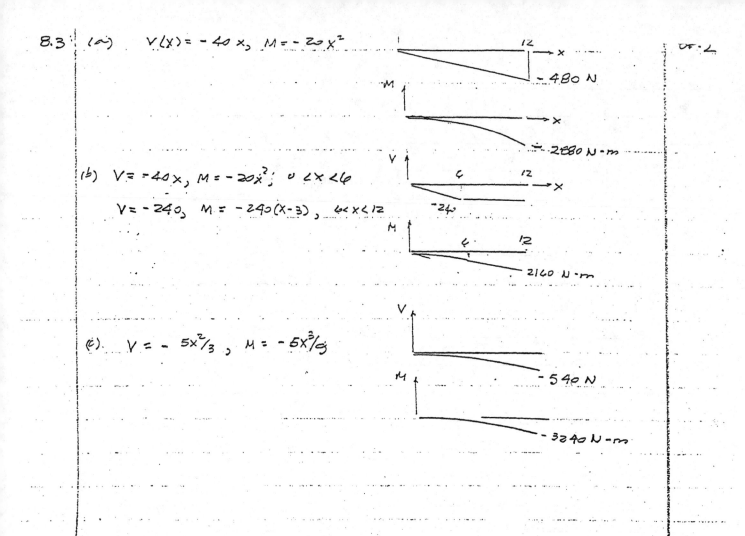

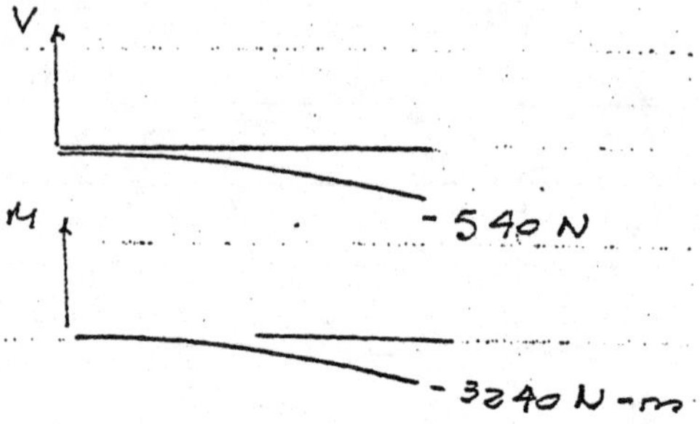

(d) $V = 1200 - 300x$

$M = 1200x - 150x^2$

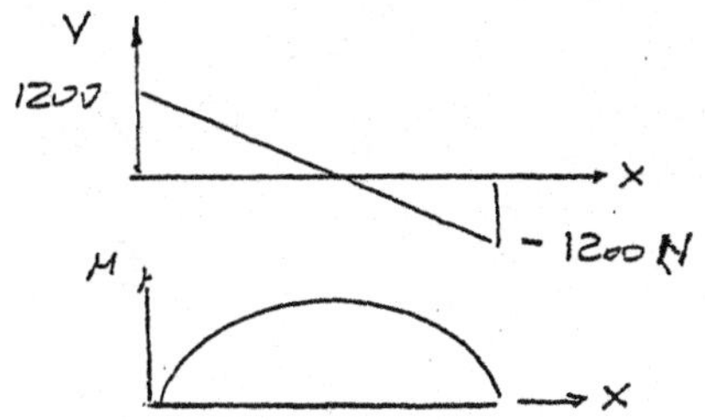

(e)

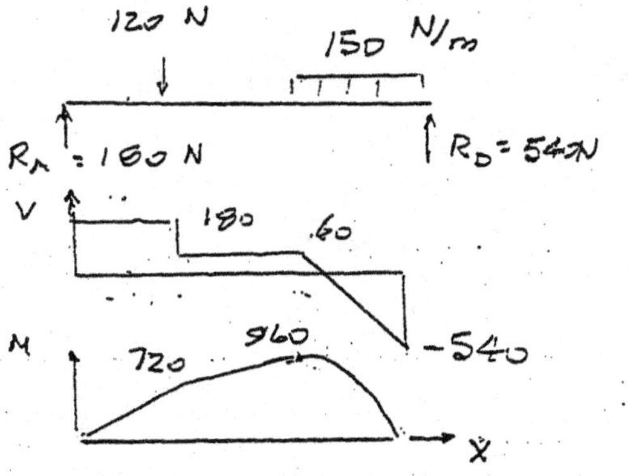

$V = 180, \; M = 180x, \quad 0 < x < 4$

$V = 60, \quad M = 60x + 480, \; 4 \leq x \leq 8$

$V = -150x + 1260, \; M = -75x^2 + 1260x - 4320$

$8 \leq x \leq 12$

(f)

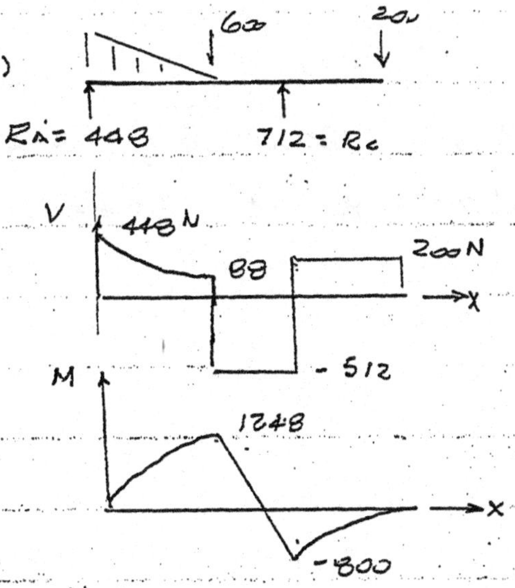

$V = 10x^2 - 120x + 448$

$M = 10x^3/3 - 60x^2 + 448x$ } $0 \leq x \leq 6$

$V = -512$; $M = -512x + 4320, \quad 6 < x < 10$

$V = 200, \quad M = -200(14 - x), \quad 10 < x < 14$

(g)

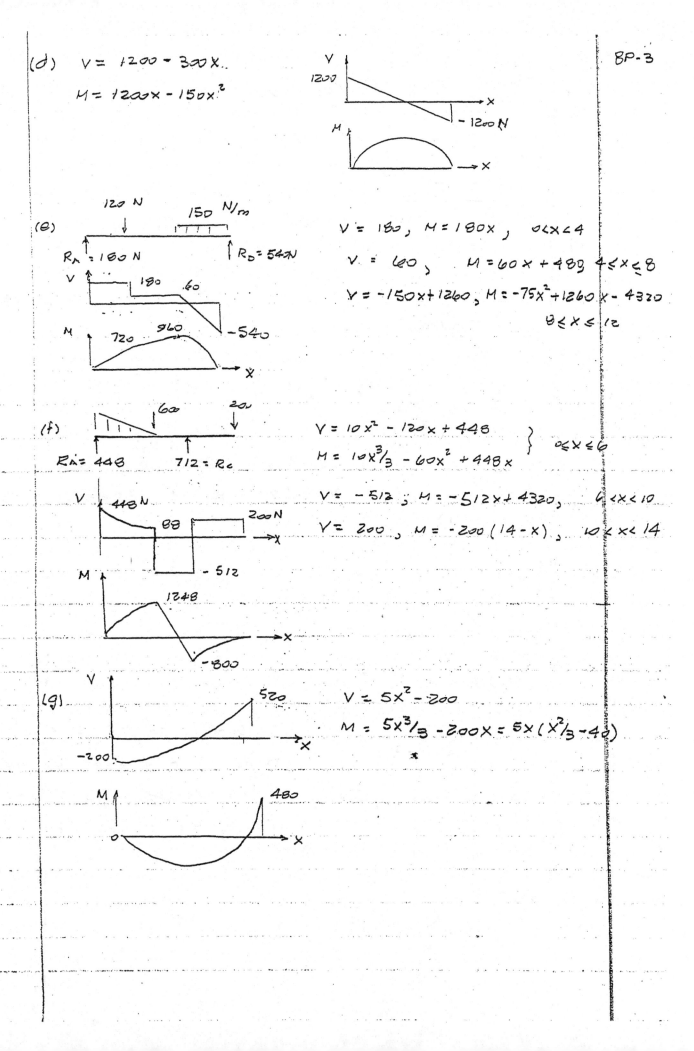

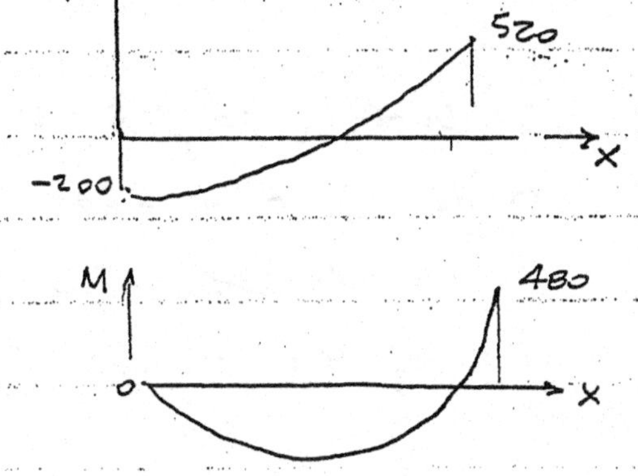

$V = 5x^2 - 200$

$M = 5x^3/3 - 200x = 5x(x^2/3 - 40)$

(h)

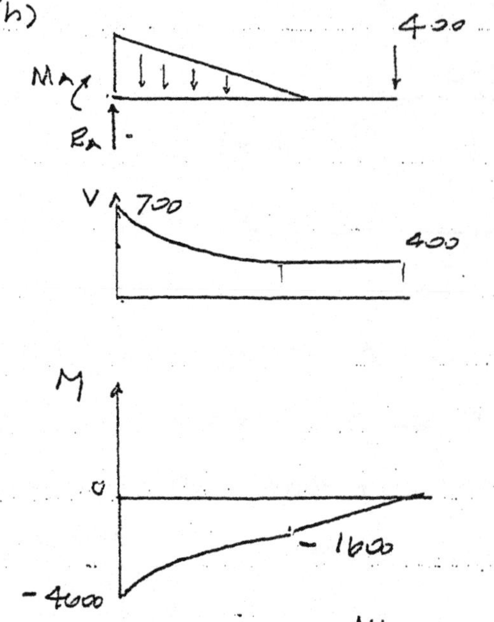

$M_A = -4000 - 600 = -4600$ N-m $\circlearrowleft$

$R_A = 700$ N

$V = 25x^2/3 - 100x + 700$

$M = 25x^3/9 - 50x^2 + 700x - 4600$ $\Big\}$ $0 \le x \le 6$

$V = 400,\quad M = +400(x - 10),\quad 6 < x \le 10$

(i)   60 N/m

$M_A = -720$ N-m $\circlearrowleft$

$R_A = 120 \qquad R_L = 120$

$V = 120,\quad M = 120x - 720,\quad 0 < x < 6$

$V = -60x + 480,\quad M = -30x^2 + 480x - 1800$

(j)   80 N/m   200

$R_A = 160 \qquad 220 \qquad 60$

$V = 160 - 80x,\quad M = -40x^2 + 160x,\quad 0 < x < 4$

$V = -160,\quad M = -160x + 640,\quad 4 < x < 8$

$V = 260,\quad M = 260x - 2720,\quad 8 < x < 10$

$V = 60,\quad M = 60x - 720,\quad 10 < x < 12$

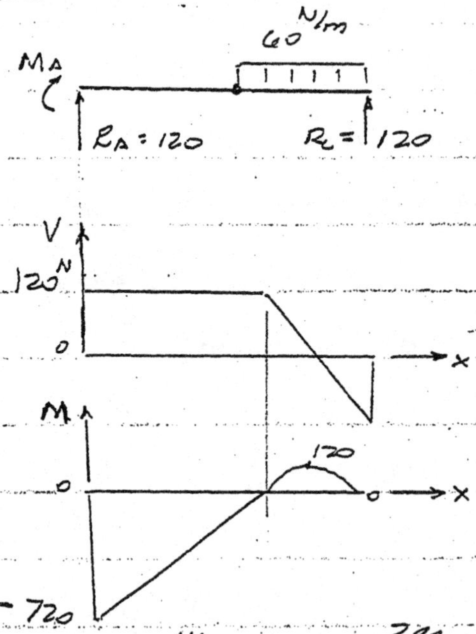

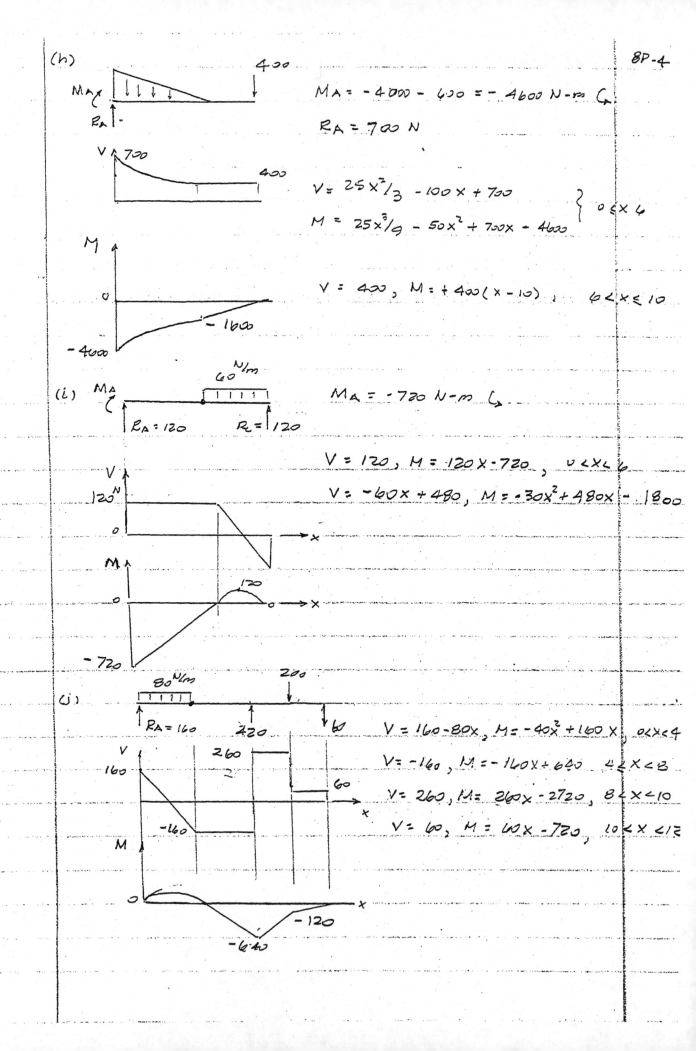

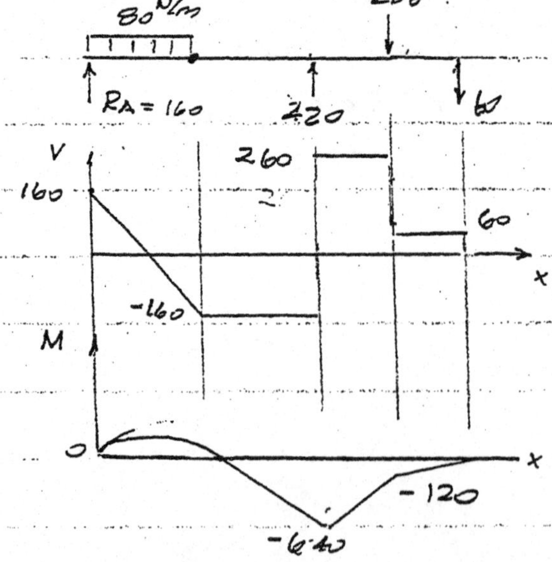

8.4

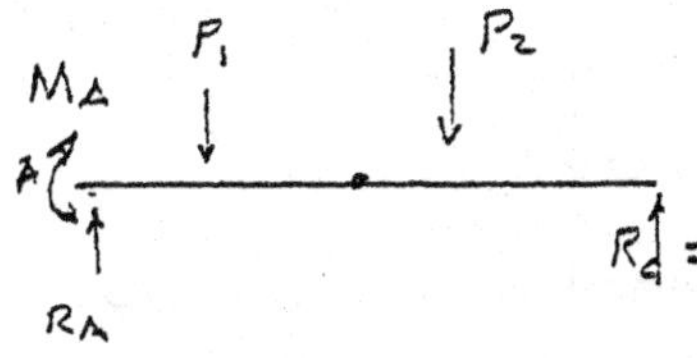

$$R_D = \frac{P_2(a - L/2)}{L/2} = 2P_2\left(a_2/L - 1/2\right)$$

$$R_A = P_1 + P_2 - 2P_2\left(a/L - 1/2\right)$$

$$= P_1 + 2P_2\left(1 - a_2/L\right)$$

$$M_A = -R_A L/2 + P_1\left(L/2 - a_1\right)$$

$$= -\left[P_1 + 2P_2\left(1 - a_2/L\right)\right] L/2 + P_1\left(L/2 - a_1\right)$$

$$= -P_1 a_1 - P_2\left(L - a_2\right)$$

$$\left.\begin{aligned} V &= P_1 + 2P_2\left(1 - a_2/L\right), \\ M &= -P_1 a_1 - P_2(L - a_2) + \left[P_1 + 2P_2\left(1 - a_2/L\right)\right]x \\ &= P_1(x - a_1) + P_2\left[2(1 - a_2/L)x - (L - a_2)\right] \end{aligned}\right\} \quad 0 < x < a_1$$

$$\left.\begin{aligned} V &= 2P_2\left(1 - a_2/L\right) \\ M &= -P_1 a_1 - P_2(L - a_2) + \left[P_1 + 2P_2\left(1 - a_2/L\right)\right]x - P_1(x - a_1) \\ &= -P_1 a_1 - P_2(L - a_2) + P_1 a_1 + 2P_2\left(1 - a_2/L\right)x \end{aligned}\right\} \quad a_1 < x < a_2$$

$$= P_2\left[-L + a_2 + 2x - 2a_2 x/L\right]$$

$$\left.\begin{aligned} V &= -2P_2\left(a_2/L - 1/2\right) \\ M &= 2P_2\left(a_2/L - 1/2\right)(L - x) \end{aligned}\right\} \quad a_2 < x < L$$

8.5

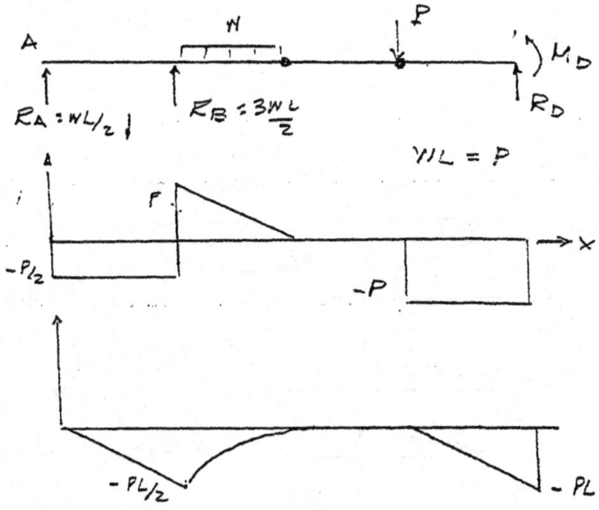

$$M_D = PL$$
$$R_D = P$$

$$V = -wL/2, \quad M = -wLx/2, \quad 0 < x < L$$

$$\left.\begin{aligned} V &= wL - w(x - L) = -wx + wL \\ M &= \frac{w}{2}\left(-x^2 + 4Lx - L^2\right) \end{aligned}\right\} \quad L < x < 2L$$

$$V = M = 0, \quad 2L < x < 3L$$

$$V = -P, \quad M = -P(x - 3L)$$

8.6

a)

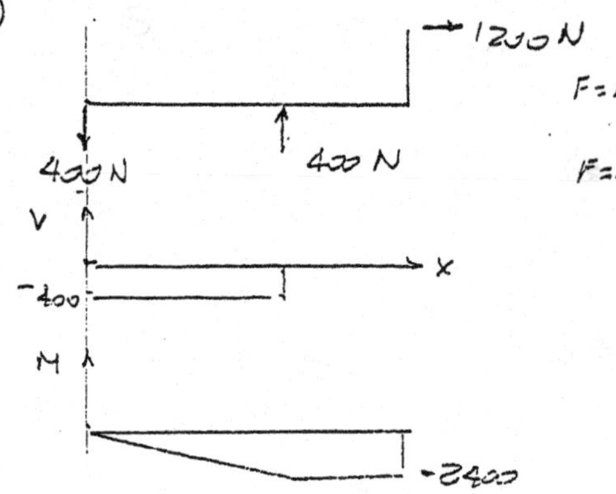

$F = 1200, V = -400, M = -400x, \quad 0 < x < 6$

$F = 1200, V = 0, M = -2400 \text{ N·m} \quad 6 < x < 10$

b)

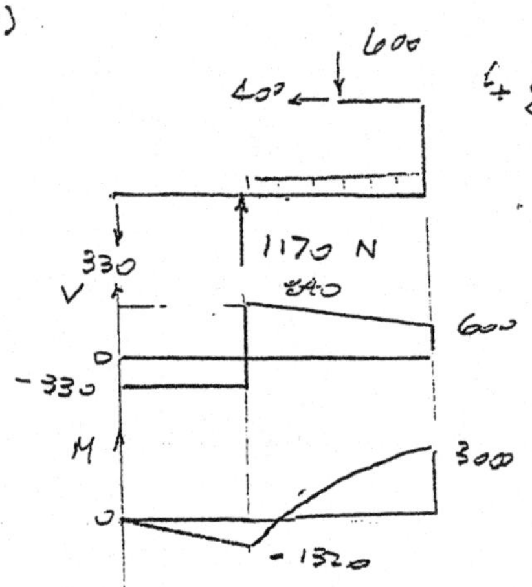

$\circlearrowleft + \sum M_A = 0 \Rightarrow 4R_B - 7(600) - (240)(7) + 400(3) = 0$

$F = -400, V = -330; M = -330x \quad 0 \le x < 4$

$F = -400, V = 1000 - 40x \quad 4 < x < 10$

$M = -20x^2 + 1000x - 5000$

c)

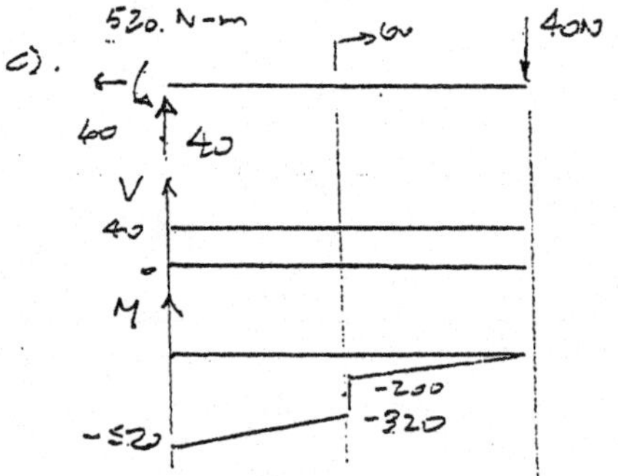

$F = 60 \text{ N}, V = 40 \text{ N}, M = -520 + 40x, \quad 0 < x < 5$

$F = 0, V = 40 \text{ N}, M = -400 + 40x \quad 5 < x < 10$

d)

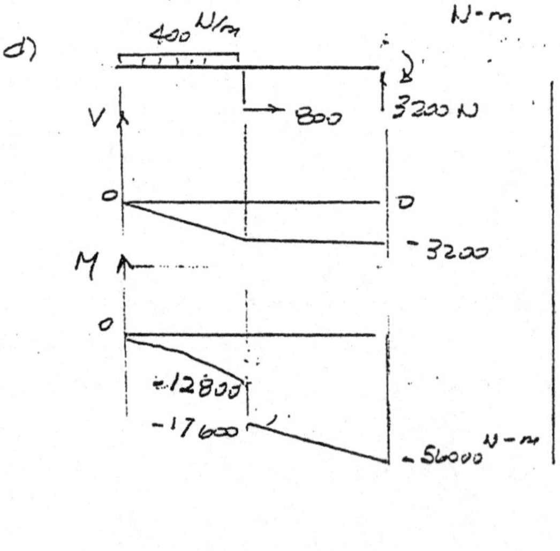

$F = 0, V = -400x, M = -200x^2, \quad 0 < x < 8$

$F = 0, V = -3200, M = -3200x + 8000, \quad 8 < x < 16$

e)

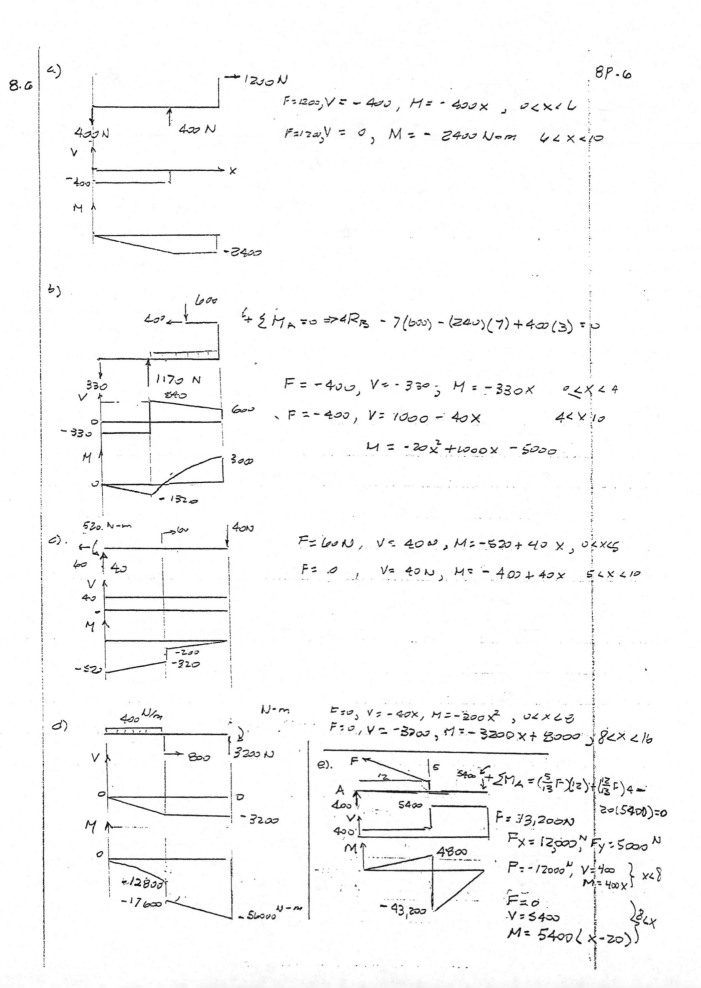

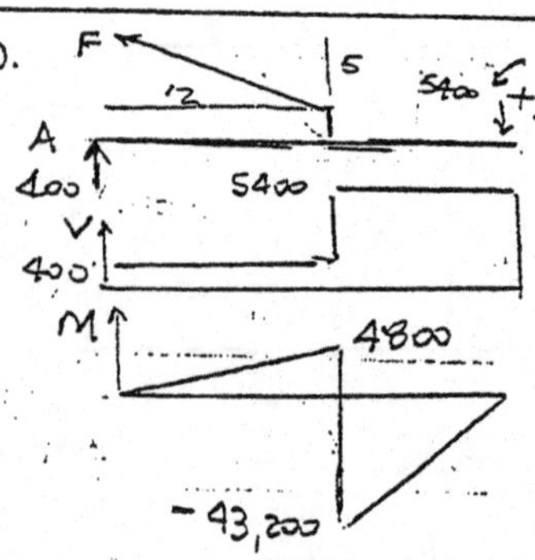

$\circlearrowleft + \sum M_A = \left(\frac{5}{13}F\right)(12) + \left(\frac{12}{13}F\right)4 - 20(5400) = 0$

$F = 13{,}200 \text{ N}$

$F_x = 12{,}000 \text{ N}; F_y = 5000 \text{ N}$

$F = -12000 \text{ N}, V = 400, M = 400x \quad \} \; x < 8$

$F = 0, V = 5400, M = 5400(x - 20) \quad \} \; 8 < x$

$P < wL^2/8e$

8.7

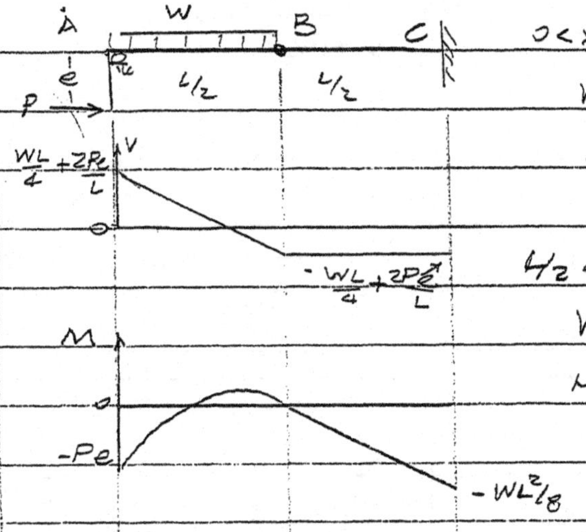

$0 < x < L/2$:

$V(x) = wL/4 - wx + 2Pe/L$

$M(x) = wLx/4 - wx^2/2 + 2Pe/L\,x - Pe$

$L/2 < x < L$:

$V = -wL/4 + 2Pe/L$

$M = +wLx/4 + 2Pex/L - Pe - \frac{w}{2}(x - L/4)^2$

AB: $M_{max} \Rightarrow V = 0$

$x = L/4 + 2Pe/wL$

$M_{max} = (wL/4 + 2Pe/L)(L/4 + 2Pe/wL) - w/2(L/4 + 2Pe/wL)^2 - Pe$

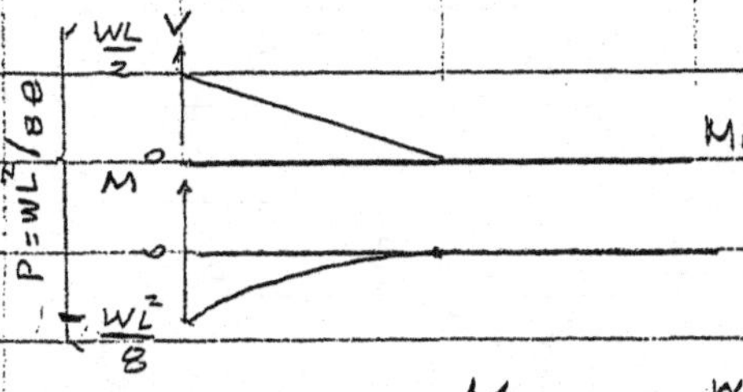

$M_{max} = w/2\,(L/4 + 2Pe/wL)^2 - Pe$

$= w/2\,[L^2/16 + Pe/w + 4P^2e^2/w^2L^2] - Pe$

$= wL^2/32 + Pe/2 + 2P^2e^2/wL^2 - Pe$

$M_{max} = wL^2/32 + 2P^2e^2/wL^2 - Pe/2$

$dM_{max}/dP = 0 \Rightarrow 4Pe^2/wL^2 - e/2 = 0 \Rightarrow P = wL^2/8e$.

$(M_{max})_{min} = wL^2/32 + 2(wL^2/64)\frac{1}{wL^2} - wL^2/16 = 0$

8.8

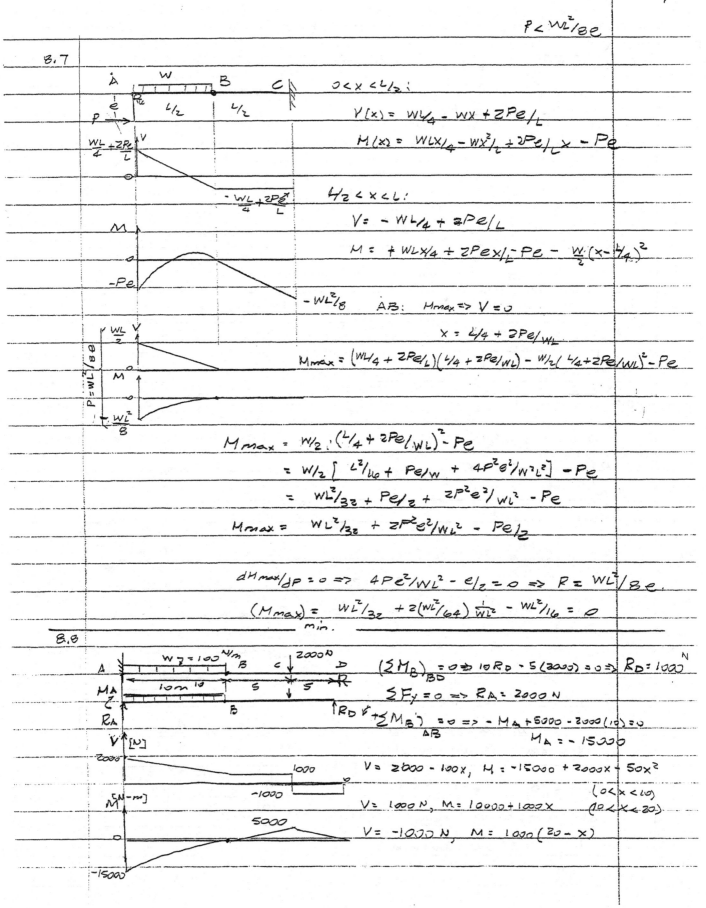

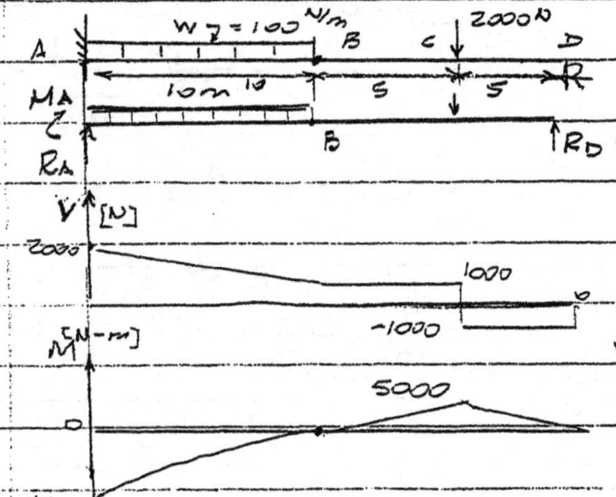

$(\sum M_B)_{BD} = 0 \Rightarrow 10R_D - 5(2000) = 0 \Rightarrow R_D = 1000$ N

$\sum F_y = 0 \Rightarrow R_A = 2000$ N

$+\sum(M_B)_{AB} = 0 \Rightarrow -M_A + 5000 - 2000(10) = 0$

$M_A = -15000$

$V = 2000 - 100x$, $M = -15000 + 2000x - 50x^2$ $(0 < x < 10)$

$V = 1000$ N, $M = 10000 + 1000x$ $(10 < x < 20)$

$V = -1000$ N, $M = 1000(20 - x)$

8.9

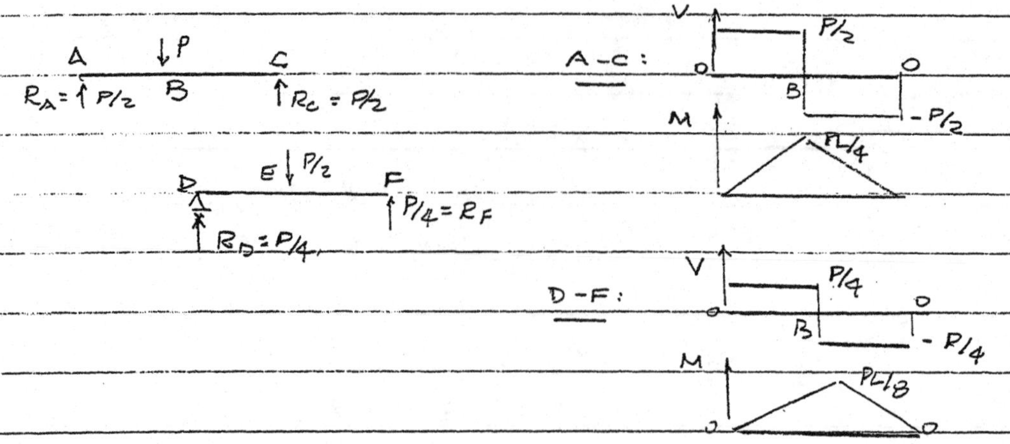

8.10

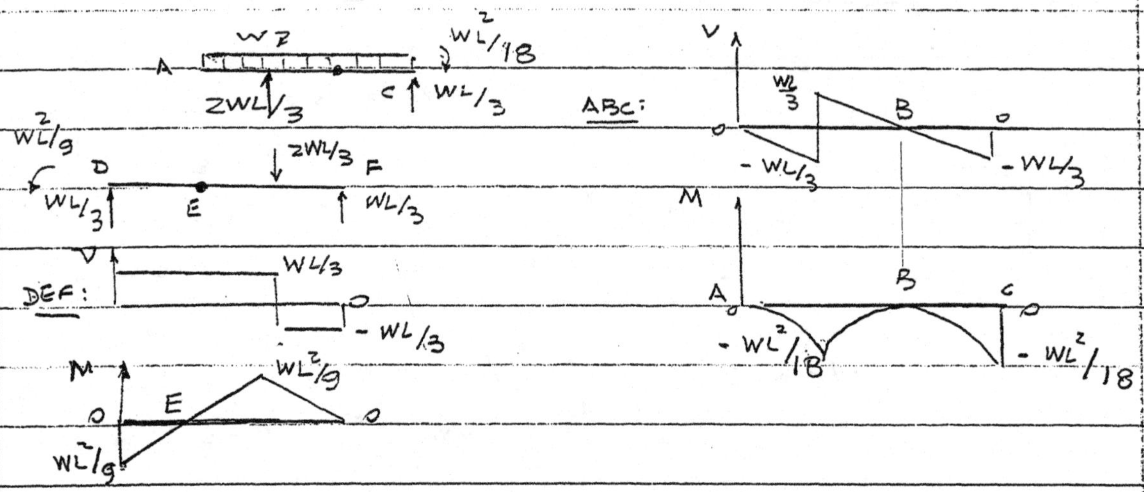

8.11 a) $0 < x < 8 : dV/dx = 0 \Rightarrow q = 0$

$8 < x < 12 : dV/dx = -250 \Rightarrow q = 250\ N/m$

$\Delta V|_B = -2000 \Rightarrow$ Concentrated Force, $P = 2000$ acts at B.

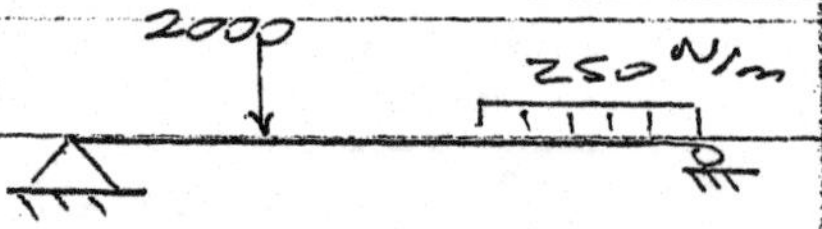

b) $0 < x < 5 : dV/dx = 0 \Rightarrow q = 0$

$5 < x < 10 : dV/dx = 500 \Rightarrow q = -500\ N/m$

$M_A = 2000 \Rightarrow$ couple acts at A; $V_A = -1000 \Rightarrow P = 1000\ N$ acts at A.

$\Delta M = -3000 \Rightarrow$ couple acts at C.

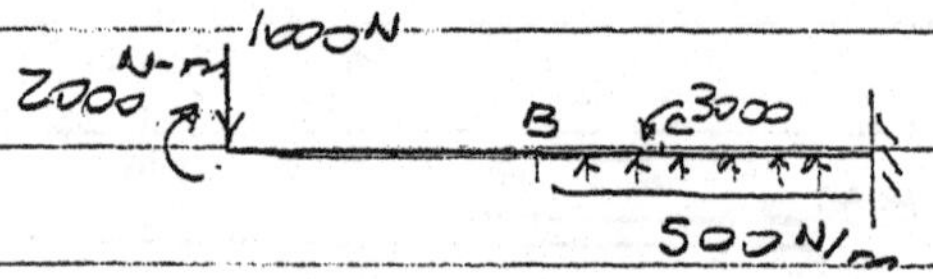

c) $0 \le x \le 18 : dV/dx = 0 \Rightarrow q = 0$

$\Delta V|_B = -1000 \Rightarrow P = 1000\ N$ acts at B

$\Delta M|_C = +3000 \Rightarrow$ Couple acts at C.

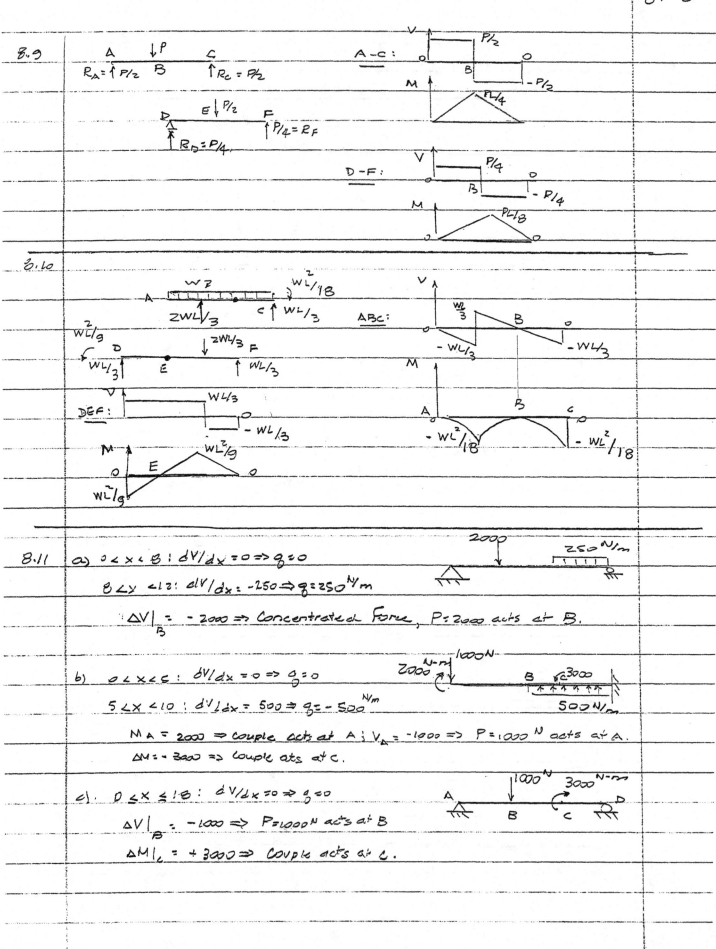

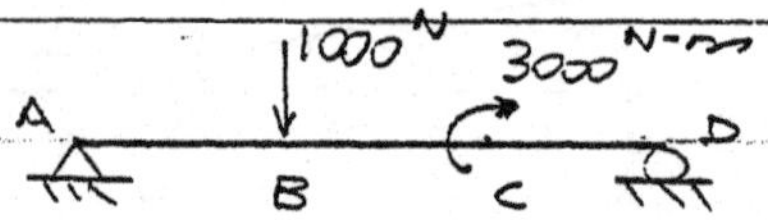

8.12 $q(x) = q_0 (x/L)^n$

$$V(x) = -\int q(\xi)\,d\xi = -\frac{q_0}{L^n}\int_0^x \xi^n\,d\xi = -\frac{q_0 x^{n+1}}{(n+1)L^n} = -\frac{q_0}{n+1}(x/L)^n \cdot x$$

$$M(x) = -\int_0^x q(\xi)\cdot(x-\xi)\,d\xi = -\frac{q_0}{L^n}\int_0^x (x-\xi)\,\xi^n\,d\xi = -q_0/L^n\left[\frac{x^{n+2}}{n+1} - \frac{x^{n+2}}{n+2}\right] = -\frac{q_0}{L^n}\frac{x^{n+2}}{(n+1)(n+2)}$$

$\therefore R_B = V_B = -q_0 L/_{n+1}$, $M_B = -\dfrac{q_0 L^2}{(n+1)(n+2)}$

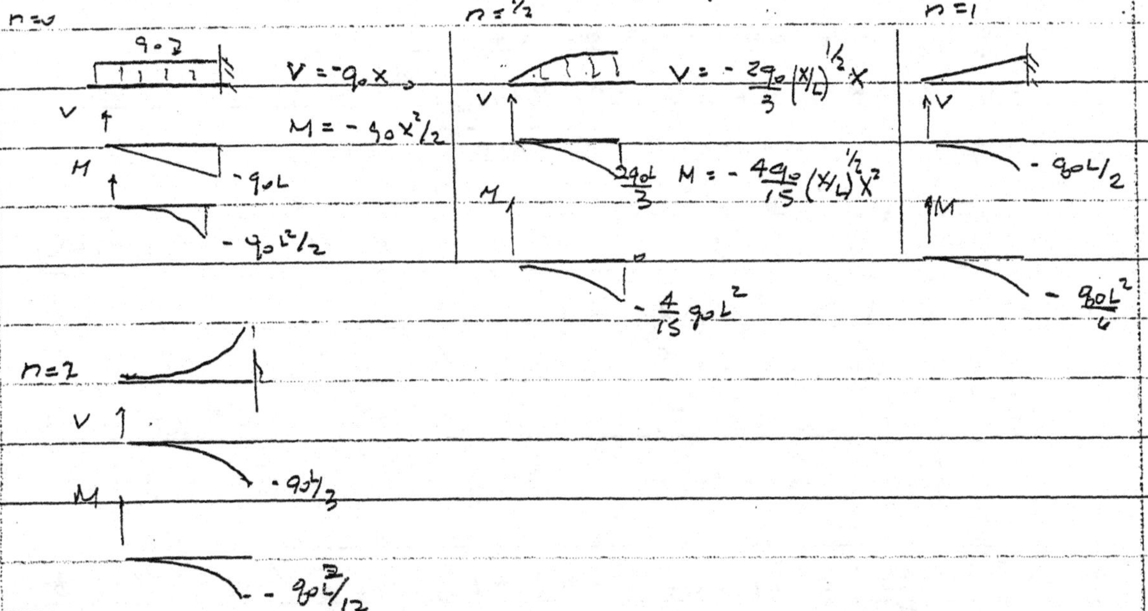

8.13 $M(0) = M(L) = 0 \Rightarrow V(0)L = \int_0^L q(\xi)\cdot(x-\xi)\,d\xi$

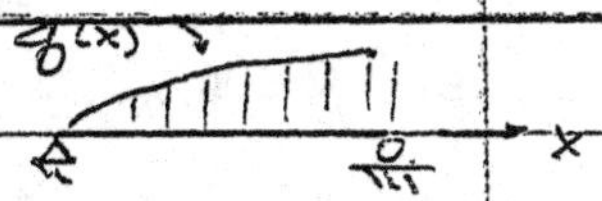

$R_A = V(0) = q_0/L \int_0^L (L-\xi)\sin \pi\xi/2L\,d\xi$

$= q_0/L\left\{-2L^2/\pi \cos \pi\xi/2L\Big|_0^L - \left[-\xi(2L/\pi)\cos \pi\xi/2L\Big|_0^L + (2L/\pi)\int \cos \pi\xi/2L\,d\xi\right]\right\}$

$= q_0/L\left\{2L^2/\pi - (2L/\pi)^2\right\} = (2q_0 L/\pi^2)(\pi - 2)$

$V(x) = V(0) - \int_0^x q(\xi)\,d\xi = (2q_0 L/\pi^2)(\pi+2) - q_0\int_0^x \sin \pi\xi/2L\,d\xi$

$= 2(\pi+2)q_0 L/\pi^2 + 2q_0 L/\pi\,(\cos \pi\xi/2L)\Big|_0^x$

$= 2(\pi-2)q_0 L/\pi^2 + 2q_0 L/\pi\,(\cos \pi x/2L - 1) = \frac{2}{\pi^2}(-2 + \pi\cos \pi x/2L)\,q_0 L$

$M(x) = V(0)\cdot x - \int_0^x (x-\xi)\,q(\xi)\,d\xi$

$= 2/\pi^2\,[-2x + 2L\sin \pi x/2L]\,q_0 L = 4/\pi^2\,[-x + L\sin \pi x/2L]\,q_0 L$

$R_B = -V(L) = (4/\pi^2)\,q_0 L$

8.14 a). $M(0)=M(L) \Rightarrow R_A = V(0) = q_0/L \int_0^L (L-\xi)(1-e^{-\alpha\xi/L})d\xi$

$R_A = q_0/L \{ (L\xi - \xi^2/2)|_0^L - L\int_0^L e^{-\alpha\xi/L} + \int_0^L \xi e^{-\alpha\xi/L} d\xi \}$

$= q_0/L \{ L^2/2 + L^2/\alpha\, e^{-\alpha\xi/L}|_0^L + [-L/\alpha\, \xi e^{-\alpha\xi/L}|_0^L + L/\alpha \int_0^L e^{-\alpha\xi/L} d\xi \}$

$= q_0/L \{ L^2/2 + L^2/\alpha (e^{-\alpha L/L} - 1) - L^2/\alpha\, e^{-\alpha L/L} - L^2/\alpha^2\, e^{-\alpha\xi/L}|_0^L \}$

$= q_0 L \{ 1/2 + 1/\alpha (e^{-\alpha L/L} - 1) - 1/\alpha\, e^{-\alpha/L} - 1/\alpha^2 (e^{-\alpha} - 1) \}$

$= q_0 L \{ (1/2 - 1/\alpha + 1/\alpha^2) - \frac{1}{\alpha^2} e^{-\alpha} \} = \frac{q_0 L}{\alpha^2} [(\alpha^2/2 - \alpha + 1) - e^{-\alpha}]$

b)

$V(x) = R_A - \int q(\xi)d\xi = \frac{q_0 L}{\alpha^2}[\alpha^2/2 - \alpha + 1 - e^{-\alpha}] - q_0 \int_0^x (1 - e^{-\alpha\xi/L})d\xi$

$= (q_0 L/\alpha^2)(\alpha^2/2 - \alpha + 1 - e^{-\alpha}) - q_0(\xi + L/\alpha\, e^{-\alpha\xi/L})|_0^x$

$= (q_0 L/\alpha^2)(\alpha^2/2 - \alpha + 1 - e^{-\alpha}) - q_0(x + L/\alpha\, e^{-\alpha x/L} - L/\alpha)$

$= \frac{q_0}{\alpha^2}[(\alpha^2/2 - \alpha + 1 - e^{-\alpha})L - \alpha^2 x + \alpha L e^{-\alpha x/L} + L\alpha]$

$= q_0/\alpha^2\, [(1 + \alpha^2/2 - e^{-\alpha} - \alpha e^{-\alpha x/L})L - \alpha^2 x]$

$\alpha \to \infty \therefore V(x) = q_0(L/2 - x)$

$M(x) = V(0)x - \int_0^x (x-\xi)\, q(\xi)d\xi$

$= q_0 \{ \frac{L}{\alpha^2}[\alpha^2/2 - \alpha + 1 - e^{-\alpha}]x - [x\int_0^x (1 - e^{-\alpha\xi/L})d\xi - \int_0^x \xi(1 - e^{-\alpha\xi/L})d\xi]\}$

$= q_0 \{ \frac{Lx}{\alpha^2}(\alpha^2/2 - \alpha + 1 - e^{-\alpha}] - [(x\xi + Lx/\alpha\, e^{-\alpha\xi/L}]_0^x + [\xi^2/2 - (-L/\alpha\, \xi e^{-\alpha\xi/L} + L/\alpha \int e^{-\alpha\xi/L} d\xi)]_0^x \}$

$= q_0 \{ \frac{Lx}{\alpha^2}(\alpha^2/2 - \alpha + 1 - e^{-\alpha}) - (x^2 + \frac{Lx}{\alpha} e^{-\alpha x/L} - \frac{Lx}{\alpha}) + [x^2/2 + \frac{Lx}{\alpha} e^{-\alpha x/L} + \frac{L^2}{\alpha^2}(e^{-\alpha x/L} - 1) \}$

$M(x) = q_0 \{ \frac{Lx}{\alpha^2}(\alpha^2/2 + 1 - \alpha e^{-\alpha}) - x^2/2 + L^2/\alpha^2 (1 - e^{-\alpha x/L}) \}$ $\quad \alpha \to \infty \quad M(x) = q_0 x(L-x)/2$

c). $(R_A)_b/(R_A)_a = q_b/q_a$

$q_b = q_0 \tanh(\alpha x/L) = q_0 \left(\frac{e^{\alpha x/L} - e^{-\alpha x/L}}{e^{\alpha x/L} + e^{-\alpha x/L}} \right) = q_0 \left(\frac{1 - e^{-2\alpha x/L}}{1 + e^{-2\alpha x/L}} \right)$

$q_b/q_a = \frac{1 - e^{-2\alpha x/L}}{(1 + e^{-2\alpha x/L})(1 - e^{-\alpha x/L})} = \frac{(1 + e^{-\alpha x/L})(1 - e^{-\alpha x/L})}{(1 + e^{-2\alpha x/L})(1 - e^{-\alpha x/L})} = \frac{1 + e^{-\alpha x/L}}{1 + e^{-2\alpha x/L}} \begin{cases} > 1 & \alpha \text{ finite} \\ 1 & \alpha \to \infty \end{cases}$

$\therefore (R)_b \geq (R)_a$

8.15 a). Assume radius to N.A. $= R$ & R_m; $\varepsilon_t = \frac{R_m + d/2 - R}{R}$, $\varepsilon_c = \frac{R_m - d/2 - R}{R}$

$\therefore R\varepsilon_t = R_m + d/2 - R$ (i) $\quad$ (i) + (ii) $\Rightarrow 2R_m = R(2 + \varepsilon_t + \varepsilon_c)$; (i) − (ii) $\Rightarrow R = d/(\varepsilon_t - \varepsilon_c)$

$R\varepsilon_c = R_m - d/2 - R$ (ii) $\qquad \therefore R_m = d/2 \left(\frac{2 + \varepsilon_t + \varepsilon_c}{\varepsilon_t - \varepsilon_c} \right)$

b) $d \to 0 \Rightarrow \varepsilon_m = (R_m - R)/R$

c). $R = R_m \Rightarrow \varepsilon_t = d/2R$, $\varepsilon_c = -d/2R$ $\therefore \Rightarrow \varepsilon_c = -\varepsilon_t$

8.16 a). $M = EI/R \qquad I = \pi d^4/64 = \pi(3\times10^{-3})^4/64 = 3.976\times10^{-12}\ m^4$

$M = (120\times10^9)(3.976\times10^{-12})/0.75 = 0.636\ N\text{-}m = 63.6\ N\text{-}cm$

b). $\sigma_x = Mc/I = (0.636)(1.5\times10^{-3})/(3.976\times10^{-12}) = 240\times10^6\ N/m^2 = 240\ MPa$

8.17

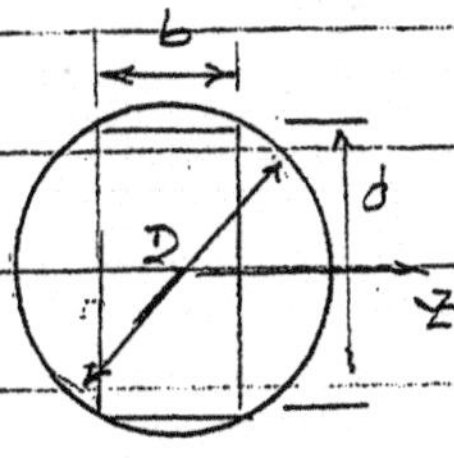

$D^2 = d^2 + b^2 \qquad b = (D^2 - d^2)^{1/2}$

(a). $k = M/EI$; $\quad I = bd^3/12 = b/12\,(D^2 - b^2)^{3/2}$

Maximize $I \Rightarrow dI(b)/db = 0$

$\Rightarrow (D^2 - b^2)^{3/2} - \frac{3b}{2}(D^2 - b^2)^{1/2}\cdot 2b = 0$

$(D^2 - b^2)^{1/2}[D^2 - b^2 - 3b^2] = 0 \Rightarrow 4b^2 = D^2 \Rightarrow b = D/2$

$d = (D^2 - b^2)^{1/2} = D(1 - b^2/D^2)^{1/2} = \sqrt{3}D/2 \Rightarrow d/b = \sqrt{3}$

b). $\sigma = \dfrac{Md/2}{bd^3/12} = \dfrac{6M}{bd^2} = \dfrac{6M}{b(D^2 - b^2)} = \dfrac{6M}{\alpha(b)}$

Maximize α: $\quad d\alpha/db = D^2 - 3b^2 = 0 \Rightarrow b = \sqrt{3}/3\,D$

$d = (D^2 - b^2)^{1/2} = \sqrt{2/3}\,D = \sqrt{2}\cdot\sqrt{3}/3\,D \Rightarrow d/b = \sqrt{2}$

8.18 $\varepsilon = y/R \Rightarrow R = c/\varepsilon \qquad A = d^2 = 64\ mm^2 \Rightarrow d = 8\ mm \Rightarrow c = 4\ mm.$

a) $R = (4\times10^{-3})/(16.00\times10^{-4}) = 1/4\times10 = 2.5\ m$

b) $M = EI/R = \dfrac{(200\times10^9)(8\times10^{-3})^4/12}{2.5} = 27.31\ N\text{-}m$

c). $\sigma = E\varepsilon = Mc/I = 320\ MPa$

8.19 $M = \sigma I/c \qquad I = b^4/12 - \pi R^4/4 = \dfrac{(50\times10^{-3})^4}{12} - \dfrac{\pi(10^{-2})^4}{4} = 5.123\times10^{-7}\ m^4$

$M = \dfrac{(120\times10^6)(5.123\times10^{-7})}{25\times10^{-3}} = 2.462\ kN\text{-}m$

8.20 $\bar{y} = 4R/3\pi \qquad \sigma = Mc/I, \quad I_{z'z'} = \pi R^4/8, \quad c = -R(1 - 4/3\pi)$

$I_{zz} = I_{z'z'} - (\pi R^2/2)(4R/3\pi)^2 = \pi R^4/8 - 8R^4/9\pi$

$I_{zz} = \pi R^4(1/8 - 8/9\pi^2)$

$\sigma_c = -\dfrac{24M}{R^3}\left(\dfrac{3\pi - 4}{9\pi^2 - 64}\right)$; $\quad \sigma_t = \dfrac{96M}{R^3}\left(\dfrac{1}{9\pi^2 - 64}\right)$

8.21 a). $M = \iint_A y\,\sigma_x\,dA$; $\sigma = \alpha\varepsilon^n$, $1<n$, n odd

$\varepsilon = y/R \Rightarrow M = b\alpha\int_{-d/2}^{d/2} y\,\varepsilon^n\,dy = \frac{b\alpha}{R^n}\int_{-d/2}^{d/2} y^{n+1}\,dy$

$M = (b\alpha/R^n)\left.\frac{y^{n+2}}{n+2}\right|_{-d/2}^{d/2} = \frac{2\alpha b}{R^n(n+2)}(d/2)^{n+2} \Rightarrow \alpha = \frac{R^n(n+2)2^{n+2}}{2bd^{n+2}}\cdot M$

$\alpha = \frac{R^n(n+2)2^{n+1}}{bd^{n+2}}\cdot M \Rightarrow \sigma_x = \alpha\varepsilon^n = \alpha y^n/R^n$

$\sigma_x = \left[\frac{(n+2)2^{n+1}}{bd^{n+2}}\right] M y^n$

b). $1/R^n = \left[\frac{(n+2)2^{n+1}}{\alpha bd^{n+2}}\right] M \Rightarrow \kappa = 1/R = \frac{2[2(n+2)]^{1/n}}{d(\alpha bd^2)^{1/n}}\cdot M^{1/n}$

c). $n=1$, $\alpha \to E \Rightarrow \sigma_x = \frac{12}{bd^3}\cdot My = My/I$; $1/R = \frac{12M}{Ebd^3} = M/EI$.

8.22

$\sigma = \begin{cases}\alpha\varepsilon^n, & \varepsilon>0\\ -\alpha|\varepsilon|^n, & \varepsilon\le 0\end{cases}$, $0<n\le 1 \Rightarrow \sigma_x = \begin{cases}\alpha y^n/R^n, & y\ge 0\\ -(\alpha/R^n)|y|^n, & y\le 0\end{cases}$

$M = \iint \sigma\cdot y\,dA = \frac{b\alpha}{R^n}\left\{-\int_{-d/2}^{0} y|y|^n dy + \int_0^{d/2} y^{n+1}dy\right\}$

Let $z = -y$ $(z>0)$

$\therefore \int_{-d/2}^{0} y|y|^n dy = \int_{d/2}^{0}(-z)|-z|^n(-dz) = \int_{d/2}^{0} z|-z|^n dz = \int_{d/2}^{0} z^{n+1}dz = -\int_0^{d/2} y^{n+1}dy$

$\therefore M = 2b\alpha/R^n\int_0^{d/2} y^{n+1}dy = \frac{2b\alpha}{R^n}\left.\frac{y^{n+2}}{(n+2)}\right|_0^{d/2} = \frac{2b\alpha(d/2)^{n+2}}{R^n(n+2)}$, $0<n\le 1$

Note: Result is same as in problem 8.21 with $n = 1, 3, 5 \cdots$
Results for (b) & (c) are same.

8.23 Conclusion

8.24 $\sigma_a = E_a \varepsilon_a$, $\sigma_b = E_b \varepsilon_b$, $\varepsilon_a = \varepsilon_b = y/R$

Assume $\bar{y}$ is measured from interface.

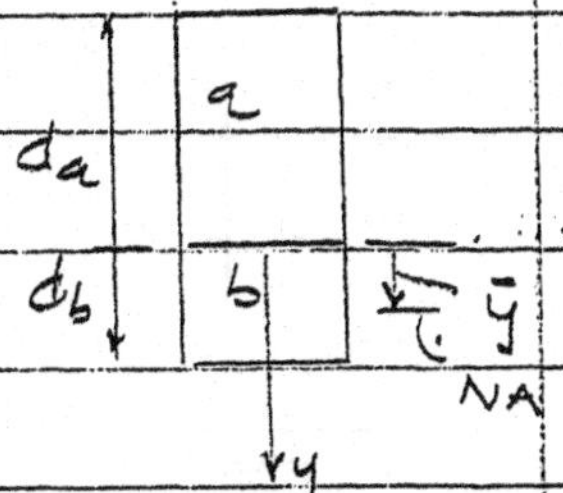

$$F = \iint \sigma \, di = b\int \sigma \, dy = 0$$

$$\int_{-(d_a+\bar{y})}^{-\bar{y}} E_a y \, dA + E_b\left[\int_{-\bar{y}}^{0} y \, dA + \int_{0}^{d_b-\bar{y}} y \, dA\right] = 0$$

$$E_a \int_{-(d_a+\bar{y})}^{-\bar{y}} y \, dA + E_b \int_{-\bar{y}}^{d_b-\bar{y}} y \, dA = 0 \Rightarrow \tfrac{1}{2}\left\{ E_a y^2 \Big|_{-(d_a+\bar{y})}^{-\bar{y}} + E_b y^2 \Big|_{-\bar{y}}^{(d_b-\bar{y})} \right\} = 0$$

$$E_a[\bar{y}^2 - (d_a+\bar{y})^2] + E_b[(d_b-\bar{y})^2 - \bar{y}^2] = 0$$

$$E_a[-d_a^2 - 2d_a\bar{y}] + E_b[d_b^2 - 2d_b\bar{y}] = 0 \Rightarrow \bar{y} = \tfrac{1}{2}\,\frac{E_b d_b^2 - E_a d_a^2}{E_b d_b + E_a d_a}$$

Note: If $E_a = E_b \Rightarrow \bar{y} = (d_b - d_a)/2$

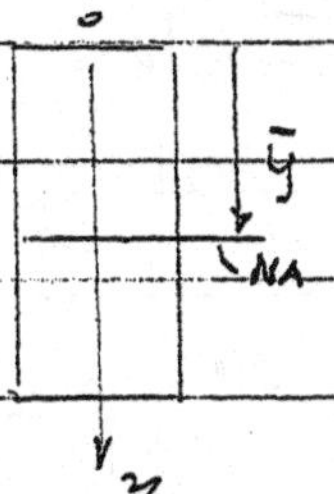

8.25 $E(\eta) = E_0[1 + \beta(\eta/d)]$, $\varepsilon = y/R \Rightarrow \sigma = E(\eta)\,y/R$

$$y = \eta - \bar{y} \Rightarrow \sigma(\eta) = E(\eta)(\eta - \bar{y})/R$$

a) $F = \iint \sigma(\eta)\,dA = \dfrac{bE_0}{R}\displaystyle\int_0^d [1 + \beta/d\,\eta](\eta - \bar{y})\,d\eta = 0$

$$\int_0^d [-\bar{y} + (1 - \beta\bar{y}/d)\eta + (\beta/d)\eta^2]\,d\eta = 0$$

$$\left\{-\bar{y}\eta + [1 - \beta\bar{y}/d]\eta^2/2 + (\beta/d)\eta^3/3\right\}\Big|_0^d = 0$$

$$-\bar{y}d + \tfrac{1}{2}[d^2 - \beta d\bar{y}] + \beta d^2/3 = 0$$

$$-\bar{y} - \beta\bar{y}/2 + d/2 + \beta d/3 = 0 \Rightarrow (1 + \beta/2)\bar{y} = d/6\,(3 + 2\beta)$$

$$\bar{y} = d/3\left[\frac{3+2\beta}{2+\beta}\right]$$

b) Since M is the same about all horizontal axes:

$$M = \iint_A \eta\,\sigma(\eta)\,dA = \frac{bE_0}{R}\int_0^d [-\bar{y}\eta + (1 - \beta\bar{y}/d)\eta^2 + (\beta/d)\eta^3]\,d\eta$$

$$= bE_0/R\left\{-\bar{y}d^2/2 + (1 - \beta\bar{y}/d)d^3/3 + \beta d^3/4\right\}$$

Substituting for $\bar{y}$

$$M = \frac{bd^3 E_0[\beta^2 + 6\beta + 6]}{36(2+\beta)R} = \frac{E_0 I}{R}\,\frac{\beta^2 + 6\beta + 6}{3(\beta+2)}$$

Note: $\beta = 0 \Rightarrow M = E_0 I/R$.

8.26

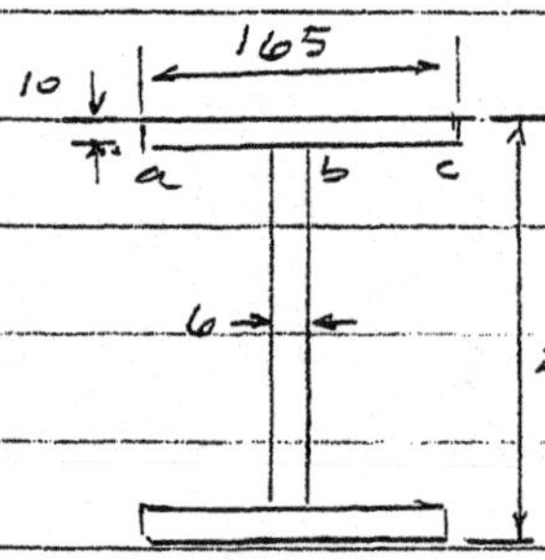

a) $I = \frac{(180)(6)^3}{2} + 2\left[\frac{165(10)^3}{12} + (1650)(95)^2\right] = 29.81 \times 10^6\ mm^4$

$S = I/c = \frac{29.81\times10^6}{100} = 298.1 \times 10^3\ mm^3$

W 203 x 36 $I = 34.5\times10^6\ mm^4$, $S = 342\times10^3\ mm^3$

Note: Difference due mainly to neglect of fillets.

b) $M_{max} = \sigma_{all} \cdot S = (150\times10^6)(298.1\times10^{-6}) = 44{,}715\ N\text{-}m = 44.72\ kN\text{-}m$

c). $\sigma = My/I = \frac{(44.72\times10^6)}{(29.81\times10^6)} \cdot y = 1.5\,y$ [N/mm^2 = MPa], y in mm.

$F_{flange} = 165(1.5)\int_{-100}^{-90} y\,dy = +247.5\, y^2/2 \Big|_{-100}^{-90} = -235.1\ kN$ (comp.)

(d) F_{web} (hd) $= 6(1.5)\int_{-90}^{0} y\,dy = 4.5\,y^2\Big|_{-90}^{0} = -36.45\ kN$

$F_{web}/F_{fl} = 15.5\%$

e). $M_{flange} = (-235.1\times10^3)(-190\times10^{-3}) = 44.67\ kN\text{-}m$

$M|_{web} = (F_{web})(\frac{2}{3}\cdot 90) = (-36.45\times10^3)(60\times10^{-3}) = 4.37\ kN$

$M|_{fl}/M_{TOT} = 99.9\%$,

8.27

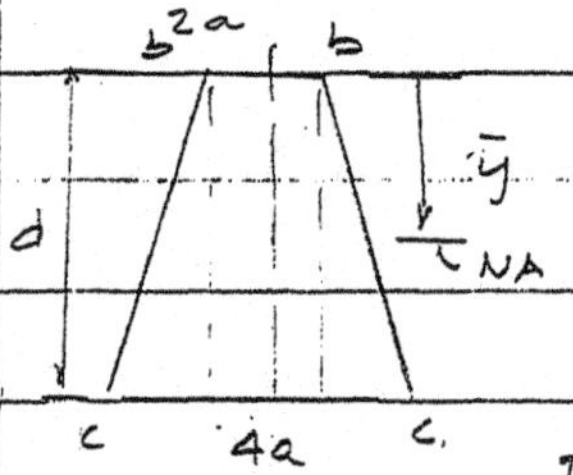

$A\bar{y} = (2ad)(d/2) + 2[(ad/2)(2d/3)] = ad^2 + 2ad^2/3 = 5ad^2/3$

$\bar{y} = \frac{5ad^2/3}{3ad} = 5d/9$

$I = \left[\frac{(2a)d^3}{12} + (2ad)(5d/9 - d/2)^2\right] + 2\left[ad^3/36 + \left(\frac{ad}{2}\right)\left(\frac{2d}{3} - \frac{5d}{9}\right)^2\right]$

$= ad^3\{[1/6 + 2(1/18)^2] + 2[1/36 + \frac{1}{2}(1/9)^2]\} = \frac{13ad^3}{54} = 0.24074\,ad^3$

$\sigma = Mc/I$, $c = y_{max} = 5d/9 \Rightarrow \sigma = \frac{30M}{13ad^2}$

$M_E = 13ad^2\sigma_0/30$

8.28

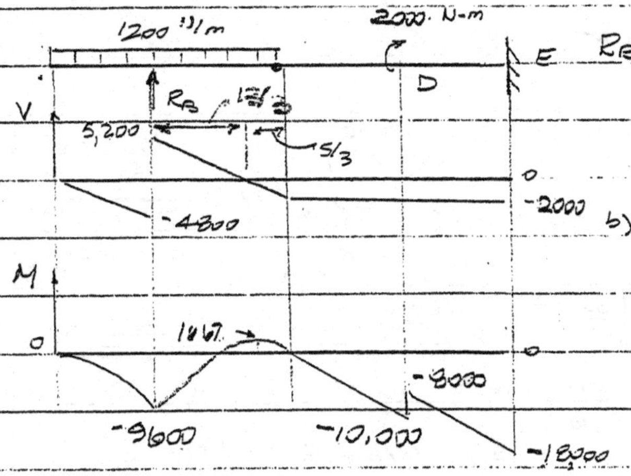

$R_B = 10{,}000\,N$

$M_{max} = (2000/2)(5/3) = 1667\ N\text{-}m$

b). $\sigma_T = My/I = \frac{(-18000)(-4\times10^{-2})}{720\times10^{-8}} = 100\ MPa$

$\sigma_C = \frac{(-18000)(6\times10^{-2})}{720\times10^{-8}} = -150\ MPa$

σ_{xT} & σ_{xc} at E; σ_{xT} at a-a, σ_{xc} at b-b

$\tau_{c\text{-}c} = VQ/Ib$ $\quad V_{max} = 5200$, $Q = (2\times6)(5) = 60\ cm^3$

$\tau = \frac{(5200)(60)}{720\times3} = 145.5\ N/cm^2 = 1.455\ MPa.$

8.29

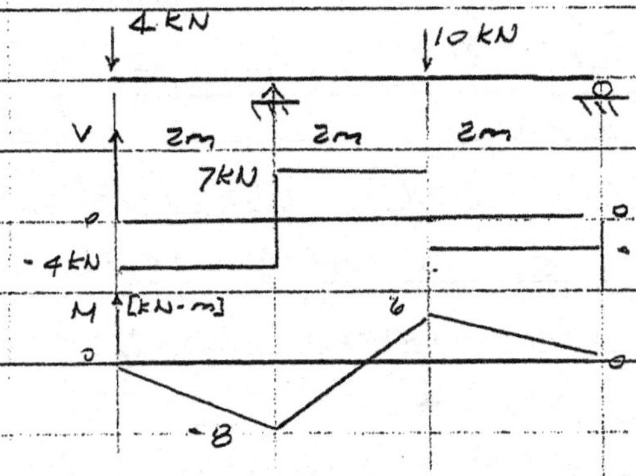

$R_B = 11\ kN$

$\bar{y} = \frac{(48\times2) + (48\times10)}{24\times4} = 6\ cm.$

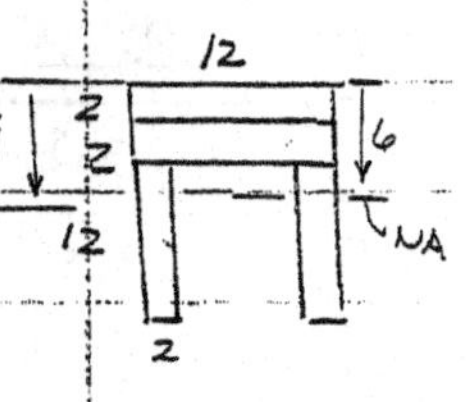

$I = \left[\frac{12\times4^3}{12} + (48\times4^2)\right] + 2\left[\frac{2\times12^3}{12} + 24\times4^2\right]$

$= 2176\ cm^4$

(a). Tension: $\sigma = My/I = \frac{(6\times10^5)(10)}{2176} = 2.76\times10^3\ N/cm^2 = 27.5\ MPa$

(b) Comp.: $\sigma = \frac{(6\times10^5)(10)}{2176} = -3.68\times10^3\ N/cm^2 = -36.8\ MPa$

c). $Q = (2\times10)(3) = 60\ cm^3$

$\tau = \frac{(7\times10^3)(60)}{(2176)(2)} = 96.5\ N/cm^2 = 965\ kPa$

d). $Q = (12\times4)(-4) = -192\ cm^3$ $\quad \tau = \frac{(-7\times10^3)(192)}{(2176)2} = -308\ N/cm^2 = 3.08\ MPa.$

8.30

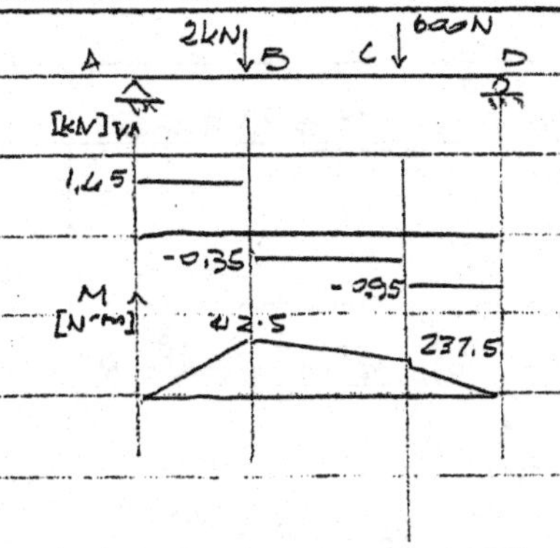

$\bar{y} = 25\ mm.$ $\quad I_{NA} = 5.125\times10^5\ mm^4$

$R_A = 1.65\ kN$ (a) $\sigma_x = \frac{(412.5\times10^3)(25\times10^{-3})}{5.125\times10^{-3}} = 2.012\ MPa$ (tens.)

$\sigma_x = \frac{(412.5\times10^3)(-35\times10^{-3})}{5.125\times10^{-3}} = -2.817\ MPa$ (comp.)

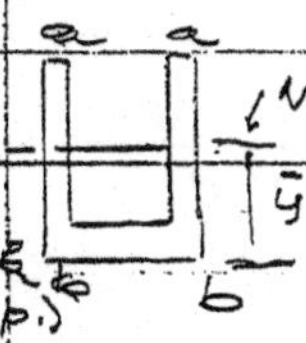

(c)

$\tau = VQ/Ib$ $\quad Q = (60)(-5) = -300\ mm^3$

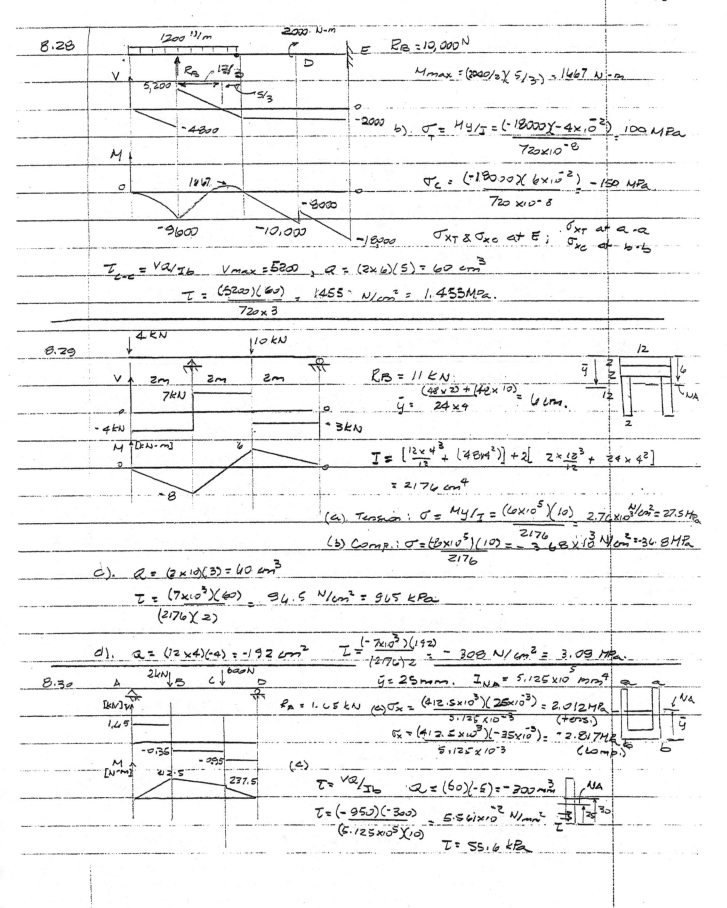

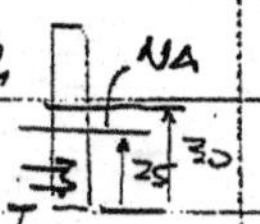

$\tau = \frac{(-950)(-300)}{(5.125\times10^5)(10)} = 5.56\times10^{-2}\ N/mm^2$

$\tau = 55.6\ kPa$

8.31

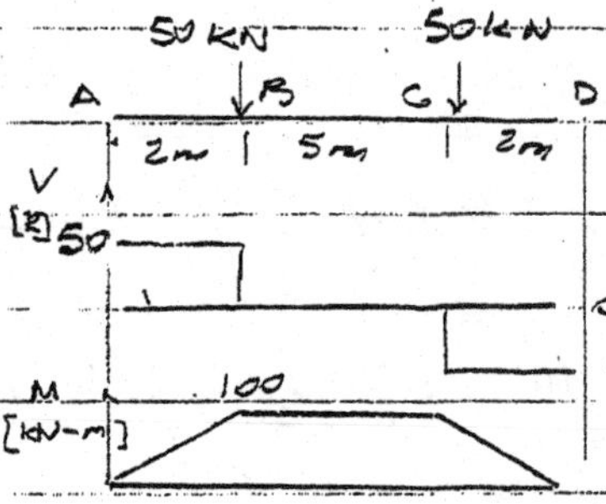

S305 x 47 : $S = 596 \times 10^3\ mm^3$, $I = 90.7 \times 10^6\ mm^4$

a) $\sigma = M/S = \frac{100 \times 10^6}{596 \times 10^3} = 167.8\ N/mm^2 = 167.8\ MPa$

b) $Q|_{cc} = (13.8 \times 59.05)(-145.5)^{13.8}$

$= -1.186 \times 10^5\ mm^3$

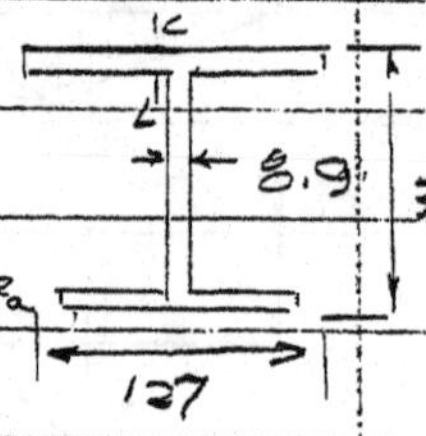

$\tau = \frac{(50 \times 10^3)(-1.186 \times 10^5)}{(13.8)(90.7 \times 10^6)} = 4.74\ MPa$

c) (i) $Q|_{y=0} = -2(1.186 \times 10^5) + (\frac{152.4^2}{2} \times 8.9) = 3.405 \times 10^{10}\ mm^3$.

$\tau = \frac{(50 \times 10^3)(3.405 \times 10^{10})}{(90.7 \times 10^6)(8.9)} = 21.1\ N/mm^2 = 21.1\ MPa$

(ii) $\tau_{ave} = V/A_W$; $A_W = (277.2)(8.9) = 2467\ mm^2 \Rightarrow \tau_{ave} = 20.27\ MPa$

% diff. = 3.9%

d) $R = EI/M = (200 \times 10^9)(90.7 \times 10^{-6})/(100 \times 10^3) = 181.4\ m$

8.32

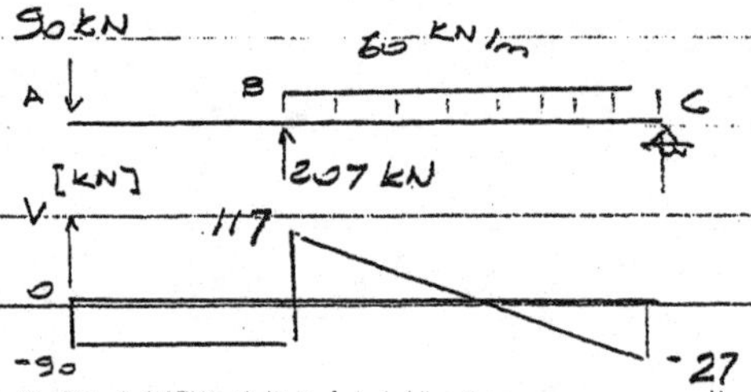

$I = (12)^3/12 + 4(4.5)(1)^3 + 4(4.5 \times 1)(5.5)^2 = 706.5\ cm^4$

τ_{a-a} : $Q_{a-a} = (4.5 \times 1)(5.5) = 24.75\ cm^3$

$\tau_{a-a} = (117 \times 10^3)(24.75)/(706.5 \times 1) = 4.099 \times 10^3\ N/cm^2$

$\tau_{a-a} = 41\ MPa$

τ_{c-c} : $Q_{cc} = -2(4.5 \times 1)(5.5) + (6 \times 1)(3) = -67.5\ cm^3$

$\tau_{c-c} = \frac{117 \times (-67.5)}{(706.5 \times 1)} = 11.18\ N/cm^2 = 112\ MPa$

8.33

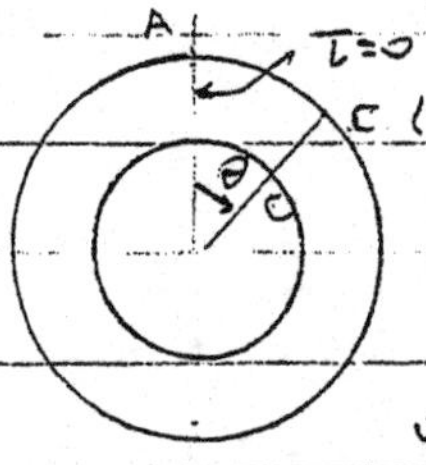

(b) $I = \frac{\pi}{4}(b^4 - a^4)$

$Q = \iint_A y\,dA$, $dA = (r\,d\phi)\,dr$

$y = r\cos\phi$

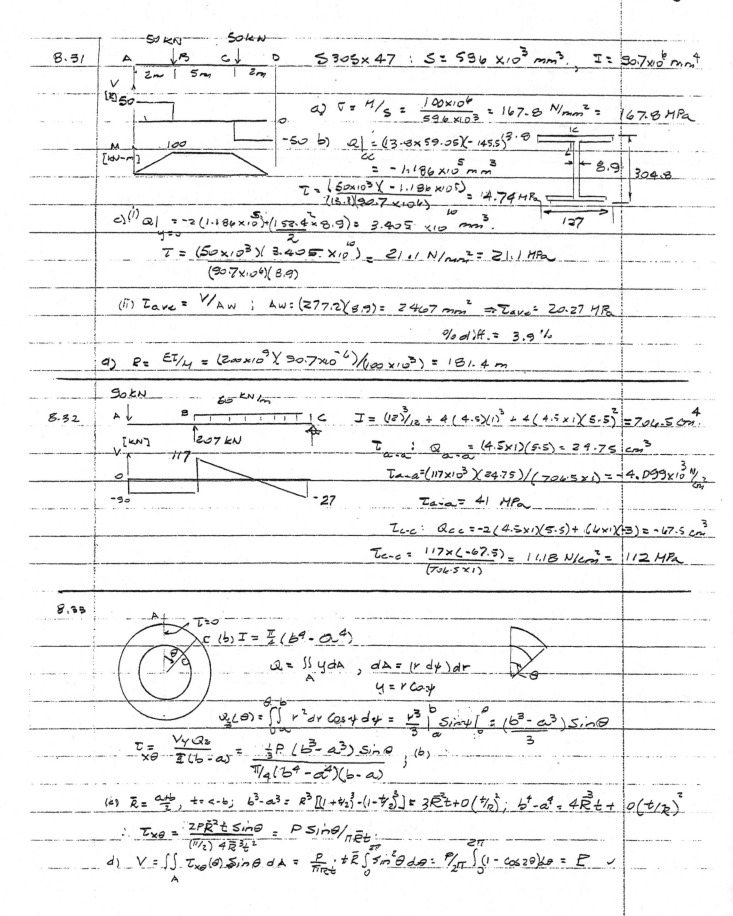

$Q_z(\theta) = \int_0^{\theta}\int_a^{b} r^2 dr \cos\phi\, d\phi = \frac{r^3}{3}\Big|_a^b \sin\phi\Big|_0^\theta = \frac{(b^3 - a^3)}{3}\sin\theta$

$\tau_{x\theta} = \frac{V_y Q_z}{I(b-a)} = \frac{\frac{1}{3}P(b^3 - a^3)\sin\theta}{\pi/4(b^4 - a^4)(b-a)}$; (b)

(c) $\bar{R} = \frac{a+b}{2}$, $t = a-b$; $b^3 - a^3 = \bar{R}^3[(1+t/2)^3 - (1-t/2)^3] = 3\bar{R}^2 t + O(t/R)^2$; $b^4 - a^4 = 4\bar{R}^3 t + O(t/R)^2$

$\therefore \tau_{x\theta} = \frac{2P\bar{R}^2 t \sin\theta}{(\pi/2)\,4\bar{R}^3 t^2} = P\sin\theta/\pi\bar{R}t$

d) $V = \iint_A \tau_{x\theta}(\theta)\sin\theta\, dA = \frac{P}{\pi\bar{R}t} \cdot t\bar{R}\int_0^{2\pi}\sin^2\theta\, d\theta = P/2\pi\int_0^{2\pi}(1 - \cos 2\theta)d\theta = P$ ✓

8.34

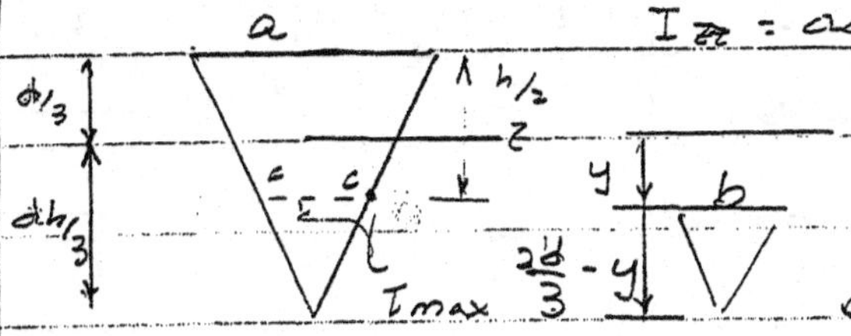

$I_{zz} = ad^3/36$

$b = \frac{a}{d}(2d/3 - y) = a(2/3 - y/d)$

$Q(y) = \frac{b}{2}(2d/3 - y)[y + \frac{2d/3 - y}{3}]$

$= \frac{b}{27}(2d - 3y)(3y + d)$

$\tau = \frac{VQ}{Ib} = \frac{V}{I} \cdot \frac{(2d - 3y)(3y - d)}{27} = \frac{V}{27I}(2d^2 + 3dy - 9y^2)$

$d\tau/dy = 0 \Rightarrow 3d - 18y = 0 \Rightarrow y_{cr} = d/6. \Rightarrow \tau_{max} = \frac{9Vd^2}{4 \cdot 27I} = \frac{9Vd^2}{4 \cdot 27(ad^3/36)} = \frac{3V}{ad}$

$\tau_{max} = \frac{3}{2}\frac{V}{A}$, $A = ad/2$; τ_{max} at $d/3 + d/6 = d/2$

$k = 3/2$

8.35

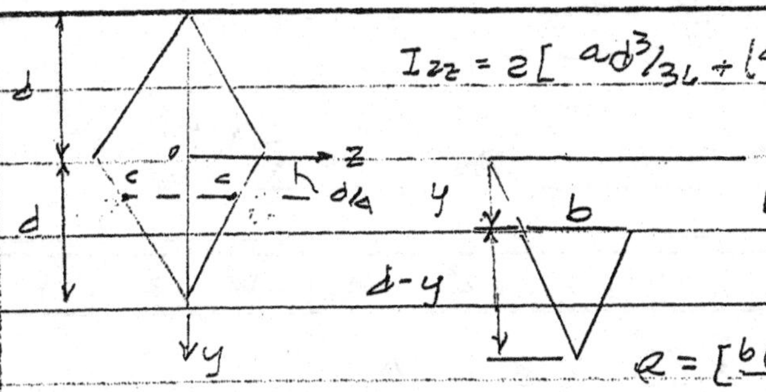

$I_{zz} = 2[ad^3/36 + (ad/2)(d/3)^2] = ad^3/6$

$b/a = \frac{d-y}{d} \Rightarrow b = a(1 - y/d)$

$Q = [\frac{b(d-y)}{2}][y + (\frac{d-y}{3})] = \frac{b}{6}(d-y)(2y+d)$

$Q/b = 1/6(d^2 + dy - 2y^2) \Rightarrow d\tau/dy = 0 \Rightarrow \frac{\partial Q/b}{\partial y} = 0 \Rightarrow d - 4y = 0 \Rightarrow y = d/4$

$\tau_{max} = \frac{V}{I}(Q/b)_{max} = \frac{V}{6I}(9d^2/8) = \frac{3Vd^2}{16I} = \frac{3Vd^2}{16 \cdot ad^3/6} = \frac{9V}{8ad} = \frac{9}{8}V/A$, $k = 9/8$

8.36

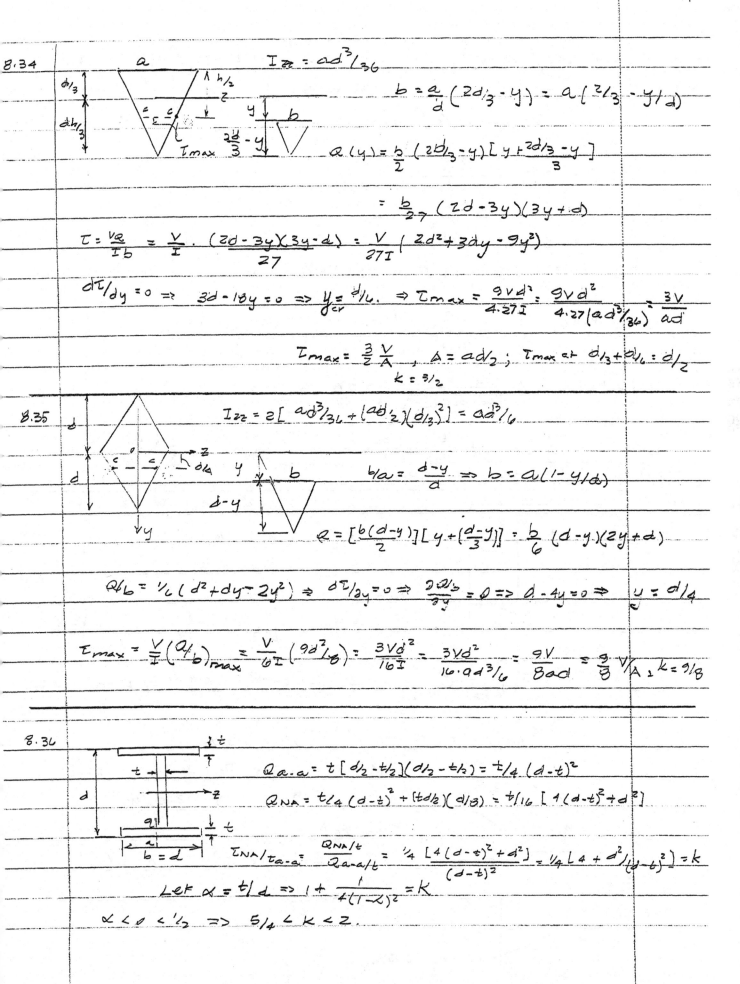

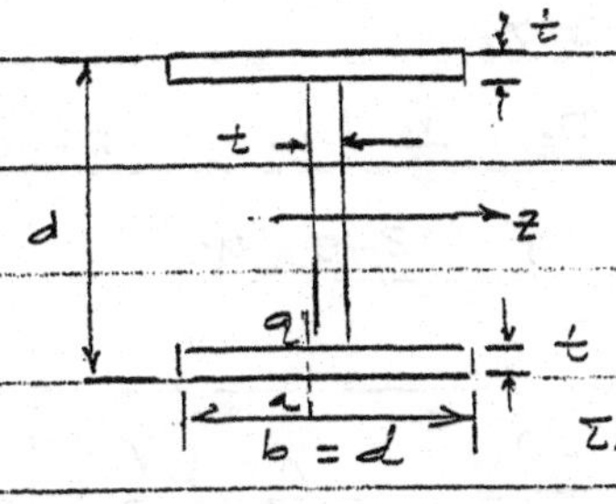

$Q_{a\text{-}a} = t[d/2 - t/2](d/2 - t/2) = t/4\,(d-t)^2$

$Q_{NA} = t/4\,(d-t)^2 + (td/2)(d/8) = t/16\,[4(d-t)^2 + d^2]$

$\tau_{NA}/\tau_{a\text{-}a} = \frac{Q_{NA}/t}{Q_{a\text{-}a}/t} = 1/4\,\frac{[4(d-t)^2 + d^2]}{(d-t)^2} = 1/4\,[4 + d^2/(d-t)^2] = k$

Let $\alpha = t/d \Rightarrow 1 + \frac{1}{4(1-\alpha)^2} = K$

$\alpha < 0 < 1/2 \Rightarrow 5/4 < K < 2.$

8.37 V_y: $I_{zz} = \frac{1}{12}[b_o d_o^3 - b_i d_i^3]$ $Q_z = [b_o(d_o - d_i)/2][\frac{d_o}{2} - (\frac{d_o - d_i}{4})]$

$= \frac{1}{8} b_o(d_o - d_i)(d_o + d_i).$

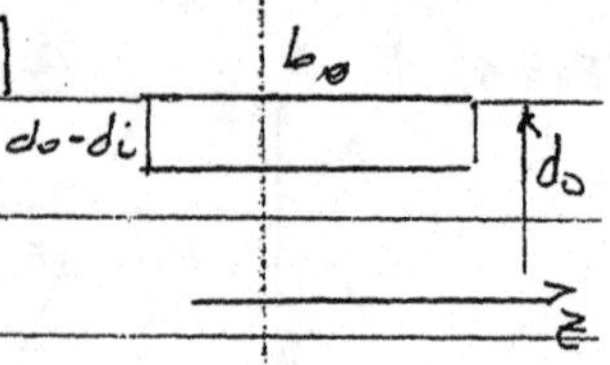

$Q_z / b = Q/(b_o - b_i) = \frac{1}{8} b_o(d_o^2 - d_i^2)/(b_o - b_i)$

$(\tau)_y = V_y Q / I b = \frac{3V_y}{2} \frac{b_o(d_o^2 - d_i^2)}{(b_o - b_i)(b_o d_o^3 - b_i d_i^3)}$

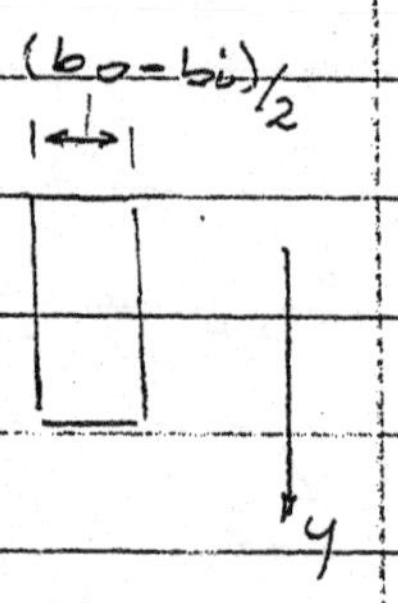

V_z: $I_{yy} = \frac{1}{12}[d_o b_o^3 - d_i b_i^3]$

$Q_y = [d_i(b_o - b_i)/2][b_o/2 - (\frac{b_o - b_i}{4})] = \frac{1}{8} d_i(b_o^2 - b_i^2)$

$(\tau)_z = V_z Q_y / I_{yy} b = \frac{3V_z}{2} \frac{d_i(b_o^2 - b_i^2)}{(d_o b_o^3 - d_i b_i^3)(b_o - b_i)} = \frac{3V_z(b_o + b_i)d_i}{2(d_o b_o^3 - d_i b_i^3)}$

$b_o = d_o$, $b_i = d_i$ $\Rightarrow (\tau)_y = \frac{3V_y}{2} \cdot \frac{b_o(b_o^2 - b_i^2)}{(b_o - b_i)(b_o^4 - b_i^4)} = \frac{3V_y}{2} \frac{b_o}{(b_o - b_i)(b_o^2 + b_i^2)}$

$(\tau)_z = \frac{3V_z}{2} \frac{(b_o + b_i)b_i}{b_o^4 - b_i^4}$ $(\tau)_y/(\tau)_z = (V_y/V_z)(\frac{b_o}{b_i})(\frac{b_o^2 + b_i^2}{b_o^2 - b_i^2})$

→ PROBLEM 8.38 FOUND ON P. BP-19

8.39 a)

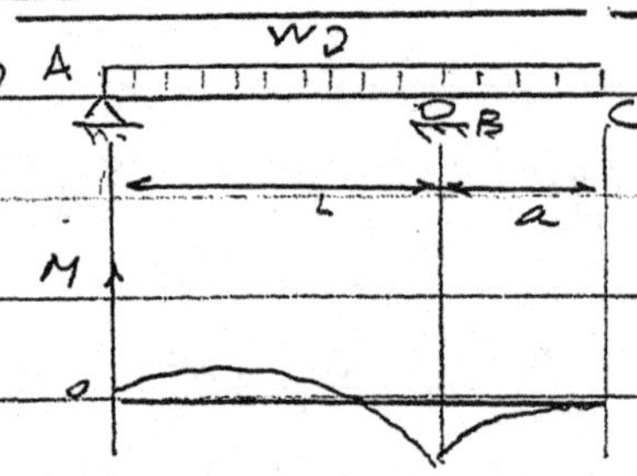

$\Sigma M_B = 0$ $R_A = \frac{1}{L}(wL^2/2 - wa^2/2) = \frac{w}{2L}(L^2 - a^2)$

A-B: $V = w/2L(L^2 - a^2) - wx$

$M = \frac{w}{2L}(L^2 - a^2)x - wx^2/2$

$V = 0 \Rightarrow x_{cr} = (L^2 - a^2)/2L$

$\Rightarrow M_{max} = \frac{w}{8L^2}(L - a)^2$, $M_B = -wa^2/2$

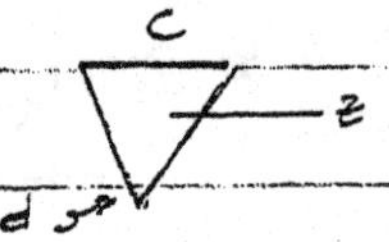

$\sigma_b|_{M>0} = \sigma_c|_{M<0} \Rightarrow 2[\frac{w}{8L^2}(L - a)^2] = wa^2/2 \Rightarrow L^2 - a^2 = \sqrt{2}aL$

$a^2 + \sqrt{2}La - L^2 = 0 \Rightarrow a/L = (\sqrt{6} - \sqrt{2})/2 = 0.5176$

b)

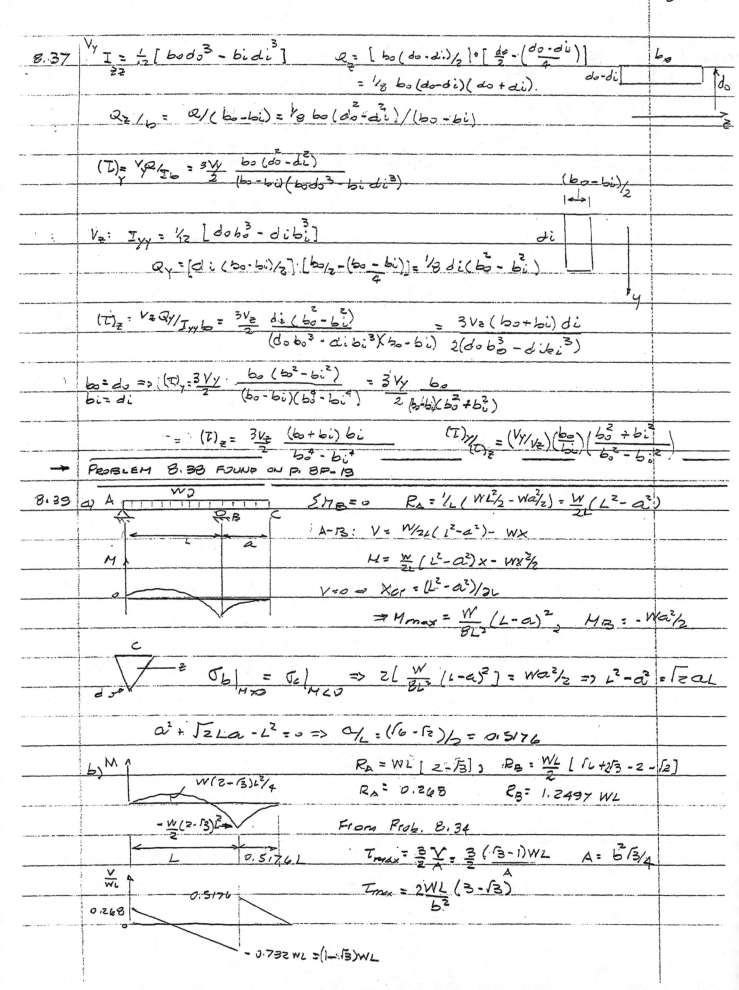

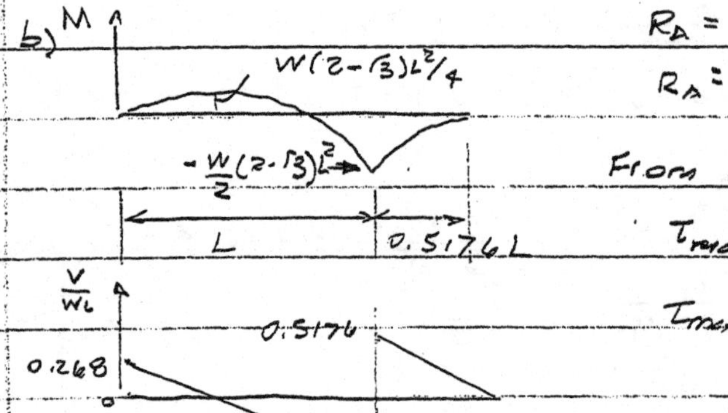

$R_A = wL[2 - \sqrt{3}]$, $R_B = \frac{wL}{2}[\sqrt{6} + 2\sqrt{3} - 2 - \sqrt{2}]$

$R_A = 0.268$ $R_B = 1.2497\, wL$

From Prob. 8.34

$\tau_{max} = \frac{3}{2}\frac{V}{A} = \frac{3}{2}\frac{(\sqrt{3} - 1)wL}{A}$ $A = b^2\sqrt{3}/4$

$\tau_{max} = \frac{2wL}{b^2}(3 - \sqrt{3})$

8.38

From Prob. 8.37: $I_{zz} = \frac{1}{12}[b_o d_o^3 - b_i d_i^3]$

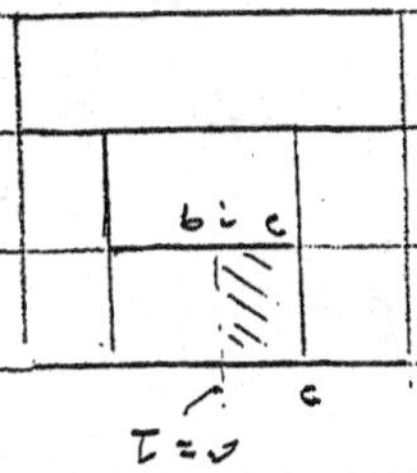

Along c-c: $\frac{Q_z}{b_{c-c}} = \frac{1}{2}[b_i \cdot (\frac{d_o - d_i}{2})][\frac{d_i}{2} + (\frac{d_o - d_i}{4})] / b_{c-c}$

$b_{c-c} = \frac{d_o - d_i}{2}$ $= \frac{b_i(d_o^2 - d_i^2)}{16(d_o - d_i)/2} = \frac{b_i}{8}(d_o + d_i)$

$\tau_{c-c} = VQ/Ib = \frac{3V}{2}\frac{b_i(d_o + d_i)}{b_o d_o^3 - b_i d_i^3}$

$b_o = d_o,\ b_i = d_i \Rightarrow \tau_{c-c} = \frac{3V}{2}\left[\frac{d_i(d_o + d_i)}{d_o^4 - d_i^4}\right] = \frac{3V}{2A}\left(\frac{d_i}{d_o - d_i}\right)$

→ PROBLEM 8.39 FOUND ON P. 8P-18

8.40

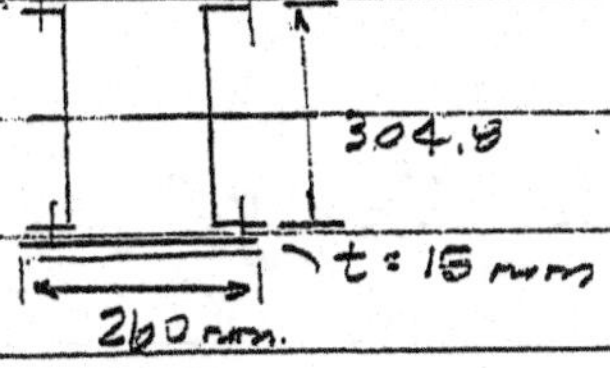

C 305×45 $I_{zz} = 67.4\times10^6\ mm^4$ $\tau_{all} = 150\ MPa,\ s = 800\ mm$

$I = 2(67.4\times10^6) + 2[(260)(15)^3/12 + (260\times15)(159.9)^2]$

$= 2(67.4\times10^6) + 2(99.716\times10^6) = 334.8\times10^6\ mm^4$

$Q = (260\times15)(159.9) = 62.36\times10^4\ mm^3$

$b\tau_{c-c} = VQ/I = \frac{62.36\times10^4 V}{(199.43\times10^6)} = 1.866\times10^{-3} V\ [N/mm]$

$V_{equiv} = (800\times b)\tau_{cc} = 1.4926\ V\ [N]$

Bolts: $F_{bolt} = 450\ N \Rightarrow 1.4926 V = 2\times450 = 900 \Rightarrow V = 603\ N$

$P/2 = V \Rightarrow P = 1206\ N.$

8.41

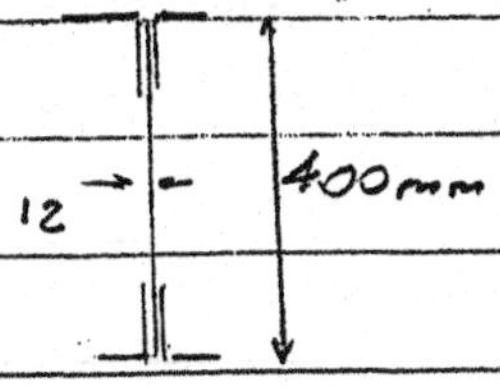

L 102×102×9.5 $I = 1.81\times10^6,\ A = 1845\ mm^2$ 29 mm

$I = 12(400)^3/12 + 4[(1.81\times10^6) + (1845)(171)^2] = 287\times10^6\ mm^4$

$Q = (1845)(171) = 3.155\times10^5\ mm^3$

$b\tau = VQ/I = \frac{(P)(3.155\times10^5)}{(287\times10^6)} = \frac{(240\times10^3)(3.155\times10^5)}{287\times10^6} = 263.8\ N/mm$

$s = 100\ mm$

$b\tau s = V = 263.8\times10^2\ [N]$

$A_b = \pi/4\,(20)^2 = 314.2\ mm^2;\ \tau = F/A = \frac{263.8\times10^2}{314.2\times10^2} = 83.97\ N/mm^2 = 83.97\ MPa$

8.42

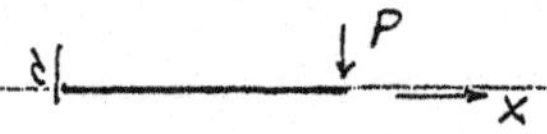

$M(x) = -P(L-x)$

a)

$$\sigma = \frac{Mh/2}{I(x)} = \frac{6M}{bh^2(x)} = \frac{-6P(L-x)}{bh^2(x)}\ ;\ \sigma_0 = \sigma(0) = \frac{-6PL}{bh^2(0)} = -\frac{6PL}{bh_0^2}$$

$$\sigma(x) = \sigma_0 \Rightarrow \frac{6P(L-x)}{bh^2(x)} = \frac{-6PL}{bh_0^2} \Rightarrow h^2(x) = \frac{h_0^2}{L}(L-x)$$

$$h(x) = h_0(1-x/L)^{1/2}$$

$$dh(x)/dx = -\frac{h_0}{2L}(1-x/L)^{-1/2} \leq \tan\theta$$

$$\therefore (1-x/L)^{1/2} \geq -\frac{h_0}{2L\tan\theta} \Rightarrow 1 - x/L \geq (h_0/2L)^2 \frac{1}{\tan^2\theta}$$

$$x/L \leq 1 - (h_0/2L)^2 \frac{1}{\tan^2 5^\circ} = 0.855$$

$$0 < x/L < 0.855$$

8.43

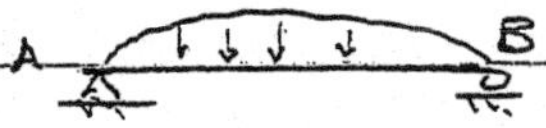

$q(x) = q_0 \sin \pi x/L$. $\quad R_A = q_0/2 \int_0^L \sin \pi x/L\, dx = q_0 L/\pi$

$M(0) = 0$

$$\therefore M(x) = \frac{q_0 L x}{\pi} - q_0 \int_0^x (x-\xi)\sin \pi\xi/L\, d\xi = \frac{q_0 L^2}{\pi^2}\sin \pi x/L \qquad M_{max}: M(L/2) = q_0 L^2/\pi^2$$

a). $\sigma_0 = \frac{6M(L/2)}{bh_0^2} \Rightarrow h_0^2 = \frac{6q_0 L^2}{\pi^2 b\sigma_0} \Rightarrow h_0 = L/\pi\left(\frac{6q_0}{\sigma_0 b}\right)^{1/2}$

b).

$$\sigma = \frac{M(x)h(x)/2}{bh^3(x)/12} = \frac{6M(x)}{bh^2(x)} = \sigma(x = L/2) = \frac{6q_0 L^2}{\pi^2 bh_0^2}$$

$$\therefore h^2(x) = \frac{M(x)\pi^2}{q_0 L^2} h_0^2 = h_0^2 \sin \pi x/L$$

$$h(x) = h_0(\sin \pi x/L)^{1/2}.$$

c). $dh/dx = (h_0/2)(\pi/L)(\cos \pi x/L)^{-1/2}$; $\quad dh/dx\big|_{\substack{max \\ (x=0)}} = h_0\pi/2L$

$$h_0\pi/2L < \tan 5^\circ \Rightarrow h_0^2 \leq \frac{4L^2}{\pi^2}\tan^2 5^\circ \Rightarrow \frac{6q_0L^2}{b\sigma_0} \leq \frac{4L^2}{\pi^2}\tan^2 5^\circ$$

$$q_0 \leq \frac{2b\sigma_0}{3\pi^2}\tan^2 5^\circ$$

d). $b = 10\text{ cm}$, $\sigma_0 = 10\text{ MPa} \Rightarrow q_0 = \frac{2(0.1)(10\times10^6)}{3\pi^2}\tan^2 5^\circ = 517\text{ N/m}$

$L = 3\text{ m} \quad h_0^2 = 6q_0L^2/b\sigma_0 = 2.79\times10^{-2} \Rightarrow h_0 = 0.167\text{ m} = 16.7\text{ cm}.$

8.44

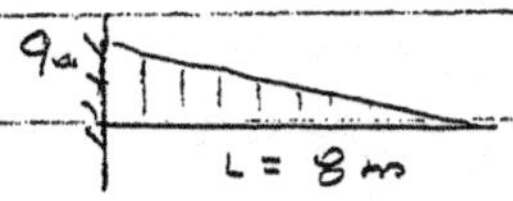

$V_{max} = W, \quad M_{max} = (q_o L/2)(L/3) = q_o L^2/6 = WL/3$

254W45: $S = 531 \times 10^3\ mm^3,\ A_W = [266 - 2(13)](6.6) = 1824\ mm^2$

$V = \tau A_W = (80)(1824) = 145{,}920\ N \qquad W = 145.9\ kN$

$M = \sigma S = (150)(531 \times 10^3) = 79.65 \times 10^6\ N\text{-}mm = 79.65 \times 10^3\ N\text{-}m$

$W = 3M_{max}/L = 3(79.65 \times 10^3)/8 = 29.88\ kN$

Flexure governs : $W_{all} = 29.88\ kN$.

8.45 a) $V_{max} = W/2, \quad M_{max} = qL^2/8 = WL/8$

$V_{max} = (d_w t_w)\tau_{allow} \Rightarrow W_s = 2A_w\tau$

$M_{max} = \sigma_{all} \cdot S \Rightarrow W_f = 8M/L = 8S\sigma_{all}/L$

b) $W_f < W_s \Rightarrow 8S\sigma_{all}/L < 2A_w\tau \Rightarrow L > 4S\sigma_{all}/A_w\tau_{all}$ (Flex governs)

$W_f > W_s \Rightarrow L < 4S\sigma_{all}/A_w\tau_{all}$ (Shear governs)

c). S203 × 34: $S = 265 \times 10^3\ mm^3,\ d = 203.2\ mm,\ t_f = 10.8,\ t_w = 11.2\ mm.$

Flex. governs: $W_f = 8(265 \times 10^3)(200)/(5 \times 10^3) = 84.8\ kN$

8.46 $\tau b = VQ/I$; At a: $Q = (10 \times 26)(17) = 4420\ mm^3$; $V_{BC} = 480\ N$

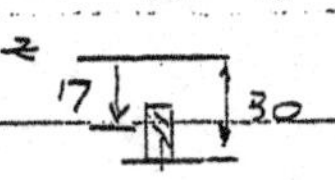

$\tau b = (480)(4420)/(2.0 \times 10^6) = 1.0608\ N/mm = 10.608\ N/cm$

$f = \tau b s \Rightarrow s = f/\tau b = 300/10.608 = 28.3\ cm$

At b: $Q = (4420) + (80 \times 10)(35) = 3.242 \times 10^4\ mm^3$

$\tau b = (480)(3.242 \times 10^4)/(2 \times 10^6) = 7.78\ N/mm = 77.8\ N/cm$

$s = 2f/\tau b = 2 \cdot 300/77.8 = 7.71\ cm.$

8.47

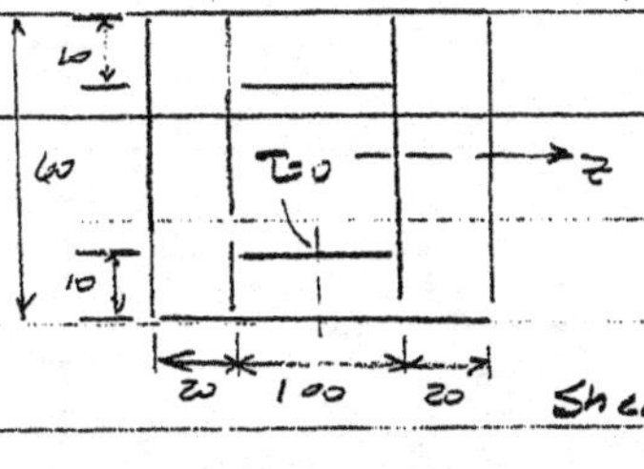

Flexure: $M = q_o L^2/8 \qquad I = [(140)(60)^3 - (100)(40)^3]/12 = 1.9867 \times 10^6\ mm^4$

$M_{max} = \sigma I/c = \dfrac{(25)(1.9867 \times 10^6)}{30} = 1.655 \times 10^6\ N\text{-}mm$

$M_{max} = 1655\ N\text{-}m$

$q_o = 8M/L^2 = 8(1655)/4 = 3311\ N/m \Rightarrow W = 6622\ N$

Shear (in glue):

$Q = (50 \times 10)(25) = 12.5 \times 10^3\ mm^3$

$V = \dfrac{\tau b I}{Q} = \dfrac{2(10)(1.9867 \times 10^6)}{12.5 \times 10^3} = 3178\ N$; $V = W/2$

$W = 2V = 6357\ N \qquad W_{all} = 6357\ N$

8.48

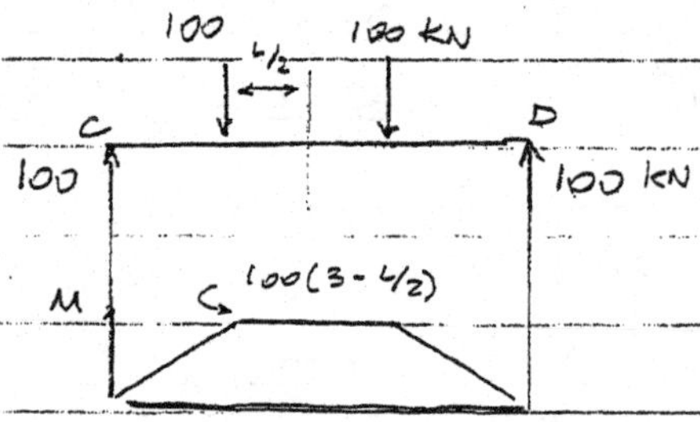

C-D: $V_{max} = 100$ kN

$M_{max} = 100(3 - L/2) = 300 - 50L$

W 254 x 67 $S = 805 \times 10^3$ mm^3

$d = 257$, $t_f = 15.7$, $t_w = 8.9$ mm.

$A_w = (257 - 31.4)(8.9) = 2008$ mm^2

$V_{all} = \tau A_w = (85)(2008) = 170$ kN ok

$M_{all} = \sigma S = (140)(805 \times 10^3) = 1.127 \times 10^8$ N-mm $= 112.7 \times 10^3$ N-m $= 112.7$ kN-m

$300 - 50 \cdot L = 112.7 \Rightarrow L = 187.3/50 = 3.75$ m.

AB: $M_{max} = PL/4 = 93.75$ kN-m $= 93.75 \times 10^6$ N-mm

$S = M/\sigma_{all} = 93.75 \times 10^6/140 = 669.6 \times 10^3$ mm^3 Choose: W356x45

8.49 $I = I_0 + 2[bt^3/12 + bt(\frac{d+t}{2})^2] = I_0 + 2btd^2[(t/d)^2/12 + (1+t/d)^2/4] \approx I_0 + btd^2/2$ $t/d \ll 1$

$S = \frac{I_0 + btd^2/2}{d/2 + t} = \frac{I_0 + btd^2/2}{d/2(1 + 2t/d)} \approx \frac{I_0 + btd^2/2}{d/2} = S_0 + btd$

8.50

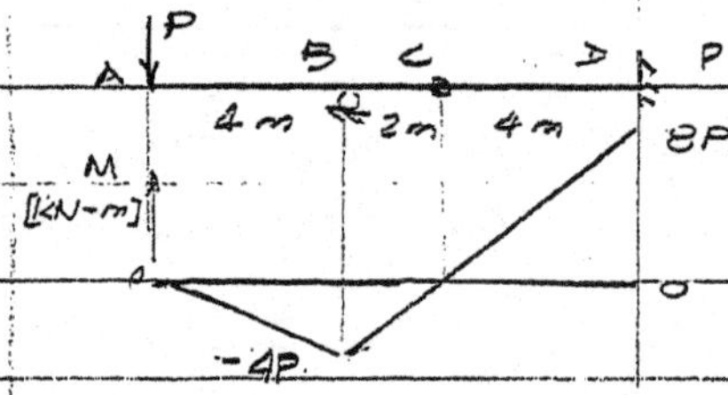

$P = 60$ kN: $|M_{max}| = 480$ kN-m $S = M/\sigma_{all}$

$\sigma = 160$ MPa; $S_0 = \frac{480 \times 10^3}{160 \times 10^6} = 3 \times 10^{-3}$ m^3 $= 3000 \times 10^3$ mm^3

Choose: S 610 x 149: $S_0 = 3260 \times 10^3$ mm^3; $d = 609.6$ mm

$b_f = 184$, $t_f = 22.1$, $t_w = 18.9$]

C-D: $M_{max} = 960$ kN-m. $\Rightarrow$ Require $S = 6000 \times 10^3$ mm^3

$S = S_0 + btd \Rightarrow \Delta S = btd$. Let $b = b_f \Rightarrow t = \Delta S / b_f d = \frac{2740 \times 10^3}{(184)(609.6)} = 24.4$ mm

Use $t = 25$ mm^2

Check: $I = 995 \times 10^6 + 2[\frac{(184)(25)^3}{12} + (184 \times 25)(327.3)^2] = 1981 \times 10^6$ mm^4

$S = I/c = \frac{1981 \times 10^6}{329.8} = 6005 \times 10^3$ mm^3.

$P = 120$ kN:

$V_{BD} = 240$ kN $Q = (25)(184)[329.8 - 12.5] = 1460 \times 10^3$ mm^3

$b\tau = VQ/I = \frac{(240 \times 10^3)(1460 \times 10^3)}{1981 \times 10^6} = 176.88$ N/mm.

$F_{bolt} = \tau A_b = (\frac{\pi d^2}{4})\tau = (100\pi)(50) = 5000\pi$ N

$(b\tau)s = 2F = 10^4\pi \Rightarrow s = 10^4\pi/176.88 = 177.6$ mm $= 17.8$ cm.

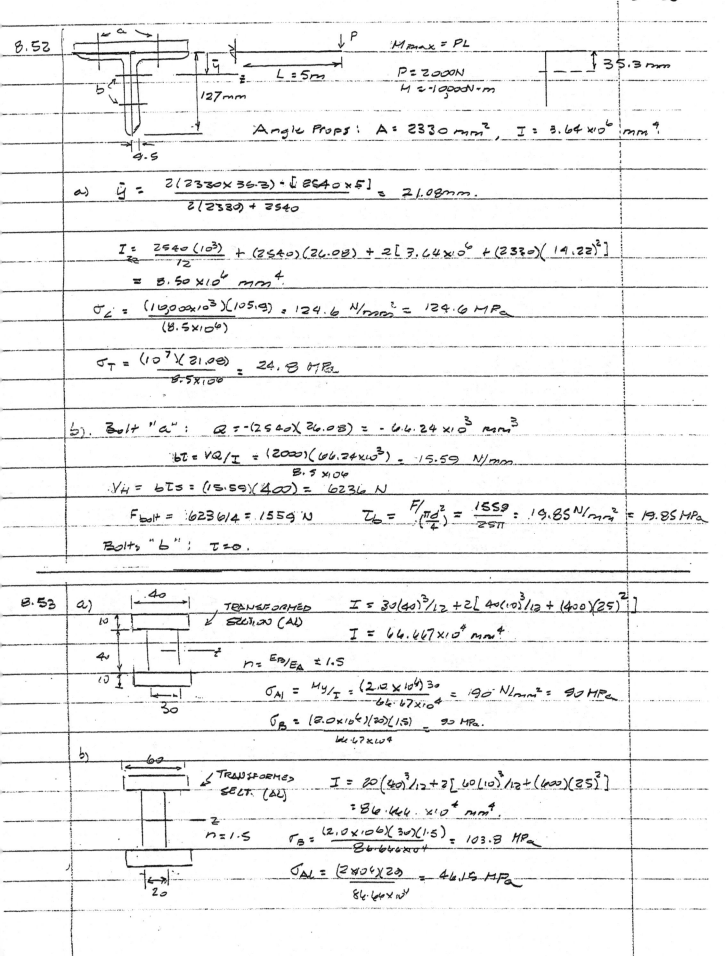

8.52

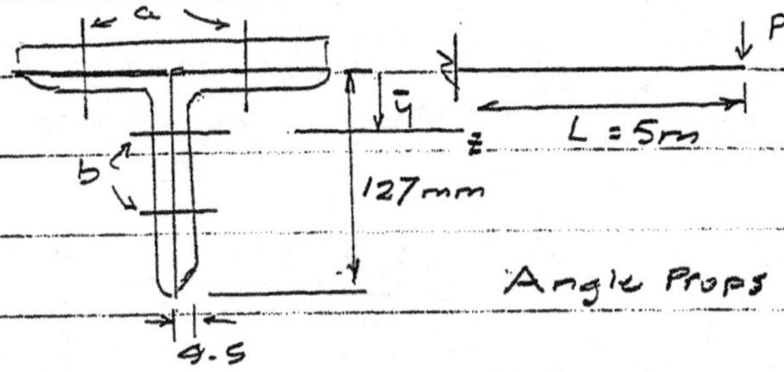

$M_{max} = PL$

$P = 2000N$

$M = 10{,}000N\cdot m$

Angle Props: $A = 2330\ mm^2$, $I = 3.64 \times 10^6\ mm^4$

a) $\bar{y} = \dfrac{2(2330 \times 35.3) - [2540 \times 5]}{2(2330) + 2540} = 21.08mm.$

$I_{zz} = \dfrac{2540(10^3)}{12} + (2540)(26.08) + 2[3.64\times10^6 + (2330)(14.22)^2]$

$= 8.50 \times 10^6\ mm^4$

$\sigma_C = \dfrac{(10{,}000\times10^3)(105.9)}{(8.5\times10^6)} = 124.6\ N/mm^2 = 124.6\ MPa$

$\sigma_T = \dfrac{(10^7)(21.08)}{8.5\times10^6} = 24.8\ MPa$

b). Bolt "a": $Q = -(2540)(26.08) = -66.24\times10^3\ mm^3$

$b\tau = VQ/I = \dfrac{(2000)(66.24\times10^3)}{8.5\times10^6} = 15.59\ N/mm$

$V_H = b\tau s = (15.59)(400) = 6236\ N$

$F_{bolt} = 6236/4 = 1559\ N \qquad \tau_b = \dfrac{F}{(\frac{\pi d^2}{4})} = \dfrac{1559}{25\pi} = 19.85\ N/mm^2 = 19.85\ MPa$

Bolts "b": $\tau = 0$.

8.53 a)

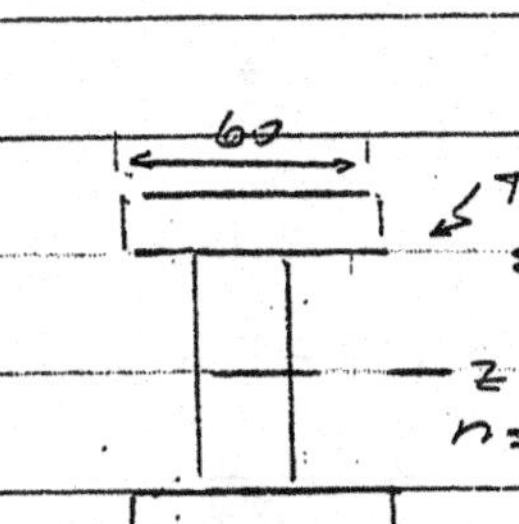

TRANSFORMED SECTION (AL)

$I = 30(40)^3/12 + 2[40(10)^3/12 + (400)(25)^2]$

$I = 66.667\times10^4\ mm^4$

$n = E_B/E_A = 1.5$

$\sigma_{Al} = My/I = \dfrac{(2.0\times10^6)30}{66.67\times10^4} = 190\ N/mm^2 = 90\ MPa$

$\sigma_B = \dfrac{(2.0\times10^6)(20)(1.5)}{66.67\times10^4} = 90\ MPa.$

b)

TRANSFORMED SECT. (AL)

$I = 20(40)^3/12 + 2[60(10)^3/12 + (600)(25)^2]$

$= 86.664\times10^4\ mm^4.$

$n = 1.5 \qquad \sigma_B = \dfrac{(2.0\times10^6)(30)(1.5)}{86.664\times10^4} = 103.8\ MPa$

$\sigma_{AL} = \dfrac{(2\times10^6)(20)}{86.66\times10^4} = 46.15\ MPa$

8.54 $n = E_S / E_A = 200/70 = 2.857$

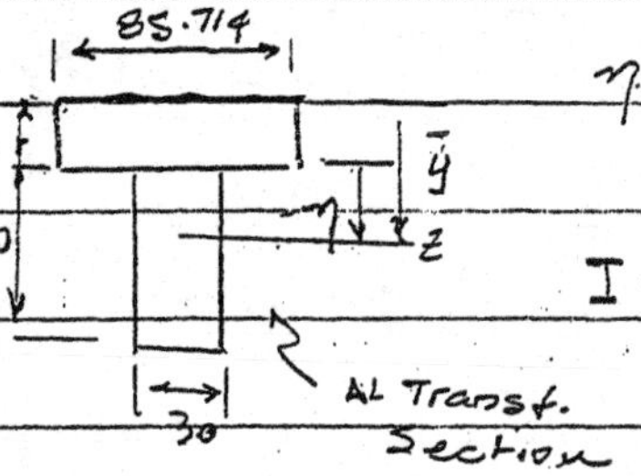

$$\eta = \frac{(85.714)(20)(-10) + (1200)(20)}{1714.29 + 1200} = 2.353 \Rightarrow \bar{y} = 22.353 \text{ mm.}$$

$$I = \frac{(30)(40)^3}{12} + (1200)(17.647)^2 + \frac{85.714 \times 20^3}{12} + (1714.3)(12.353)^2$$

$$= 85.244 \times 10^4 \text{ mm}^4.$$

a). $$\sigma_A = \frac{(1200 \times 10^3)(37.747)}{85.244 \times 10^4} = 53.13 \text{ N/mm}^2 = 53.13 \text{ MPa (tens.)}$$

$$\sigma_S = \frac{(1200 \times 10^3)(-22.353)(2.857)}{85.244 \times 10^4} = -89.9 \text{ MPa (comp.)}$$

b) $$R = EI/M = \frac{(70 \times 10^9)(85.244 \times 10^{-8})}{1200} = 49.7 \text{ m.}$$

8.55 $n = 200/12 = 16.67$

$$I = \frac{(150)(250)^3}{12} + 2(16.67)\left[125 \times 10^3/12 + (1250)(130)^2\right] = 89.997 \times 10^7 \text{ mm}^4$$ WOOD TRANSFORM SECT.

$M = \sigma I / c$

Wood: $$M = \frac{(12)(89.997 \times 10^7)}{125} = 8.64 \times 10^7 \text{ N-mm} = 86.4 \text{ kN-m.}$$

Steel: $$M = \frac{(200)(89.997 \times 10^7)}{130} \cdot \frac{1}{16.67} = 8.305 \times 10^7 \text{ N-mm} = 83.06 \text{ kN-m.}$$

$M = 83.06$ kN-m (steel governs)

$$R = EI/M = \frac{(12 \times 10^9)(89.997 \times 10^{-5})}{83.06 \times 10^3} = 130 \text{ m.}$$

8.56 $n = 200/10 = 20$

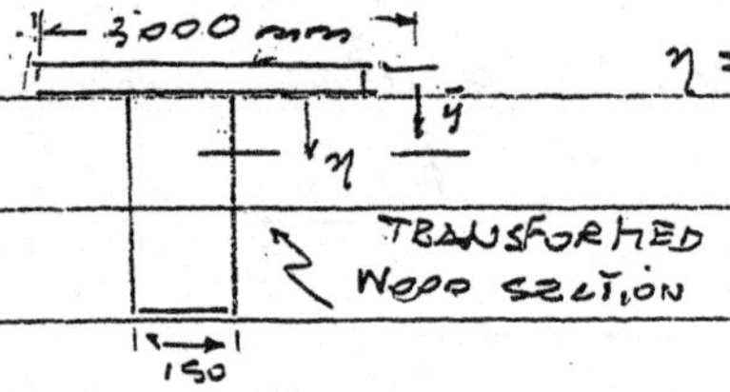

$$\eta = \frac{(3000 \times 15)(-7.5) + (350 \times 150)(175)}{(4.5 \times 10^4) + (5.25 \times 10^4)} = \frac{8.85 \times 10^6}{9.75 \times 10^4} = 90.769$$

$\bar{y} = 105.769$

$$\frac{(\sigma_W)_{max}}{(\sigma_S)_{max}} = \frac{350 - \eta}{n \bar{y}} = \frac{1}{20} \cdot \frac{259.23}{105.769} = 0.1225$$

$(\sigma_W)_{max} = 0.1225\ \sigma_S = 9.19$ MPa.

8.57

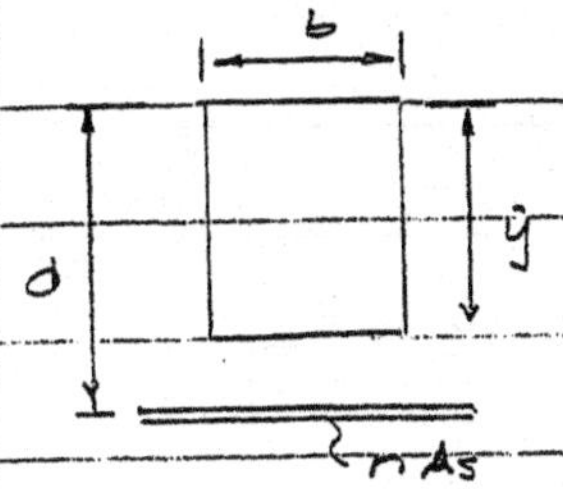

$$\bar{y} = \frac{b\bar{y}^2/2 + nA_s d}{b\bar{y} + nA_s} \Rightarrow b\bar{y}^2/2 + nA_s\bar{y} - nA_s d = 0$$

$$\bar{y} = \frac{nA_s}{b}\left[\left(1 + 2bd/nA_s\right)^{1/2} - 1\right]$$

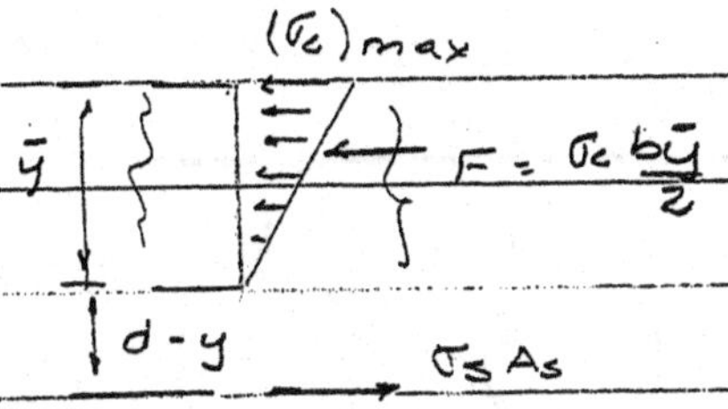

$$\Sigma F = 0 \Rightarrow \sigma_c b\bar{y}/2 = \sigma_s A_s \Rightarrow \sigma_s = (b\bar{y}/2A_s)\sigma_c$$

$$M = F(2\bar{y}/3) + \sigma_s A_s(d-\bar{y})$$

$$= (b\bar{y}^2/3)\sigma_c + A_s(d-\bar{y})\sigma_s$$

$$M = b\bar{y}\left[\bar{y}/3 + (d-\bar{y})/2\right]\sigma_c$$

$$\sigma_c = \frac{6M}{b\bar{y}}\left[3d - \bar{y}\right]^{-1}; \quad \sigma_s = \frac{3M}{A_s}(3d-\bar{y})^{-1}$$

8.58 $M = PL/4 = 20\ \text{kN-m} = 20\times10^6\ \text{N-mm}.$

$n = 200/15 = 13.33$

$$\bar{y} = \frac{nA_s}{b}\left[\left(1 + 2bd/nA_s\right)^{1/2} - 1\right]$$

$A_s = 4(\pi d^2/4) = \pi d^2 = 225\pi$; $d = 275$ mm, $b = 200$ mm.

$$\bar{y} = \frac{(13.33)(225\pi)}{200}\left[\left(1 + \frac{(400)(275)}{(13.33)(225\pi)}\right)^{1/2} - 1\right] = 120.622\ \text{mm}.$$

$$\sigma_c = \frac{6M}{b\bar{y}}(3d-\bar{y})^{-1} = \frac{6(20\times10^6)}{(200)(120.622)}\left[3(275) - 120.622\right]^{-1} = 7.06\ \text{N/mm}^2 = 7.06\ \text{MPa}$$

$$\sigma_s = \frac{3M}{A_s}(3d-\bar{y})^{-1} = \frac{3(20\times10^6)}{(225\pi)}\left[3(275) - 120.622\right]^{-1} = 120.5\ \text{MPa}$$

8.59 $\varepsilon = y/R \Rightarrow 1/R = \varepsilon_c/\bar{y} = \varepsilon_s/(d-\bar{y}) \Rightarrow \dfrac{\sigma_c}{E_c\bar{y}} = \dfrac{\sigma_s}{(d-\bar{y})E_s} \Rightarrow \dfrac{d-\bar{y}}{\bar{y}} = \dfrac{E_c\sigma_s}{E_s\sigma_c}$

$$d/\bar{y} = 1 + E_c\sigma_s/E_s\sigma_c$$

$$\therefore \bar{y} = \frac{d}{1 + E_c\sigma_s/E_s\sigma_c}$$

8.60 $M_z = -(750)(180) = -135\times10^3$ N-m, $M_y = (250)(220) = 55\times10^3$ N-m

$\sigma_x = P/A + M_z y/I_{zz} + M_y z/I_{yy}$

$A = 1920$ mm², $I_{zz} = 60(32)^3/12 = 163.84\times10^3$ mm⁴; $I_{yy} = (32)(60)^3/12 = 576\times10^3$ mm⁴.

(a) $\sigma_x = \dfrac{500}{1920} + \dfrac{(-135\times10^3)(-16)}{163.8\times10^3} = 13.447$ N/mm² $= 13.45$ MPa

(b) $\sigma_x = \dfrac{500}{1920} + \dfrac{(-135\times10^3)(-16)}{163.8\times10^3} + \dfrac{(55\times10^3)(15)}{576\times10^3} = 14.88$ MPa

(c) $\sigma_x = \dfrac{500}{1920} + \text{"} + \dfrac{(55\times10^3)(30)}{576\times10^3} = 16.31$ N/mm² $= 16.31$ MPa

8.61 $(\tau)_V = VQ/Ib$; $Q_{a\text{-}a} = \left(\dfrac{\pi R^2}{2}\right)(4R/3\pi) = 2R^3/3\pi$

$I = \pi R^4/4$

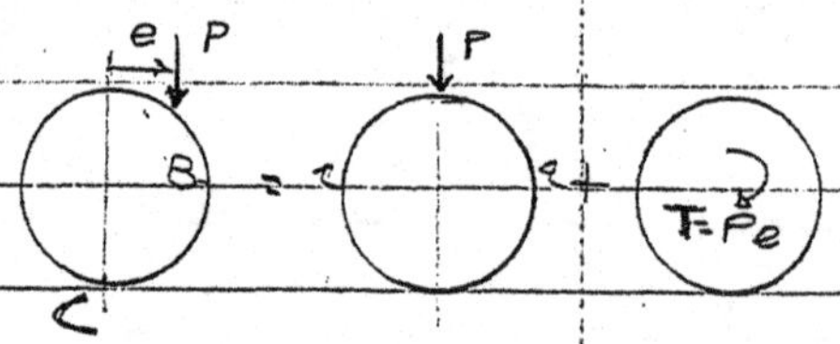

$(\tau_{a\text{-}a})_V = \dfrac{P(2R^3/3\pi)}{(\pi R^4/4)(2R)} = \dfrac{4P}{3\pi^2R^2}$ $\sigma_x = 0$

$(\tau)_T = \dfrac{2TR}{\pi R^4} = \dfrac{2T}{\pi R^3} = \dfrac{2Pe}{\pi R^3}$. At B: $\tau_B = \dfrac{P}{\pi R^2}\left[\dfrac{4}{3\pi} + \dfrac{2e}{R}\right] = \dfrac{P}{3\pi^2R^3}(4R + 2\pi e)$

At C:

$(\tau_B)_{max\,(e=R)} = \dfrac{P}{3\pi^2R^2}(4+2\pi) = \dfrac{2P}{3\pi^2R^2}(2+\pi)$

$\sigma_x = \dfrac{4MR}{\pi R^4} = \dfrac{4PL}{\pi R^3}$, $\tau_V = 0$

$\tau_{max} = [(\sigma_x/2)^2 + \tau^2]^{1/2} = \left[4\left(\dfrac{PL}{\pi R^3}\right)^2 + 4\left(Pe/\pi R^3\right)^2\right]^{1/2}$

$\tau_{max} = \dfrac{2P}{\pi R^3}\left[L^2 + e^2\right]^{1/2} = \dfrac{2P}{\pi R^2}\left(L^2/R^2 + 1\right)^{1/2} > \tau_{B\,max}$

$e = R$, $L/R = 12$; $\tau_{max} = 2\sqrt{145}\,P/\pi R^2$

8.62 a). $\sigma_x| = \dfrac{P\cos\alpha}{A} + \dfrac{[(P\cos\alpha)e + (P\sin\alpha)L]\,y_0}{I} = 0$

$R^2\cos\alpha + 4[(\cos\alpha)e + (\sin\alpha)L]\,y_0 = 0 \Rightarrow y_0 = -\dfrac{R^2\cos\alpha}{4[e\cos\alpha + L\sin\alpha]}$

$y_0 = -\dfrac{R^2}{4(e + L\tan\alpha)}$

b) $\alpha = 0$ $y|_C = -R \Rightarrow e = R/4$

c). $\sigma_x|_D = P/A + PR\,e/I = \dfrac{P}{\pi R^4}(R^2 + R^2) = 2P/\pi R^2$; $\tau = TR/J = \dfrac{PR^2/4}{\pi R^4/2} = P/2\pi R^2$

$\tau_{max} = [(\sigma_x/2)^2 + \tau_{xy}^2]^{1/2} = \dfrac{P}{\pi R^2}[1 + 1/4]^{1/2} = \dfrac{\sqrt{5}P}{2\pi R^2}$ $\tan 2\theta = -\dfrac{\sigma_x}{2\tau_{xy}} = -2 \Rightarrow \theta = -31.8° + 90$

$\theta = 58.3°$

8.63

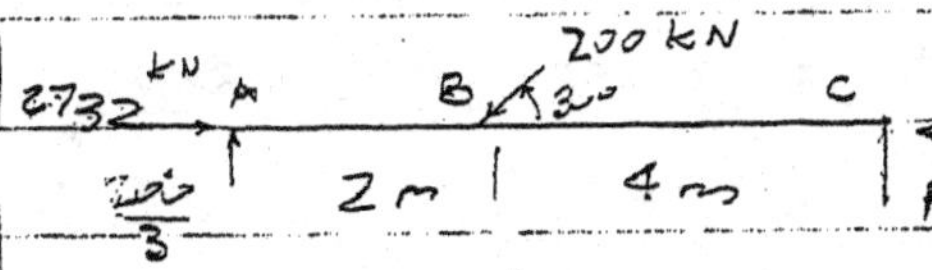

$F = 100(1+\sqrt{3}) = 2.732$ kN, $M = \frac{400}{3}$ kN-m $= \frac{400\times10^6}{3}$ N-mm

$\sigma_x = -\frac{2.732\times10^3}{A} \pm \frac{400\times10^6/3}{S}$

W305 × 97 $A = 12{,}325\ mm^2$, $S = 1440\times10^3\ mm^3$

B-C:

$\sigma_x = -\frac{100\times10^3}{12.325\times10^3} \pm \frac{400\times10^6}{3(1440\times10^3)} = -8.1136 \pm 92.59$

A-B: $\sigma_x = -(2732\times10^3)/12.732\times10^3) \pm 92.59 = -22.17 \pm 92.59$

At $x = 2-\varepsilon$: $\sigma_x = -114.75$ MPa (top), $+70.43$ MPa (bott.)

$x = 2+\varepsilon$: $\sigma_x = -100.71$ MPa (top), $+84.48$ MPa (bott.)

8.64

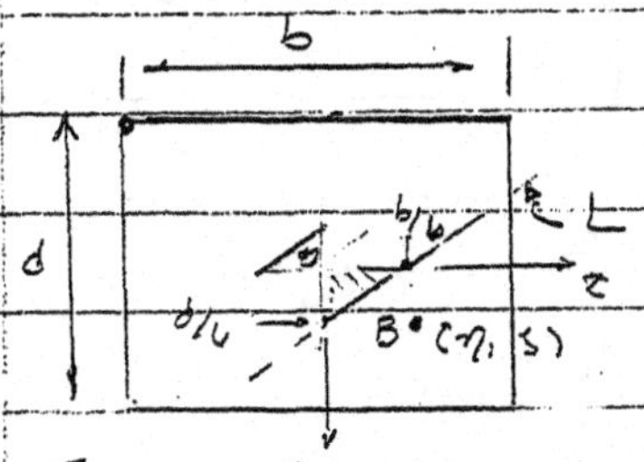

$\sigma_x = -\frac{P}{A} + \frac{M_z y}{I_{zz}} + \frac{M_y z}{I_{yy}}$

P comp. at pt. B: $M_z = -P\eta$, $M_z = -P\zeta$.

$\sigma_x \le 0 \Rightarrow y$ $\therefore \sigma_x = -P\left(\frac{1}{A} + \eta y/I_{zz} + \zeta z/I_{yy}\right) \le 0$

$\Gamma = \frac{1}{A} + \eta y/I_{zz} + \zeta z/I_{yy} \ge 0$. Smallest value of Γ: $y = -d/2$, $z = -b/2$

$\therefore \frac{1}{A} - \eta d/2I_{zz} - \zeta b/2I_{yy} \ge 0 \Rightarrow \frac{1}{bd} - 6\eta/bd^2 - 6\zeta/db^2 \ge 0$

$\therefore$

$6\eta/d + 6\zeta/b \le 1 \Rightarrow \eta/d + \zeta/d - 1/6 = 0$ is eq. of line L

$\therefore$ core lies in core [shaded]. By Symmetry other parts are like Fig (8.12.6)

8.65

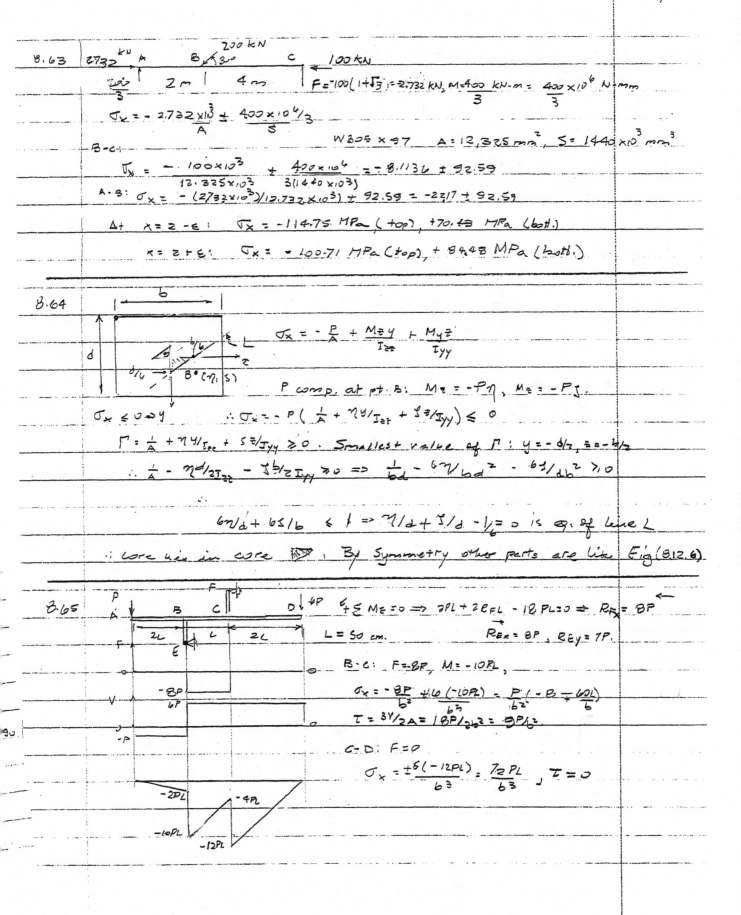

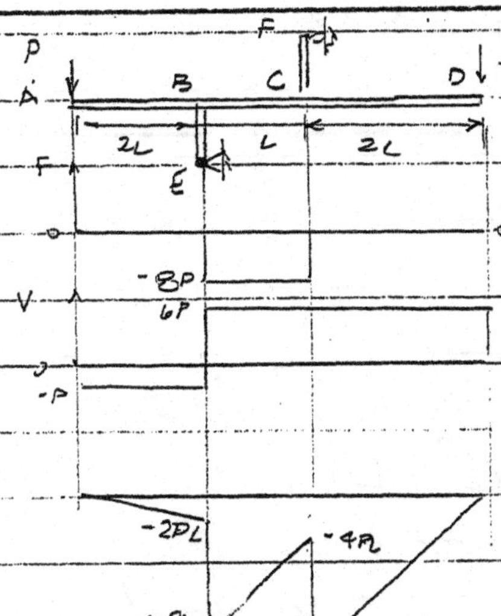

$\circlearrowleft + \sum M_E = 0 \Rightarrow 2PL + 2R_{FL} - 18PL = 0 \Rightarrow R_{Fx} = 8P$

$L = 50$ cm. $R_{Ex} = 8P$, $R_{Ey} = 7P$.

B-C: $F = -8P$, $M = -10PL$,

$\sigma_x = -\frac{8P}{b^2} \pm \frac{6(-10PL)}{b^3} = \frac{P}{b^2}\left(-8 \mp \frac{60L}{b}\right)$

$\tau = 3V/2A = 18P/2b^2 = 9P/b^2$.

C-D: $F = 0$

$\sigma_x = \pm\frac{6(-12PL)}{b^3} = \frac{72PL}{b^3}$, $\tau = 0$

8.66 $AE = (4^2+3^2+1.5^2)^{1/2} \times 10^2 = 522\,mm$. $P_x = -150\,P/AE = 0.2873P$; $P_y = -300P/AE = -0.5747P$; $P_z = -0.7663\,P$

$\underset{\sim}{M} = (150\,\underset{\sim}{i} - 75\,\underset{\sim}{j}) \times (-1.5\underset{\sim}{i} - 3\underset{\sim}{j} - 4\underset{\sim}{k}) \times 100\,P/AE = (57.47\,\underset{\sim}{i} + 114.94\,\underset{\sim}{j} - 107.76\,\underset{\sim}{k})P$ (Vector conven)

$B: \sigma_x = -0.2873P/A - 107.76\,\dfrac{Pd}{2I_{xx}}$ ‖ $C: \sigma_x = -0.2873P/A + 114.9\,Pd/2I_{yy}$

$\tau = 4/3\,P_z/A + \dfrac{M_x d/2}{J}$ ‖ $\tau = 4/3\,\dfrac{P_y}{A} + \dfrac{M_x d/2}{J}$

$(M_x = T)$ $P = 500\,N$; $d = 20\,mm$, $A = \pi d^2/4 = 314.16$, $I_{xx} = I_{yy} = \pi d^4/64 = 7{,}854\,mm^4$

$\therefore$ At B: $\sigma_x = -68.06\,MPa$; $\tau = 19.92\,MPa$ $J = 15{,}708\,mm^4$

At C: $\sigma_x = 72.72\,MPa$, $\tau = 19.52\,MPa$

8.67

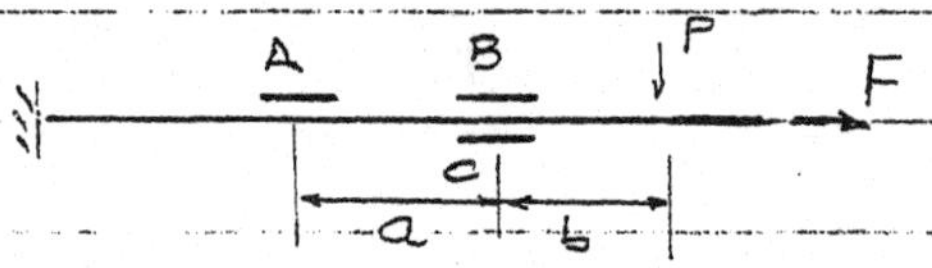

$\sigma_x|_A = F/A + \dfrac{P(a+b)c}{I} = \varepsilon_A E$

$\sigma_x|_{B,C} = F/A \pm \dfrac{Pbc}{I} = (\varepsilon_B, \varepsilon_C)E$

$c = d/2$, $a = 3\,cm$, $E = 70\,GPa = 70\times10^5\,N/cm^2$, $I = d^4/12$.

$A: F/d^2 + \dfrac{6P(3+b)}{d^3} = (70\times10^5)(550\mu) \Rightarrow Fd + 6P(3+b) = (38.5\times10^2)d$ (i)

$B: F/d^2 + 6Pb/d^3 = (70\times10^5)(400\mu) \Rightarrow Fd + 6Pb = (28.0\times10^2)d$ (ii)

$C: F/d^2 - 6Pb/d^3 = (70\times10^5)(-300\mu) \Rightarrow Fd - 6Pb = -(21.0\times10^2)d$ (iii)

$(ii) + (iii) \Rightarrow 2F = 7\times10^2 \Rightarrow F = 3.5\times10^2 = 350\,N$

$d = 1\,cm$.

$(i) - (ii) \Rightarrow 18P = (10.5\times10^2)d = 10.5\times10^2 = 1050\,N. \Rightarrow P = 58.3\,N$

$\therefore \Rightarrow b = \dfrac{(28\times10^2 - F)d}{6P} = \dfrac{[28\times10^2 - 3.5\times10^2]d}{6\cdot58.3} = 7.0\,cm$

8.68

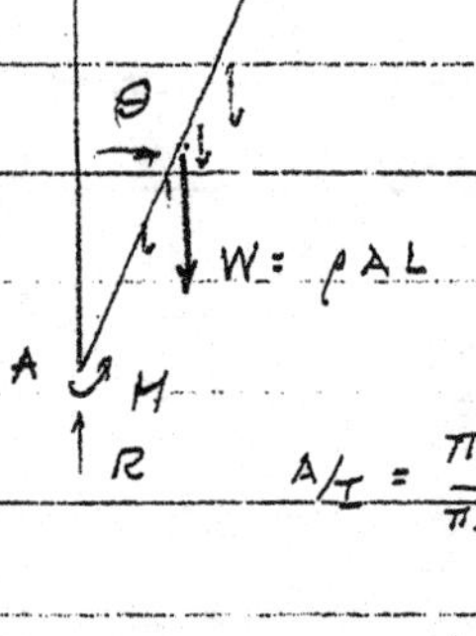

$M = W\dfrac{L}{2}\sin\theta = \dfrac{\rho AL^2}{2}\sin\theta$

$\sigma_x = -\dfrac{W}{A}\cos\theta + \dfrac{Md/2}{I} = -\dfrac{\rho AL}{A}\cos\theta + \dfrac{\rho AL^2 d}{4I}\sin\theta$

$= \rho L\left[-\cos\theta + AdL/4I\,\sin\theta\right]$

$A/I = \dfrac{\pi d^2/4}{\pi d^4/64} = 16/d^2$ $\sigma_x = \rho L\left[-\cos\theta + 4L/d\,\sin\theta\right] = 0$

$\tan\theta = d/4L \Rightarrow \theta = \tan^{-1}(d/4L)$

8.69 $V = P, \quad T = PR$

$$\tau = \frac{kP}{A} + \frac{Td}{2J} = \frac{kP}{A} + \frac{PRd}{2J} = P\left[k\frac{4}{\pi d^2} + \frac{16Rd}{\pi d^4}\right] = \frac{4P}{\pi d^3}(kd + 4R)$$

$$k: \quad Q = (4R/3\pi)(\pi R^2/2) = 2R^3/3 \;;\; Q/I = \frac{2R^3/3}{\pi R^4/4} = 8/3\pi R$$

$$\tau = VQ/2IR = \frac{4}{3}\frac{V}{\pi R^2} \Rightarrow k = 4/3$$

$$\therefore \quad \tau = \frac{16P}{\pi d^3}(d/3 + R) = \frac{16P}{3\pi d^3}(d + 3R) = \frac{16RP}{3\pi d^3}(3 + d/R).$$

8.70 a) $M_E = \frac{\sigma_0 I}{2h/3} = \frac{3\sigma_0 I}{2d} \qquad I_{NA} = ad^3/36$

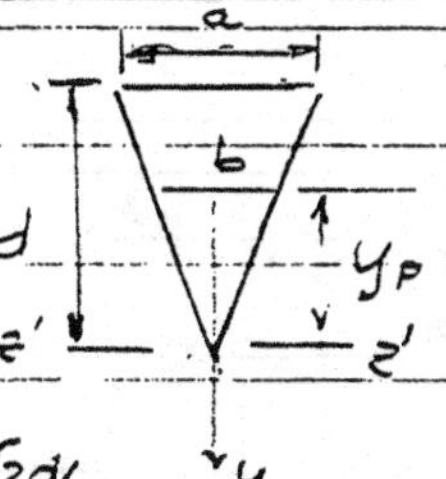

$$M_E = \frac{3\sigma_0 ad^3}{72d} = \frac{\sigma_0 ad^2}{24}$$

b). $b = \frac{y_p a}{d} \qquad A_T = A_C \Rightarrow \frac{by_p}{2} = \frac{ad}{2} \Rightarrow y_p = d/2 \Rightarrow y_p = \sqrt{2}d/2$

$$M_p\Big|_{z'z'} = \left(\frac{\sigma_0 ad}{2}\right)(2d/3) - 2\left(\frac{\sigma_0 by}{2}\right)\left(\frac{2y_p}{3}\right) = \frac{\sigma_0}{3}[ad^2 - 2by_p^2] = \frac{\sigma_0}{3}(ad^2 - bd^2)$$

$$M_p = \frac{\sigma_0 d^2}{3}(a - \sqrt{2}a/2) = \frac{\sigma_0 d^2 a}{6}(2 - \sqrt{2}) = 0.0976\ \sigma_0 ad^2$$

c) $M_p/M_E = 4(2 - \sqrt{2}) = 2.343$

8.71

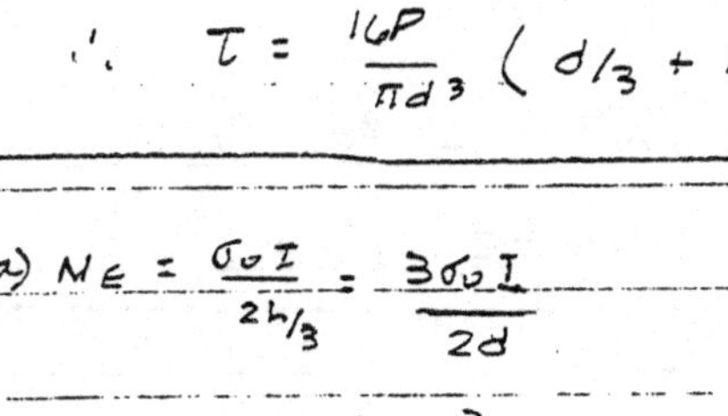

$\ell/\varepsilon_0 = 5/\varepsilon \Rightarrow \ell = 5\varepsilon_0/\varepsilon \;;\; \varepsilon_0 = \sigma_0/E = 10^{-3}$

$\ell = 5\times 10^{-3}/4\times 10^{-3} = 1.25$ cm.

$dy = 3.75$ cm.

$$M = b\int_{-5}^{5}\sigma(y)\cdot y\,dy = b\left\{-\sigma_0\int_{-5}^{-\ell} y\,dy + \frac{\sigma_0}{\ell}\int_{-\ell}^{\ell} y^2\,dy + \sigma_0\int_{\ell}^{5} y\,dy\right\}$$

$$= b\sigma_0\left\{\frac{y^2}{2}\Big|_{-5}^{-\ell} + \frac{1}{\ell}\frac{y^3}{3}\Big|_{-\ell}^{\ell} + \frac{y^2}{2}\Big|_{\ell}^{5}\right\}$$

$$= b\sigma_0\left\{-\frac{1}{2}(\ell^2 - 25) + \frac{2}{3\ell}\ell^3 + \frac{1}{2}(25 - \ell^2)\right\}$$

$$= b\sigma_0\{25 - \ell^2 + 2\ell^2/3\} = b\sigma_0(25 - \ell^2/3)$$

$b = 10,\ \ell = 1.25 \qquad M = 10\sigma_0[25 - 0.5208] = 244.48\,\sigma_0$

8.72 W762×196: $\sigma_0 = 250$ MPa; $S = 6225\times10^3\ mm^3$

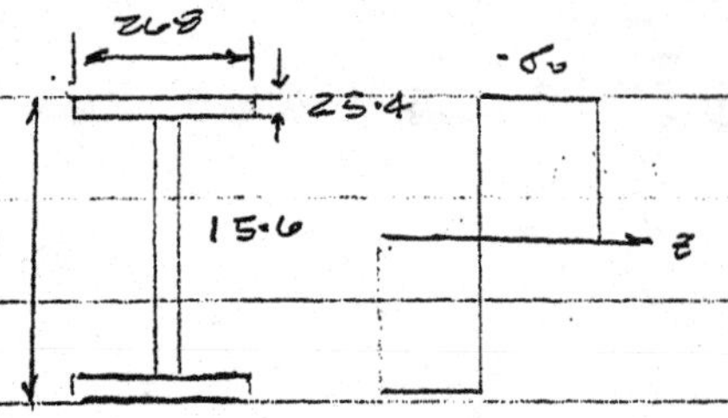

$M_E = (250)(6225\times10^3) = 1.556\times10^9$ N-mm

$M_E = 1556$ kN-m

$M_P = 2[M_F + M_W]$

$M_P = 2\{[\sigma_0(248\times25.4)(385-12.7)] + \frac{1}{2}[(\frac{770-50.8}{2})^2\times(15.6)]\sigma_0\}$

$= 2\sigma_0\{(2.534\times10^6) + (2.2819\times10^6)\}$

$= 2\sigma_0\{4.8162\times10^6\} = 2(250)(4.8162\times10^6) = 2.408\times10^9$ N-mm

$M_P = 2408$ kN-m.

8.73

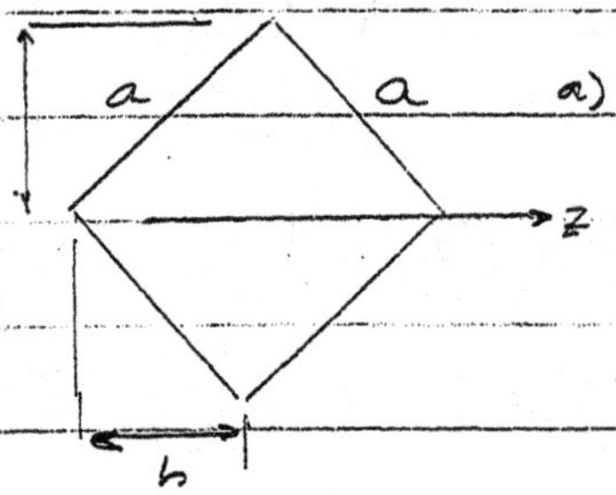

a) $M_E = \sigma_0 I/c$, $c = h = \sqrt{2}a/2$

$I = 4\left(\frac{h^4}{12}\right) = h^4/3$

$M_E = \frac{\sigma_0 h^4/3}{h} = \frac{\sigma_0 h^3}{3} = \frac{\sqrt{2}a^3\sigma_0}{12}$

b). $M_P = 4[\sigma_0(h^2/2)(h/3)] = 2\sigma_0 h^3/3 = \sqrt{2}a^3\sigma_0/6$

8.74

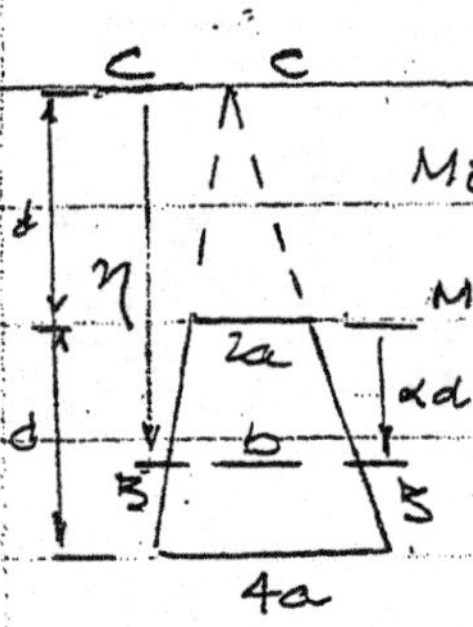

$M_E = \frac{13}{30}\sigma_0 a d^2$ [See Problem 8.27]

M_P: Locate BB which bi-sects area.

$\eta/b = 2d/4a \Rightarrow b = \frac{2a}{d}(1+\alpha)d = 2a(1+\alpha)$, $0 \le \alpha \le 1$

$\eta = (1+\alpha)d$ $A_T = A_C \Rightarrow (2a+b)rd = (b+4a)(1-r)d$

$\Rightarrow (2+r)r = (3+r)(1-r) \Rightarrow 2r^2 + 4r - 3 = 0$

$r = \frac{1}{2}(\sqrt{10}-2)$

$M_P|_{c-c} = \sigma_0\left[\{[(4a)(2d)/2](4d/3) - (\eta b/2)(2\eta/3)\} - \{(\frac{\eta b}{2})(2\eta/3) - (2ad/2)(2d/3)\}\right]$

$= \sigma_0\{16ad^2/3 + \eta^2 b(2/3) + 2ad^2/3\} = \sigma_0\{16ad^2 - \frac{10}{4}d^2(\sqrt{10}a)2 + 2ad^2\}\frac{2}{3}$

$= 2\sigma_0 ad^2/3\{8 - 10\sqrt{10}/4 + 1\} = \frac{\sigma_0 ad^2}{3}(18 - 5\sqrt{10})$

$M_P/M_E = \frac{(18-5\sqrt{10})/3}{13/30} = \frac{10}{13}(18-5\sqrt{10}) = 1.6835$

8.75 a)

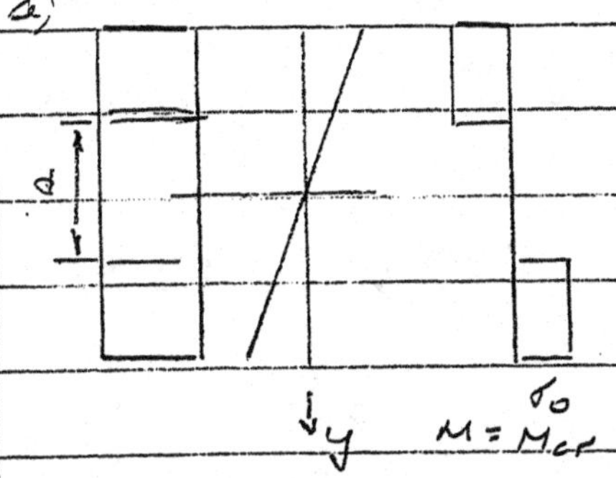

$$M = \iint \sigma(y)\cdot y\,dy = 2b\int_{-d(1/2+\alpha)}^{0} \sigma(y)\,dy$$

At $M = M_{cr}$

$\sigma(y) = 0, \quad -d/2 < y < 0$

$\sigma(y) = -\sigma_0, \quad -d(1/2+\alpha) < y < -d/2$

a)

$$\therefore M_{cr} = 2b\int_{-d(1/2+\alpha)}^{-d/2} -\sigma_0 y\,dy = -2b\sigma_0\left[y^2/2\right]_{-d(1/2+\alpha)}^{-d/2}$$

$$M_{cr} = -b\sigma_0\left[d^2/4 - d^2(1/2+\alpha)^2\right]$$

$$= -bd^2\sigma_0\left[1/4 - (1/2+\alpha)^2\right] = (\alpha^2+\alpha)bd^2\sigma_0 = \alpha(\alpha+1)bd^2\sigma_0$$

b) $M > M_{cr}$: $\quad \varepsilon = y/R \Rightarrow \sigma = Ey/R \quad ; \quad -d/2 < y < 0$

$\sigma = -\sigma_0 \quad ; \quad -d(1/2+\alpha) < y < -d/2$

$$M = 2b\left\{-\sigma_0\int_{-d(1/2+\alpha)}^{-d/2} y\,dy + \frac{E}{R}\int_{-d/2}^{0} y^2\,dy\right\}$$

$$= 2b\left\{-\sigma_0 y^2/2\Big|_{-d(1/2+\alpha)}^{-d/2} + \frac{E}{3R}y^3\Big|_{-d/2}^{0}\right\} = 2b\left\{-\frac{\sigma_0 d^2}{2}\left[1/4 - (1/2+\alpha)^2\right] + \frac{Ed^3}{24R}\right\}$$

$$M = b\left\{\sigma_0 d^2(\alpha^2+\alpha) + Ed^3/12R\right\}$$

$$K = 1/R = \frac{12}{Ebd^3}\left[M - \alpha(\alpha+1)bd^2\sigma_0\right] = \frac{12}{Ebd^3}(M - M_{cr})$$

8.76

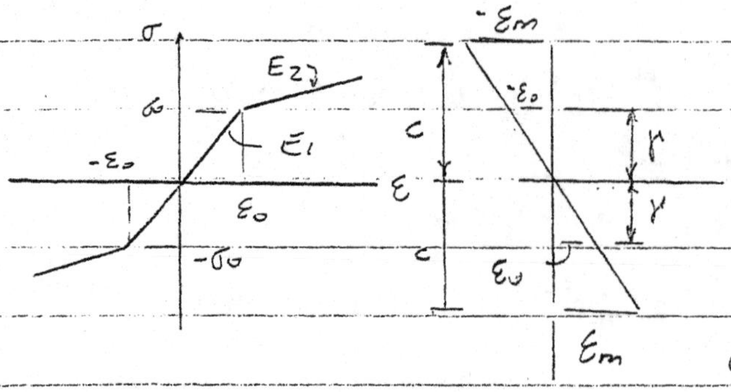

$\gamma = y\varepsilon_0/\varepsilon = c\,\varepsilon_0/\varepsilon_m = \sigma_0 c/E_1\varepsilon_m$ (I)

$\sigma = E_1\varepsilon = \left(\frac{E_1\varepsilon_m}{c}\right)y \;,\quad 0 \le |y| \le \gamma$

$\sigma = \sigma_0(1 - E_2/E_1) + \left(\frac{E_2\varepsilon_m}{c}\right)y \;,\quad \gamma \le |y| \le c$ eq. (II)

$\sigma = A + By$

$$M = \iint_A y\,\sigma(y)\,dA = 2b\left\{\int_0^{\gamma} \frac{E_1\varepsilon_m}{c}y^2\,dy + \int_{\gamma}^{c}(A+By)\,y\,dy\right\}$$

$$= 2b\left\{\frac{E_1\varepsilon_m}{3c}y^3\Big|_0^{\gamma} + (Ay^2/2 + By^3/3)\Big|_{\gamma}^{c}\right\}$$

$$= 2b\left\{\frac{E_1\varepsilon_m}{3c}\gamma^3 + A/2\,(c^2-\gamma^2) + B/3\,(c^3-\gamma^3)\right\}$$

$$M = b/3\left\{2\left[\frac{E_1\varepsilon_m}{c} - B\right]\gamma^3 + (3A + 2Bc)c^2 - 3A\gamma^2\right\}.$$

Substituting for A, B, from eq (I) and noting that $\sigma_0 = E_1\varepsilon_0$, upon rearranging terms:

$$M = \frac{bd^2}{12}\sigma_0\left\{\left[3 - (\varepsilon_0/\varepsilon_m)^2\right](1 - E_2/E_1) + 2E_2/E_1\,(\varepsilon_m/\varepsilon_0)\right\}.$$

b). $b = 20\text{ mm}^2$, $d = 30$ mm, $\sigma_0 = 120$ MPa, $E_1 = 100$ GPa, $E_2 = 50$ GPa, $\varepsilon_m = 3.6\times10^{-3}$

$$M = \frac{(20)(30)^2}{12}\cdot 120\left\{\left[3 - (1/3)^2\right](1-0.5) + (1)(3)\right\} = 8\times10^5\text{ N-mm} = 800\text{ N-m}$$

c). Eq. (I): $\gamma = \frac{(120)(15)}{(100\times10^3)(3.6\times10^{-3})} = 5$ cm.

Eq. II : $\sigma_m = \pm 240$ MPa.

Unloading: $\sigma = -Mc/I$, $I = bd^3/12 = 4.5\times10^{-4}$ mm^4. $\sigma = -\frac{(8\times10^5)(15)}{I} = -266.7$ MPa

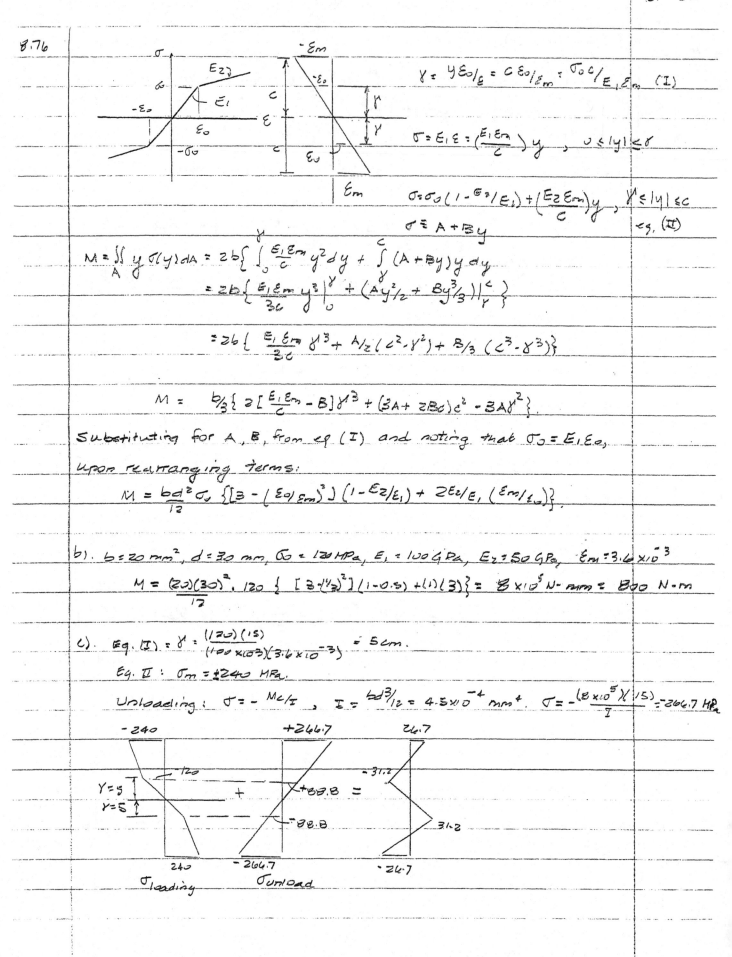

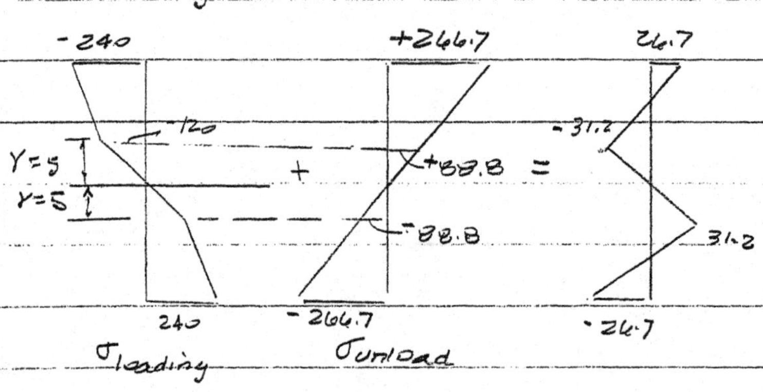

8.77

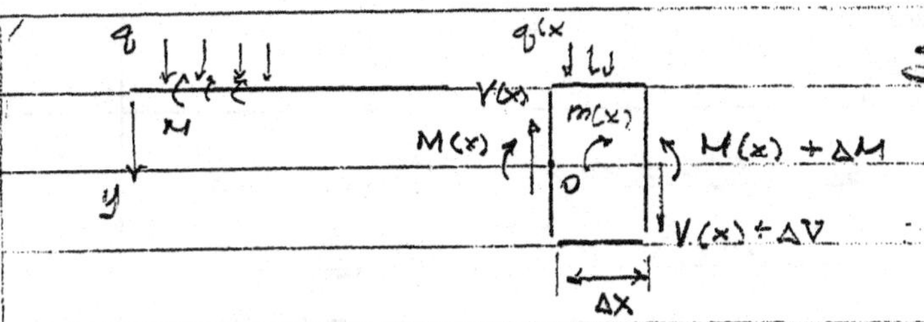

$\Sigma F_y = 0$:
$dV/dx = -q$ [as in text]

$\curvearrowleft \Sigma M_o = 0 \Rightarrow [M(x) + \Delta M] - M(x) - [V(x) + \Delta V] \cdot \Delta x - [q(x)\Delta x](\alpha \Delta x) + m(x)\Delta x = 0$
$0 \le \alpha \le 1$

$$\frac{\Delta M(x)}{\Delta x} = V(x) + \Delta V + q(x) \cdot \alpha \Delta x + m(x)$$

Taking limit as $\Delta x \to 0$ [$\Delta V \to 0$] $\Rightarrow dM/dx = V(x) + m(x)$

$\therefore d^2M/dx^2 = -q(x) + m(x)$

8.78

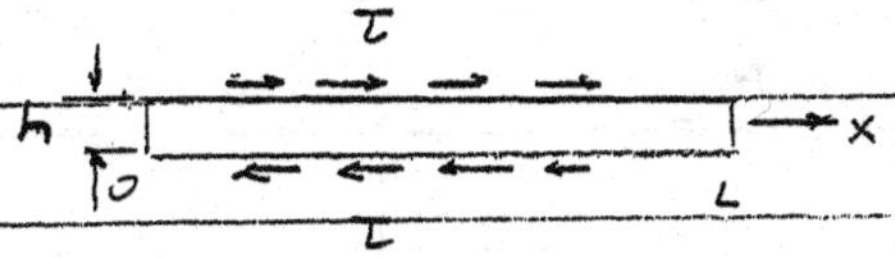

a) $$\frac{dM(x)}{dx} = m(x) = h\tau(x) = hA\sin(2\pi x/L)$$

$M(x) = \frac{-AhL}{2\pi}\cos\left(\frac{2\pi x}{L}\right) + C \qquad M(0) = 0 \Rightarrow C = +Ah/2\pi L$

$M(x) = AhL/2\pi\,(1 - \cos 2\pi x/L)$

b) $1/R = \frac{M}{EI}$; $I = (1)h^3/12 \Rightarrow R(x)^{-1} = \frac{6AL}{Eh^2\pi}(1 - \cos 2\pi x/L)$

8.79 $I = \frac{(3R)(2R)^3}{12} - \frac{\pi R^4}{4} = \frac{R^4}{4}(8 - \pi)$

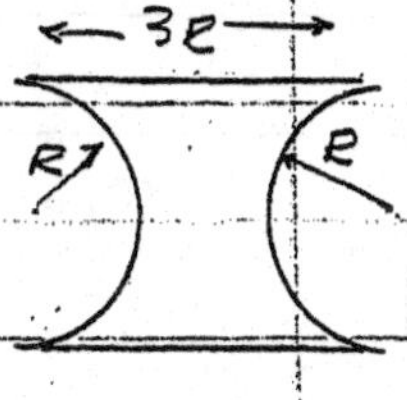

a). $M_E = \frac{\sigma_0 I}{R} = \frac{\sigma_0 R^3(8-\pi)}{4} = 1.2146\,\sigma_0 R^3$.

b). $M_P = 2\sigma_0\left[(3R^2)(R/2) - \left(\frac{\pi R^2}{2}\right)\left(\frac{4}{3\pi}\right)\right] = \frac{5\sigma_0 R^3}{3}$

$M_P/M_E = 1.317$

c). $\tau_{max}\Big|_{NA} = VQ/Ib \qquad V = \tau_0 I b/Q_{NA} \qquad Q_{NA} = \frac{5}{6}R^3$

$$\therefore V = \frac{\tau_0[(8-\pi)R^4/4]R}{5R^3/6} = \frac{3}{10}(8-\pi)\tau_0 R^2 = 1.4575\,\tau_0 R^2$$

7.80

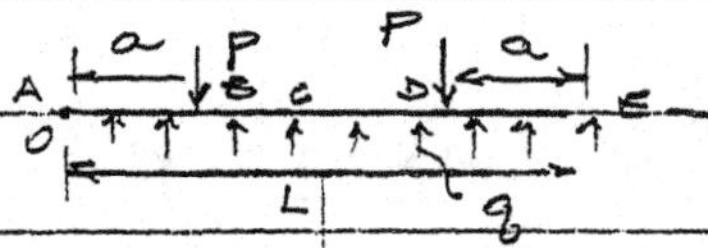

$q = 2P/L$

a)

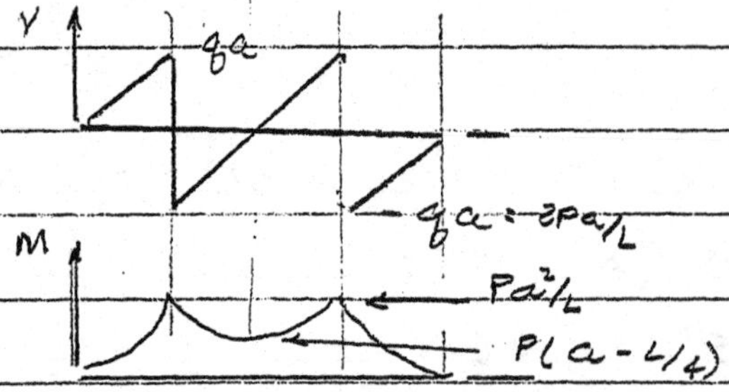

AB: $V(x) = qx = 2Px/L$

$M(x) = qx^2/2 = Px^2/L$

B-D: $V(x) = 2Px/L - P = P(2x/L - 1)$

$M(x) = Px^2/L - P(x-a) = \frac{P}{L}(x^2 - Lx + aL)$

b).

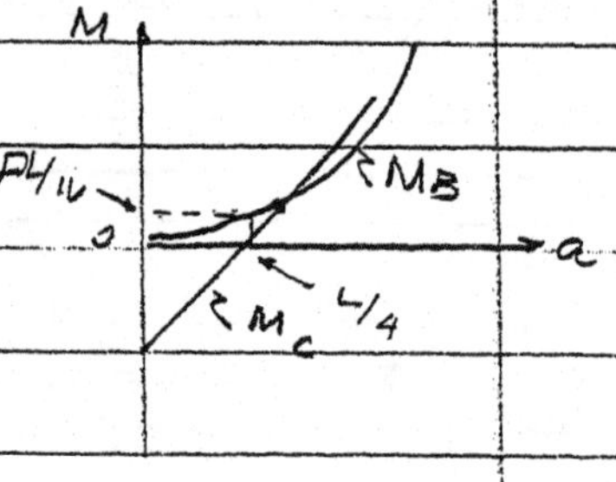

Require $M_{BD} \geq 0$ $\quad M_{min}(L/2) = 0 \Rightarrow a = L/4$

$M_B = M_{max} = M(x=a) = Pa^2/L$

$M_B = PL/16$

$M = \sigma_0 I/c = \sigma_0 b^3/6$

$\sigma_0 b^3/6 = PL/16 \Rightarrow b = (3PL/8\sigma_0)^{1/3}$

7.81 $\quad M(x) = Px/2 \quad 0 \leq x \leq L/2 \qquad d_0 > d_s.$

$d_s(x) = d_s + \frac{2(d_0 - d_s)}{L}x \qquad I(x) = \frac{bd^3}{12} = \frac{b}{12}\left[d_s + \frac{2(d_0 - d_s)}{L}x\right]^3$

$= b/12\,[d_s + \gamma x]^3, \quad \gamma = 2(d_0 - d_s)/L$

$\sigma = Mc/I = \frac{(Px/2)(d/2)}{bd^3/12} = \frac{3Px}{bd^2(x)} = \frac{3Px}{b\,[d_s + \gamma x]^2}$

$\sigma_{max} \rightarrow \frac{d\sigma(x)}{dx} = 0 \Rightarrow \frac{d}{dx}\left[x(d_L + \gamma x)^{-2}\right] = 0$

$(d_L + \gamma x)^{-2} - 2x(d_L + \gamma x)^{-3}\cdot\gamma = 0$

$(d_s + \gamma x) - 2\gamma x = 0 \Rightarrow x_{cr} = \frac{d_s}{\gamma} = \frac{d_s L}{2(d_0 - d_s)} = \frac{L}{2}\frac{1}{(d_0/d_s - 1)} < L/2$

$\therefore \sigma(x_{cr}) = \frac{3Px_{cr}}{b[d_s + \gamma x_{cr}]^2} = \frac{3Pd_sL}{2b(d_0 - d_s)[2d_s]^2} = \frac{3PL}{8b(d_0 - d_s)d_s}$

7.82 $W = q_0 L/2$

a) 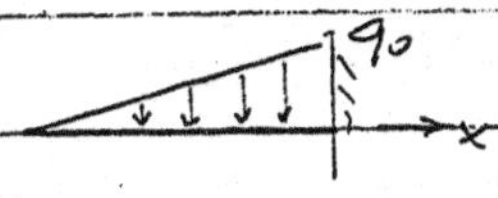 $M = -q_0 x^3/6L = -\frac{Wx^3}{3L^2}$; $\sigma_0 = \frac{M(L) d_0/2}{b h^3/12} = \frac{6M(L)}{b h_0^2}$

(i)

$$|M(L)| = q_0 L^2/6 = WL/3$$

$$h_0^2 = \frac{6|M(L)|}{b\sigma_0} = \frac{2WL}{b\sigma_0} \qquad h_0 = (2WL/b\sigma_0)^{1/2}$$

(ii)

$$h^2(x) = \frac{2Wx^3}{bL^2\sigma(x)} = \frac{2WL}{b\sigma(x)}\left(\frac{x}{L}\right)^3 \Rightarrow \sigma(x) = \frac{2WL}{bh^2(x)}\left(\frac{x}{L}\right)^3$$

$$\sigma(x) = \sigma_0 \Rightarrow h^2(x) = \frac{2WL}{b\sigma_0}(x/L)^3 = h_0^2 (x/L)^3 \Rightarrow h(x) = h_0 (x/L)^{3/2}$$

b).

$$\frac{dh(x)}{dx} = \tfrac{3}{2L} h_0 (x/L)^{1/2} < \tan 5^\circ$$

$$(x/L)^{1/2} \le \frac{2L \tan 5^\circ}{3h_0} \qquad W = 5\,kN,\ \sigma_0 = 10\,MPa,\ L/b = 40,\ L = 2m.$$

$$h_0 = \left[\frac{(10\times10^3)(40)}{(10\times10^6)}\right]^{1/2} = 0.2\ m$$

$\therefore$ $x/L \le 0.7637$

c). $x/L \le 1 \Rightarrow$ Require $L = \frac{3h_0}{2\tan 5^\circ} = 3.43\ m.$

8.83 $V = q_0 x^2/2L = Wx^2/L^2.$

$$\tau = \frac{3}{2}\frac{V}{A} = \frac{3Wx^2}{2L^2 b h(x)} \Rightarrow \text{At } x = L \Rightarrow h_0 = \frac{3W}{2b\tau} = \frac{3(5\times10^3)}{2(5\times10^{-2})(2\times10^6)} = 0.075\ m.$$

$$h(x) = h_0 (x/L)^2$$

8.84 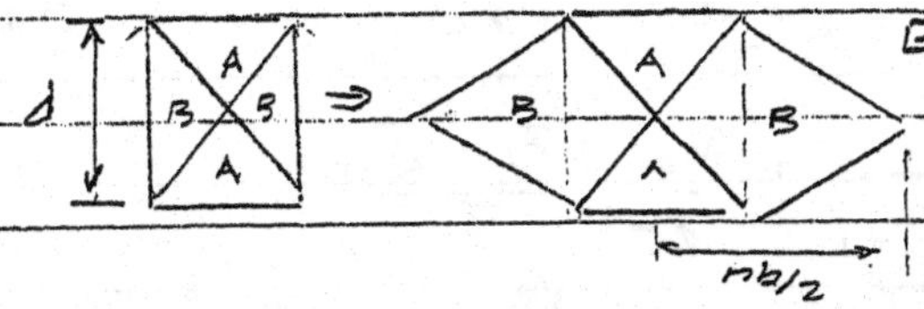

$E_B = nE_A$, $n > 0$ Transformed as material A;

$$I = 2[b(d/2)^3/36 + (bd/4)(d/3)^2] + 4[(nb/2)(d/2)^3/12]$$

$$I = (n+3)bd^3/48;\ \sigma_{A\,max} = \frac{M(d/2)}{I} = \frac{24M}{(n+3)bd^3}$$

$$\sigma_{B\,max} = 24Mn/(n+3)bd^3.$$

8.85 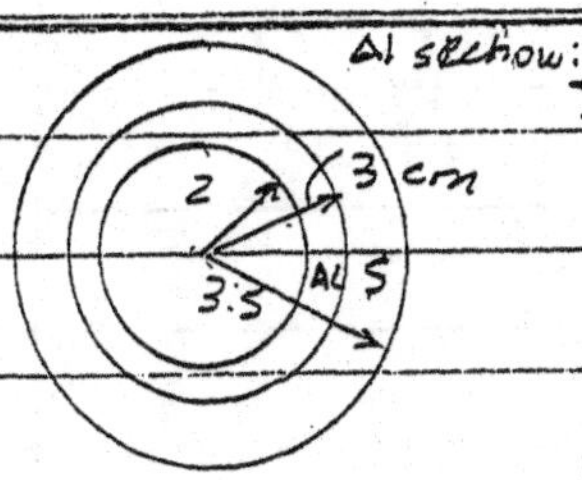

Al section:

$$I = \pi/4\,[(3^4 - 2^4) + n(3.5^4 - 3^4)] = 204.03\ cm^4,\quad n = E_S/E_{Al} = 200/70$$

$$\sigma_{Al} = Mc/I = \frac{(2500\times10^2)3}{206.03} = 3640\ N/cm^2 = 36.40\ N/mm^2 = 36.4\ MPa$$

$$\sigma_{St} = nMc/I = \frac{(2500\times10^2)(3.5)n}{206.03} = 12,13 4\ N/cm^2 = 121.3\ MPa$$

$$R = EI/M = \frac{(70\times10^9)(204.03\times10^{-8})}{2500} = 57.7\ m$$

8.86 a)

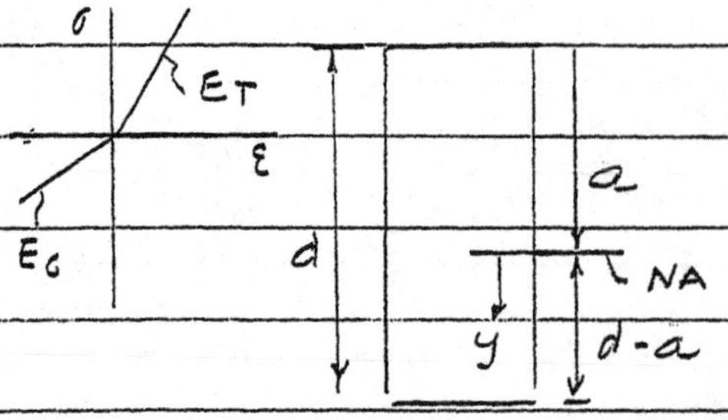

$$F = \iint \sigma\, dA = b\int \sigma(y)\,dy = 0$$

$$b\left\{E_c\int_{-a}^{0} y\,dy + E_T\int_{0}^{d-a} y\,dy\right\} = 0$$

$$\frac{b}{2}\left\{E_c y^2\Big|_{-a}^{0} + E_T y^2\Big|_{0}^{d-a}\right\} = 0 \Rightarrow -E_c a^2 + E_T(d-a)^2 = 0 \quad (i)$$

$$a = (E_T/E_c)^{1/2}(d-a) \quad (ii)$$

$$M = \frac{1}{R}\iint_A y\sigma\, dA = \frac{b}{R}\left[E_c\int_{-a}^{0} y^2 dy + E_T\int_{0}^{d-a} y^2 dy\right\} = \frac{b}{3R}\left[E_c a^3 + E_T(d-a)^3\right]$$

Substituting (ii):

$$M = \frac{b}{3R}\left[E_c(E_T/E_c)^{3/2} + E_T\right](d-a)^3 = \frac{E_T b}{3R}\left[1+(E_T/E_c)^{1/2}\right](d-a)^3 \quad (iii)$$

Let $\beta = (E_T/E_c)^{1/2}$: $M = E_T b/3R\,[1+\beta](d-a)^3$.

From (i): $(d-a)^2 = 1/\beta^2\, a^2$

$$d^2 - 2ad + a^2 = 1/\beta^2\, a^2 \Rightarrow \beta^2 d^2 - 2a\beta^2 d + (\beta^2-1)a^2 = 0$$

$$d = \frac{1}{2\beta^2}\left\{2a\beta^2 \pm \left[4a^2\beta^4 - 4\beta^2(\beta^2-1)a^2\right]^{1/2}\right\} = \frac{1}{2\beta^2}\left[2a\beta^2 \pm 2a\beta\right]$$

$$d = a(1 + 1/\beta) = \left(\frac{\beta+1}{\beta}\right)a \Rightarrow a = \left(\frac{\beta}{\beta+1}\right)d$$

$$(d-a) = \left(1 - \frac{\beta}{\beta+1}\right)d = \left(\frac{1}{\beta+1}\right)d$$

$$(iii) \Rightarrow M = \frac{E_T b}{3R}\left[1+\beta\right]\frac{d^3}{(\beta+1)^3} = \frac{bd^3}{3R}E_T\frac{1}{(1+\beta)^2}$$

$$M = \frac{bd^3}{3R}E_T\frac{1}{\left[1+(E_T/E_c)^{1/2}\right]^2} = \frac{bd^3}{3R}\frac{E_T E_c}{(E_T^{1/2}+E_c^{1/2})^2}$$

$$I = bd^3/12 \Rightarrow M = \frac{E^* I}{R},\quad E^* = \frac{4E_T E_c}{(\sqrt{E_T}+\sqrt{E_c})^2}$$

b)

$$(\sigma_x)_{T\,max} = \frac{E_T y_{max}}{R} = \frac{E_T(d-a)}{R} = \frac{ME_T(d-a)}{E^* I} = \frac{ME_T(\sqrt{E_T}+\sqrt{E_c})^2}{4E_T E_c I}\left(\frac{1}{\beta+1}\right)d$$

$$= \frac{Md}{4E_c I}(\sqrt{E_T}+\sqrt{E_c})^2 \cdot \frac{1}{(E_T/E_c)^{1/2}+1} = \frac{Md}{4I}\frac{(\sqrt{E_T}+\sqrt{E_c})}{\sqrt{E_c}}$$

8.87

$$I_{zt} = \pi R^4/8 - \left(\frac{\pi R^2}{2}\right)\left(\frac{4R}{3\pi}\right)^2 = \frac{\pi R^4}{8} - \frac{8R^4}{9\pi} = \frac{R^4}{72\pi}\left(9\pi^2 - 64\right)$$

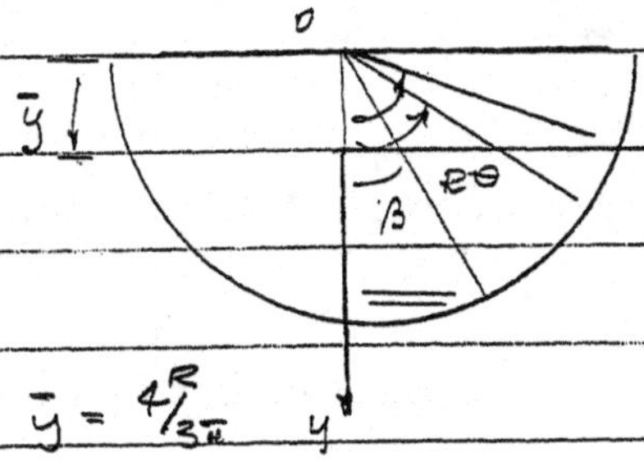

$\bar{y} = 4R/3\pi$

$$dQ(y) = 2\{R\sin\beta\, d(R\cos\beta)[R\cos\beta - \bar{y}]\}$$
$$= 2R^2\sin^2\beta\,[R\cos\beta - \bar{y}]\,d\beta$$
$$Q(\theta) = 2R^3\int_0^\theta \sin^2\beta\cos\beta\,d\beta - 2R^2\bar{y}\int_0^\theta \sin^2\beta\,d\beta$$

$$Q(\theta) = \frac{2R^3}{3}\sin^3\beta\Big|_0^\theta - \frac{2R^2\bar{y}}{2}\int_0^\theta (1-\cos 2\beta)\,d\beta$$

$$Q(\theta) = \frac{2R^3}{3}\sin^3\theta - R^2\bar{y}\left[\beta - \frac{\sin 2\beta}{2}\right]_0^\theta = \frac{2R^3}{3}\sin^3\theta - R^2\bar{y}\left(\theta - \frac{\sin 2\theta}{2}\right)$$
$$= \frac{2R^3}{3}\left[\sin^3\theta - \frac{2}{\pi}\left(\theta - \frac{\sin 2\theta}{2}\right)\right]$$

$$\tau(\theta) = \frac{VQ}{Ib} \qquad \frac{Q(\theta)}{b(\theta)} = \frac{Q(\theta)}{2R\sin\theta} = \frac{R^2}{3}\left\{\sin^2\theta - 2/\pi\left[\frac{\theta}{\sin\theta} - \cos\theta\right]\right\}$$

$$\frac{d[Q/b]}{d\theta} = 0 \Rightarrow 2\sin\theta\cos\theta - 2/\pi\left[\frac{1}{\sin\theta} - \frac{\theta\cos\theta}{\sin^2\theta} + \sin\theta\right] = 0$$

$$2\sin\theta\cos\theta - \left(2/\pi\sin^2\theta\right)\left[\sin\theta - \theta\cos\theta + \sin^3\theta\right] = 0$$

$$\pi\sin^2\theta\cos\theta - [\sin\theta - \theta\cos\theta + \sin^3\theta] = 0$$

$$\sin\theta[(\pi\cos\theta - 1)\sin^2\theta - 1] + \theta\cos\theta = 0$$

or

$$[(1-\pi\cos\theta)\sin^2\theta + 1]\tan\theta = \theta \Rightarrow \theta_{cr}$$

In general: $\tau = \frac{VQ}{Ib} = \frac{24V\pi}{R^2(9\pi^2-64)\sin\theta}\left[\sin^3\theta - \frac{1}{\pi}(2\theta - \sin 2\theta)\right]$

FOR PROBLEM 8.97:

$\theta_{cr} = 60.775430^\circ = 1.060732$ rad

$\tau_{max} = 0.0995576\ VR^2/I = 0.907073\ V/R^2 = 1.4248\ V/A$

$\tilde{y} = 0.48823R > \bar{y}$

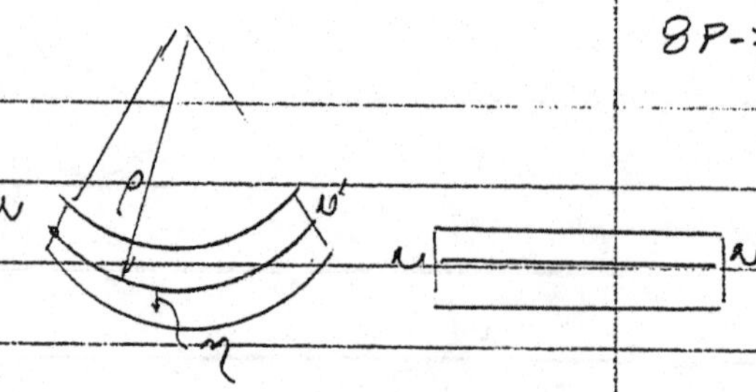

8.38 $\varepsilon_x(P) = \frac{\Delta x - (\rho+\eta)\Delta\theta}{(\rho+\eta)\Delta\theta}$; $\Delta x = \rho\Delta\theta$

a)

$\therefore \varepsilon = -\frac{\eta}{\rho+\eta}$, $\sigma = -\frac{E\eta}{\rho+\eta}$

b)

$F = \iint_A \sigma(\eta)\,dA = -bE\int_{-c_1}^{c_2}\frac{\eta}{\rho+\eta}\,d\eta = 0$; $\int_{-c_1}^{c_2}\frac{\eta}{\rho+\eta}\,d\eta = \frac{1}{\rho}\int_{-c_1}^{c_2}\frac{\eta}{1+\eta/\rho}\,d\eta = 0$ NOT DEFINITION OF CENTROID

c)(i)

$\int_{-c_1}^{c_2}\frac{\eta}{1+\eta/\rho}\,d\eta = \rho^2\{1+\eta/\rho - \ln[1+\eta/\rho]\}_{-c_1}^{c_2} = 0 \Rightarrow d/\rho + \ln(1-c_1/\rho) - \ln(1+c_2/\rho) = 0$

$\therefore d/\rho = \ln(1+c_2/\rho) - \ln(1-c_1/\rho) = \ln\left(\frac{1+c_2/\rho}{1-c_1/\rho}\right)$ (I)

$e^{d/\rho} = \frac{1+c_2/\rho}{1-c_1/\rho} \Rightarrow (\rho+c_2-d)e^{d/\rho} = \rho + c_2$ $[d = c_1+c_2]$

Let $\alpha = d/\rho \Rightarrow c_2/d = \frac{(1-\alpha)e^{\alpha}-1}{\alpha(1-e^{\alpha})}$

(ii)

$M = \iint_A \eta\sigma(\eta)\,dA = -bE\int_{-c_1}^{c_2}\frac{\eta^2}{\rho+\eta}\,d\eta = -bE\left[\tfrac12(\rho+\eta)^2 - 2\rho(\rho+\eta) + \rho^2\ln(\rho+\eta)\right]_{-c_1}^{c_2}$

$= -bE\rho^2\left\{-d/\rho + \frac{d(c_2-c_1)}{2\rho^2} + \ln[\rho(1+c_2/\rho)] - \ln[\rho(1-c_1/\rho)]\right\}$

d)

$= -bE\rho^2\{-d/\rho + d(c_2-c_1)/2\rho^2 + \ln(1+c_2/\rho) - \ln(1-c_1/\rho)\}$ (II)

(i) $d/\rho \ll 1 \Rightarrow c_i/\rho \ll 1$:

$\ln(1+c_2/\rho) - \ln(1-c_1/\rho) = [c_2/\rho - \tfrac12 c_2^2/\rho^2 + \tfrac13 c_2^3/\rho^3 \ldots] - [-c_1/\rho - \tfrac12 c_1^2/\rho^2 - \tfrac13 c_1^3/\rho^3 + \ldots]$

$= \frac{c_1+c_2}{\rho} + \frac{c_1^2-c_2^2}{2\rho^2} + \tfrac13\frac{c_2^3+c_1^3}{\rho^3} = \frac{d}{\rho} + \frac{d(c_1-c_2)}{2\rho^2} + \frac{c_2^3+c_1^3}{3\rho^3} + O(c^4)$ (III)

$\therefore$ (I) $\Rightarrow \frac{d(c_1-c_2)}{2} + \frac{c_1^3+c_2^3}{3\rho} = 0$ Note: $c_1^3+c_2^3 = d(d^2-3c_1c_2)$ (IV)

$\therefore \frac{3\rho}{2}(c_1-c_2) + d^2 - 3c_1c_2 = 0 \Rightarrow 3c_2^2 - 3(d+\rho)c_2 + d(d+3\rho/2) = 0$ (V)

since $c_1+c_2 = d$

Sol. $c_2 = \rho/6\{3(1+d/\rho) \pm \sqrt{3}(3-d^2/\rho^2)^{1/2}\} \equiv \rho\beta/6$

relevant root

$\therefore c_2/d = \frac{1}{6\alpha}[3(1+\alpha) \pm 3(1-\alpha^2/3)^{1/2}] = \frac{1}{2\alpha}[(1+\alpha) - (1-\alpha^2/6 \ldots)]$

$\therefore c_2/d = \tfrac12(1+\alpha/6) + O(\alpha^4)$, $\alpha = d/\rho$ [see p. 8P-39 for simpler solution]

(ii) (II) & (III) $\Rightarrow M = -bE\rho^2 \left\{ \dfrac{c_1^3 + c_2^3}{3\rho^3} + O(c^4) \right\} = -\dfrac{bE}{3\rho} d(d^2 - 3c_1c_2)$ by (IV)

Note: $c_1c_2 = (d - c_2)c_2 = dc_2 - c_2^2 = d\rho\beta/6 - (d+\rho)\rho\beta/6 + \dfrac{d}{3}(d + 3\rho/2)$ by (V)

$= -\rho^2\beta/6 + \dfrac{d}{3}(d + 3\rho/2)$

$\therefore c_1c_2 = -\rho^2/2 \pm \rho^2/2\,(1 - d^2/3\rho^2)^{1/2} + d^2/3$

$\therefore$

$$M = -\frac{bE}{\rho}\left[d^3/3 - c_1c_2 d\right] = -\frac{bE\rho d}{2}\left[1 \mp (1 - d^2/3\rho^2)^{1/2}\right]$$

$$= -\frac{bE\rho d}{2}\left[1 \mp (1 - d^2/6\rho^2 \cdots)\right] = -\frac{bd^3E}{12\rho}$$

$$M = -EI/\rho$$

LOCATION OF NEUTRAL AXIS: (SIMPLER SOLUTION)

$$c_2/d = \frac{(1-\alpha)e^\alpha - 1}{\alpha(1 - e^\alpha)} = \frac{[(1-\alpha)(1 + \alpha + \alpha^2/2 + \alpha^3/6 + \cdots)] - 1}{\alpha[1 - (1 + \alpha + \alpha^2/2 + \alpha^3/6 + \cdots]}$$

$$= \frac{[1 + \alpha + \alpha^2/2 + \alpha^3/6 + \cdots) - (\alpha + \alpha^2 + \alpha^3/2 + \cdots)] - 1}{-\alpha[\alpha + \alpha^2/2 + \alpha^3/6 \cdots]} = \frac{-\alpha^2/2 - \alpha^3/3}{-\alpha^2 - \alpha^3/2}$$

$$c_2/d = \frac{1/2 + \alpha/3}{1 + \alpha/2} = 1/2(1 + 2\alpha/3)(1 - \alpha/2) = 1/2(1 + \alpha/6) + O(\alpha^2)$$

8.89 a). $R_A = R_0 - c = M_A/EI_0$, $R_B = R_0 + c = M_B/EI_0$

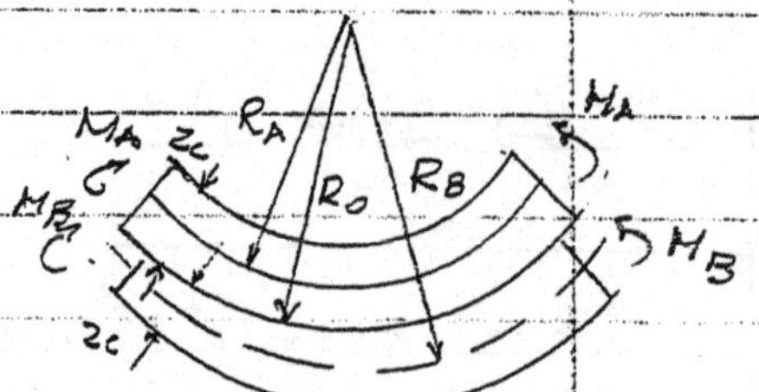

$I_0 = 2bc^3/3$.

$$M_B/M_A = \frac{R_0+c}{R_0-c} = \frac{1+c/R_0}{1-c/R_0} =$$

$$= (1+c/R_0)[1 + c/R_0 + (c/R_0)^2 + \dots]$$

$$= 1 + 2c/R_0 + 2(c/R_0)^2 + \dots \; ; \; c/d << 1 \Rightarrow M_B \cong M_A = M$$

b)(i) By superposition, apply $M = 2M_A = 2M_B$.

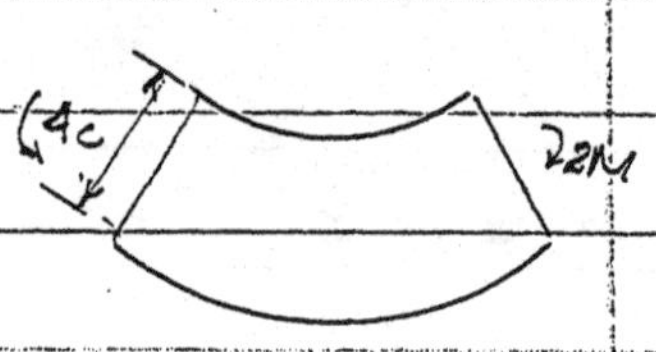

Combined strips: $I = b(4c)^3/12 = 8I_0$

Initial stress: $\sigma_0 = \pm \frac{M}{I_0}\eta$ in "A" and "B"

Due to M: $\sigma = \mp\frac{(2M)(2\eta)}{I} = \mp\frac{4M\eta}{8I_0} = \mp\frac{M\eta}{2I_0}$.

In A: Top $(\eta = -c)$: $\sigma = -\frac{Mc}{I_0} - \frac{M(-c)}{2I_0} = -Mc/2I_0$ (comp.)

Bottom $(\eta = c)$ $\sigma = +Mc/2I_0$ (tension)

In B: Top $(\eta = -c)$: $\sigma = -Mc/I - \frac{M(-c)}{2I} = -Mc/2I_0$ (comp.)

Bottom $(\eta = c)$ $\sigma = Mc/2I_0$ (tension).

(ii) Distribution:

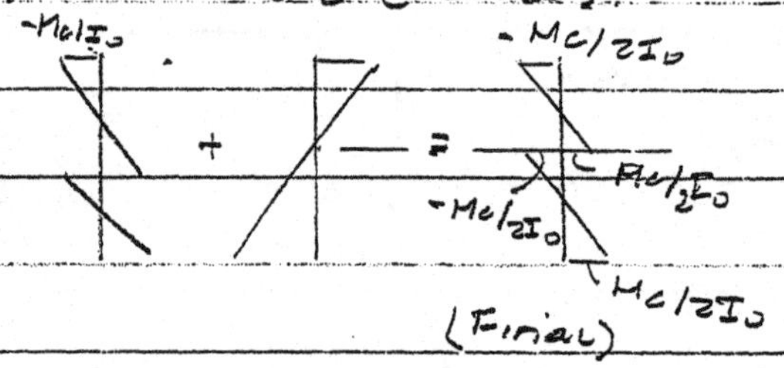

c).

$$1/R = \frac{M}{EI_0} + \frac{(-2M)}{EI} = \frac{M}{EI_0} - \frac{2M}{8EI_0} = \frac{M}{EI_0}(3/4) \Rightarrow \frac{1/R}{1/R_0} = 3/4 \Rightarrow R = \frac{4}{3}R_0$$

8.90

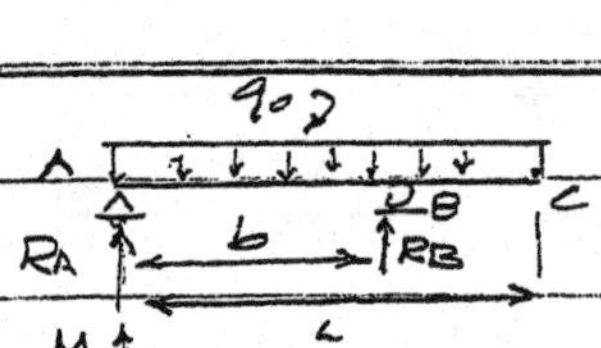

$\Sigma M_A = 0 \Rightarrow R_B = q_0L^2/2b$; $R_A = q_0L(1 - L/2b)$ Note: $R_A > 0$ if $b > L/2$

$M_B = -q_0L^2/2(1 - b/L)^2$

$M(x) = R_Ax - qx^2/2 = q_0[(1 - L/2b)Lx - x^2/2]$, $0 \le x \le b$

M_{max}: $dM/dx = 0 \Rightarrow L(1 - L/2b) - x = 0 \Rightarrow x_{cr} = L(1 - L/2b)$

$M_{max} = (q_0L^2/2)(1 - L/2b)^2$, $0 < x < b$

Setting:

$|M_{max}| = |M_B| \Rightarrow q_0L^2/2(1 - L/2b)^2 = q_0L^2/2(L-b)^2 \Rightarrow \frac{L^2}{4b^2}(2b - L)^2 = (L-b)^2$

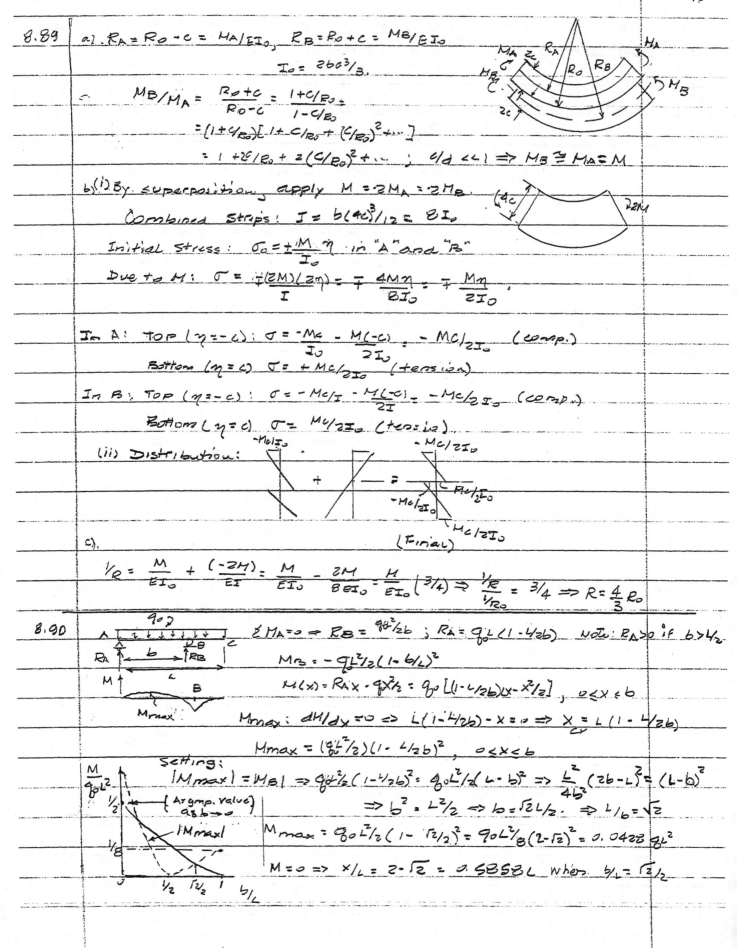

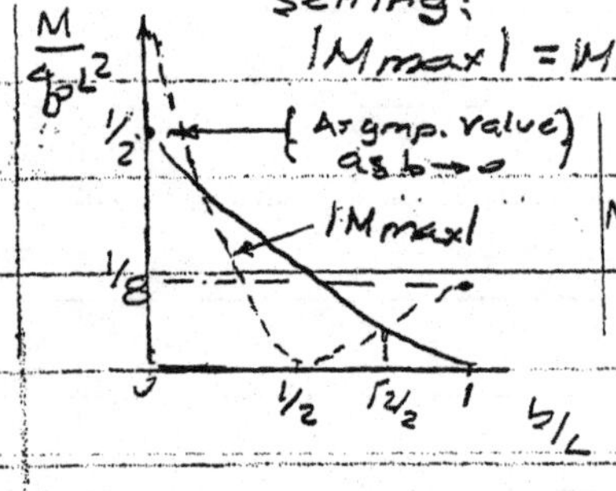

$\Rightarrow b^2 = L^2/2 \Rightarrow b = \sqrt{2}L/2$. $\Rightarrow L/b = \sqrt{2}$

$M_{max} = q_0L^2/2(1 - \sqrt{2}/2)^2 = q_0L^2/8(2-\sqrt{2})^2 = 0.0428\, q_0L^2$

$M = 0 \Rightarrow x/L = 2 - \sqrt{2} = 0.5858L$ when $b/L = \sqrt{2}/2$

8.91. $I = a^4/12$ about all centroidal axes.

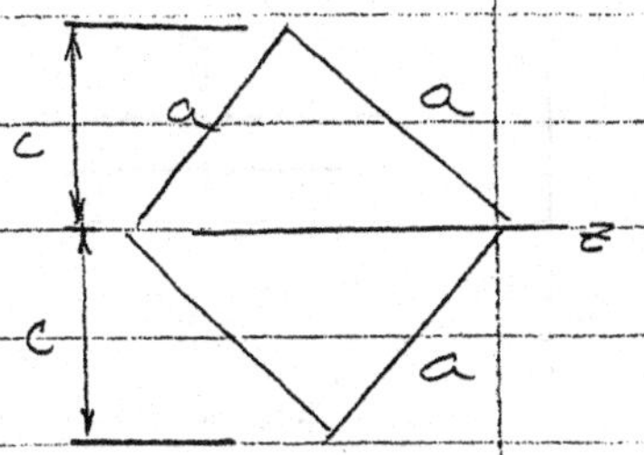

$a = \sqrt{2}\,c \Rightarrow I_{zz} = c^4/3$

a). $(\sigma_x)_a = \pm Mc/I = 3M/c^3 = \pm 6\sqrt{2}\,M/a^3$

b).

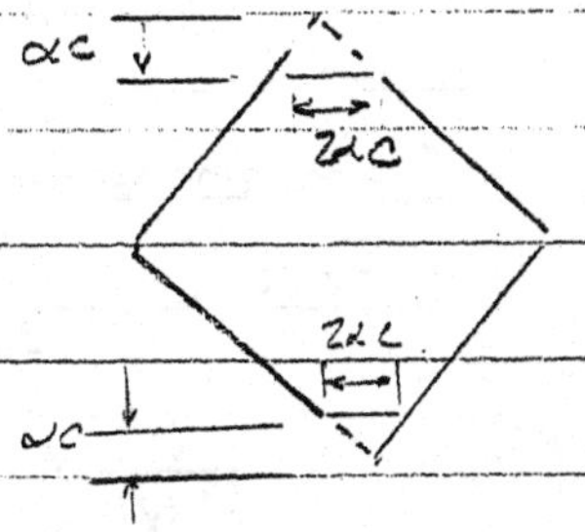

$I = c^4/4 - 2\cdot[(2\alpha c)(\alpha c)^3/36 + (\alpha c)^2[c - \frac{2}{3}(\alpha c)]^2]$

$= c^4/3\,[1 - \alpha^2(3\alpha^2 - 8\alpha + 6)]$

$(\sigma_x)_b = \dfrac{Mc(1-\alpha)}{I} = \dfrac{Mc(1-\alpha)}{c^4/3\,[1-\alpha^2(3\alpha^2-8\alpha+6)]}$

$(\sigma_x)_b = \dfrac{3M}{c^3}\cdot\dfrac{(1-\alpha)}{1-\alpha^2(3\alpha^2-8\alpha+6)}$

$R_\sigma(\alpha) = \dfrac{(\sigma_x)_a}{(\sigma_x)_b} = \dfrac{1-\alpha^2(3\alpha^2-8\alpha+6)}{1-\alpha} = 3\alpha^3 - 5\alpha^2 + \alpha + 1 = (\alpha-1)^2(3\alpha+1)$

c). $R_\sigma = 1 \Rightarrow 3\alpha^3 - 5\alpha^2 + \alpha + 1 = 1 \Rightarrow \alpha(3\alpha^2 - 5\alpha + 1) = 0 \Rightarrow \alpha = 0.2324$

$R_\sigma > 0 \Rightarrow 0 < \alpha < 0.2324$, $R_\sigma < 1 \Rightarrow 0.2324 \le \alpha \le 1$

Implication: Removing material with $\alpha < 0.2324$ reduces stress σ_x.

8.92 $\sigma(y) = \begin{cases} y\,\sigma_0/\ell & 0 \le |y| < \ell \\ \sigma_0 & \ell < |y| < d/2 \end{cases}$, $d = 2c$

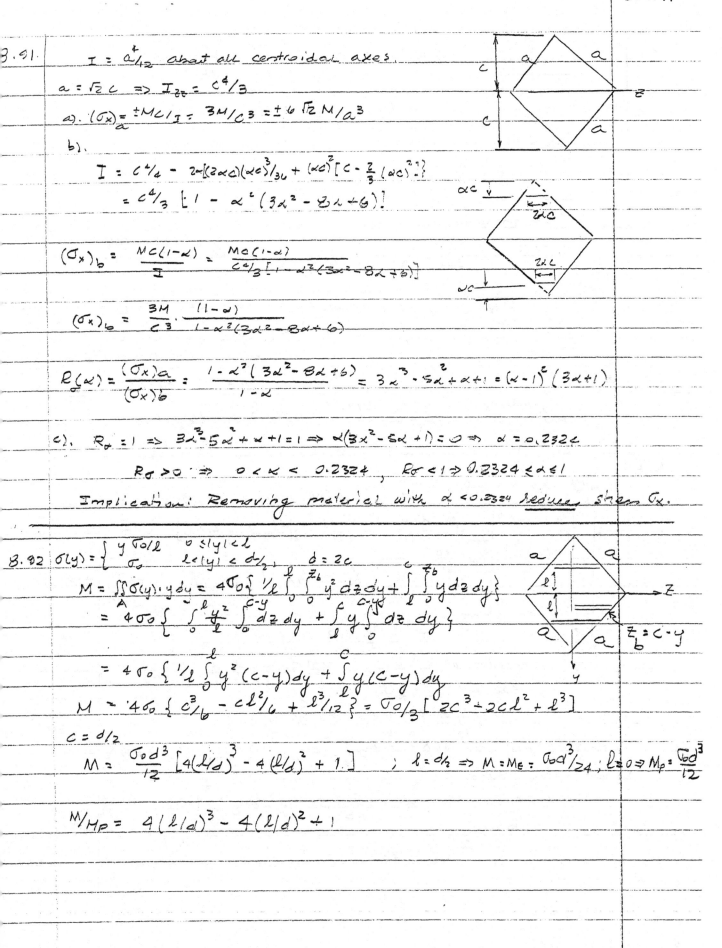

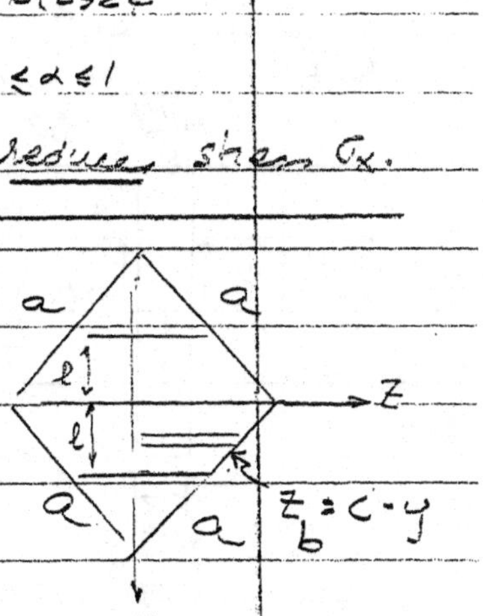

$M = \iint_A \sigma(y)\cdot y\,dy = 4\sigma_0\{1/\ell \int_0^\ell \int_0^{z_b} y^2\,dz\,dy + \int_\ell^c \int_0^{z_b} y\,dz\,dy\}$

$= 4\sigma_0\{\int_0^\ell \frac{y^2}{\ell}\int_0^{c-y} dz\,dy + \int_\ell^c y\int_0^{c-y} dz\,dy\}$

$= 4\sigma_0\{1/\ell \int_0^\ell y^2(c-y)\,dy + \int_\ell^c y(c-y)\,dy$

$M = 4\sigma_0\{c^3/6 - c\ell^2/6 + \ell^3/12\} = \sigma_0/3\,[2c^3 - 2c\ell^2 + \ell^3]$

$c = d/2$

$M = \dfrac{\sigma_0 d^3}{12}[4(\ell/d)^3 - 4(\ell/d)^2 + 1]$; $\ell = d/2 \Rightarrow M = M_E = \sigma_0 d^3/24$; $\ell = 0 \Rightarrow M_p = \dfrac{\sigma_0 d^3}{12}$

$M/M_p = 4(\ell/d)^3 - 4(\ell/d)^2 + 1$

8.93

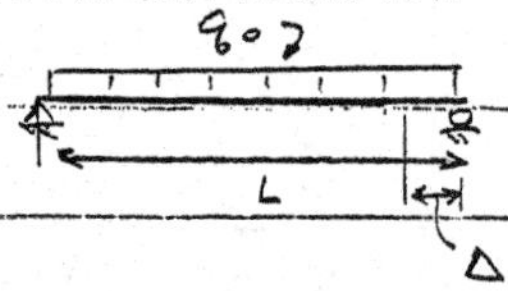

$$M = \frac{q_0 x(L-x)}{2} \qquad \sigma_x = \frac{Mc}{I} = \frac{q_0 x(L-x)\, d/2}{2bd^3/12}$$

a). $$\sigma_x = \frac{3 q_0 x(L-x)}{bd^2}\ ; \quad \sigma_y = q_0/b$$

$$|\sigma_y/\sigma_x| = \frac{q_0/b}{[3q_0 x(L-x)]/bd^2} = \frac{d^2}{3x(L-x)}$$

At Δ from the end B:

$$\therefore |\sigma_y/\sigma_x| = \frac{d^2}{3\Delta(L-\Delta)} = \frac{d^2}{3L^2(\Delta/L)(1-\Delta/L)} = \frac{1}{3}\left(\frac{d}{L}\right)^2 \frac{[1+\Delta/L+(\Delta/L)^2 \cdots]}{\Delta/L}$$

$$\therefore |\sigma_y/\sigma_x| \le \frac{1}{3(\Delta/L)}\left(d/L\right)^2$$

Let $d/L = 1/20$, $\Delta/L = 0.05 \Rightarrow |\sigma_y/\sigma_x| < 0.0167 = 1.7\%$

$d/L = 1/100$ $\Delta/L = 0.05 \Rightarrow |\sigma_y/\sigma_x| < 0.07\%$

$d/L = 1/20$ $\Delta/L = 0.10 \Rightarrow |\sigma_y/\sigma_x| < 0.8\%$

$d/L = 1/100$ $\Delta/L = 0.10 \Rightarrow |\sigma_y/\sigma_x| < 0.03\%$

b) $\sigma_x|_{max} = \dfrac{6q_0L^2/8}{bd^2}$; $|\sigma_y/\sigma_x| = \dfrac{4}{3}(d/L)^2 \Rightarrow d/L = 1/20 \Rightarrow |\sigma_y/\sigma_x| = 0.3\%$

$d/L = 1/100 \Rightarrow |\sigma_y/\sigma_x| = 0.01\%$

8.94

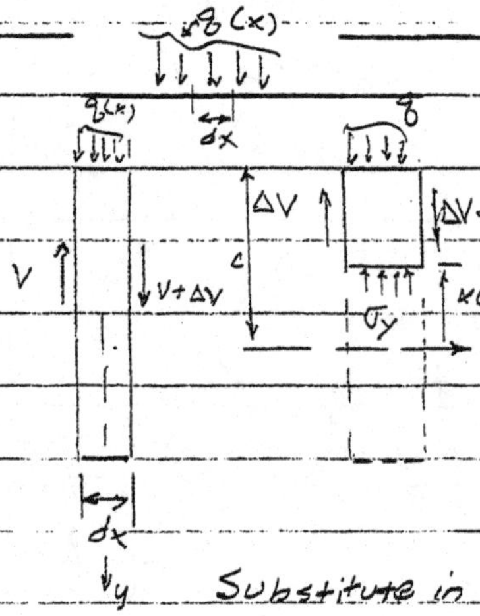

$$\downarrow + \Sigma F_y = 0 \Rightarrow [\Delta V + d(\Delta V)] - \Delta V + q(x)dx - \sigma_y b\, dx = 0$$

$$\sigma_y = \frac{1}{b}\left[\frac{d(\Delta V)}{dx} + q(x)\right] \qquad (I)$$

$$\Delta V = \iint_A d\tau_{xy}\, dA = b\int_{-c}^{-\alpha c} d\tau_{xy}(y)\, dA = b\, dV \int_{-c}^{-\alpha c} \frac{Q(y)}{Ib}\, dy$$

$$[y = -\alpha c]$$

$$\Delta V = \frac{dV}{I}\int_{-c}^{-\alpha c} Q(y)\, dy \qquad II$$

Substitute in (I) $\Rightarrow$ $\sigma_y = 1/b\left[q(x) + \frac{1}{I}\frac{dV}{dx}\int_{-c}^{-\alpha c} Q(y)\, dx\right]$

But $dV/dx = -q(x)$

$$\therefore \sigma_y = \frac{q(x)}{b}\left[1 - \frac{1}{I}\int_{-c}^{-\alpha c} Q(y)\, dy\right]$$

$$Q(y) = \frac{b}{2}(c^2 - y^2) \Rightarrow \int_{-c}^{-\alpha c} Q(y)\, dy = b/2\int_{-c}^{-\alpha c}(c^2-y^2)\, dy = \frac{bc^3}{2}\left(\alpha^3/3 - \alpha + 2/3\right)$$

$I = 2bc^3/3$. Substituting:

$$\therefore \sigma_y = \frac{q}{4b}[2 + 3\alpha - \alpha^3] = \frac{q}{4b}\left(2 - 3y/c + (y/c)^3\right)$$

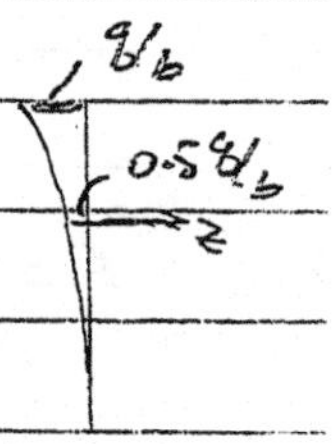

$\sigma_y > 0 \leftarrow$ compression

$$\sigma_y|_{NA} = q/2b.$$

8.96 See Problem 8.14 for analytic solution.

8.97 See Problem 8.37 for derivation of transcendental eq. and solution.

8.98

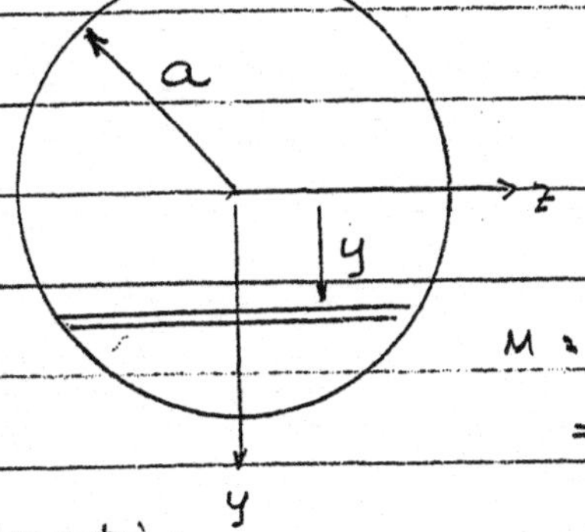

$$\varepsilon = y/R \qquad \sigma_x = \begin{cases} (y/\ell)\,\sigma_0 , & 0 \le |y| \le \ell \\ \sigma_0 , & \ell \le y \le a \end{cases}$$

$$M = \iint \sigma_x(y) \cdot y \, dA , \quad dA = 2(a^2 - y^2)^{1/2} dy$$

$$M = 2\left[\sigma_0 \int_0^{\ell} \frac{2y^2}{\ell}(a^2-y^2)^{1/2} dy + \sigma_0 \int_{\ell}^{a} 2(a^2-y^2)^{1/2} y \, dy\right]$$

$$= 4\sigma_0 \left[1/\ell \int y^2 (a^2-y^2)^{1/2} dy + \int_{\ell}^{a} y(a^2-y^2)^{1/2} dy \right]$$

Integrating

$$M = 4\sigma_0 \left\{ \left(-\frac{y}{4\ell}(a^2-y^2)^{3/2} + \frac{a^2}{8\ell}\left[y(a^2-y^2)^{1/2} + a^2 \sin^{-1}(\ell/a) \right] \right|_0^{\ell} - \frac{1}{3}(a^2-y^2)^{3/2} \Big|_{\ell}^{a} \right\}$$

$$= 4\sigma_0 \left\{ -\tfrac{1}{4}(a^2-\ell^2)^{3/2} + \frac{a^2}{8\ell}\left[\ell(a^2-\ell^2)^{1/2} + a^2 \sin^{-1}(\ell/a)\right] + \tfrac{1}{3}(a^2-\ell^2)^{3/2} \right\}$$

Recombining:

$$M = \frac{\sigma_0 a^3}{6} \left\{ (5 - 2\ell^2/a^2)(1-\ell^2/a^2)^{1/2} + 3(a/\ell)\sin^{-1}(\ell/a) \right\}$$

Note: $\ell = a \rightarrow M = \sigma_0 \pi a^3/4 = \frac{\sigma_0 I}{c} = M_E$ ✓

$\ell = 0 \rightarrow M = \frac{\sigma_0 a^3}{6}\left[5 + 3a\left[1/\ell \sin^{-1}(\ell/a)\right]\right\}$ as $\ell \rightarrow 0$

$$\lim_{\ell \to 0} 1/\ell \sin^{-1}(\ell/a) = \lim \frac{1}{\sqrt{1-\ell^2/a^2}} \cdot \frac{1}{a} = 1/a$$

$M(\ell \rightarrow 0) = 4\sigma_0 a^3/3 = M_P$.

Note: $M_P/M_E = 16/3\pi$

$$M/M_P = 1/8\left[(5 - 2\ell^2)(1-\ell^2/a^2)^{1/2} + 3(a/\ell)\sin^{-1}(\ell/a) \right]$$

Solve numerically for ℓ/a.

$$1/R = \varepsilon_0/\ell = \sigma_0/E\ell = \frac{(3M_P/4a^3)(1/a)}{E\ell \cdot 1/a} = \frac{3M_P}{4a^4 E} \cdot \frac{1}{\ell} , \quad \ell = \ell(M/M_P)$$

$$EI/R = \frac{(3\pi/16)M_P}{(\ell/a)} \quad \text{where } \ell/a = f(M/M_P).$$

Plots:

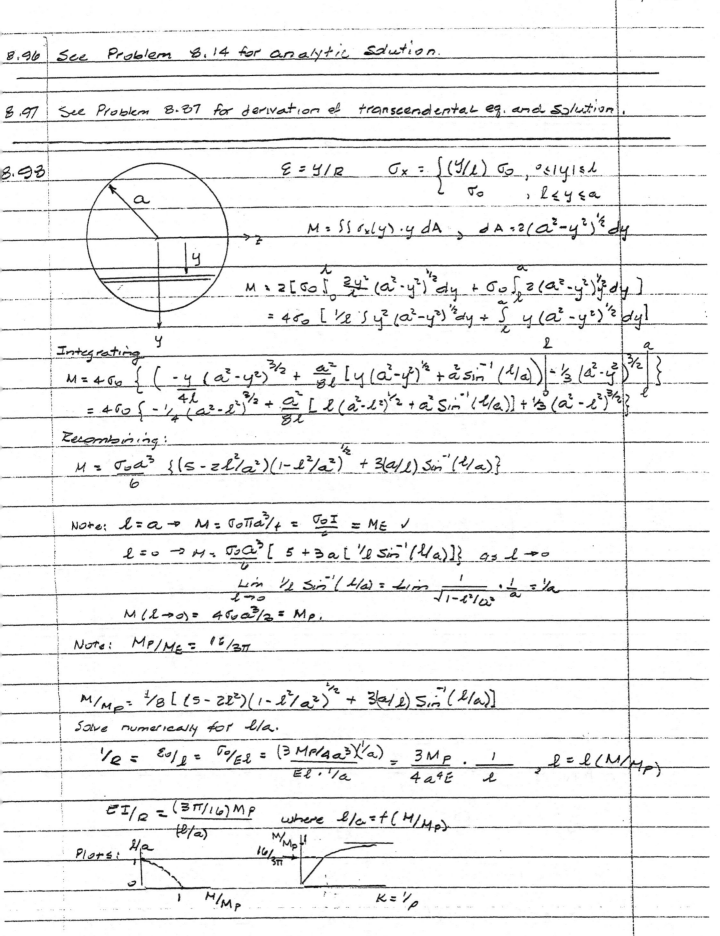

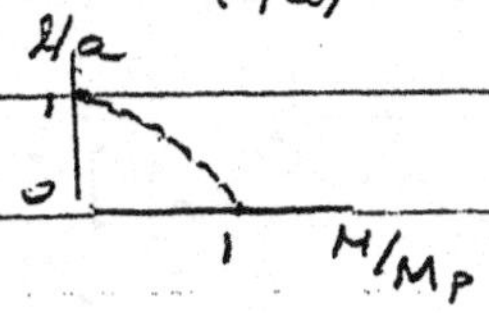

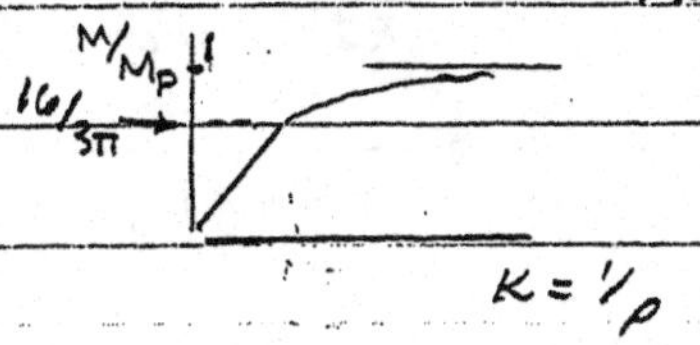

9.1

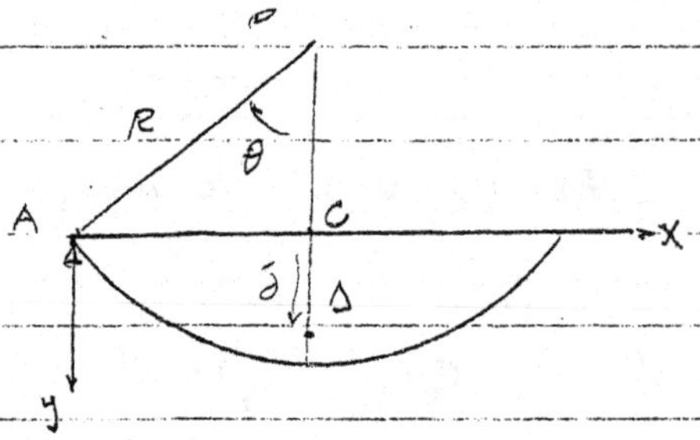

$M_E = \sigma_0 I/c = 6\sigma_0 bd^2$

$\therefore \frac{M_E}{EI} = \frac{2\sigma_0}{Ed}$

$|OC| = R\cos\theta \qquad R\theta = L/2 \Rightarrow \theta = L/2R$

$|OC| = R\cos(L/2R)$

$R = \frac{EI}{M_E} = Ed/2\sigma_0 \Rightarrow |OC| = \frac{Ed}{2\sigma_0}\cos\left(\frac{L\sigma_0}{Ed}\right)$

$|OC| = \frac{L}{2}\left(\frac{E}{\sigma_0}\cdot d/L\right)\cos\left(\frac{\sigma_0}{E}\frac{L}{d}\right)$

$|AC| = R\sin(L/2R) = L/2\left(\frac{E}{\sigma_0}\cdot d/L\right)\sin\left(\frac{\sigma_0}{E}\cdot\frac{L}{d}\right)$

$O: \left[\frac{L}{2}\left(\frac{E}{\sigma_0}\cdot\frac{d}{L}\right)\cos\left(\frac{\sigma_0 L}{Ed}\right), -\frac{L}{2}\left(\frac{E}{\sigma_0}\cdot\frac{d}{L}\right)\sin\left(\frac{\sigma_0}{E}\cdot\frac{L}{d}\right)\right]$

Linear Theory:

$\delta = M_E L^2/8EI = \frac{\sigma_0 L^2}{4Ed} \qquad R_L\Big|_c = |v''|^{-1} = \frac{EI}{M_E} = Ed/2\sigma_0$ same as R.

Y coord. of O': $-(R-\delta) = -\left[\frac{Ed}{2\sigma_0} - \frac{\sigma_0 L^2}{4Ed}\right] = -\left\{\frac{Ed}{2\sigma_0}\left[1 - \left(\frac{\sigma_0}{E}\right)^2 (L/d)^2\right]\right\}$

$O': \left[L/2, -\left\{\frac{L}{2}\left(\frac{E}{\sigma_0}\right)(d/L)\left[1 - (\sigma_0/E)^2 (L/d)^2\right]\right\}\right]$

Steel: $E/\sigma_0 = 10^3 \qquad d/L = 100 \Rightarrow O[0.4975L, 4.992L]$

$O_L[0.5L, 4.950L]$

9.2

$R_{ex}(L/2) = \frac{EI}{M(L/2)} = 2EI/PL \qquad v(x) = \frac{P}{6EI}(3Lx^2 - x^3)\ ;\ v' = \frac{P}{2EI}(2Lx - x^2),\ v'' = \frac{P}{EI}(L-x)$

$v'(L/2) = 3PL^2/8EI,\ v''(L/2) = PL/2EI.$

$K = \frac{v''}{[1+(v')^2]^{3/2}} \qquad K(L/2) = \frac{PL/2EI}{[1+(3PL^2/8EI)^2]^{3/2}} \Rightarrow R = 1/K = (2EI/P)[1+(3PL^2/8EI)^2]^{3/2}$

$\sigma_{max} = \sigma_0 \Rightarrow M_E = \frac{2\sigma_0 I}{d} = PL \Rightarrow \frac{PL^2}{EI} = \frac{2\sigma_0}{E}(L/d)$

$\therefore$

$R = 2EI/PL\left[1 + \frac{9\cdot 4}{64}\left(\frac{\sigma_0}{E}\right)^2 (L/d)^2\right]^{3/2} = R_{ex}\left[1 + \frac{9}{16}(\sigma_0/E)^2 (L/d)^2\right]^{3/2}$

eg. STEEL: $\sigma_0/E = 10^{-3}$ For $(L/d) = 20$ $R/R_{ex} = 1.00034$; $L/d = 100$, $R/R_{ex} = 1.0085$

$R/R_{ex} = [1 + 9/16(\sigma_0/E)^2 (L/d)^2]^{3/2} \simeq 1 + 27/32(\sigma_0/E)^2 (L/d)^2$

9.3 a) 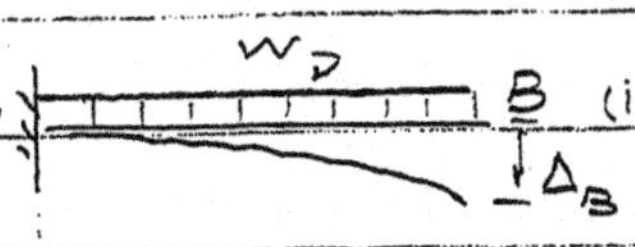

(i) $EIv'' = -M(x) = +\frac{wL^2}{2} - wLx + wx^2/2$

B.C. $v(0) = v'(0) = 0$

$EIv'(x) = wL^2x/2 - wLx^2/2 + wx^3/6 + A$; $v'(0) = 0 \Rightarrow A = 0$

$EIv(x) = wL^2x^2/4 - wLx^3/6 + wx^4/24 + B$; $v(0) = 0 \Rightarrow B = 0$

(iii) $\Delta_B = v(L) = \frac{wL^4}{8EI}$

b). 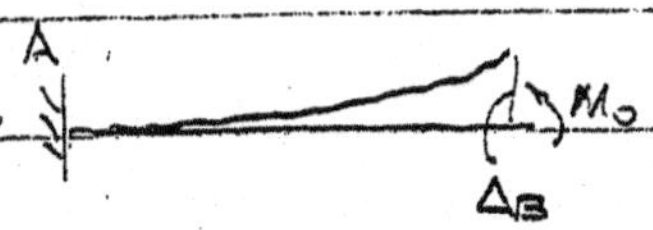

$EIv'' = -M_0$

B.C. $v'(0) = v(0) = 0$

$EIv'(x) = -M_0x + A$; $v'(0) = 0 \Rightarrow A = 0$

$EIv(x) = -M_0x^2/2 + B$; $v(0) = 0 \Rightarrow B = 0$

$\Delta_B = v(L) = -M_0L^2/2EI$

c). 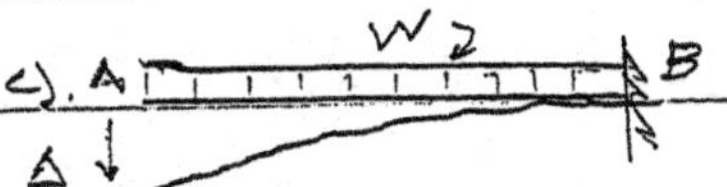

$EIv''(x) = +wx^2/2$ B.C. $v'(L) = v(L) = 0$

$EIv'(x) = +wx^3/6 + A$; $v'(L) = 0 \Rightarrow A = -wL^3/6$

$EIv(x) = +wx^4/24 - wL^3x/6 + B$; $v'(L) = 0 \Rightarrow B = +wL^4/8$

$v(x) = \frac{w}{24EI}[x^4 - 4L^3x + 3L^4]$; $\Delta_A = v(0) = wL^4/8EI$

d). 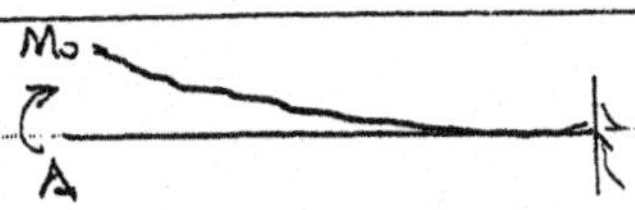

$EIv''(x) = -M_0$; B.C. $v'(L) = v(L) = 0$

$EIv'(x) = -M_0x + A$; $v'(L) = 0 \Rightarrow A = M_0L$

$EIv(x) = -M_0x^2/2 + M_0Lx + B$; $v(L) = 0 \Rightarrow B = -M_0L^2/2$

$v(x) = -\frac{M_0}{2EI}(x^2 - 2Lx + L^2) = -\frac{M_0}{2EI}(L-x)^2$

$\Delta_A = v(0) = -M_0L^2/2EI$

e). 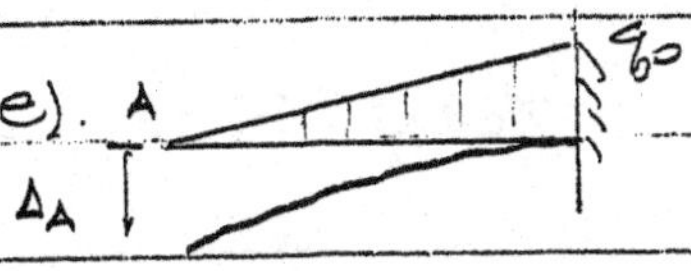

$M(x) = -q_0x^3/6L$ $EIv''(x) = q_0x^3/3L$; B.C. $v'(L) = v(L) = 0$

$EIv'(x) = q_0x^4/24L + A$; $v'(L) = 0 \Rightarrow A = -q_0L^3/24$

$EIv(x) = q_0x^5/120L - q_0L^3x/24 + B$

$v(L) = 0 \Rightarrow B = q_0L^4/30$

$v(x) = \frac{q_0}{120EI}(x^5/L - 5L^3x + 4L^4)$

$\Delta_A = v(0) = \frac{q_0L^4}{30EI}$

f).

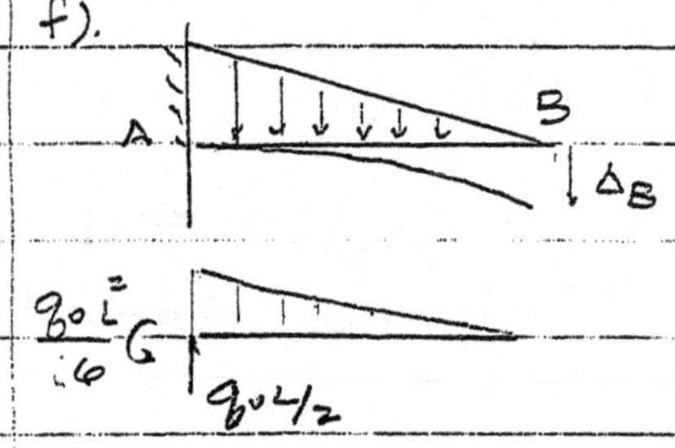

$$M(x) = -q_0(L-x)^3/6L$$

$$EIv'' = q_0(L-x)^3/6L \qquad B.C.\ v'(0) = v(0) = 0$$

$$EIv'(x) = -q_0(L-x)^4/24L + A \Rightarrow v'(0) = 0 \Rightarrow A = q_0L^3/24$$

$$EIv(x) = +q_0(L-x)^5/120L + q_0L^3x/24 + B$$

$$v(0) = 0 \Rightarrow B = -q_0L^4/120$$

$$v(x) = \frac{q_0}{120EI}\left[\frac{1}{L}(L-x)^5 + 5L^3x - L^4\right]$$

$$\Delta_B = v(L) = q_0L^4/30EI.$$

g).

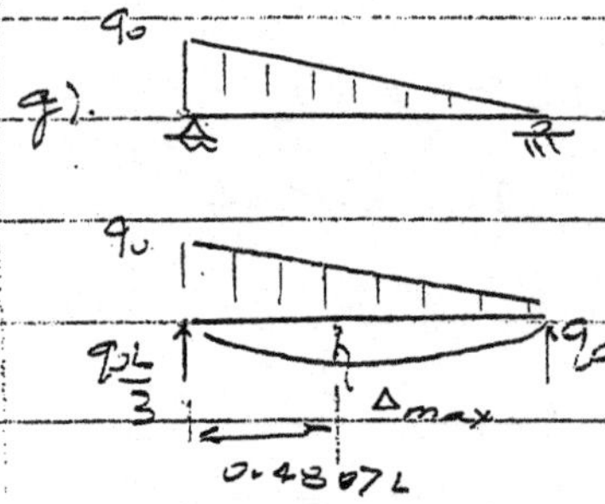

$$M(x) = \frac{q_0L}{6}(L-x) - q_0(L-x)^3/6L$$

$$EIv''(x) = \frac{q_0}{6L}\left[-L^2(L-x) + (L-x)^3\right] \qquad B.C.\ v(0) = v(L) = 0$$

$$EIv'(x) = \frac{q_0}{6L}\left[L^2(L-x)^2/2 - (L-x)^4/4\right] + A$$

$$EIv(x) = q_0/6L\left[-L^2(L-x)^3/6 + (L-x)^5/20\right] + Ax + B$$

$$v(0) = 0 \Rightarrow q_0/6L\left(-L^5/6 + L^5/20\right) + B = 0 \Rightarrow B = 7q_0L^4/360$$

$$v(L) = 0 \Rightarrow B + AL = 0 \Rightarrow A = -7q_0L^3/360$$

$$v(x) = \frac{q_0}{360EIL}\left[-10L^2(L-x)^3 + 3(L-x)^5 - 7L^4x + 7L^5\right]$$

$$v_{max} \Rightarrow v'(x_{cr}) = 0 \Rightarrow \frac{q_0}{6L}\left[L^2(L-x)^2/2 - (L-x)^4/4\right] - 7q_0L^3/360 = 0$$

$$\Rightarrow q_0L^3/6\left[\tfrac{1}{2}(1-x/L)^2 - \tfrac{1}{4}(1-x/L)^4 - 7/60\right] = 0$$

$$\text{Let } \eta = (1-x/L)^2 \Rightarrow \eta^2 - 2\eta + 7/15 = 0 \Rightarrow \eta = 1-(8/15)^{1/2}$$

$$x_{cr}/L = 1 - \left[1-(8/15)^{1/2}\right]^{1/2} = 0.48067$$

$$v_{max} = \frac{2.1347986\,q_0L^4}{360\,EI} = 0.005922\,\frac{q_0L^4}{EI}.$$

h).

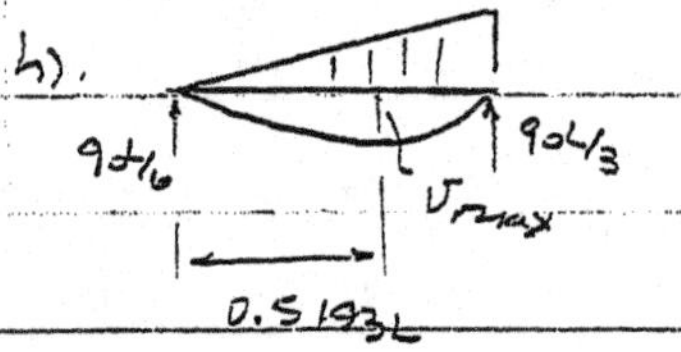

$$M: \frac{q_0}{6}Lx - q_0x^3/6L$$

$$EIv''(x) = \frac{q_0}{6L}(-L^2x + x^3] \qquad B.C.\ v(0) = v(L) = 0$$

$$EIv'(x) = \frac{q_0}{6L}(-L^2x^2/2 + x^4/4] + A$$

$$EIv(x) = \frac{q_0}{6L}(-L^2x^3/6 + x^5/20] + Ax + B \Rightarrow v(0) = 0 \Rightarrow B = 0$$

$$v(L) = 0 \Rightarrow A = 7q_0L^3/360$$

$$v(x) = \frac{q_0}{360EIL}\left[-10L^2x^3 + 3x^5 + 7L^4x\right]$$

$$v_{max} \Rightarrow v'(x_{cr}) = 0 \Rightarrow \frac{q_0}{6L}\left(-L^2x^2/2 + x^4/4 + 7L^4/60\right) = 0$$

$$x = L\left[1-(8/15)^{1/2}\right]^{1/2} = 0.5193\,L$$

$$v_{max} = 0.005922\,qL^4/EI$$

i). 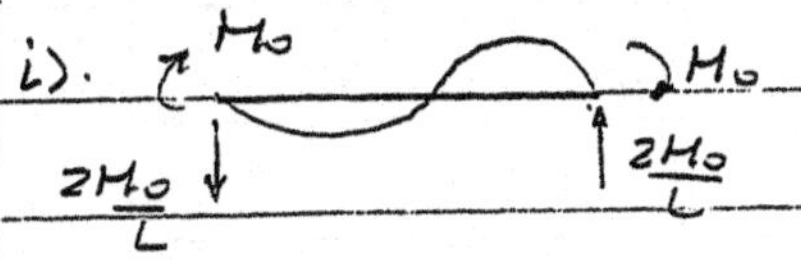

$M(x) = M_0 - \frac{2M_0}{L}x = M_0(1 - 2x/L)$

$EIv''(x) = -M_0(1 - 2x/L)$ B.C. $v(0) = v(L) = 0$

$EIv'(x) = -M_0(x - x^2/L) + A$

$EIv(x) = -M_0(x^2/2 - x^3/3L) + Ax + B$; $v(0) = 0 \Rightarrow B = 0$

$v(L) = 0 \Rightarrow A = +M_0L/6$

$$v(x) = \frac{M_0}{6EI}[2x^3/L - 3x^2 + Lx]$$

$v_{max} \Rightarrow v'(x_{cr}) = 0 \Rightarrow M_0[-x + x^2/L + L/6] = 0 \Rightarrow x/L = \frac{1}{2}[1 \pm \frac{1}{3}\sqrt{3}]$

$x_{cr}/L = 0.21133,\ 0.788675$

$$|v_{max}| = \pm\frac{\sqrt{3}M_0L^2}{108EI} = \pm 0.01604\, M_0L^2/EI$$

9.4 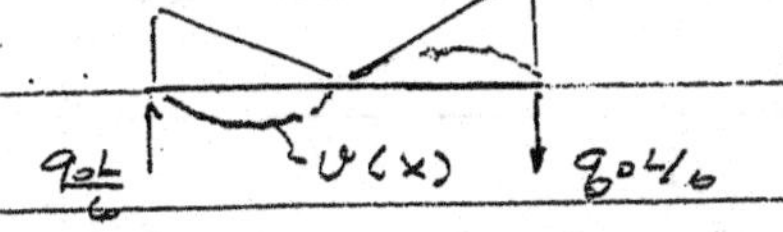

$q(x) = q_0(1 - 2x/L)$ $0 \le x \le L$

$M(x) = \frac{q_0L}{6} - \int_0^x q(\xi)\cdot(x - \xi)d\xi$ [see eq. (8.5.7)]

$M(x) = \frac{q_0}{6L}[2x^3 - 3Lx^2 + L^2x]$

$EIv''(x) = -M = -\frac{q_0}{6L}(2x^3 - 3Lx^2 + L^2x)$ B.C. $v(0) = v(L) = 0$

$EIv'(x) = -\frac{q_0}{6L}[x^4/2 - Lx^3 + L^2x^2/2] + A$

$EIv(x) = -\frac{q_0}{6L}[x^5/10 - Lx^4/4 + L^2x^3/6] + Ax + B$

$v(0) = 0 \Rightarrow B = 0$; $v(L) = 0 \Rightarrow A = +q_0L^3/360$

$$v(x) = -\frac{q_0}{360EIL}[6x^5 - 15Lx^4 + 10L^2x^3 - L^3x]$$

Note: $v(L/2) = 0$

9.5 a)

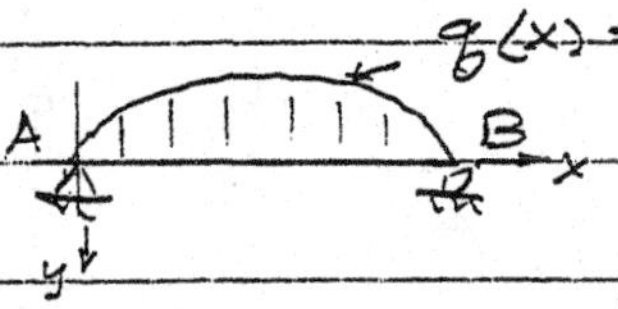

$q(x) = \frac{4q_0}{L^2}(Lx - x^2)$

$R_A = \frac{1}{L}\int_0^L q(\zeta)(L-\zeta)\,d\zeta = \frac{4q_0}{L^3}\int_0^L (L\zeta-\zeta^2)(L-\zeta)\,d\zeta = q_0L/3$

$M(x) = R_A x - \frac{4q_0}{L^2}\int_0^x (L\zeta-\zeta^2)(x-\zeta)\,d\zeta$

$= q_0Lx/3 - \frac{4q_0}{L^2}\left\{ x\int_0^x (L\zeta-\zeta^2)\,d\zeta - \int_0^x (L\zeta^2-\zeta^3)\,d\zeta \right\}$

$M(x) = \frac{q_0}{3L^2}(x^4 - 2Lx^3 + L^3x)$

$EIv''(x) = -M(x) = -\frac{q_0}{3L^2}(x^4 - 2Lx^3 + L^3x)$ B.C. $v(0) = v(L) = 0$

$EIv'(x) = -\frac{q_0}{3L^2}(x^5/5 - Lx^4/2 + L^3x^2/2) + A$

$EIv(x) = -\frac{q_0}{3L^2}(x^6/30 - Lx^5/10 + L^3x^3/6) + Ax + B$; $v(0) = 0 \Rightarrow B = 0$

$v(L) = 0 \Rightarrow A = q_0L^3/30$

$v(x) = -\frac{q_0x}{90EIL^2}(x^5 - 3Lx^4 + 5L^3x^2 - 3L^5)$

By symmetry: $v_{max} = v(L/2) = \frac{23q_0L^4}{1920\,EI}$

b)

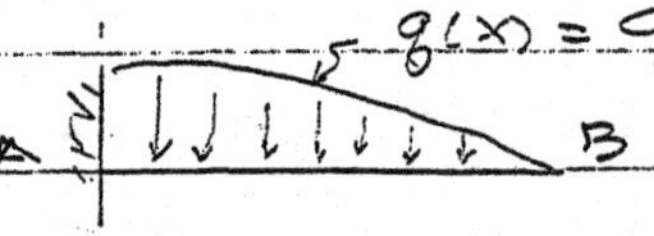

$q(x) = q_0\cos\frac{\pi x}{2L}$

$R_A = q_0\int_0^L \cos\frac{\pi x}{2L}\,dx = \frac{2q_0L}{\pi}\sin\frac{\pi x}{2L}\Big|_0^L = 2q_0L/\pi$

$M_A = -\int_0^L x\,q(x)\,dx = -q_0\int_0^L x\cos\frac{\pi x}{2L}\,dx = -\frac{2q_0L^2}{\pi^2}(\pi-2)$

$M(x) = M_A + R_Ax - q_0\int_0^x (x-\zeta)\cos\frac{\pi\zeta}{2L}\,d\zeta$

$= M_A + R_Ax - q_0\left\{ x\int_0^x \cos\pi\zeta/2L\,d\zeta - \int_0^x \zeta\cos\pi\zeta/2L\,d\zeta \right\}$

$= M_A + R_Ax - q_0\left\{ \frac{2Lx}{\pi}\sin\pi x/2L - (2L/\pi)^2\left[\cos\frac{\pi x}{2L} + \frac{\pi x}{2L}\sin\frac{\pi x}{2L} - 1\right]\right\}$

$= M_A + R_Ax - (2L/\pi)^2 q_0(1 - \cos\pi x/2L)$

$M(x) = \frac{2q_0L^2}{\pi^2}\left\{-\pi + \pi x/L + 2\cos\pi x/2L\right\}$

$EIv''(x) = \frac{2q_0L^2}{\pi^2}\left[\pi - \pi x/L - 2\cos\pi x/2L\right]$ B.C. $v'(0) = v(0) = 0$

$EIv'(x) = \frac{2q_0L^2}{\pi^2}\left[\pi x - \pi x^2/2L - \frac{4L}{\pi}\sin\pi x/2L\right] + A$; $v'(0) = 0 \Rightarrow A = 0$

$EIv(x) = \frac{2q_0L^2}{\pi^2}\left[\pi x^2/2 - \pi x^3/6L + \frac{8L^2}{\pi^2}\cos\frac{\pi x}{2L}\right] + B$ $v(0) = 0 \Rightarrow B = -16\frac{q_0L^4}{\pi^4}$

$v(x) = \frac{2q_0L^2}{\pi^4EI}\left[\pi^3x^2/2 - \pi^3x^3/6L + 8L^2\cos\pi x/2L - 8L^2\right]$

$v(L) = \frac{2q_0L^4(\pi^3 - 24)}{3\pi^4EI}$

c).

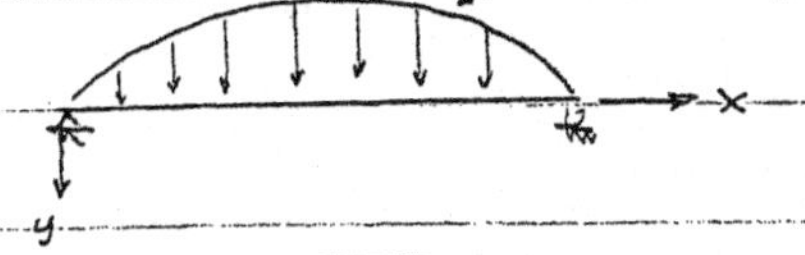

$q(x) = q_0 \sin \pi x/L$

$$2R = q_0 \int_0^L \sin\frac{\pi x}{L}\, dx = -q_0 L/\pi \cos\frac{\pi x}{L}\Big|_0^L = 2 q_0 L/\pi$$

$$R = q_0 L/\pi$$

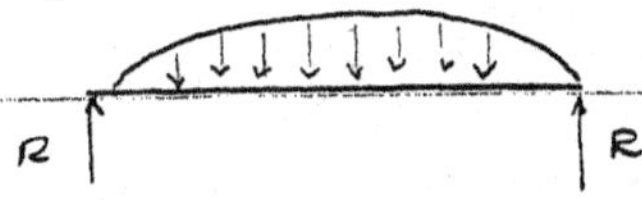

$$M(x) = q_0 L x/\pi - q_0 \int_0^x (x-\zeta) \sin\frac{\pi \zeta}{L}\, d\zeta$$

$$= q_0 L x/\pi - q_0 \left\{ -\frac{Lx}{\pi} \cos\frac{\pi \zeta}{L}\Big|_0^x - \int_0^x \zeta \sin\frac{\pi\zeta}{L}\, d\zeta \right\}$$

$$= q_0 L x/\pi - q_0 \left\{ -\frac{Lx}{\pi}\left(\cos\frac{\pi x}{L} - 1\right) - \frac{L}{\pi}\left(\frac{L}{\pi}\sin\frac{\pi x}{L} - x\cos\frac{\pi x}{L}\right)\right\}$$

$$M(x) = q_0 L^2/\pi^2 \sin \pi x/L$$

$EI\, v''(x) = -q_0 L^2/\pi^2 \sin \pi x/L$ B.C. $v(0) = v(L) = 0$

$EI\, v'(x) = q_0 (L/\pi)^3 \cos \pi x/L + A$

$EI\, v(x) = q_0 (L/\pi)^4 \sin \pi x/L + Ax + B$; $v(0) = 0 \Rightarrow B = 0$; $v(L) = 0 \Rightarrow A = 0$

$v_{max} = v(L/2) = q_0 L^4/\pi^4 EI$

9.6

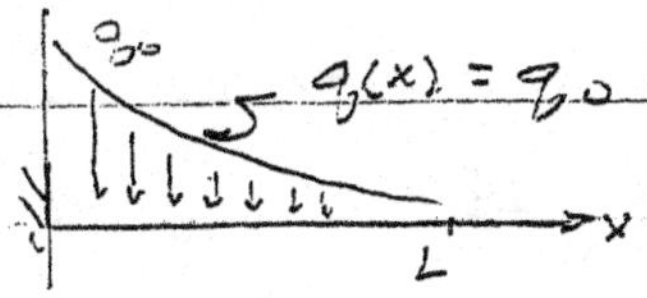

$q(x) = q_0 e^{-\alpha x/L}$

$$R = q_0 \int_0^L e^{-\alpha x/L} dx = -\frac{q_0 L}{\alpha} e^{-\alpha x/L}\Big|_0^L = \frac{q_0 L}{\alpha}(1 - e^{-\alpha})$$

$$M_A = -q_0 \int_0^L x e^{-\alpha x/L} dx = -q_0 \left\{ -\frac{L}{\alpha} x e^{-\alpha x/L}\Big|_0^L + \frac{L}{\alpha}\int_0^L e^{-\alpha x/L} dx \right\}$$

$$= \frac{q_0 L^2}{\alpha^2}\left[e^{-\alpha}(1+\alpha) - 1\right]$$

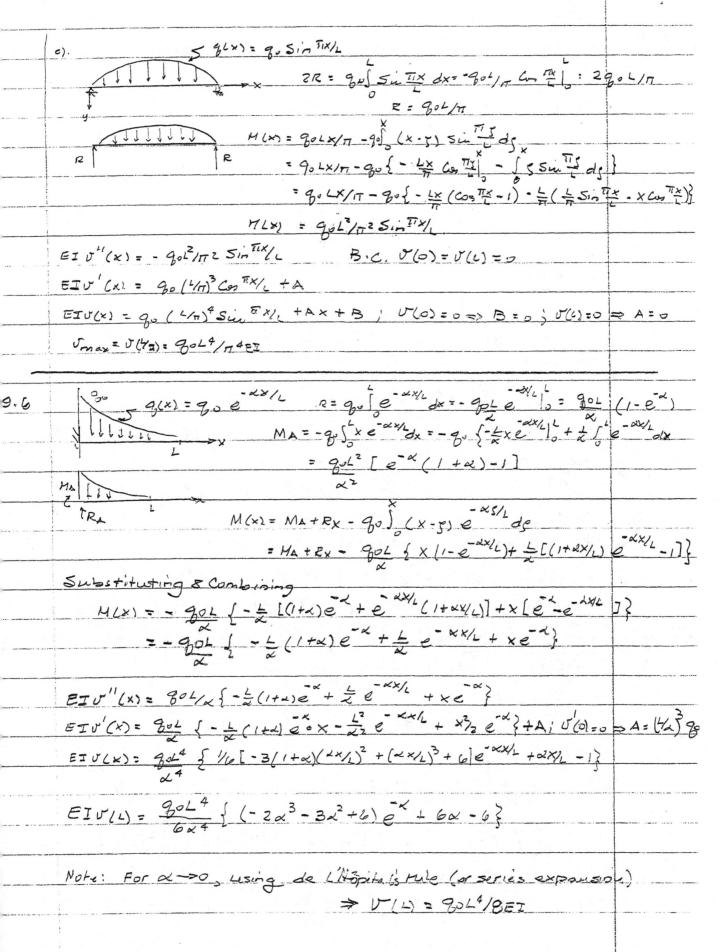

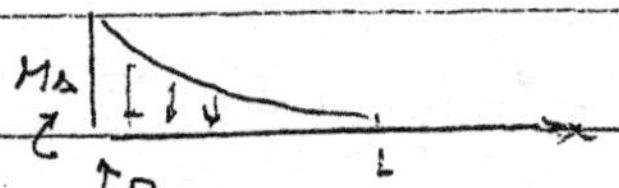

$$M(x) = M_A + Rx - q_0 \int_0^x (x-\zeta) e^{-\alpha\zeta/L} d\zeta$$

$$= M_A + Rx - \frac{q_0 L}{\alpha}\left\{ x(1 - e^{-\alpha x/L}) + \frac{L}{\alpha}\left[(1 + \alpha x/L) e^{-\alpha x/L} - 1\right]\right\}$$

Substituting & Combining

$$M(x) = -\frac{q_0 L}{\alpha}\left\{ -\frac{L}{\alpha}\left[(1+\alpha)e^{-\alpha} + e^{-\alpha x/L}(1 + \alpha x/L)\right] + x\left[e^{-\alpha} - e^{-\alpha x/L}\right]\right\}$$

$$= -\frac{q_0 L}{\alpha}\left\{ -\frac{L}{\alpha}(1+\alpha)e^{-\alpha} + \frac{L}{\alpha} e^{-\alpha x/L} + x e^{-\alpha}\right\}$$

$$EI\, v''(x) = q_0 L/\alpha \left\{ -\frac{L}{\alpha}(1+\alpha) e^{-\alpha} + \frac{L}{\alpha} e^{-\alpha x/L} + x e^{-\alpha}\right\}$$

$$EI\, v'(x) = \frac{q_0 L}{\alpha}\left\{ -\frac{L}{\alpha}(1+\alpha) e^{-\alpha} x - \frac{L^2}{\alpha^2} e^{-\alpha x/L} + x^2/2\, e^{-\alpha}\right\} + A;\ v'(0) = 0 \Rightarrow A = (L/\alpha)^3 q_0$$

$$EI\, v(x) = \frac{q_0 L^4}{\alpha^4}\left\{ 1/6\left[-3(1+\alpha)(\alpha x/L)^2 + (\alpha x/L)^3 + 6\right] e^{-\alpha x/L} + \alpha x/L - 1\right\}$$

$$EI\, v(L) = \frac{q_0 L^4}{6\alpha^4}\left\{ (-2\alpha^3 - 3\alpha^2 + 6) e^{-\alpha} + 6\alpha - 6 \right\}$$

Note: For $\alpha \to 0$, using de L'Hôpital's rule (or series expansion)

$$\Rightarrow v(L) = q_0 L^4/8EI$$

9.7

$M(x) = -Px$, $b(x) = x b_0 / L$ $\quad I(x) = b_0 d^3 x / 12$

$EI v''(x) = +Px \Rightarrow E \dfrac{b_0 d^3 x}{12L} v''(x) = Px$; B.C. $v'(L) = v(L) = 0$

$E v''(x) = \dfrac{12PL}{b_0 d^3} \Rightarrow E v'(x) = \dfrac{12PL}{b_0 d^3} x + A$; $v'(L) = 0 \Rightarrow A = -12PL^2 / b_0 d^3$

$E v(x) = 6PLx^2 / b_0 d^3 - 12PL^2 x / b_0 d^3 + B$; $v(L) = 0 \Rightarrow B = +6PL^3 / b_0 d^3$.

$E v(x) = \dfrac{PL}{b_0 d^3}(6x^2 - 12Lx + 6L^2)$; $v(0) = 6PL^3 / E b_0 d^3$.

9.8

General case:

$I(x) = I_0 \sin \frac{\pi x}{L}$

$q = q_0 \sin \dfrac{n\pi x}{L}$

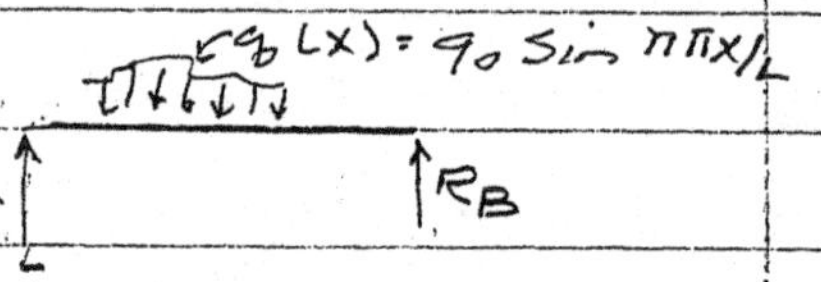

$(\Sigma M)_A = 0 \Rightarrow R_B = \dfrac{q_0}{L} \int_0^L x \sin \dfrac{n\pi x}{L} dx$; $n = 1, 2, \ldots$

$R_B = \dfrac{q_0}{L}\left\{ -\dfrac{L}{n\pi} x \cos \dfrac{n\pi x}{L} \Big|_0^L + \dfrac{L}{n\pi} \int \cos \dfrac{n\pi x}{L} dx \right\} = \dfrac{q_0}{L}\left\{ -\dfrac{L^2}{n\pi} \cos n\pi - \dfrac{L^2}{n^2\pi^2} \sin \dfrac{n\pi x}{L} \Big|_0^L \right\}$

$R_B = -\dfrac{q_0 L \cos n\pi}{n\pi}$; $R_A = \int_0^L q(x) dx - R_B$

$q_0 \int_0^L \sin \dfrac{n\pi x}{L} dx = -q_0 L / n\pi \cos \dfrac{n\pi x}{L} \Big|_0^L = -\dfrac{q_0 L}{n\pi}[\cos n\pi - 1]$.

$R_A = q_0 L / n\pi \, [(1 - \cos n\pi) + \cos n\pi] = q_0 L / n\pi$

$M(x) = R_A x - \int_0^x (x - \xi) q(\xi) d\xi \Rightarrow M(x) = \dfrac{q_0 L^2}{n^2\pi^2} \sin \dfrac{n\pi x}{L}$

$n = 1$ $\quad EI_0 \sin \dfrac{\pi x}{L} v''(x) = -\dfrac{q_0 L^2}{\pi^2} \sin \dfrac{\pi x}{L} \Rightarrow EI_0 v''(x) = -q_0 L^2 / \pi^2$

$EI_0 v'(x) = -q_0 L^2 x / \pi^2 + A$; $EI_0 v(x) = -q_0 L^2 x^2 / 2\pi^2 + Ax + B$; $v(0) = 0 \Rightarrow B = 0$

$v(L) = 0 \Rightarrow A = q_0 L^3 / 2\pi^2 \Rightarrow v(x) = (q_0 L^4 / 2EI_0 \pi^2)(x/L - x^2/L^2)$; $v_{max} = 3 q_0 L^4 / 8\pi^2 EI_0$

NOTE: SEE GENERAL CASE IN PROB. 9.8

9.9 $n = 2$: $\quad M(x) = \dfrac{q_0 L^2}{4\pi^2} \sin \dfrac{2\pi x}{L}$

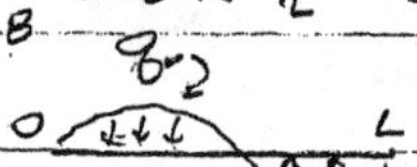

$EI(x) v''(x) = -M(x)$

$EI_0 \sin \dfrac{\pi x}{L} v''(x) = -\dfrac{q_0 L^2}{4\pi^2} \sin \dfrac{2\pi x}{L} = -\dfrac{q_0 L^2}{2\pi^2} \sin \dfrac{\pi x}{L} \cos \dfrac{\pi x}{L}$.

$EI_0 v''(x) = -\dfrac{q_0 L^2}{2\pi^2} \cos \pi x / L \Rightarrow EI_0 v'(x) = -\dfrac{q_0 L^3}{2\pi^3} \sin \dfrac{\pi x}{L} + A$

$EI_0 v(x) = \dfrac{q_0 L^4}{2\pi^4} \cos \dfrac{\pi x}{L} + Ax + B$; $v(0) = 0 \Rightarrow B = -q_0 L^4 / 2\pi^4$

$v(L) = 0 \Rightarrow \dfrac{q_0 L^4}{2\pi^4}(-1 - 1) + AL = 0 \Rightarrow A = q_0 L^3 / \pi^4$

$v(x) = \dfrac{q_0 L^4}{2\pi^4 EI_0} \{\cos \pi x / L + 2x/L - 1\}$;

Note: $v(L/2) = 0$; $v'(L/2) = -\dfrac{q_0 L^3}{2\pi^4 EI_0}(\pi - 2)$

$v_{max} \Rightarrow v'(x_{cr}) = 0$

$\sin \dfrac{\pi x}{L} = 2/\pi$

$x/L = \dfrac{1}{\pi} \sin^{-1}(2/\pi) = 0.2197$

$v_{max} = (1.08 \times 10^{-3}) q_0 L^4 / EI_0$

9.10

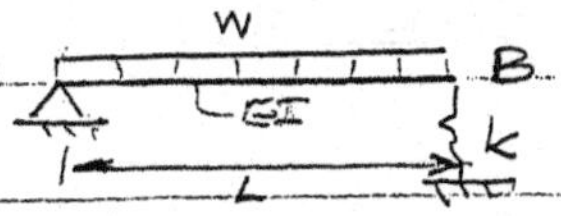

$M(x) = wLx/2 - wx^2/2$

$EIv''(x) = -w[Lx/2 - x^2/2]$ B.C. $v(0) = 0$, $v(L) = wL/2k$

$R_A = R_B = wL/2$

$EIv'(x) = -w[Lx^2/4 - x^3/6] + A$

$EIv(x) = -w(Lx^3/12 - x^4/24) + Ax + B$

$v(0) = 0 \Rightarrow B = 0$; $v(L) = wL/2k \Rightarrow -\dfrac{wL^4}{24EI} + AL = \dfrac{wL}{2k}$

$A = wL^3/24EI + w/2k$

$v(x) = \dfrac{w}{24EI}[2Lx^3 + x^4 + (L^3 + 2EI/k)x]$

$v(x) = \dfrac{wL^4}{24EI}[(x/L)^4 - 2(x/L)^3 + (1 + 12\alpha)(x/L)]$, $\alpha = EI/kL^3$

$k \to \infty$; $v(L/2) = 5wL^4/384EI$; $EI \to \infty$ $v(x) = wx/k$

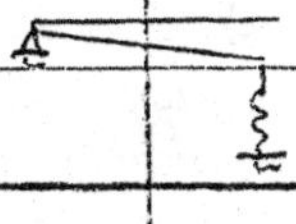

9.11

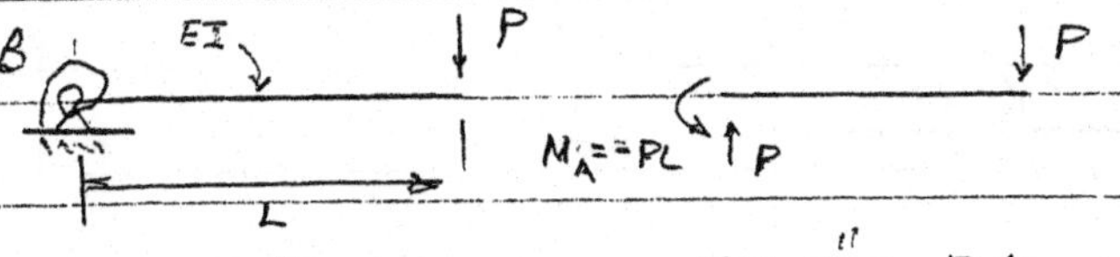

$M(x) = -PL + Px$; $EIv'' = P(L - x)$ B.C. $v(0) = 0$,

$M_A = -\beta\theta_A = -\beta v'(0) \Rightarrow v'(0) = \dfrac{PL}{\beta}$

$EIv'(x) = P(Lx - x^2/2) + A$

$EIv'(0) = A = PL/\beta$

$EIv(x) = P(Lx^2/2 - x^3/6) + (PL/\beta)x + B_{\to 0}$ $[v(0) = 0]$

$v(x) = \dfrac{P}{6EI}\left[3Lx^2 - x^3 + \dfrac{6EIL}{\beta}x\right] = \dfrac{PL^3}{6EI}[3(x/L)^2 - (x/L)^3 + 6\gamma(x/L)]$

$\gamma = EI/\beta L$

b) $\beta \to \infty \Rightarrow \gamma = 0 \Rightarrow v(L) = PL^3/3EI$

c). $EI \to \infty \Rightarrow v(x) = PLx/\beta$

9.12 a) $\Delta_x = \frac{1}{2}\int_0^L (v')^2 dx \qquad v(x) = \frac{M_E}{2EI}(Lx - x^2) \Rightarrow v'(x) = \frac{M_E}{2EI}(L-2x)$

$\Delta_x = \frac{M_E^2}{4(EI)^2}\int_0^L (L-2x)^2 dx = -\frac{M_E^2}{24(EI)^2}(L-2x)^3\Big|_0^L = \frac{M_E^2 L^3}{12(EI)^2}$

$M_E = \sigma_0 I/\alpha d \Rightarrow \Delta_x = \left(\frac{\sigma_0}{E}\right)^2 \frac{L^2}{(\alpha d)^2}\cdot L \simeq \gamma^2 L/12$

b). $\Delta x - (\Delta x)_{ex} = \gamma^2 L/12 - L\left[1 - \frac{2}{\gamma}\sin\gamma/2\right] = L\left\{\frac{\gamma^2}{12} - \left[1 - \frac{2}{\gamma}\left(\frac{\gamma}{2} - \frac{\gamma^3}{8\cdot 6} + \frac{\gamma^5}{120\cdot 32} \ldots\right)\right]\right.$

$\frac{\Delta x - (\Delta x)_{ex}}{(\Delta x)_{ex}} = \frac{\gamma^2 L/24}{L\left[1 - \frac{2}{\gamma}\sin\gamma/2\right]} = \frac{\gamma^2/24}{1 - (1 - \gamma^2/24 + \gamma^4/60\cdot 32)} = \frac{\gamma^2/24}{\gamma^2/24 - \gamma^4/60\cdot 32}$

$= \frac{1}{1 - \gamma^2/80} = 1 + \gamma^2/80$

9.13 

a) $v'(x) = \frac{P}{EI}(Lx - x^2/2)$

$u(x) = \frac{1}{2}\int_0^x (v')^2 dx \quad \frac{P^2}{2(EI)^2}\int_0^x (L\xi - \xi^2/2)^2 d\xi = \frac{P^2}{2(EI)^2}\left(L^2x^3/3 - Lx^4/4 + x^5/20\right)$

$\Delta_x = U(L) = P^2L^5/15(EI)^2$

b) $\frac{\Delta_x}{L} = P^2L^4/15(EI)^2 = \frac{(9\times10^6)(16)}{15(10^{11})^2(10^{-10})} = 9.6\times10^{-6}$

9.14 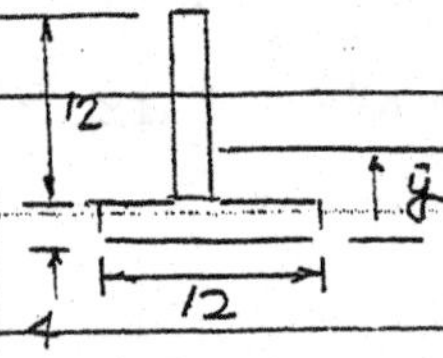

$\bar{y} = \frac{48\times2 + 48\times10}{2(12\times4)} = 6\ cm. \qquad I = \left(\frac{12\times4^3}{12} + 48\times4^2\right) + \left(\frac{4\times12^3}{12} + 48\times4^2\right)$

$I = 2176\ cm^4$

$\Delta_x = P^2L^5/15(EI)^2$ [see Prob. 9.11]

$M_{max} = PL = \sigma I/c \Rightarrow P = \sigma I/cL = \frac{(10\times10^4)(2176\times10^{-8})}{(10\times10^{-2})(2)} = 1088\ N$

$\Delta_x = \frac{(1.088\times10^3)^2(2)^5}{15(15\times10^9)^2(2.176\times10^{-3})^2} = 2.37\times10^{-9}\ m.$

$\Delta x/L = 1.19\times10^{-9}$

9.15 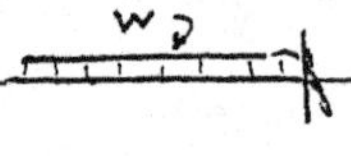

$V = -wx \qquad \tau = \frac{3}{2}\frac{V}{bd} \qquad \gamma = \frac{\tau}{G} = \frac{3}{2}\frac{V}{Gbd}$

$v_s = \frac{3w}{2bdG}\int_0^L x\,dx = \frac{3wL^2}{4bdG}; \quad v_f = \frac{wL^4}{8EI}; \quad \frac{v_s}{v_f} = \frac{1}{2}\frac{G}{E}\left(\frac{d}{L}\right)^2$

$v_s/v_f = \frac{1}{4}\left(\frac{d}{L}\right)^2\left(\frac{1}{1+\nu}\right)$

9.16 

$V = P/2 \quad \gamma = \frac{3}{2}\frac{P}{2}\frac{1}{Gbd} = \frac{3P}{4Gbd}; \quad v_s = \frac{3P}{4Gbd}(L/2) = \frac{3PL}{8Gbd},$

$v_f = PL^3/48EI \Rightarrow v_s/v_f = \frac{3PL/8Gbd}{PL^3/4Ebd^3} = \frac{3}{2}(G/E)(d/L)^2 = \frac{3}{4}(d/L)^2\frac{1}{1+\nu}.$

9.17

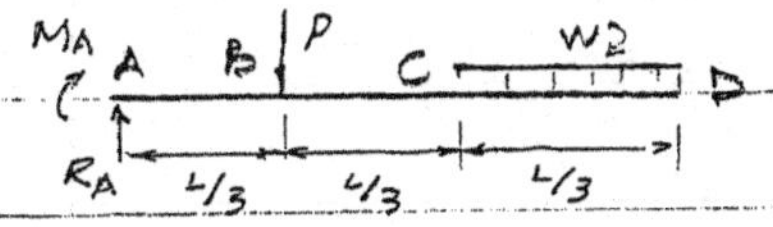

$R_A = P + WL/3$ $\qquad$ B.C. $v'(0) = v(0) = 0$

$M_A = -PL/3 - 5WL^2/18$

$M(x) = -PL/3 - 5WL^2/18 + (P + WL/3)x - P\langle x - L/3\rangle - \frac{W}{2}\langle x - 2L/3\rangle^2; \quad 0 \le x \le L$

$EIv''(x) = PL/3 + 5WL^2/18 - (P + WL/3)x + P\langle x - L/3\rangle + W/2\langle x - 2L/3\rangle^2$

$EIv'(x) = PLx/3 + 5WL^2x/18 - 1/2(P + WL/3)x^2 + P/2\langle x - L/3\rangle^2 + \frac{W}{6}\langle x - 2L/3\rangle^3 + A$ ($A = 0$)

$EIv(x) = PLx^2/6 + 5WL^2x^2/36 - 1/6(P + WL/3)x^3 + P/6\langle x - L/3\rangle^3 + \frac{W}{24}\langle x - 2L/3\rangle^4 + B$ ($B = 0$)

$$v_c = v(L) = \frac{4PL^3}{81EI} + \frac{163WL^4}{1944EI}.$$

9.18

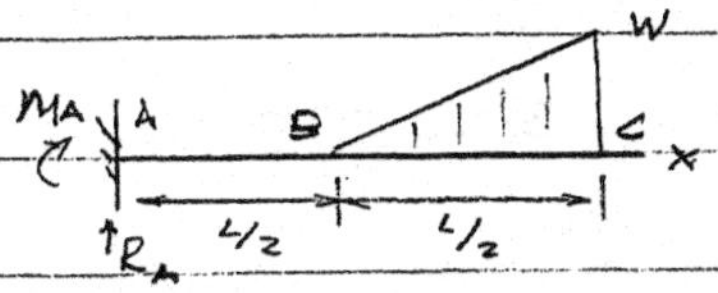

$R_A = WL/4 \qquad M_A = -5WL^2/24$

$M(x) = -5WL^2/24 + WLx/4 - \frac{W}{3L}\langle x - L/2\rangle^3$

$EIv''(x) = 5WL^2/24 - WLx/4 + \frac{W}{3L}\langle x - L/2\rangle^3$

$EIv'(x) = 5WL^2x/24 - WLx^2/8 + \frac{W}{12L}\langle x - L/2\rangle^4 + A$ ($A = 0$)

$EIv(x) = 5WL^2x^2/48 - WLx^3/24 + \frac{W}{60L}\langle x - L/2\rangle^5 + B$ ($B = 0$)

b) $v_B = v(L/2) = WL^4/48EI, \quad v_C = \frac{7wL^4}{640EI}; \quad v_C/v_B = 0.525$

c). $W = 6$ kN/m, $L = 4$ m, WF 203×22 $I = 20\times10^6$ mm^4,

$$v_C = \frac{7(6\times10^3)(4)^4}{640(200\times10^9)(20\times10^{-6})} = 4.20\times10^{-3}\text{ m} = 4.2\text{ mm}.$$

9.19

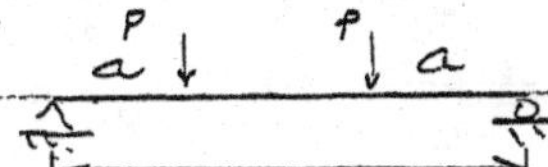

$M(x) = Px - P\langle x - a\rangle - P\langle x - (L-a)\rangle$

$EIv''(x) = -Px + P\langle x - a\rangle + P\langle x - (L-a)\rangle$

$EIv'(x) = -Px^2/2 + P/2\langle x - a\rangle^2 + \frac{P}{2}\langle x - (L-a)\rangle^2 + A$

$EIv(x) = -Px^3/6 + P/6\langle x - a\rangle^3 + P/6\langle x - (L-a)\rangle^3 + Ax + B$

$v(0) = 0 \Rightarrow B = 0; \quad v(L) = 0 \Rightarrow -PL^3/6 + P/6(L-a)^3 + Pa^3/6 + AL = 0$

$\therefore A = \frac{Pa}{2}(L-a)$

$EIv(L/2) = P/6EI\left[-L^3/8 + \langle L - a\rangle^3 + 3a(L-a)(L/2)\right] = \frac{Pa}{6EI}(3L^2/4 - a^2)$

$v(L/2)\big|_{a=L/2} = PL^3/24EI.$

9.20

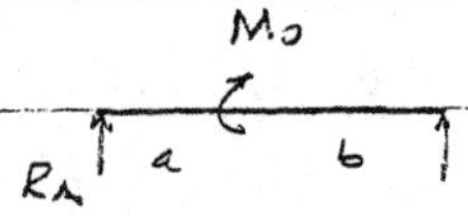

$R_A = -M_0/L \qquad M(x) = -M_0x/L + M_0\langle x-a\rangle^0$

$EIv''(x) = +M_0x/L - M_0\langle x-a\rangle^0$

$EIv'(x) = M_0x^2/2L - M_0\langle x-a\rangle + A$

$EIv(x) = M_0x^3/6L - M_0\langle x-a\rangle^2/2 + Ax + B \;;\; v(0)=0 \Rightarrow B=0$

$v(L) = 0 \Rightarrow M_0L^2/6 - M_0(L-a)^2/2 + AL = 0$

$A = M_0/6L\,[3(L-a)^2 - L^2]$

$v(x) = \dfrac{M_0}{6EI}\left[x^3/L - 3\langle x-a\rangle^2 + (2L - 6a + 3a^2/L)x\right]$

$v(a) = \dfrac{M_0a}{3EIL}(2a^2 - 3aL + L^2)$

Values for which $v(x_{cr}) = v_{max}$ occurs for $x<a$:

$v'(x_{cr}) = M_0x^2/2L + A = 0 \Rightarrow M_0x^2/2L + M_0/6L\,[3(L-a)^2 - L^2] = 0$

Require $3(L-a)^2 - L^2 < 0$

$2L^2 - 6La + 3a^2 < 0 \Rightarrow 3(a/L)^2 - 6(a/L) + 2 < 0$

Let $f(a/L) = 3(a/L)^2 - 6(a/L) + 2 \qquad f(a/L) < 0$

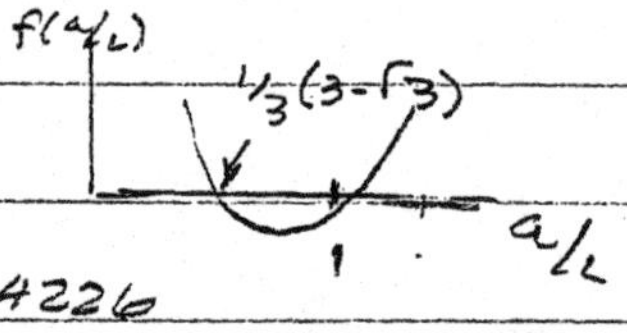

$f(a/L) = 0 \Rightarrow a/L = 1/6\,[6 \pm (36-24)^{1/2}]$

$a/L = 1 - \sqrt{12}/6 = 1 - \sqrt{3}/3 = 1/3(3-\sqrt{3}) = 0.4226$

v_{max} occurs for $x<a$ only if $a/L > 0.4226$.

$x_{cr}^2 = -1/3\,[2L^2 - 6La + 3a^2] = -L^2/3\,[2 - 6(a/L) + 3(a/L)^2]$

Note: $x_{cr} = x_{cr}(a/L)$

9.21

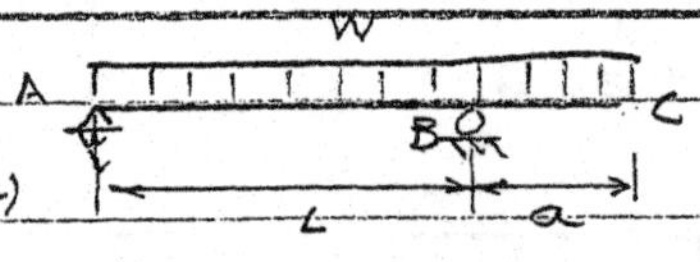

$(\Sigma M)_A = 0 \Rightarrow R_B = W(L+a)^2/2L \,,\; \Sigma F_y = 0 \Rightarrow R_A = \dfrac{W(L^2-a^2)}{2L}$

a) $M(x) = \dfrac{W}{2}(L^2-a^2)x - Wx^2/2 + \dfrac{W(L+a)^2}{2L}\langle x-L\rangle$

$EIv''(x) = -\dfrac{W}{2L}(L^2-a^2)x + Wx^2/2 - \dfrac{W(L+a)^2}{2L}\langle x-L\rangle \Rightarrow EIv'(x) = -\dfrac{W}{4L}(L^2-a^2)x^2 + \dfrac{Wx^3}{6} - \dfrac{W(L+a)^2}{4L}\langle x-L\rangle^2 + A$

$EIv(x) = -\dfrac{W}{12L}(L^2-a^2)x^3 + Wx^4/24 - \dfrac{W(L+a)^2}{12L}\langle x-L\rangle^3 + Ax + B$

$v(0) = 0 \Rightarrow B = 0 \;;\; v(L) = 0 \Rightarrow A = \dfrac{WL}{24}(L^2 - 2a^2)$

$v(x) = \dfrac{W}{24EI}\left\{-2(L^2-a^2)x^3/L + x^4 - \dfrac{2(L+a)^2}{L}\langle x-L\rangle^3 + (L^2-2a^2)Lx\right\}.$

$= \dfrac{WL^4}{24EI}\left\{-2(1-a^2/L^2)(x/L)^3 + (x/L)^4 - 2(1+a/L)^2\langle x/L - 1\rangle^3 + (1-2a^2/L^2)(x/L)\right\}$

b) $v_C = 0 \Rightarrow a/L = 1/6\,[\sqrt{13} - 1] = 0.4343$

9.22

$R_A = -Pa/L, \quad R_B = P(a+L)/L$

$M(x) = -Pa/L + \frac{P}{L}(a+L)\langle x-L\rangle$

$EI\,v''(x) = +Pax/L - \frac{P}{L}(L+a)\langle x-L\rangle$

$EI\,v'(x) = Pax^2/2L - \frac{P}{2L}(L+a)\langle x-L\rangle^2 + A$

$EI\,v(x) = Pax^3/6L - \frac{P}{6L}(L+a)\langle x-L\rangle^3 + Ax + B \qquad v(0)=0 \Rightarrow B=0$

$v(L)=0 \Rightarrow A = -PaL/6$

$v(x) = \frac{P}{6EI}\left[ax^3/L - (L+a)\langle x-L\rangle^3/L - aLx\right]$

$v_{max} \Rightarrow v'(x_{cr}) = 0 \Rightarrow x_{cr} = \sqrt{3}L/3 \Rightarrow |v|_{max} = PaL^2\sqrt{3}/27EI$

$v_c = \frac{Pa^2(L+a)}{3EI} \qquad |v_{max}| = v_c \Rightarrow a/L = 1/6\left[(9+4\sqrt{3})^{1/2} - 3\right]$

$|v_{max}| = v_c = 0.010596\, PL^3/EI$

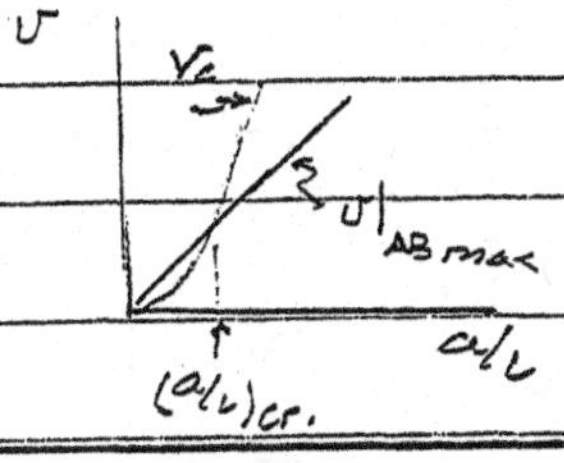

9.23

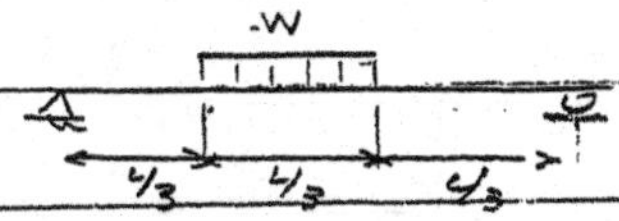

$M(x) = \frac{wL}{6}x - \frac{w}{2}\langle x-L/3\rangle^2 + \frac{w}{2}\langle x-2L/3\rangle^2$

$EI\,v''(x) = -wLx/6 + \frac{w}{2}\langle x-L/3\rangle^2 - \frac{w}{2}\langle x-2L/3\rangle^2$

$EI\,v'(x) = -wLx^2/12 + \frac{w}{6}\langle x-L/3\rangle^3 - \frac{w}{6}\langle x-2L/3\rangle^3 + A$

$EI\,v(x) = -wLx^3/36 + \frac{w}{24}\langle x-L/3\rangle^4 - \frac{w}{24}\langle x-2L/3\rangle^4 + Ax + B$

$v(0)=0 \Rightarrow B=0; \quad v(L)=0 \Rightarrow A = 13wL^3/18\cdot 36$

$v(L/2) = \frac{205\, wL^4}{31{,}104\, EI}$

9.24

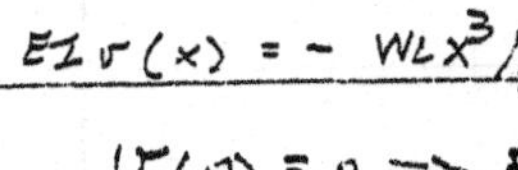

$R_C = wL/4, \quad R_A = P + wL/4, \quad M_A = -PL/4 - wL^2/8$

Note: Due to discontinuity in slope at B, must consider two domains: $0 < x < L/2$; $L/2 \le x_2 \le L$

$M(x) = -PL/4 - wL^2/8 + (P + wL/4)x - P\langle x-L/4\rangle \;; \quad 0 \le x \le L/2$

$M(x) = -PL/4 - wL^2/8 + (P+wL/4)x - P\langle x-L/4\rangle - \frac{w}{2}\langle x-L/2\rangle^2; \quad L/2 \le x \le$

$EI\,v_1''(x) = PL/4 + wL^2/8 - (P+wL/4)x + P\langle x-L/4\rangle$

$EI\,v_2''(x) = PL/4 + wL^2/8 - (P+wL/4)x + P\langle x-L/4\rangle + \frac{w}{2}\langle x-L/2\rangle^2$

B.C. $v_1'(0) = v_1(0) = 0, \quad v_2(L) = 0, \quad v_1(L/2) = v_2(L/2)$

$EIv_1'(x) = (PL + \frac{wL^2}{2})x/4 - (P + wL/4)x^2/2 + P\langle x - L/4\rangle^2/2 + A_1 \quad [v'(0) = 0]$

$EIv_1(x) = (PL + \frac{wL^2}{2})x^2/8 - (P + wL/4)x^3/6 + P\langle x - L/4\rangle^3/6 + B_1 \quad [v(0) = 0]$

$EIv_2'(x) = (PL + \frac{wL^2}{2})x/4 - (P + wL/4)x^2/2 + P\langle x - L/4\rangle^2/2 + w\langle x - L/2\rangle^3/6 + A_2$

$EIv_2(x) = (PL + \frac{wL^2}{2})x^2/8 - (P + wL/4)x^3/6 + P\langle x - L/4\rangle^3/6 + w\langle x - L/2\rangle^4/24 + A_2x + B_2$

$v_1(L/2) = v_2(L/2) \Rightarrow B_2 = -A_2L/2$

$v_2(L) = \frac{11}{384}PL^3 + \frac{3wL^4}{128} + A_2L + B_2 = 0$

$\therefore A_2 = -\frac{11PL^2}{192} - 3wL^3/64$

$v_B = \frac{5PL^3}{384EI} + \frac{wL^4}{96EI}$

$v(3L/4) = \frac{5PL^3}{768EI} + \frac{37wL^4}{256 \cdot 24}$

$\Delta\theta_B = \theta_B^+ - \theta_B^- = \frac{A_2}{EI} = -\frac{11PL^2}{192EI} - \frac{3wL^3}{64EI}$

9.25

$M(x) = M_A + R_Ax$

$EIv''(x) = -M_A - R_Ax \quad ; \quad B.C. \quad v'(0) = v(0) = v(L) = 0$

$EIv'(x) = -M_Ax - R_Ax^2/2 + A \;(=0), \quad EIv(x) = -M_Ax^2/2 - R_Ax^3/6 + B \;(=0)$

$v(L) = 0 \Rightarrow R_A = -3M_A/L$

$\circlearrowleft (\sum M_i)_B = 0 \Rightarrow R_AL + M_A = M_0 \Rightarrow -3M_A = M_0 \Rightarrow M_A = -M_0/3, \quad R_A = 3M_0/2L, \quad R_B = -3M_0/2L$

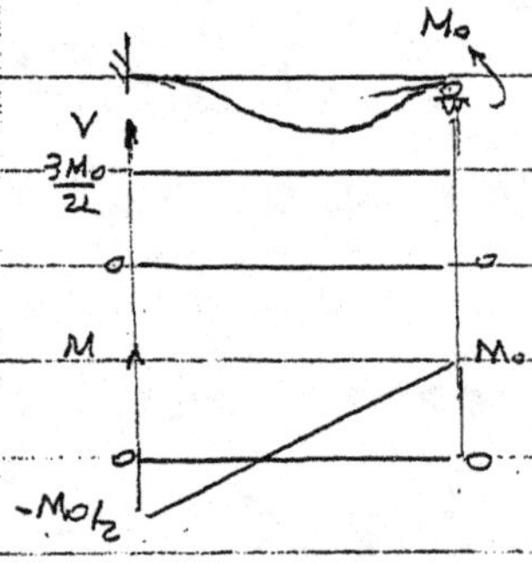

$M(x) = \frac{M_0}{2}(3x - 1) \qquad v(x) = \frac{M_0}{4EI}[x^2 - x^3/L]$

$x_{cr} = 2L/3 \qquad |v_{max}| = +\frac{M_0L^2}{27EI}$

9.26

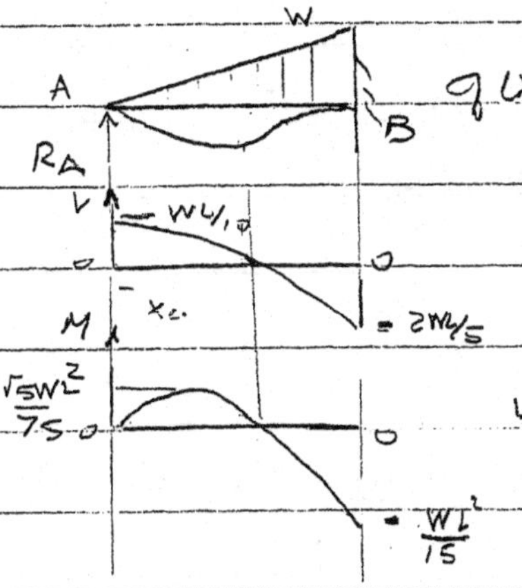

$q(x) = Wx/L \qquad M = R_A x - Wx^3/6L$

$EI v''(x) = -R_A x + Wx^3/6L \qquad$ B.C. $v(0) = v'(L) = v(L) = 0$

$EI v'(x) = -R_A x^2/2 + Wx^4/24L + A$

$EI v(x) = -R_A x^3/6 + Wx^5/120L + Ax + B \qquad [v(0)=0]$

$v'(L) = 0 \Rightarrow A = R_A L^2/2 - WL^4/24L$

$v(L) = 0 \Rightarrow -R_A L^3/6 + WL^4/120 + R_A L^3/2 - WL^3/24 = 0$

$R_A = WL/10 \Rightarrow R_B = 2WL/5$

$M_B = R_A L - WL^2/6 = WL^2/15$

$M_{max} = M(x = \sqrt{5}/5) = \sqrt{5}WL^2/75; \quad M_B = -2WL^2/15$

$v' = 0 \Rightarrow 5x^4 - 6L^2x^2 + L^4 = 0 \Rightarrow x = \sqrt{5}L/5 \qquad v_{max} = \frac{8\sqrt{5}\,WL^4}{7500} = \frac{2\sqrt{5}\,WL^4}{1875}$

9.27

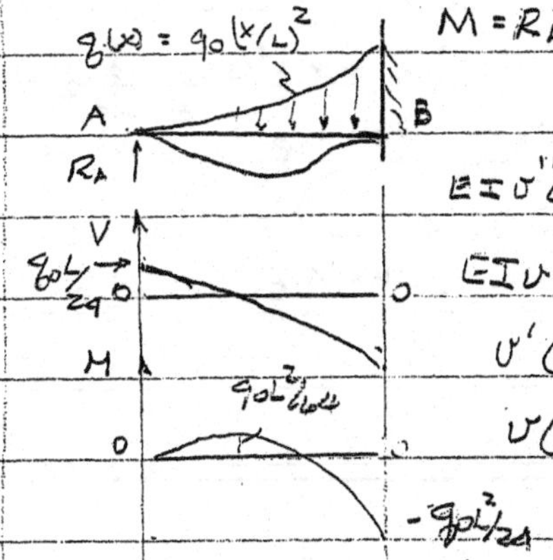

$M = R_A x - \frac{q_0}{L^2}\int_0^x (x-\xi)\xi^2\, d\xi = R_A x - \frac{q_0}{L^2}\left[x\cdot\frac{\xi^3}{3} - \frac{\xi^4}{4}\right]_0^x = R_A x - q_0 x^4/12L^2$

$EI v''(x) = -R_A x + q_0 x^4/12L^2; \quad EI v'(x) = -R_A x^2/2 + q_0 x^5/60L^2 + A$

$EI v(x) = -R_A x^3/6 + q_0 x^6/360L^2 + Ax + B; \quad v(0) = 0 \Rightarrow B = 0$

$v'(L) = 0 \Rightarrow A = R_A L^2/2 - q_0 L^3/60$

$v(L) = 0 \Rightarrow -R_A L^3/6 + q_0 L^4/360 + R_A L^3/2 - q_0 L^4/60 = 0$

$R_A = q_0 L/24 \qquad R_B = \frac{q_0}{L^2}\int x^2\, dx - R_A = 7q_0 L/24$

$M_{max} = M(x = L/2) = q_0 L^2/64$

9.28

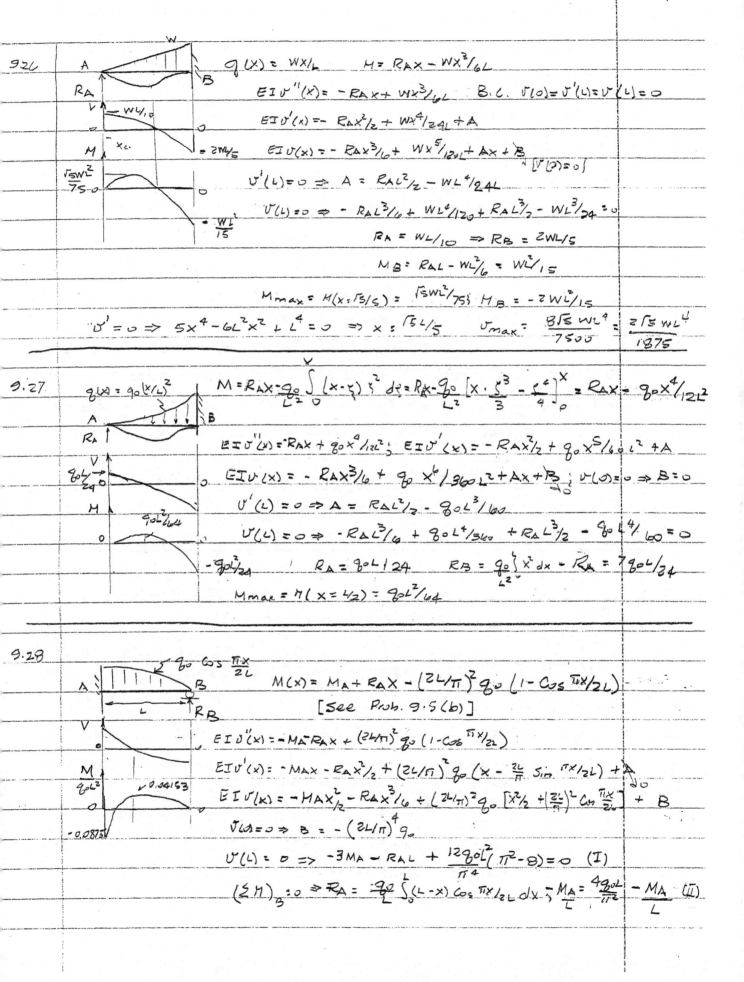

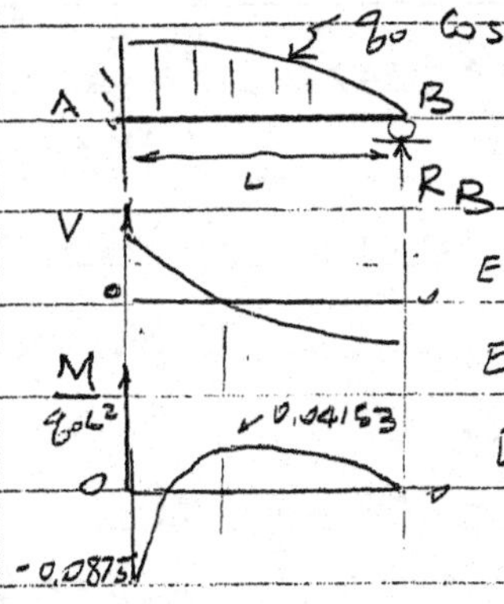

$M(x) = M_A + R_A x - (2L/\pi)^2 q_0 (1 - \cos \pi x/2L)$

[See Prob. 9.5(b)]

$EI v''(x) = -M_A - R_A x + (2L/\pi)^2 q_0 (1 - \cos \pi x/2L)$

$EI v'(x) = -M_A x - R_A x^2/2 + (2L/\pi)^2 q_0 (x - \frac{2L}{\pi} \sin \pi x/2L) + A$ ($A = 0$)

$EI v(x) = -M_A x^2/2 - R_A x^3/6 + (2L/\pi)^2 q_0 \left[x^2/2 + \left(\frac{2L}{\pi}\right)^2 \cos\frac{\pi x}{2L}\right] + B$

$v(0) = 0 \Rightarrow B = -(2L/\pi)^4 q_0$

$v(L) = 0 \Rightarrow -3M_A - R_A L + \frac{12 q_0 L^2}{\pi^4}(\pi^2 - 8) = 0 \quad (I)$

$(\Sigma M)_B = 0 \Rightarrow R_A = \frac{q_0}{L}\int_0^L (L-x)\cos \pi x/2L\, dx - \frac{M_A}{L} = \frac{4q_0L}{\pi^2} - \frac{M_A}{L} \quad (II)$

Solving (I) & (II) $\Rightarrow M_A = \frac{4 q_0 L^2}{\pi^4}(\pi^2 - 12) = -0.0875\, q_0 L^2$

$R_A = 48 q_0 L/\pi^4 = 0.4928\, q_0 L$

$M(x) = \frac{4 q_0 L^2}{\pi^4}\{12(x/L - 1) + \pi^2 \cos \pi x/2L\}$

M_{max}: $x/L = 2/\pi \sin^{-1}(24/\pi^3) = 0.56353$, $M_{max} = 0.041525\, q_0 L^2$

$v(x) = \frac{8 q_0 L^4}{EI \pi^4}\left[-(x/L)^3 + 3(x/L)^2 + 2(\cos \frac{\pi x}{2L} - 1)\right]$

9.29

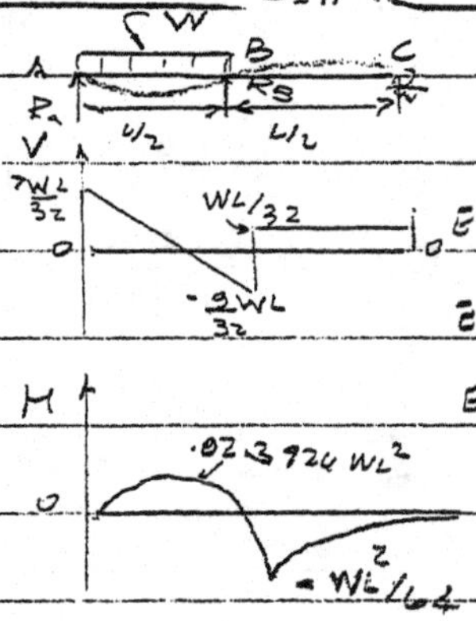

$M(x) = R_A x - wx^2/2 + R_B \langle x - L/2 \rangle + \frac{w}{2}\langle x - L/2 \rangle^2$

B.C. $v(0) = v(L/2) = v(L) = 0$

$EI v'' = -R_A x + wx^2/2 - R_B \langle x - L/2 \rangle - w/2 \langle x - L/2 \rangle^2$

$EI v'(x) = -R_A x^2/2 + wx^3/6 - R_B \langle x - L/2 \rangle^2/2 - w \langle x - L/2 \rangle^3/6 + A$

$EI v(x) = -R_A x^3/6 + wx^4/24 - R_B \langle x - L/2 \rangle^3/6 - w \langle x - L/2 \rangle^4/24 + Ax + B$

$v(0) = 0 \Rightarrow B = 0$, $v(L/2) = 0 \Rightarrow -R_A L^3/48 + wL^4/384 + AL/2 = 0$ (I)

$v(L) = 0 \Rightarrow -R_A L^3/6 - R_B L^3/48 + 5wL^4/128 + AL = 0$ (II)

$M_{max} = 49 wL^2/2048$

$\circlearrowleft (\Sigma M)_c = 0 \Rightarrow R_A L + R_B L/2 - 3wL^2/8 = 0$ (III)

(I), (III) & (II) $\Rightarrow R_A = 7wL/32$, $R_B = 5wL/16$, $A = wL^3/256$; $\Sigma F_y = 0 \Rightarrow R_C = \frac{wL}{32}\downarrow$

$v_{max} = v(x/L = 0.23216) = 5.7191 \times 10^{-4})\, wL^4/EI$

9.30

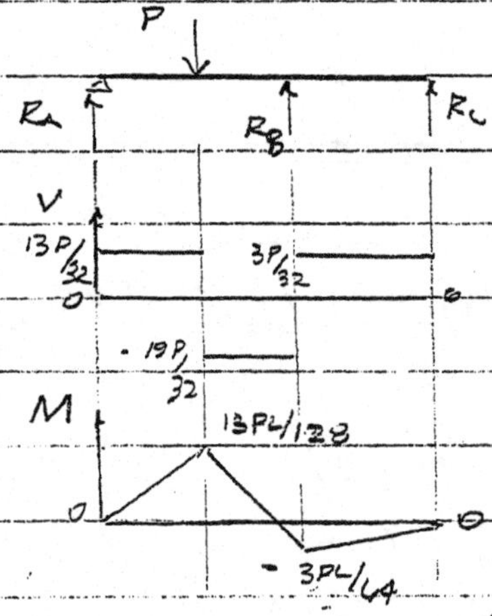

$M(x) = R_A x - P\langle x - L/4 \rangle + R_B \langle x - L/2 \rangle$

$EI v''(x) = -R_A x + P\langle x - L/4 \rangle - R_B \langle x - L/2 \rangle$

$EI v'(x) = -R_A x^2/2 + P\langle x - L/4 \rangle^2/2 - R_B \langle x - L/2 \rangle^2/2 + A$

$EI v(x) = -R_A x^3/6 + P\langle x - L/4 \rangle^3/6 - R_B \langle x - L/2 \rangle^3/6 + Ax + B$

B.C. $v(0) = 0 \Rightarrow B = 0$; $v(L/2) = 0 \Rightarrow -R_A L^3/48 + PL^3/384 + AL/2 = 0$ (I)

$v(L) = 0 \Rightarrow -R_A/6 + 9P/128 - R_B/48 + A/L^2 = 0$ (II)

$\circlearrowleft (\Sigma M)_c = 0 \Rightarrow R_A + R_B/2 - 3P/4 = 0$ (III)

$R_A = 13P/32$, $R_B = 11P/16$, $A = 3PL^2/256$, $R_C = 3P/32\downarrow$

$v_{max} = v(x/L = 0.24019) = (1.8765 \times 10^{-3})\, PL^3/EI$.

9.31

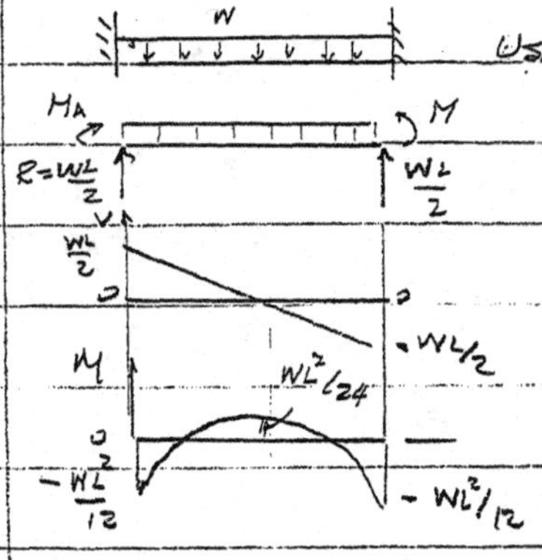

Using symmetry: $R = wL/2$; $v'(L/2) = 0$

$M(x) = M_A + wLx/2 - wx^2/2$

$EI\,v''(x) = -M_A - wLx/2 + wx^2/2$; B.C. $v(0) = v'(0) = v'(L/2) = 0$

$EI\,v'(x) = -M_A x - wLx^2/4 + wx^3/6 + A$

$v(0) = 0 \Rightarrow A = 0$; $v'(L/2) = 0 \Rightarrow M_A = -wL^2/12$

$EI\,v(x) = -M_A x^2/2 - wLx^3/12 + wx^4/24 + B$; $v(0) = 0 \Rightarrow B = 0$

$EI\,v(x) = \frac{w}{24}(x^4 - 2Lx^3 + L^2x^2)$

$v_{max} = v(L/2) = wL^4/384EI$.

Note: Without considering symmetry (R unknown) B.C's are $v(0) = v'(0) = v'(L) = v(L) = 0$, yield same solution.

9.32

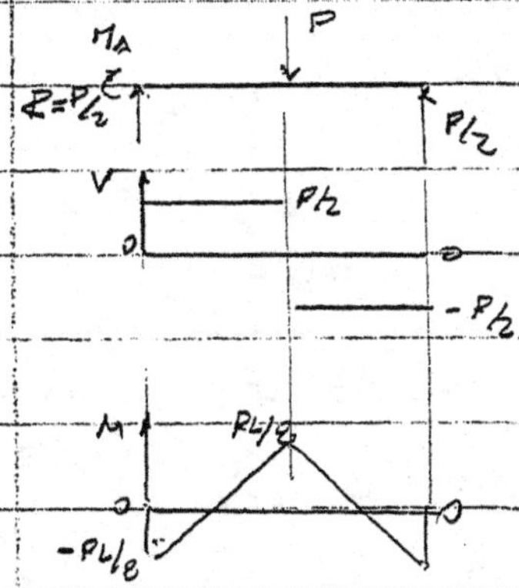

Using symmetry: $R = P/2$, $v'(L/2) = 0$

$M(x) = M_A + Px/2 - P\langle x - L/2\rangle$

$EI\,v''(x) = -M_A - Px/2 + P\langle x - L/2\rangle$

$EI\,v'(x) = -M_A x - Px^2/4 + P\langle x - L/2\rangle^2/2 + A$

$v'(0) = 0 \Rightarrow A = 0$; $v'(L/2) = 0 \Rightarrow M_A = -PL/8$

$EI\,v(x) = -M_A x^2/2 - Px^3/12 + P\langle x - L/2\rangle^3/6 + B_0$

$v(x) = \frac{P}{EI}[Lx^2/16 - x^3/12] = \frac{P}{48EI}(3Lx^2 - 4x^3)$; $0 \le x \le L/2$

$v_{max} = v(L/2) = PL^3/192EI$.

9.33

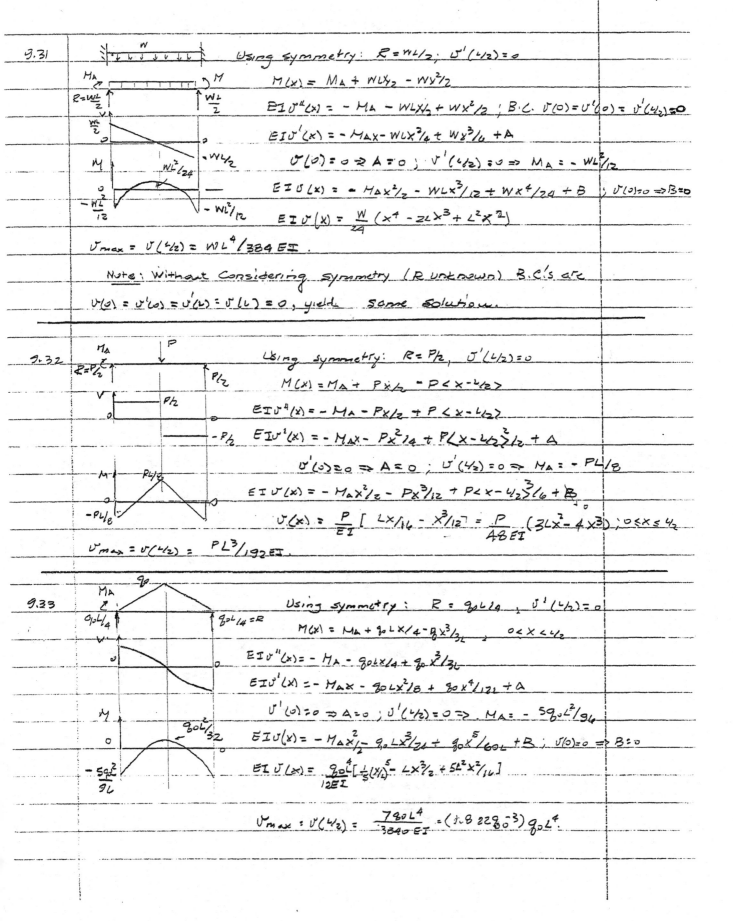

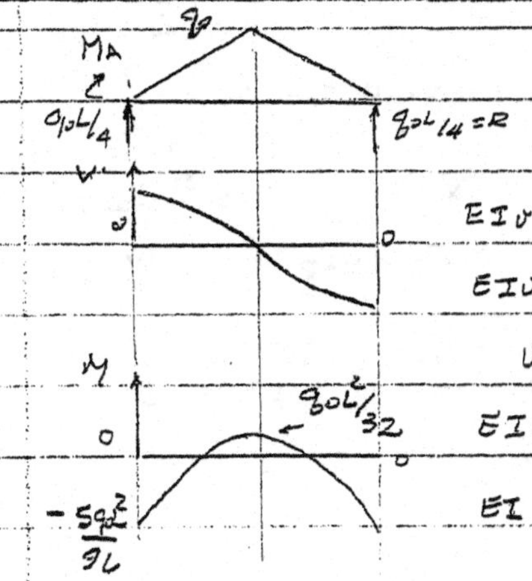

Using symmetry: $R = q_0L/4$, $v'(L/2) = 0$

$M(x) = M_A + q_0Lx/4 - q_0x^3/3L$, $0 < x < L/2$

$EI\,v''(x) = -M_A - q_0Lx/4 + q_0x^3/3L$

$EI\,v'(x) = -M_A x - q_0Lx^2/8 + q_0x^4/12L + A$

$v'(0) = 0 \Rightarrow A = 0$; $v'(L/2) = 0 \Rightarrow M_A = -5q_0L^2/96$

$EI\,v(x) = -M_A x^2/2 - q_0Lx^3/24 + q_0x^5/60L + B$; $v(0) = 0 \Rightarrow B = 0$

$EI\,v(x) = \frac{q_0L^4}{12EI}[\frac{1}{5}(x/L)^5 - Lx^3/2 + 5L^2x^2/16]$

$v_{max} = v(L/2) = \frac{7q_0L^4}{3840EI} = (1.822 \times 10^{-3})\,q_0L^4$.

9P.34

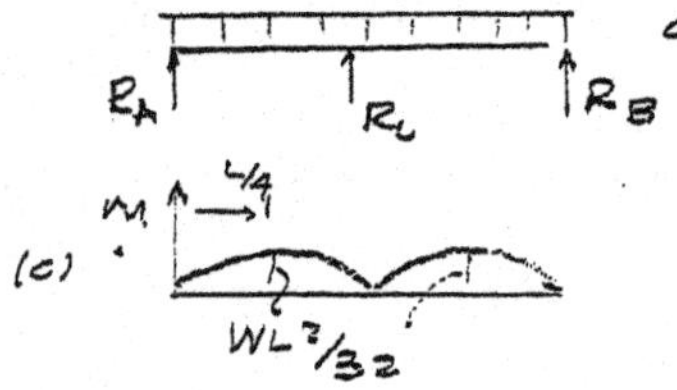

a) $M(x) = R_A x - Wx^2/2 + R_C\langle x - L/2\rangle$

$EIv''(x) = -R_A x + Wx^2/2 - R_C\langle x - L/2\rangle$

$v(0) = 0\ ;\ v(L/2) = R_C/k$

$EIv'(x) = -R_A x^2/2 + Wx^3/6 - R_C\langle x - L/2\rangle^2/2 + A$

$EIv(x) = -R_A x^3/6 + Wx^4/24 - R_C\langle x - L/2\rangle^3/6 + Ax + B$ $\quad [v(0)=0]$

$v(L) = 0 \Rightarrow A = \dfrac{R_A L^2}{6} + \dfrac{R_C L^2}{48} - WL^3/24$

$(\Sigma M)_B = 0 \Rightarrow R_A L + R_C L/2 = WL^2/2 \Rightarrow R_A = \frac{1}{2}(WL - R_C)$

$\therefore A = WL^3/24 - R_C L^2/16$

$\therefore v(x) = \dfrac{1}{48EI}\left[W(2x^4 - 4Lx^3 + 2L^3x) + R_C(4x^3 - 3L^2x) - 8R_C\langle x - L/2\rangle^3\right]$

$v(L/2) = R_C/k$

$\Rightarrow [1/k + L^3/48EI]R_C = 5WL^4/384EI \Rightarrow R_C = \dfrac{5WL}{8}\left[1 + \dfrac{48EI}{kL^3}\right]^{-1}$

b)

$R_A = (WL - R_C)/2 \Rightarrow M_C = R_A L/2 - WL^2/8 = 0 \Rightarrow R_A = WL/4$

$\Rightarrow R_C = WL - 2R_A = WL/2$.

$WL/2 = 5WL/8(1 + 48EI/kL^3)^{-1} \Rightarrow (1 + 48EI/kL^3) = 5/4$

c). See figure above for moment diagram $\quad k = 192EI/L^3$.

9.35

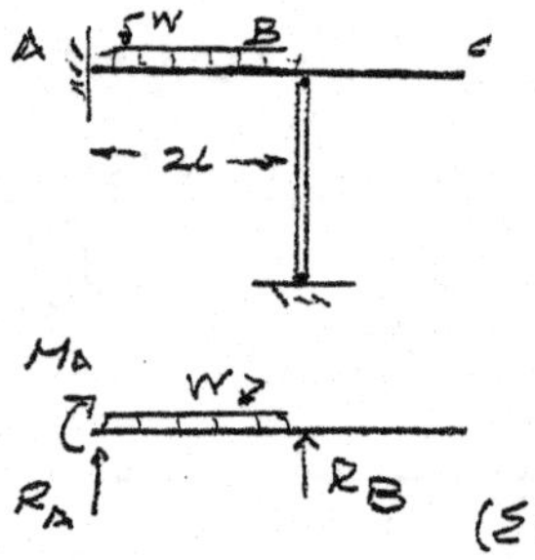

$M(x) = M_A + R_A x - Wx^2/2 + R_B\langle x - 2L\rangle$ B.C. $v'(0) = v(0) = 0$

$v(2L) = R_B h/AE$

$EIv''(x) = -M_A - R_A x + Wx^2/2 - R_B\langle x - 2L\rangle - \frac{W}{2}\langle x - 2L\rangle^2$

$EIv'(x) = -M_A x - R_A x^2/2 + Wx^3/6 - R_B\langle x - 2L\rangle^2/2 + \frac{W}{6}\langle x - 2L\rangle^3 + A$ $\quad A = 0\ [v'(0)=0]$

$EIv(x) = -M_A x^2/2 - R_A x^3/6 + Wx^4/24 - R_B\langle x - 2L\rangle^3/6 + \frac{W}{24}\langle x - 2L\rangle^4 + B$ $\quad B = 0\ [v(0)=0]$

$(\Sigma M)_B = 0 \Rightarrow M_A = 2WL^2 - 2LR_A$

$\Sigma F_y = 0 \Rightarrow R_A = 2WL - R_B \Rightarrow \therefore M_A = -2WL^2 + 2LR_B$

$v(x) = \dfrac{1}{24EI}\left\{W[x^4 - 8Lx^3 + 24L^2x^2] + (4x^3 - 24Lx^2)R_B\right]$

$\left. - R_B\langle x - 2L\rangle^3/6 + \dfrac{W}{24}\langle x - 2L\rangle^4\right\}$

$v(2L) = R_B h/AE \Rightarrow \dfrac{2WL^4}{EI} - \dfrac{8R_B L^3}{3EI} = R_B h/AE \Rightarrow R_B = \dfrac{6WL^4 A}{8AL^3 + 3hI}$,

$M_A = 2(R_B L - WL^2) = -\dfrac{2WL^2(2AL^3 + 3hI)}{8AL^3 + 3hI}$

$R_A = 2WL - R_B = \dfrac{2WL(5AL^3 + 3hI)}{8AL^3 + 3hI}$

c). $v_C = v(3L) = \dfrac{WL^4}{3EI}\left(\dfrac{30hI - 4AL^3}{8AL^3 + 3hI}\right)$

9.36

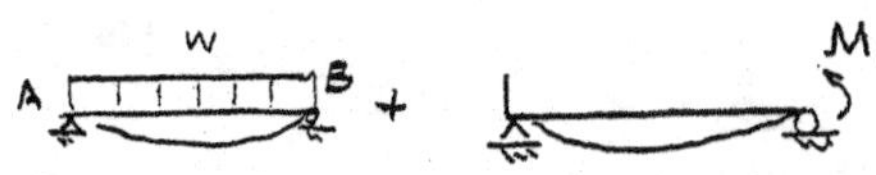

$$\theta_B = \theta_B^{w} + \theta_B^{M} = 0 \Rightarrow -\frac{wL^3}{24EI} + \left(-\frac{M_oL}{3EI}\right) = 0$$

$$M_o = -wL^2/8$$

9.37

$$\theta_B = \theta^{w} + \theta^{P} = -wL^3/24EI + \left(PL^3/16EI\right) = 0$$

$$P = 2wL/3$$

9.38

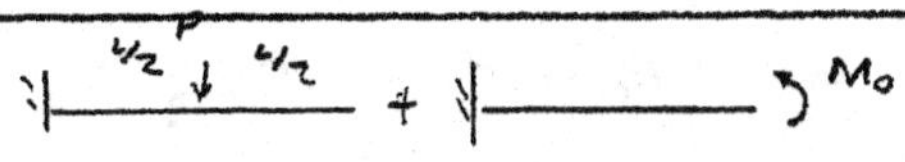

$$v(L/2) = v^{P}(L/2) + v^{M}(L/2) = PL^3/24EI + \frac{M_oL^2}{16EI} = 0$$

$$M_o = 2PL/3$$

9.39

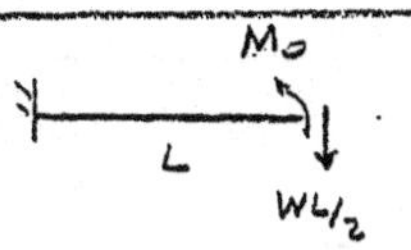

$$v(L) = v^{M}(L) + v^{w}(L) = -\frac{M_oL}{2EI} + \frac{(wL/2)L^3}{3EI} = 0$$

$$M_o = wL^2/3$$

9.40

a)

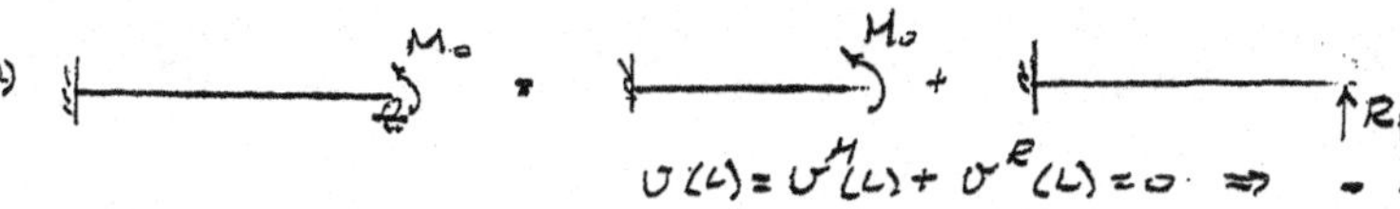

$$v(L) = v^{M}(L) + v^{R}(L) = 0 \Rightarrow -\frac{M_oL^2}{2EI} - \frac{RL^3}{3EI} = 0 \Rightarrow R_B = -3M_o/2L \downarrow$$

$$R_A = 3M_o/2L \uparrow, \quad M_A = M_o + R_BL = -M_o/2$$

b)

$$\theta_A = \theta_A^{M_o} + \theta_A^{M_A} = 0 \Rightarrow \frac{M_oL}{6EI} + \frac{M_AL}{3EI} = 0 \Rightarrow M_A = -M_o/2$$

$$\sum M_A = 0 \Rightarrow M_A - R_BL - M_o = 0 \Rightarrow R_BL = M_A + M_o = M_o/2, \quad R_B = 3M_o/2L \downarrow$$

$$R_A = 3M_o/2L \uparrow$$

9.41

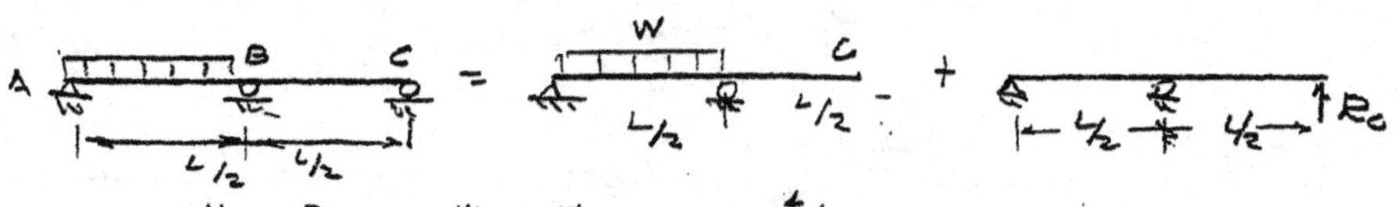

$$v_c = v_c^{w} + v_c^{R} = 0; \quad v_c^{w} = \theta_c^{w} \cdot L/2 = -wL^4/384EI.$$

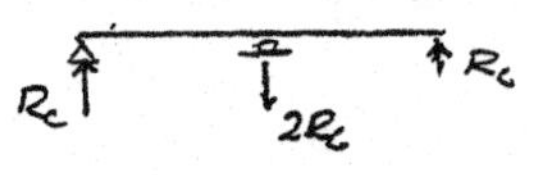

$$EIv''(x) = -M(x) = -R_cx + 2R_c\langle x - \tfrac{L}{2}\rangle; \quad EIv'(x) = -R_cx^2/2 + R_c\langle x - \tfrac{L}{2}\rangle^2 + A$$

$$EIv(x) = -R_cx^3/6 + R_c\langle x - \tfrac{L}{2}\rangle^3/3 + Ax + B; \quad v(0) = 0 \Rightarrow B = 0, \quad v(L/2) = 0 \Rightarrow A = R_cL^2/24$$

$$v(x) = \frac{R_c}{24EI}\left[-4x^3 + 8\langle x - L/2\rangle^3 + L^2x\right] \Rightarrow v(L) = -R_cL^3/12EI = v_c^{R}$$

$$-wL^4/384EI - R_cL^3/12EI = 0 \Rightarrow R_c = -wL/32; \quad (\sum M)_B = 0 \Rightarrow R_A = 7wL/32,$$

$$\sum F_y = 0 \Rightarrow R_B = 5wL/16$$

9.42

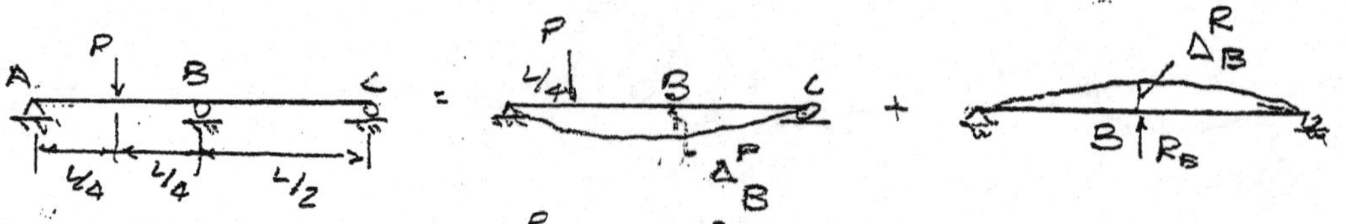

From eq (9.4.18) $\Rightarrow$ $\Delta_B^P = 11PL^3/768EI$

$\Delta_B^R = R_B L^3/48EI$

$\Delta_B = \Delta_B^P - \Delta_B^R = 0$

$11PL^3/768EI - R_B L^3/48EI \Rightarrow R_B = 11P/16$

$\Sigma M_A = 0 \Rightarrow LR_C + \frac{L}{2}R_B - PL/4 = 0 \Rightarrow LR_C = PL/4 - 11PL/32$

$R_C = -3P/32$; $\Sigma F_y = 0 \Rightarrow R_A = 13P/32$

9.43

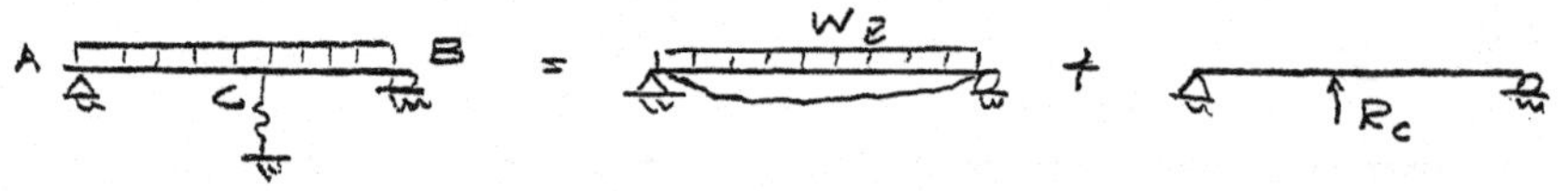

$v_C = v_C^w + v_C^R = R_C/k \Rightarrow \frac{5wL^4}{384EI} + \left(-\frac{R_C L^3}{48EI}\right) = R_C/k \Rightarrow R_C = \frac{5wL}{8}\left(1 + \frac{48EI}{kL^3}\right)^{-1}$

or

$R_C = \frac{5wL}{8}\left(\frac{\alpha}{1+\alpha}\right)$, $\alpha \equiv kL^3/48EI$

9.44 a)

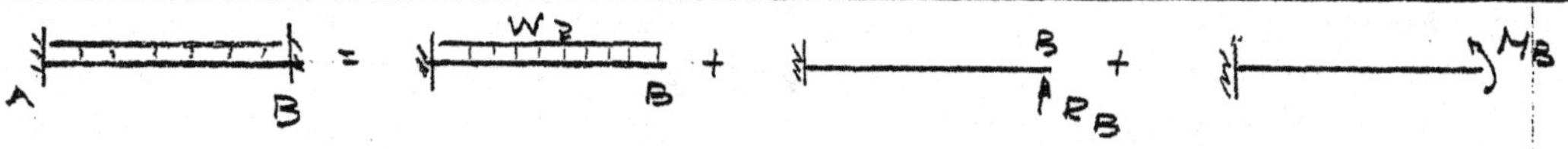

$v_B = v_B^w + v_B^R + v_B^M = 0 \Rightarrow wL^4/8EI - \frac{R_B L^3}{3EI} - \frac{M_B L^2}{2EI} = 0 \Rightarrow 8LR_B + 12M_B = 3wL^2$ (a)

$\theta_B = \theta_B^w + \theta_B^R + \theta_B^M = 0 \Rightarrow \frac{wL^3}{6EI} - \frac{R_B L^2}{2EI} - \frac{M_B L}{EI} = 0 \Rightarrow 3LR_B + 6M_B = wL^2$ (b)

Solving for (a) & (b) $\Rightarrow R_B = wL/2$, $M_B = -wL^2/12$

By symmetry: $R_A = wL/2$, $M_A = -wL^2/12$

b)

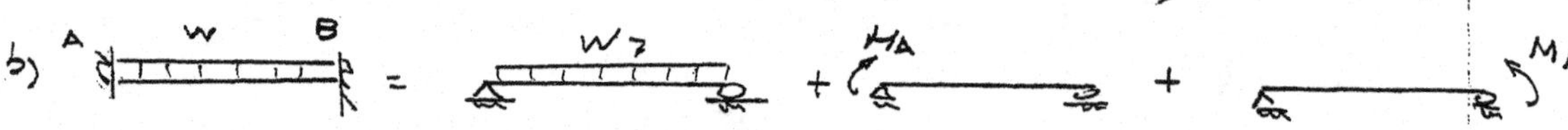

$\theta_A = \theta_A^w + \theta_A^{M_A} + \theta_A^{M_B} = 0 \Rightarrow wL^3/24EI + \frac{M_A L}{3EI} + \frac{M_B L}{6EI} = 0 \Rightarrow 8M_A + 4M_B = -wL^2$ (a)

$\theta_B = \theta_B^w + \theta_B^{M_A} + \theta_B^{M_B} = 0 \Rightarrow -wL^3/24 - \frac{M_A L}{6EI} - \frac{M_B L}{3EI} = 0 \Rightarrow 4M_A + 8M_B = -wL^2$ (b)

Solving (a) & (b) $\Rightarrow M_A = M_B \Rightarrow M_A = M_B = -wL^2/12$

$R_A = R_B = wL/2$ by symmetry.

9.45 Solution analogous with Prob. 9.44 except
in (a): $v_B^P = 5PL^3/48EI$, $\theta_B^P = PL^2/8EI$
in (b): $\theta_A^P = PL^2/16EI$, $\theta_B^P = -PL^2/16EI$.

9.46 a)

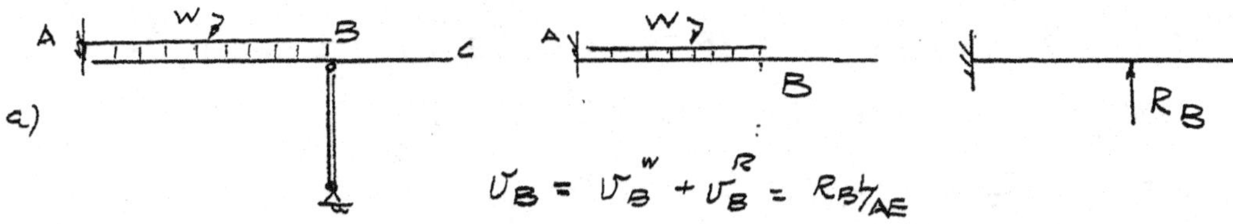

$$v_B = v_B^w + v_B^R = R_B h/AE$$

$$\frac{w(2L)^4}{8EI} + \left(-\frac{R_B(2L)^3}{3EI}\right) = R_B h/AE \Rightarrow R_B = \frac{6wL^4 A}{8AL^3 + 3hI}.$$

$$\Sigma F_y = 0 \Rightarrow R_A = \frac{2wL(5AL^3 + 3hI)}{8AL^3 + 3hI};\ \Sigma M = 0 \Rightarrow M_A = -\frac{2wL^2(2AL^3 + 3hI)}{8AL^3 + 3hI}$$

$$v_C = v_B + \theta_B L = v_B + (\theta_B^W + \theta_B^R)L = \frac{R_B L}{AE} + \left(+\frac{wL^3}{6EI} - \frac{R_B (2L)^2}{2EI}\right)L$$

Substitute for R_B above:

$$v_C = \frac{wL^4}{3EI}\left[\frac{30hI - 4AL^3}{8AL^3 + 3hI}\right].$$

9.47 a)

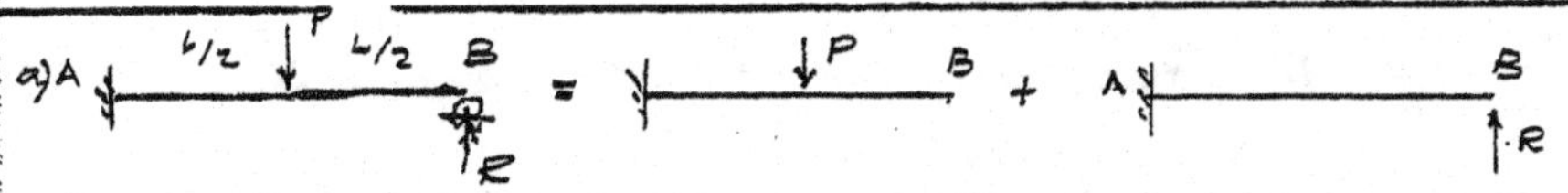

$$v_B = v_B^P + v_B^R = 5PL^3/48EI - RL^3/3EI = 0 \Rightarrow R_B = 5P/16,\ R_A = 11P/16$$
$$M_A = -3PL/16$$

b)

$$\theta_A = \theta_A^P + \theta_A^M = PL^2/16EI + M_A L^2/3EI = 0 \Rightarrow M_A = -3PL/16;\ R_A = \frac{11P}{16},\ R_B = 5P/16$$

9.48

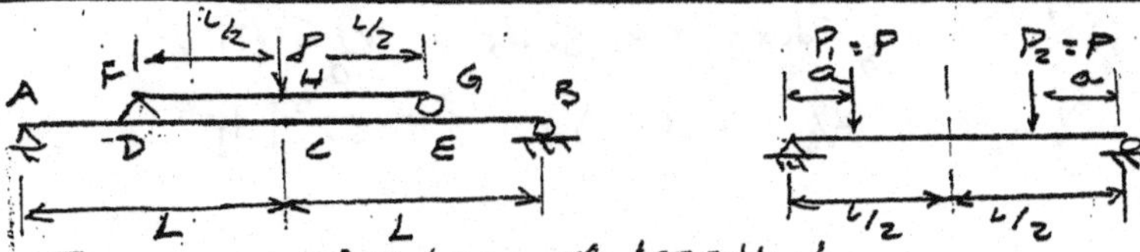

From eq. (9.4.18):

$$v_1 = v(a)\Big|_{P_1} = \frac{Pa^2(L-a)^2}{3EIL}$$

$$v_2 = v(a)\Big|_{P_2} = \frac{Pa^2(L^2 - 2a^2)}{6EIL}$$

In general for beam of length, L.

$$v(a) = v_1(a) + v_2(a) = \frac{Pa^2(3L - 4a)}{6EI} \quad (I)$$

$$v(L/2) = \frac{Pa(3L^2 - 4a^2)}{24EI} \quad (II)$$

Beam ACB: In (I) & (II): $P \to P/2$, $L \to 2L$, $a \to L/2$

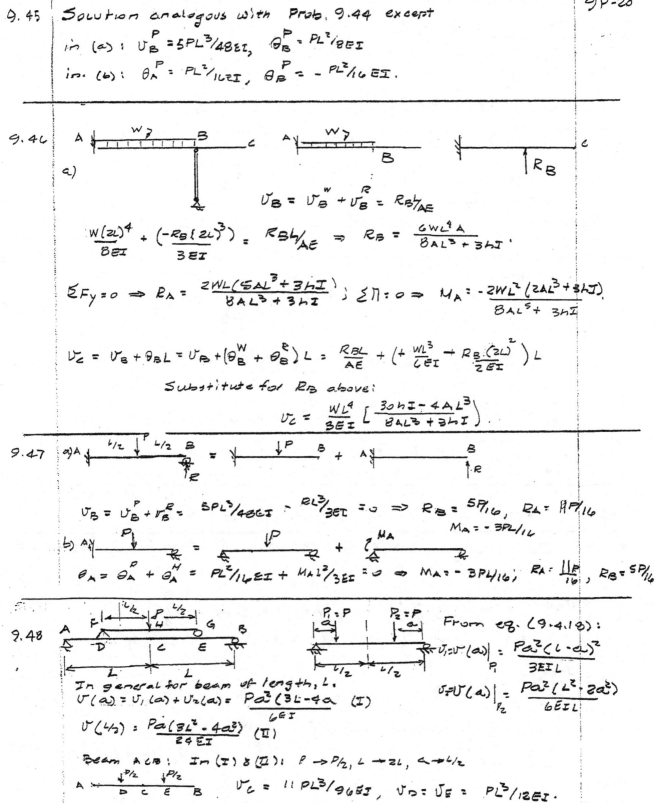

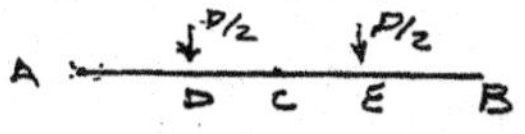

$$v_C = 11PL^3/96EI,\quad v_D = v_E = PL^3/12EI.$$

Beam FG: $v_H = v_D + PL^3/48EI = 5PL^3/48EI.$

9.49

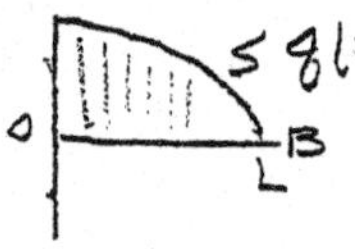

$q(x) = q_0 \cos\frac{\pi x}{2L}$

$$v_B = \frac{q_0}{6EI}\int_0^L (3L\xi^2 - \xi^3)\cos\frac{\pi\xi}{2L}\,d\xi$$

$$= \frac{q_0}{6EI}\left\{3L\left(\frac{2L}{\pi}\right)^3\left[(2z\cos z) + (z^2-2)\sin z\right] - \left(\frac{2L}{\pi}\right)^4\left[(3z^2-6)\cos z + (z^3-6z)\sin z\right]\right\}\Bigg|_0^{\pi/2}$$

$$= \frac{q_0}{6EI}(2L/\pi)^3\left\{3L\left[\pi^2/4 - 2\right] - (2L/\pi)\left[(\pi^3/8 - 3\pi) + 6\right]\right\}$$

$$v_B = \frac{2q_0L^4}{3\pi^4EI}\left[\pi^3 - 24\right] = 0.0480\, q_0L^4/EI$$

9.50

$$v_B = \frac{q_0}{6EI}\int_0^L (3L\xi^2 - \xi^3)e^{-\alpha\xi/L}\,d\xi$$

$$= \frac{q_0}{6EI}\left\{\frac{3L}{(-\alpha/L)^3}e^{-\alpha\xi/L}\left[\alpha^2/L^2\,\xi^2 + 2\alpha\xi/L + 2\right] - \frac{e^{-\alpha\xi/L}}{(\alpha/L)^4}\left[-(\alpha/L)^3\xi^3 - 3(\alpha/L)^2\xi^2 + 6\frac{\alpha}{L}\xi - 6\right]\right\}\Bigg|_0^L$$

$$= \frac{q_0}{6EI}\left\{-\frac{3L^4}{\alpha^3}\left[e^{-\alpha}(\alpha^2+2\alpha+2) - 2\right] - \frac{L^4}{\alpha^4}\left(e^{-\alpha}\left[-\alpha^3 - 3\alpha^2 - 6\alpha - 6\right] + 6\right)\right\}$$

$$v_B = \frac{q_0L^4}{6EI\alpha^4}\left\{6(\alpha-1) + e^{-\alpha}\left[-2\alpha^3 - 3\alpha^2 + 6\right]\right\}$$

b) $v_B\big|_{\alpha\to 0} = \dfrac{q_0L^4}{8EI}$ by de l'Hopital's Rule or series expansion.

9.51 $G_d(x,\xi)$ from eq (9.7.8) setting $P=1$ and $b = L-a$

9.52

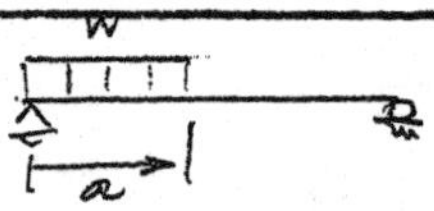

$q(x) = w[1 - \langle x-a\rangle^0] = w\langle a-x\rangle^0$ by eq. (9.7.2)

$$v(x) = \int_0^L G_d(x,\xi)q(\xi)\,d\xi = w\int_0^L G_d(x,\xi)\langle a-\xi\rangle^0 d\xi = w\int_0^a G_d(\xi,x)\,d\xi$$

$$= \frac{w}{6EI}\int_0^a\left[-(L-\xi)x^3/L - (L-\xi)^3x/L + L(L-\xi)x + \langle x-\xi\rangle^3\right]d\xi$$

Integrating

$$v(x) = \frac{w}{24EI}\left[2(L-a)^2x^3/L + (L-a)^4x/L - 2L(L-a)^2x - (x-a)^4 + x^4 - 2Lx^3 + L^3x\right]$$

$$= \frac{wL^4}{24EI}\left\{(1-a/L)\left[2(1-a/L)(x/L)^3 + \left(3(a/L)^3 - 3(a/L) - 1\right)\frac{x}{L}\right] - \left\langle\frac{x-a}{L}\right\rangle^4 + \left[(x/L)^4 - 2(x/L)^3 + (x/L)\right]\right\}$$

$$v(x)\Big|_{a=L} = \frac{w}{24EI}\left(x^4 - 2Lx^3 + L^3x\right).$$

9.53 a)

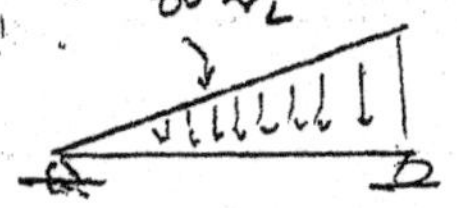

$$v(L/2) = \int_0^L G_d(L/2, \xi)\, q(\xi)\, d\xi$$

$$= \frac{q_0}{48EI} \int_0^L (\xi/L)\left[-L^3 + 9L^2\xi - 12L\xi^2 + 4\xi^3 + \langle L - 2\xi \rangle^3\right] d\xi$$

Integrating

$$v_c = v(L/2) = \frac{5 q_0 L^4}{768 EI}$$

b).

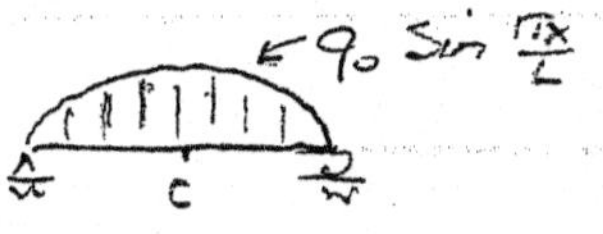

$$v(L/2) = \frac{q_0}{6EI} \int_0^L \left(\sin \frac{\pi \xi}{L}\right)\left[-(L-\xi)L^2/8 - (L-\xi)^3/2 + L^2(L-\xi)/2 + \langle L/2 - \xi \rangle^3\right] d\xi$$

Integrating : $v_c = q_0 L^4 / \pi^4$

9.54

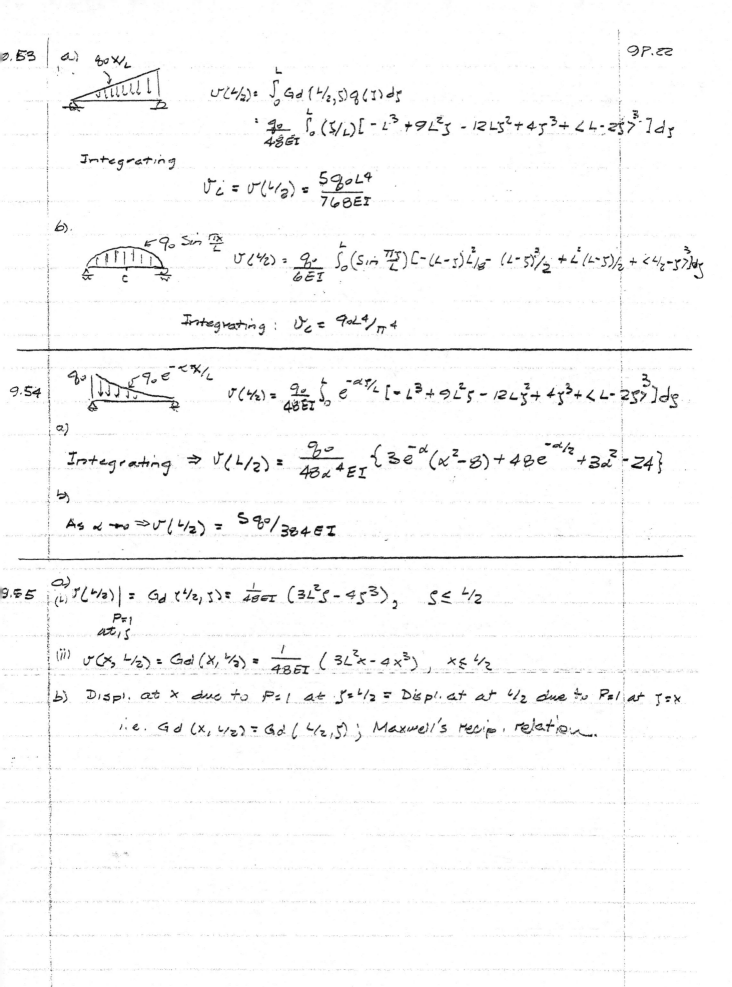

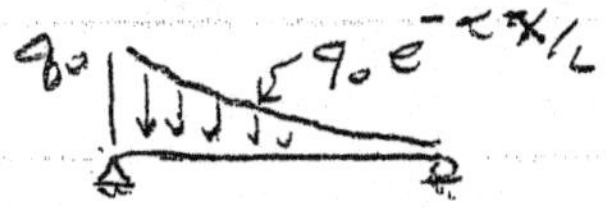

$$v(L/2) = \frac{q_0}{48EI} \int_0^L e^{-\alpha \xi/L}\left[-L^3 + 9L^2\xi - 12L\xi^2 + 4\xi^3 + \langle L - 2\xi \rangle^3\right] d\xi$$

a)

Integrating $\Rightarrow v(L/2) = \dfrac{q_0}{48\alpha^4 EI}\left\{3e^{-\alpha}(\alpha^2 - 8) + 48e^{-\alpha/2} + 3\alpha^2 - 24\right\}$

b)

As $\alpha \to 0 \Rightarrow v(L/2) = 5q_0/384EI$

9.55 a)

(i) $v(L/2)\Big|_{P=1 \text{ at } \xi} = G_d(L/2, \xi) = \frac{1}{48EI}(3L^2\xi - 4\xi^3), \quad \xi \le L/2$

(ii) $v(x, L/2) = G_d(x, L/2) = \frac{1}{48EI}(3L^2 x - 4x^3), \quad x \le L/2$

b) Displ. at x due to $P=1$ at $\xi = L/2$ = Displ. at at $L/2$ due to $P=1$ at $\xi = x$

i.e. $G_d(x, L/2) = G_d(L/2, \xi)$; Maxwell's Recip. relation.

$P = 1$

9.56 a)

$R_A = (L-\xi)/L \; , \; R_B = \xi/L$

$G_{V_C} = \begin{cases} -\xi/L, \\ (L-\xi)/L, \end{cases} \quad G_{M_C} = \begin{cases} (L-x)\xi/L & , \; \xi < x_c \\ (L-\xi)x/L & , \; x_c < \xi \end{cases}$

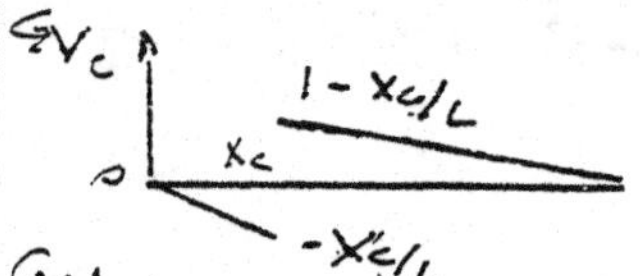

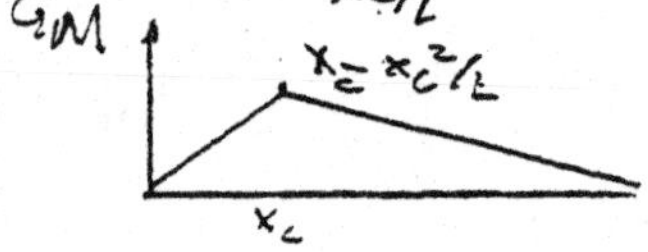

b). $V_c = w\int_0^L G_{V_C}(x, \xi)\,d\xi = \left(-\frac{x_c}{L}\right)\left(\frac{x_c}{2}\right) + (L - x_c)^2/2L$

$= (L - 2x_c)/2.$

$M_c = w\int G_{M_C}(x,\xi)\,d\xi = x_c(L - x_c)/2.$

c). $x_c = L/4$

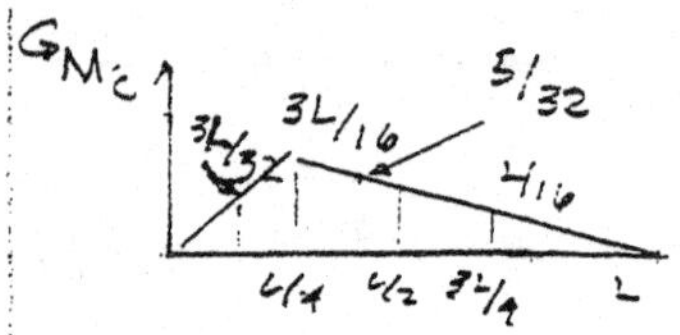

$M_c = w\int_{3L/4}^{L} G_M(L/4, \xi)\,d\xi = w(L/16)(L/4)/2 = wL^2/128$

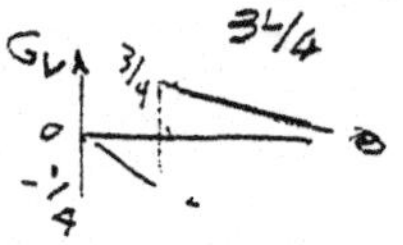

d). $M_c = \frac{1}{2}\left[\left(\frac{3L}{32} + \frac{3}{16}\right) + \left(\frac{3L}{16} + \frac{5L}{32}\right)\right](L/8) = 5wL^2/128 = 0.03906\,wL^2$

e).

$M_c = w\left[\int_a^{L/4} (3/4)\xi\,d\xi + \int_{L/4}^{L+a} \frac{1}{4}(L-\xi)\,d\xi\right] = w\left\{ \frac{3}{8}\xi^2\Big|_a^{L/4} - \frac{1}{8}(L-\xi)^2\Big|_{L/4}^{L/4+a} \right\}$

$M_c = \frac{w}{128}\left(3L^2 + 24La - 64a^2\right)$

$M_c = M_c(a)\; ; \; M_{max} \Rightarrow \partial M_c/\partial a = 0 \Rightarrow a = 3L/16$

Loading between $\frac{3L}{16} \le x \le \frac{7L}{16}$ $\quad M_{max} = 21wL^2/512 = 0.04102\,wL^2$

9.57

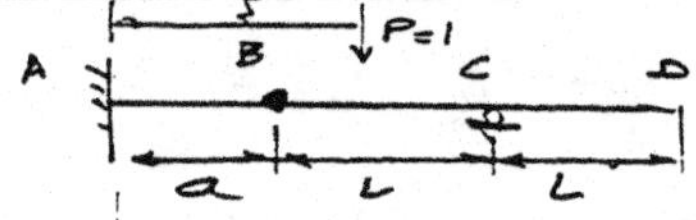

$0 \le \xi < a$:

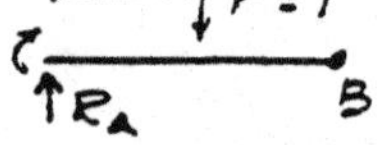

$\Sigma M_{BD} = 0 \Rightarrow R_C = 0.$

$R_A = P, \; M_A = -P\xi, \; V_C^+ = 0$

$a \le \xi \le a + 2L$:

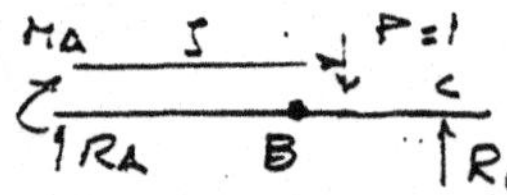

$(\Sigma M_B)_{BD} = 0 \Rightarrow R_C = \frac{\xi - a}{L}$

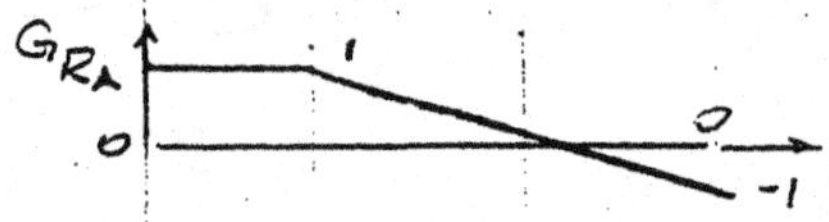

$\Sigma F_y = 0 \Rightarrow R_A = 1 - R_C = [(L+a) - \xi]/L$

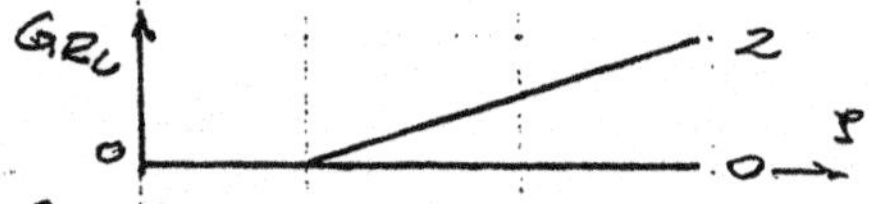

$(\Sigma M_A)_{AD} = 0 \Rightarrow M_A = [(a+L)(\frac{\xi - a}{L}) - \xi] = \frac{Pa}{L}[\xi - (L+a)]$

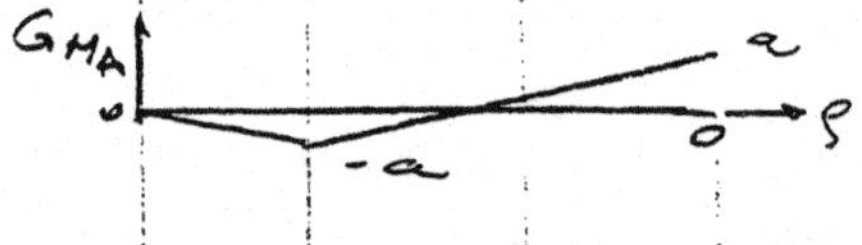

$V_C^+ = 0 \; ; \quad 0 < \xi < a + L$

$V_C^+ = 1 \; ; \; a + L < \xi < a + 2L$

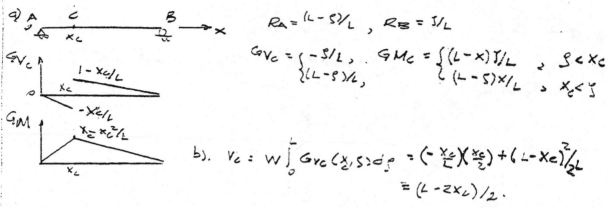

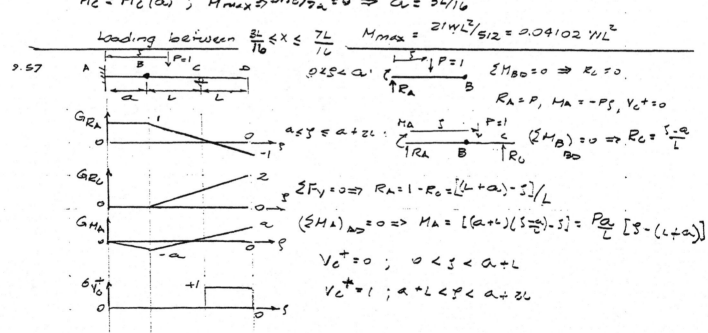

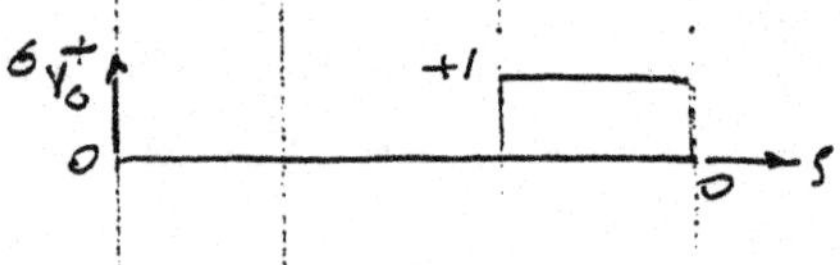

9.58

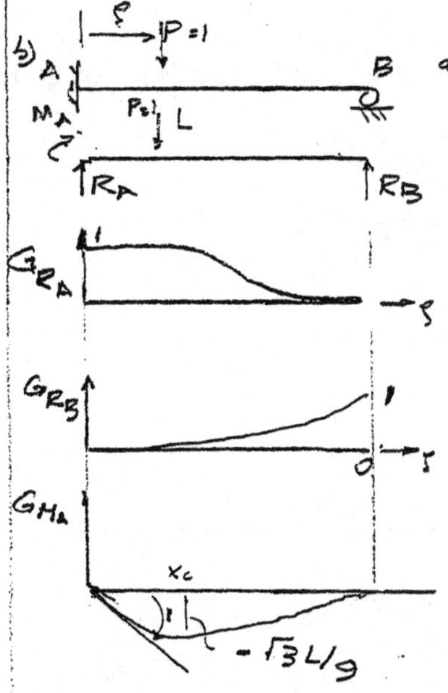

a) $R_B = \frac{P}{2L^3}(3L\xi^2 - \xi^3)$ [by eq. (3.10.15b)]

$\Rightarrow G_{RB} = \frac{1}{2L^3}(3L\xi^2 - \xi^3)$

$R_A = P - R_B \Rightarrow G_{RA} = \frac{1}{2L^3}(2L^3 - 3L\xi^2 + \xi^3)$

$M_A = R_B L - P\xi \Rightarrow G_{MA} = -\frac{\xi}{2L^2}(\xi^2 - 3L\xi + 2L^2)$

c) $dG_{MA}/d\xi = 0 \Rightarrow 3\xi^2 - 6L\xi + 2L^2 = 0 \Rightarrow \xi_{cr} = \frac{L}{3}(3 - \sqrt{3})$

$\xi_{cr} = 0.42265L \qquad |M|_{max} = \frac{\sqrt{3}}{9}PL = 384.9$ N·m

---

9.59

$EIv^{IV}(x) = W$; B.C. $v(0) = v'(0)$; $v''(L) = v'''(L) = 0$

$EIv'''(x) = Wx + A$; $v'''(L) = 0 \Rightarrow A = -WL$

$EIv''(x) = Wx^2/2 - WLx + B$; $v''(L) = 0 \Rightarrow B = +WL^2/2$

$EIv''(x) = WL^2/2 - WLx + Wx^2/2$

Note: Solution continues as in Problem 9.3a

9.60

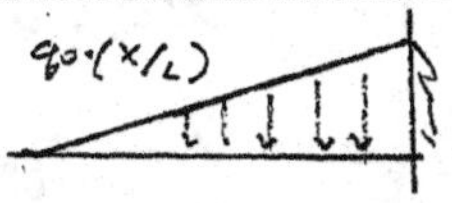

$EIv^{IV}(x) = q_0 x/L$ B.C. $v'''(0) = v''(0) = 0$, $v'(L) = v(L) = 0$

$EIv'''(x) = q_0 x^2/2L + A$, $A = 0$ $[v'''(0) = 0]$; $EIv''(x) = q_0 x^3/6L + B$, $B = 0$ $[v''(0) = 0]$

Note: Solution continues as in Problem 9.3e.

9.61

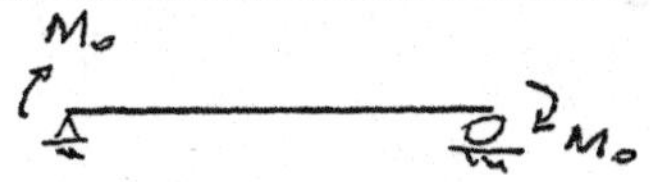

$q(x) = 0 \Rightarrow EIv^{IV}(x) = 0$; B.C. $EIv''(0) = -M_o$, $EIv''(L) = M_o$, $v(0) = v(L) = 0$

$EIv'''(x) = A$, $EIv''(x) = Ax + B$; $EIv''(0) = B = -M_o$; $EIv''(L) = M_o \Rightarrow A = 2M_o/L$

$EIv''(x) = \frac{M_o}{L}(2x - L)$

Note: Solution continues as in Prob. 9.3i.

9.62

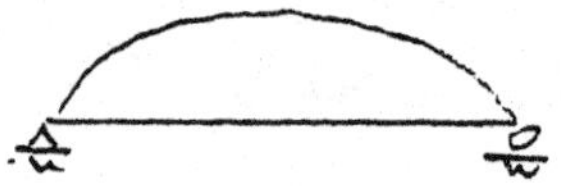

$EIv^{IV}(x) = \frac{4q_0}{L^2}(Lx - x^2)$; B.C. $v(0) = v(L) = v''(0) = v''(L) = 0$

$EIv'''(x) = 4q_0/L^2\,(Lx^2/2 - x^3/3) + A$

$EIv''(x) = \frac{4q_0}{L^2}(Lx^3/6 - x^4/12) + Ax + B \to 0$

$v''(L) = 0 \Rightarrow A = -\,q_0L/3$

$EIv''(x) = -\frac{q_0}{3L^2}(x^4 - 2Lx^3 + L^3x)$

Note: Solution continues as in Problem 9.4

9.63

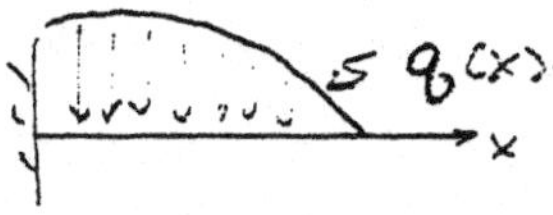

$q(x) = q_0 \cos \pi x/2L$ $\quad EIv^{IV}(x) = q_0 \cos \pi x/2L$ B.C. $v'(0) = v(0) = 0$, $v''(L) = v'''(L) = 0$

$EIv'''(x) = (2L/\pi)\,q_0 \sin \pi x/2L + A$

$v'''(L) = 0 \Rightarrow A = -\,2q_0L/\pi$

$EIv''(x) = \frac{2q_0L}{\pi}\left[-\frac{2L}{\pi}\cos \pi x/2L - x\right] + B$; $v''(L) = 0 \Rightarrow B = 2q_0L^2/\pi$

$EIv''(x) = \frac{2q_0L^2}{\pi^2}\left[\pi - \pi x/L - 2\cos \pi x/2L\right]$

Note: Solution continues as in Problem 9.5b

9.64

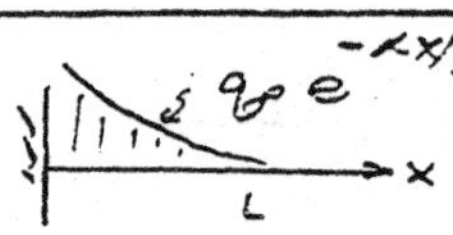

$q_0 e^{-\alpha x/L}$ $\quad EIv^{IV} = q_0 e^{-\alpha x/L}$ B.C. $v(0) = v'(0) = 0$; $v''(L) = v'''(L) = 0$

$EIv''' = -q_0L/\alpha\, e^{-\alpha x/L} + A$; $v'''(L) = 0 \Rightarrow A = q_0L/\alpha\, e^{-\alpha}$

$EIv''(x) = q_0L^2/\alpha^2\, e^{-\alpha x/L} + Ax + B$

$v''(L) = 0 \Rightarrow B = -\frac{q_0L^2}{\alpha}(1/\alpha + 1)e^{-\alpha}$

$EIv''(x) = \frac{q_0L}{\alpha}\left[-L/\alpha\,(1+\alpha)e^{-\alpha} + \frac{L}{\alpha}e^{-\alpha x/L} + x e^{-\alpha}\right]$

Note: Solution continues as in Problem 9.6

9.65

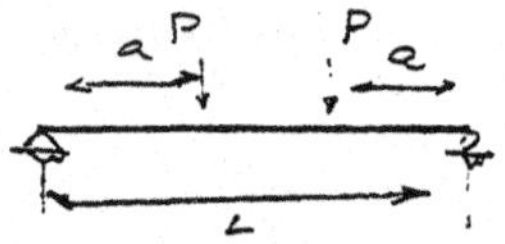

$EIv^{IV} = q(x) = P\delta(x-a) + P\delta[x-(L-a)]$ B.C. $v(0) = v''(0) = v(L) = v''(L)$

$EIv'''(x) = P\langle x-a\rangle^0 + P\langle x-(L-a)\rangle^0 + A$

$EIv''(x) = P\langle x-a\rangle + P\langle x-(L-a)\rangle + Ax + B \to 0$ $\quad v''(0) = 0$

$v''(L) = 0 \Rightarrow A = -P\{(L-a) + a\} = -PL$

$EIv''(x) = P[\langle x-a\rangle + \langle x-(L-a)\rangle] - Px$

Note: Solution continues as in Problem 9.19

9.65

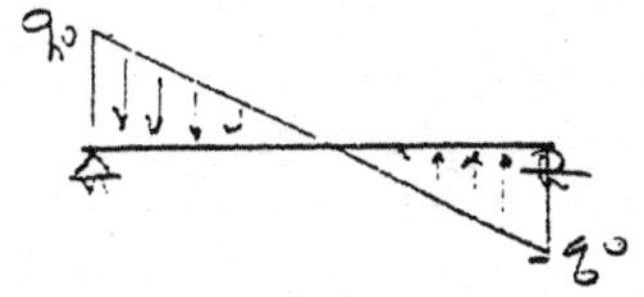

$EIv^{IV}(x) = q_0(1 - 2x/L)$, B.C. $v(0) = v(L) = v''(0) = v''(L) = 0$

$EIv'''(x) = q_0(x - x^2/L) + A$

$EIv''(x) = q_0(x^2/2 - x^3/6L) + Ax + B$ → $[v''(0) = 0]$

$v''(L) = 0 \Rightarrow A = -q_0L/6$

$EIv''(x) = q_0(x^2/2 - x^3/3L - Lx/6)$

Note: Solution continues as in Problem 9.4.

9.67

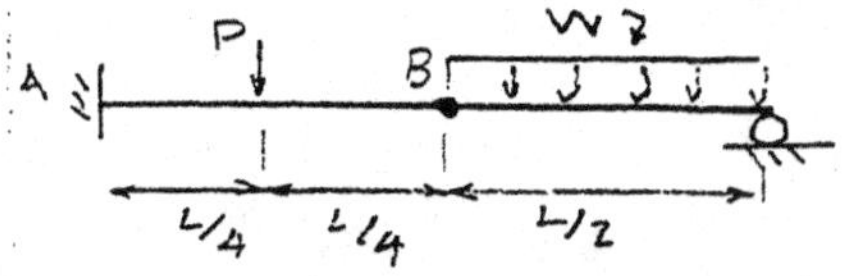

$q(x) = P\delta(x - L/4) + w\langle x - L/2\rangle^0$, $0 \le x \le L$

Note: $q(x)$ above valid over entire range of x. However, must use 2 domains since $\theta = v'$ is discontinuous at B.

D_1: $0 \le x \le L/2$

D_2: $L/2 \le x \le L$

B.C. $v_1'(0) = v_1(0) = 0$; $v_1''(L/2) = 0$; $v_2''(L/2) = 0$, $v_2(L) = 0$; $v_1'''(L/2) = v_2'''(L/2)$

$v_1(L/2) = v_2(L/2)$

$v_2''(L) = 0$

$EIv_i^{IV}(x) = P\delta(x - L/4) + w\langle x - L/2\rangle^0$, $i = 1, 2$

$EIv_i'''(x) = P\langle x - L/4\rangle^0 + w\langle x - L/2\rangle + A_i$

$EIv_i''(x) = P\langle x - L/4\rangle + w\langle x - L/2\rangle^2/2 + A_i x + B_i$

$v_1''(L/2) = 0 \Rightarrow PL/4 + A_1L/2 + B_1 = 0$ (a)

$v_2''(L/2) = 0 \Rightarrow PL/4 + A_2L/2 + B_2 = 0$ (b)

$EIv_i'(x) = P\langle x - L/4\rangle^2/2 + w\langle x - L/2\rangle^3/6 + A_i x^2/2 + B_i x + C_i$

$v_1'(0) = 0 \Rightarrow C_1 = 0$

$EIv_i(x) = P\langle x - L/4\rangle^3/6 + w\langle x - L/2\rangle^4/24 + A_i x^3/6 + B_i x^2/2 + C_i x + D_i$

$D_1 = 0$ $[v_1(0) = 0]$

$EIv_2(L) = 27PL^3/384 + wL^4/24.16 + A_2L^3/6 + B_2L^2/2 + C_2L + D_2 = 0$

$64L^2A_2 + 192LB_2 + 384C_2L = -27PL^2 - wL^3$ (c)

$v_1'''(L/2) = v_2'''(L/2) \Rightarrow A_1 = A_2 = A \Rightarrow$ (a) & (b) $\Rightarrow B_1 = B_2 = B$

$v_1(L/2) = v_2(L/2) \Rightarrow C_2L/2 + D_2 = 0 \Rightarrow D_2 = -C_2L/2$ (d)

$v_2''(L) = 0 \Rightarrow 3PL/4 + wL^2/8 + A_2L + B_2 = 0 \Rightarrow AL + B = -3PL/4 - wL^2/8$ (e)

Note: 4 unknowns (A, B, C_2, D_2) & 4 eqs. (a, c, d, e).

eq. (c) $\Rightarrow$ $64L^2A_2 + 192LB_2 + 192C_2 = -27PL^2 - wL^3$ (c')

Solve for the unknowns, A_2, B_2, C_2 using (a, c', e)

9.68 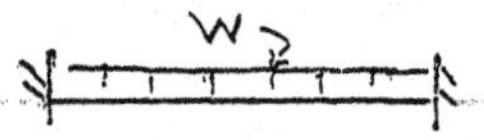

$EIv^{IV}(x) = q(x) = W$ B.C. $v(0) = v'(0) = v(L) = v'(L) = 0$

$EIv'''(x) = Wx + A$, $EIv''(x) = Wx^2/2 + Ax + B$

$EIv'(x) = Wx^3/6 + Ax^2/2 + Bx + C$; $v'(0) = 0 \Rightarrow C = 0$

$v'(L) = 0 \Rightarrow AL^2/2 + BL = -WL^3/6$ (a)

$EIv(x) = Wx^4/24 + Ax^3/6 + Bx^2/2 + D$; $v(0) = 0 \Rightarrow D = 0$

$v(L) = 0 \Rightarrow AL^2/6 + BL/2 = -WL^3/24$

Solving for A & B $\Rightarrow B = -WL^2/12$, $A = -WL/2$

$R_A = V(0) = -EIv'''(0) = -A = WL/2$; $M_A = -EIv''(0) = -B = -WL^2/12$

$R_B = -V(L) = EIv'''(L) = WL/2$, $M_B = -EIv''(L) = -WL^2/12$.

$v(x) = \frac{W}{EI}[x^4/24 - Lx^3/12 + L^2x^2/24] = \frac{Wx^2}{24EI}(x^2 - 2Lx + L^2)$

$v(L/2) = WL^4/384EI$.

9.69 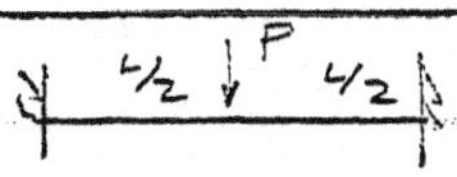

$EIv^{IV}(x) = P\delta(x - L/2)$ B.C. $v(0) = v'(0) = v(L) = v'(L) = 0$

$EIv'''(x) = P\langle x - L/2\rangle^0 + A$; $EIv''(x) = P\langle x - L/2\rangle + Ax + B$

$EIv'(x) = P\langle x - L/2\rangle^2/2 + Ax^2/2 + Bx + C$; $v'(0) = 0 \Rightarrow C = 0$

$v'(L) = 0 \Rightarrow AL + 2B = -PL/4$

$EIv(x) = P\langle x - L/2\rangle^3/6 + Ax^3/6 + Bx^2/2 + D$; $v(0) = 0 \Rightarrow D = 0$

$v(L) = 0 \Rightarrow AL + 3B = -PL/8$

Solving: $B = PL/8$, $A = -P/2$

$R_A = V(0) = -EIv'''(0) = -A = P/2$; $M_A = -EIv''(0) = -B = -PL/8$

$R_B = P/2$, $M_B = -PL/8$.

$v(x) = \frac{P}{48EI}[-4x^3 + 3Lx^2 + 8\langle x - L/2\rangle^3]$

9.70 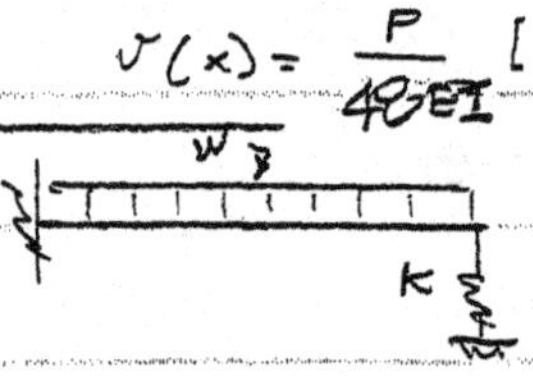

$EIv^{IV}(x) = W$ B.C. $v(0) = v'(0) = v''(L) = 0$

$EIv'''(L) = kv(L)$

$EIv'''(x) = Wx + A$; $EIv''(x) = Wx^2/2 + Ax + B \Rightarrow AL + B = -WL^2/2$ (I) $[v''(L) = 0]$

$EIv'(x) = Wx^3/6 + Ax^2/2 + Bx + C$ ($C \to 0$ $[v'(0) = 0]$); $EIv(x) = Wx^4/24 + Ax^3/6 + Bx^2/2 + D$ ($D \to 0$ $[v(0) = 0]$)

$EIv'''(L) = kv(L) \Rightarrow WL + A = \frac{k}{EI}[WL^4/24 + AL^3/6 + BL^2/2]$ (II)

Solving (I) & (II) : $\quad \alpha = kL^3/EI$

$$A = -\frac{WL}{8}\left(\frac{5\alpha+24}{\alpha+3}\right), \quad B = \frac{WL^2(\alpha+12)}{8(\alpha+3)}$$

$$v(x) = \frac{Wx^2}{48EI}\left[2x^2 - \left(\frac{5\alpha+24}{\alpha+6}\right)x + 3\left(\frac{\alpha+12}{\alpha+6}\right)L^2\right]$$

$$EIv'''(x) = -V(x) = Wx - \frac{WL}{8}\left(\frac{5\alpha+24}{\alpha+3}\right)$$

$$R_A = V(0) = \frac{WL}{8}\left(\frac{5\alpha+24}{\alpha+3}\right) \qquad R_A\Big|_{\alpha\to\infty} = 5WL/8, \quad R_A\Big|_{\alpha\to 0} = WL$$

$$R_B = -V(L) = EIv'''(L) = \frac{3\alpha WL}{8(\alpha+3)} \qquad R_B\Big|_{\alpha\to\infty} = 3WL/8, \quad R_B\Big|_{\alpha\to 0} = 0$$

$$M_A = -EIv''(0) = -\frac{\alpha+12}{8(\alpha+3)}WL^2 \qquad M_A\Big|_{\alpha\to\infty} = -WL^2/8, \quad M_A\Big|_{\alpha\to 0} = -WL^2/2$$

9.71

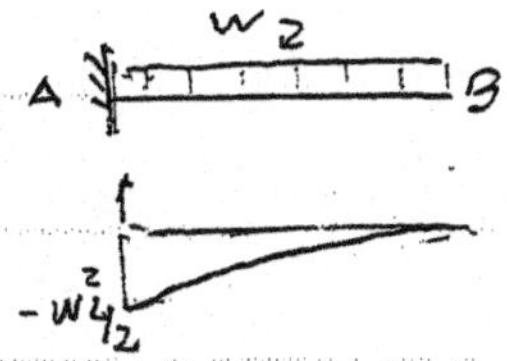

$$\theta_B = -A\Big|_0^L = -(-\tfrac{wL^2}{2})(L/3) = +wL^3/6 \;;\; \Delta_B = Q_B\Big|_0^L = +(wL^3/6)(3L/4) = wL^4/8EI.$$

9.72

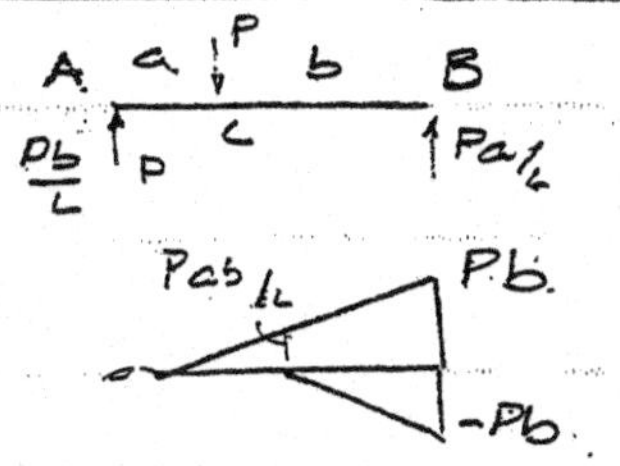

$$v_A = v_B = 0 \Rightarrow \theta_A L = Q_L\Big|_0^L = \frac{1}{EI}\left[(Pb)(L/2)(L/3) - (Pb)(b/2)(b/3)\right]$$

$$\theta_A = \frac{Pb}{6EIL}(L^2-b^2) = \frac{Pab(2L-a)}{6EIL}$$

$$v_C = \theta_A a - Q_{L/2}\Big|_0^{L/2} = \frac{Pa^2b(2L-a)}{6EIL} - \left(\frac{Pab}{L}\right)(a/2)(a/3)$$

$$v_C = \frac{P}{6EIL}\left[a^2b(2L-a) - a^3b\right] = \frac{Pa^2b(L-a)}{3EIL} = \frac{Pa^2(L-a)^2}{3EIL}$$

9.73

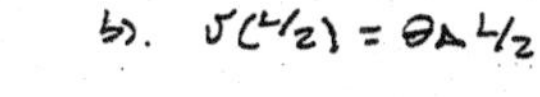

a) $v_A = v_B = 0 \Rightarrow \theta_A = \frac{Q_L}{L}\Big|_0^L = \frac{1}{EIL}\left\{(M_A L/2)(2L/3) + (M_B L/2)(L/3)\right\}$

$$= \frac{L}{6EI}\left[2M_A + M_B\right]$$

$$\theta_B = \theta_A - \frac{A}{EI}\Big|_0^L = \frac{L}{6EI}(2M_A+M_B) - \frac{L}{2EI}(M_A+M_B)$$

$$\theta_B = -\frac{L}{6EI}(M_A + 2M_B)$$

b). $v(L/2) = \theta_A L/2 - Q\Big|_0^{L/2}$

b). $$v(L/2) = \frac{L}{6EI}(2M_A+M_B)(L/2) - \frac{1}{EI}\left[\left(\frac{M_A}{2}\right)(L/2)(L/4) + \left(\frac{M_A}{2}\right)(L/4)(L/3) + \left(\frac{M_B}{2}\right)(L/4)(L/6)\right]$$

$$= \frac{L^2}{12EI}(2M_A+M_B) - \frac{L^2}{EI}\left[5M_A/48 + M_B/48\right] = \frac{L^2}{16EI}(M_A+M_B)$$

9.74

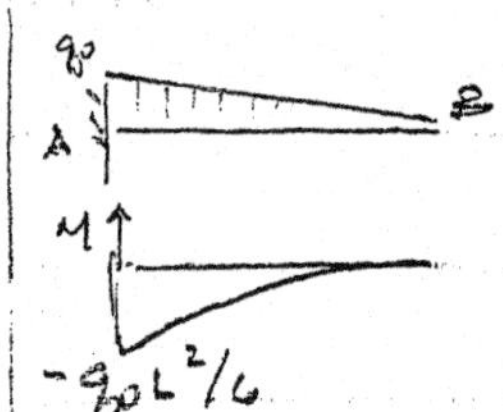

$M(x) = -q_0(L-x)^3/6L$

$v(0) = v'(0) = 0$

$\theta(L) = -A|_0^L = -[-(q_0L^2/6)(L/4)]/EI = q_0L^3/24EI$

$v(L) = -Q_L|_0^L = -[-(q_0L^3/24)(4L/5)/EI] = q_0L^4/30EI$

9.75

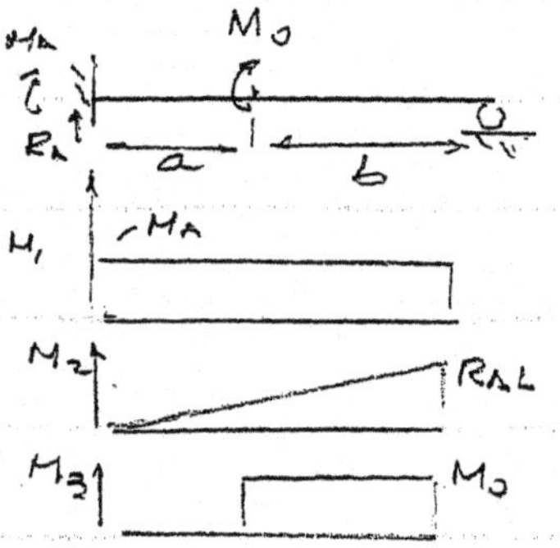

$v(0) = v'(0) = v(L) = 0 \Rightarrow Q_L|_0^L = 0$

$(M_A L)(L/2) + (R_A L^2/2)(L/3) + M_0 b^2/2 = 0$

$3M_A + R_A L = -3M_0 b^2/L^2$ (i)

$\circlearrowleft + \Sigma M_B = 0 \Rightarrow M_A + R_A L + M_0 = 0$ (ii)

(i) & (ii) $\Rightarrow M_A = M_0/2\,(1 - 3b^2/L^2)$

$R_A = -\frac{3M_0}{2L}(1 - b^2/L^2)$

$v(a) = -Q_a|_0^a = -\frac{1}{EI}[(M_A a)(a/2) + (R_A a)(a/2)(a/3)]$

$= -\frac{a^2}{6EI}[3M_A + R_A a] = -\frac{a^2}{6EI}\left\{\frac{3M_0}{2}(1 - 3b^2/L^2) - \frac{3M_0 a}{2L}(1 - b^2/L^2)\right\}$

$v(a) = -\frac{M_0 a^2}{4EI}(1 - a/L)[-2 + 4a/L - (a/L)^2]$

$v(L/2) = M_0L^2/128EI$

$v(a) = 0 \Rightarrow a/L = 2 - \sqrt{2}$; $v > 0$ for $a/L < 2 - \sqrt{2}$; $v < 0$ for $a/L > 2 - \sqrt{2}$

$M_A = 0 \Rightarrow b/L = \sqrt{3}/3 \Rightarrow a/L = 1 - \sqrt{3}/3 = \frac{1}{3}(3 - \sqrt{3})$

9.76

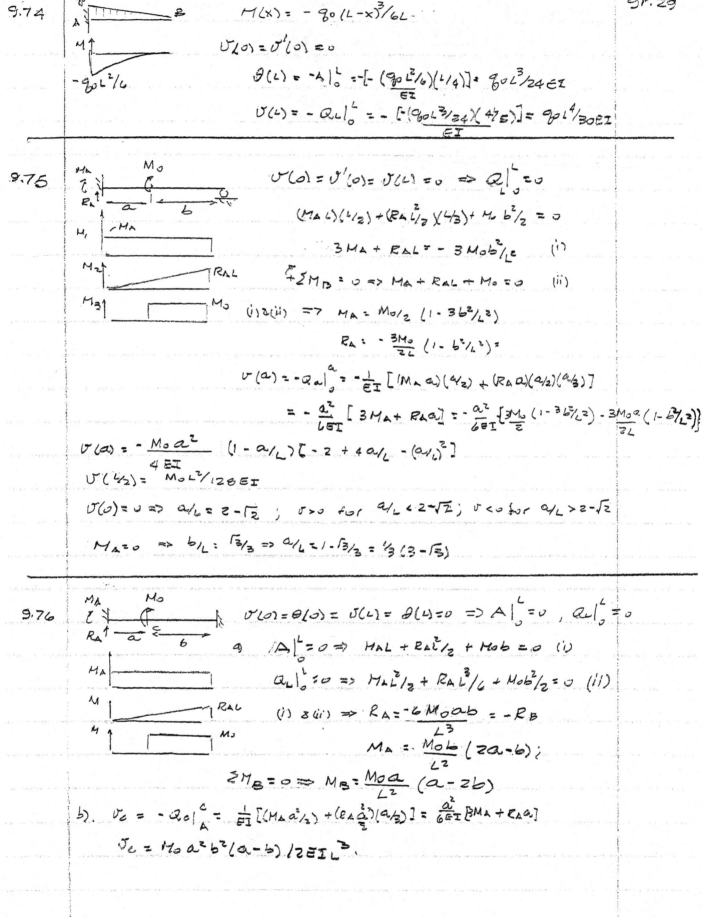

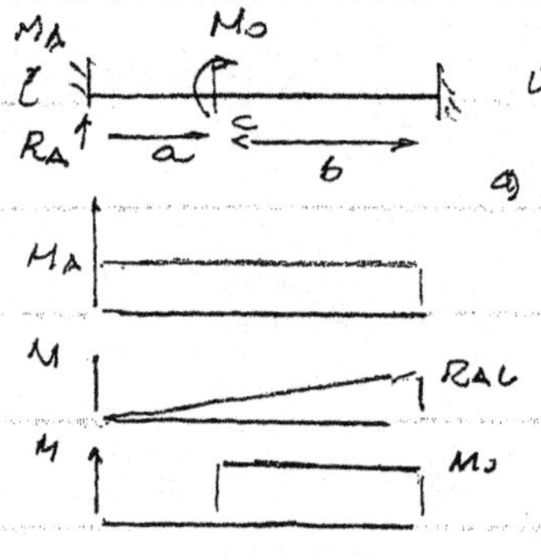

$v(0) = \theta(0) = v(L) = \theta(L) = 0 \Rightarrow A|_0^L = 0\ ,\ Q_L|_0^L = 0$

a) $A|_0^L = 0 \Rightarrow M_A L + R_A L^2/2 + M_0 b = 0$ (i)

$Q_L|_0^L = 0 \Rightarrow M_A L^2/2 + R_A L^3/6 + M_0 b^2/2 = 0$ (ii)

(i) & (ii) $\Rightarrow R_A = \frac{-6M_0ab}{L^3} = -R_B$

$M_A = \frac{M_0 b}{L^2}(2a - b)$;

$\Sigma M_B = 0 \Rightarrow M_B = \frac{M_0 a}{L^2}(a - 2b)$

b). $v_c = -Q_c|_A^c = \frac{1}{EI}[(M_A a^2/2) + (R_A \frac{a^2}{2})(a/3)] = \frac{a^2}{6EI}[3M_A + R_A a]$

$v_c = M_0 a^2 b^2 (a - b)/2EIL^3$.

9.77

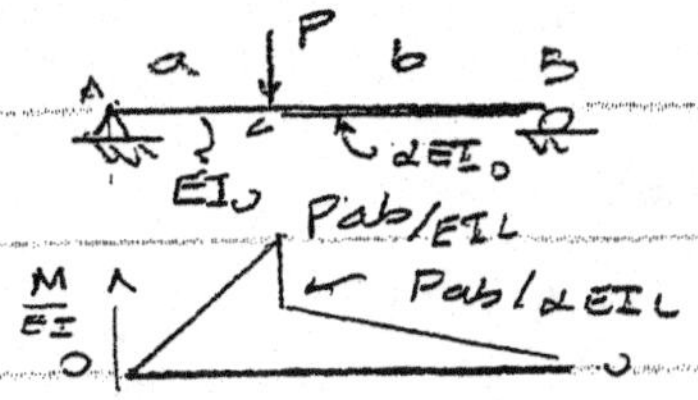

$$\theta_A = \frac{1}{L}\, Q_B\Big|_A^B$$

$$Q_B\Big|_A^B = \left(\frac{Pa^2b}{2EIL}\right)(a/3+b) + \left(\frac{Pab^2}{2EIL\alpha}\right)(2b/3)$$

$$\theta_A = \frac{Pab}{6\alpha EIL^2}\left[\alpha a(a+3b)+2b^2\right]$$

$$v_c = \theta_A a - Q_c\Big|_A^C = \theta_A a - Pa^3b/6EIL \Rightarrow v_c = \frac{Pa^2b^2(\alpha a+b)}{3\alpha EIL^2}$$

Note: $a=b=L/2$:

$\alpha = 1 \rightarrow v_c = PL^3/48EI$; $\alpha \rightarrow \infty \rightarrow v_c = PL^3/96EI$; $\alpha = 0 \Rightarrow v_c \rightarrow \infty$ unstable

---

9.78

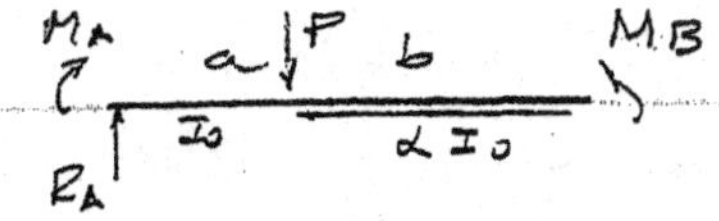

$v_A = \theta_A = v_B = \theta_B = 0 \Rightarrow A\Big|_A^B = 0,\ Q_B\Big|_A^B = 0$

$$\therefore A\Big|_A^B = 0 \Rightarrow 1/EI\left[M_A a + M_A b/\alpha + (R_A a^2/2) + R_A\cdot\frac{L+a}{2}\, b - Pb^2/2\alpha\right] = 0$$

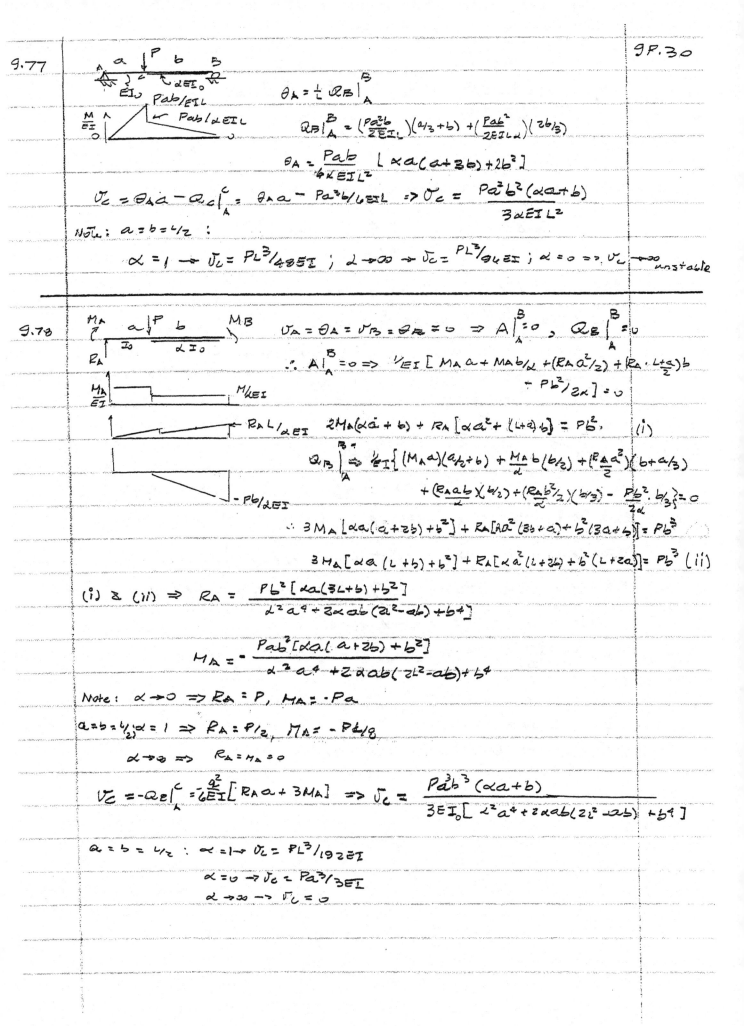

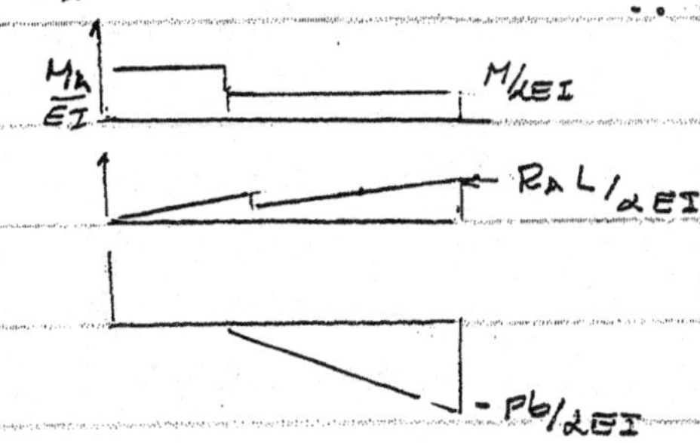

$$2M_A(\alpha a+b) + R_A\left[\alpha a^2 + (L+a)b\right] = Pb^2 \quad (i)$$

$$Q_B\Big|_A^B = \frac{1}{EI}\Big\{(M_A a)(a/2+b) + \frac{M_A}{\alpha}b(b/2) + \left(\frac{R_A a^2}{2}\right)(b+a/3) + \left(\frac{R_A ab}{\alpha}\right)(b/2) + \left(\frac{R_A b^2/2}{\alpha}\right)(b/3) - \frac{Pb^2}{2\alpha}\cdot b/3\Big\} = 0$$

$$\therefore 3M_A\left[\alpha a(a+2b)+b^2\right] + R_A\left[\alpha a^2(3b+a)+b^2(3a+b)\right] = Pb^3$$

$$3M_A\left[\alpha a(L+b)+b^2\right] + R_A\left[\alpha a^2(L+2b)+b^2(L+2a)\right] = Pb^3 \quad (ii)$$

(i) & (ii) $\Rightarrow$

$$R_A = \frac{Pb^2\left[\alpha a(3L+b)+b^2\right]}{\alpha^2a^4+2\alpha ab(2L^2-ab)+b^4}$$

$$M_A = -\frac{Pab^2\left[\alpha a(a+2b)+b^2\right]}{\alpha^2a^4+2\alpha ab(2L^2-ab)+b^4}$$

Note: $\alpha \rightarrow 0 \Rightarrow R_A = P,\ M_A = -Pa$

$a=b=L/2;\ \alpha = 1 \Rightarrow R_A = P/2,\ M_A = -PL/8$

$\alpha \rightarrow \infty \Rightarrow R_A = M_A = 0$

$$v_c = -Q_c\Big|_A^C = \frac{-a^2}{6EI}\left[R_A a + 3M_A\right] \Rightarrow v_c = \frac{Pa^3b^3(\alpha a+b)}{3EI_0\left[\alpha^2a^4+2\alpha ab(2L^2-ab)+b^4\right]}$$

$a=b=L/2$: $\alpha = 1 \rightarrow v_c = PL^3/192EI$

$\alpha = 0 \rightarrow v_c = Pa^3/3EI$

$\alpha \rightarrow \infty \rightarrow v_c = 0$

9.79

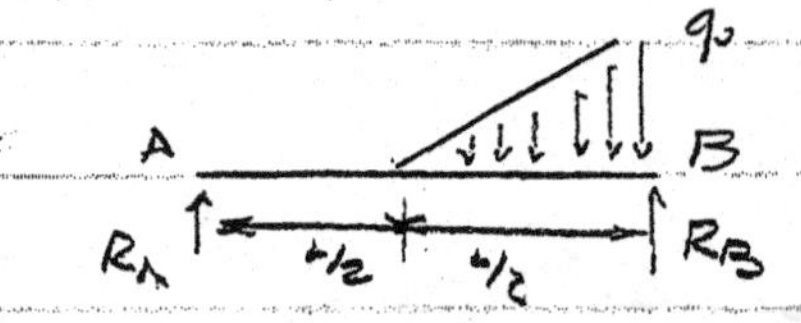

$\Sigma M_B = 0 \Rightarrow R_A = q_0 L/24$,

$M(x) = \frac{q_0}{24L}(L^2 x - 8\langle x - L/2\rangle^3)$

$EI v'' = -M(x) = -\frac{q_0}{24L}(L^2 x - 8\langle x - L/2\rangle^3)$ $\quad 0 < x < L$; B.C. $v(0) = v(L) = 0$

$EI v'(x) = -q_0/24L\,(L^2x^2/2 - 2\langle x - L/2\rangle^4) + A$

$EI v(x) = -q_0/24L\,(L^2x^3/6 - \frac{2}{5}\langle x - L/2\rangle^5) + Ax + B$ $\quad B \to 0$

$v(L) = 0 \Rightarrow A = +37L^3/5760$

$EI v(x) = -q_0/24L\,[L^2x^3/6 - 2/5\langle x - L/2\rangle^5 - \frac{37L^3 x}{240}]$

$v(x) = \frac{-q_0 L^4}{5760 EI}[40(x/L)^3 - 3\langle 2(x/L) - 1\rangle^5 - 37(x/L)]$

In AB: $v' = 0 \Rightarrow x^2/40 = 37L^2/5760$

$v(L/2) = \frac{17 q_0 L^4}{11{,}520 EI} = 1.476 \times 10^{-3}\, q_0 L^4/EI.$

$x^2 = 37L^2/120 \Rightarrow x_{cr} = 0.555L > L/2$

$\therefore x_{cr} > L/2$ (CB)

9.80

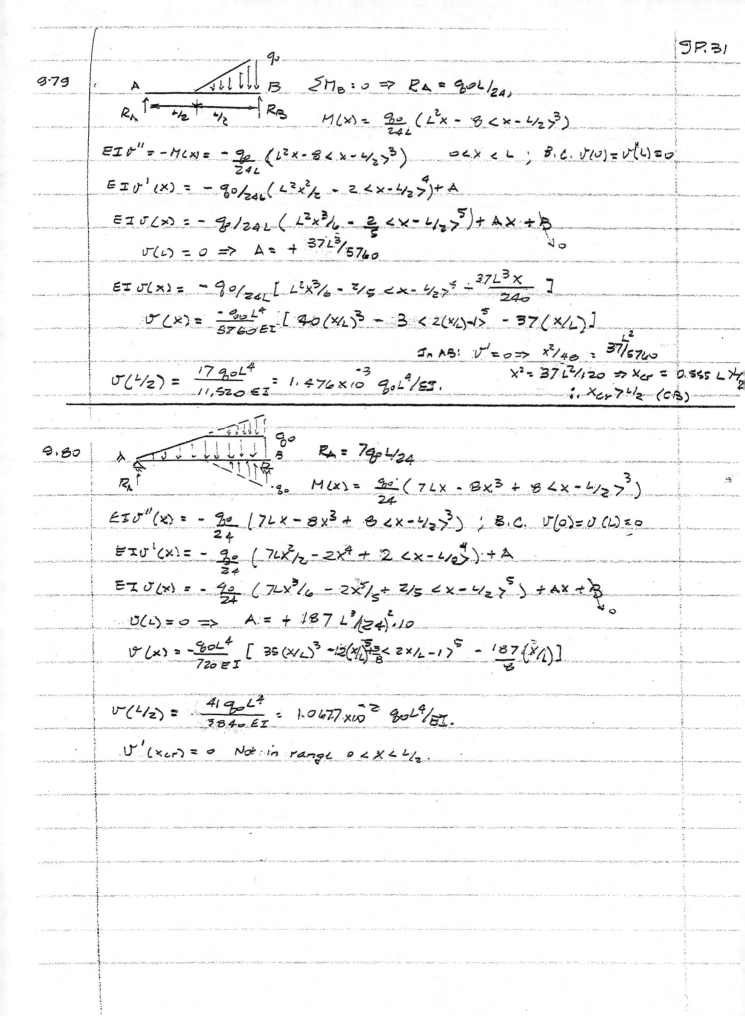

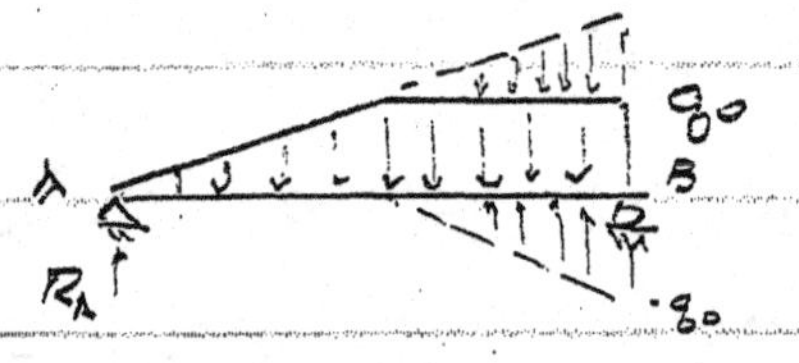

$R_A = 7q_0 L/24$

$M(x) = \frac{q_0}{24}(7Lx - 8x^3 + 8\langle x - L/2\rangle^3)$

$EI v''(x) = -\frac{q_0}{24}(7Lx - 8x^3 + 8\langle x - L/2\rangle^3)$; B.C. $v(0) = v(L) = 0$

$EI v'(x) = -\frac{q_0}{24}(7Lx^2/2 - 2x^4 + 2\langle x - L/2\rangle^4) + A$

$EI v(x) = -\frac{q_0}{24}(7Lx^3/6 - 2x^5/5 + 2/5\langle x - L/2\rangle^5) + Ax + B$ $\quad B \to 0$

$v(L) = 0 \Rightarrow A = +187L^3/(24)^2 \cdot 10$

$v(x) = \frac{-q_0 L^4}{720 EI}[35(x/L)^3 - 12(x/L)^5 + \frac{3}{8}\langle 2x/L - 1\rangle^5 - \frac{187}{8}(x/L)]$

$v(L/2) = \frac{41 q_0 L^4}{3840 EI} = 1.0677 \times 10^{-2}\, q_0 L^4/EI.$

$v'(x_{cr}) = 0$ Not in range $0 < x < L/2$

9.81 

$q(x) = q_0 \sin 2\pi x/L$ B.C. $v(0) = v''(0) = v(L) = v''(L) = 0$

$EI v''''(x) = q_0 \sin 2\pi x/L$; $EI v'''(x) = -(L/2\pi) q_0 \cos \frac{2\pi x}{L} + A$

$EI v''(x) = -(L/2\pi)^2 q_0 \sin 2\pi x/L + Ax + B$; $v''(0) = 0 \Rightarrow B = 0$

$v''(L) = 0 \Rightarrow A = 0$

$EI v'(x) = +(L/2\pi)^3 q_0 \cos 2\pi x/L + C$; $v(L) = (L/2\pi)^4 \sin 2\pi x/L + Cx + D$

$v(0) = 0 \Rightarrow D = 0$; $v(L) = 0 \Rightarrow C = 0$

$v(x) = (L/2\pi)^4 q_0 \sin 2\pi x/L$. $v(L/2) = 0$; $v(L/4) = q_0 (L/2\pi)^4$.

9.82 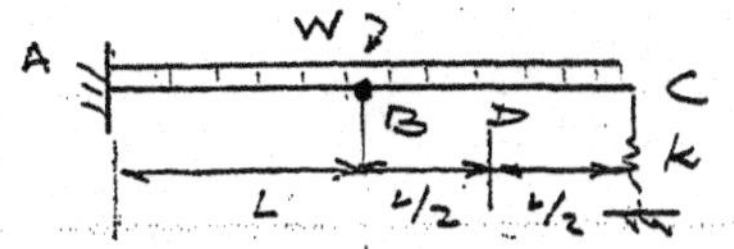

$R_C = wL/2$, $R_A = 3wL/2$

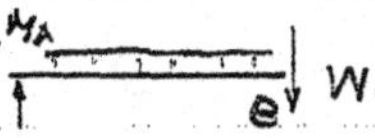

$(\Sigma M)_A = 0 \Rightarrow M_A + wL^2/2 + wL^2/2 = 0$

$M_A = -wL^2$

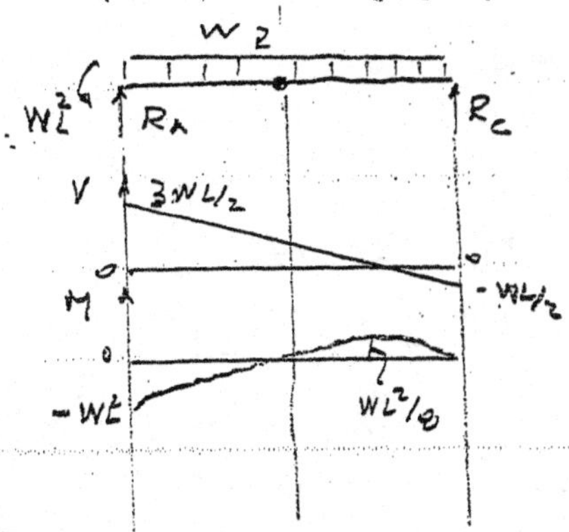

By superposition:

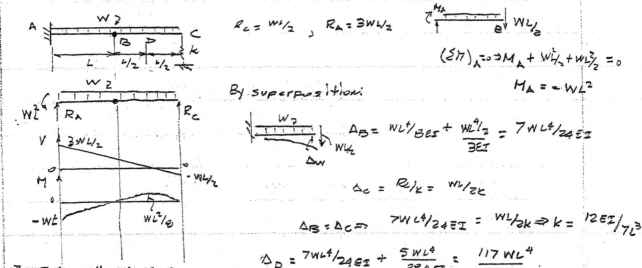

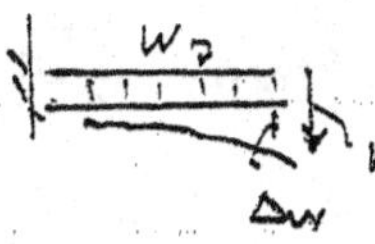

$\Delta_B = wL^4/8EI + \frac{wL^4/2}{3EI} = 7wL^4/24EI$

$\Delta_C = R_C/k = wL/2k$

$\Delta_B = \Delta_C \Rightarrow 7wL^4/24EI = wL/2k \Rightarrow k = 12EI/7L^3$

$\Delta_D = 7wL^4/24EI + \frac{5wL^4}{384EI} = \frac{117 wL^4}{384EI}$

By Integration: A-B

$EI v_1''(x) = -wL^2 - 3wLx/2 + wx^2/2$ B.C. $v_1'(0) = v(0) = 0$

$EI v_1'(x) = wL^2x - 3wLx^2/4 + wx^3/6 + A_1$ ($=0$)

$EI v_1(x) = wL^2x^2/2 - wLx^3/4 + wx^4/24 + B_1$ ($=0$) $v_1(L) = 7wL^4/24$

$v_1(x) = \frac{w}{EI}(L^2x^2/2 - Lx^3/4 - x^4/24)$; $0 < x < L$

B-C: $L < x < 2L$:

$\therefore EI v_2''(x) = wL^2 - 3wLx/2 + wx^2/2$ B.C. $v_2(L) = v_1(L)$, $v_2(2L) = wL/2k$

$EI v_2'(x) = wL^2x - 3wLx^2/4 + wx^3/6 + A_2$

$EI v_2(x) = wL^2x^2/2 - wLx^3/4 + wx^4/24 + A_2x + B_2$

$v_2(L) = v_1(L) \Rightarrow B_2 = -AL$; $v_2(2L) = v_2(L) = 7wL^4/24 \Rightarrow A_2 = -B_2/L = -wL^3/24$

$\therefore v_2(2L) = 7wL^4/24 = wL/2k \Rightarrow k = 12EI/7L^3$. (As above)

$v_B = v_C = 7wL^4/24EI$; $v_D = v(3L/2) = 117\, wL^4/384EI$

9.83 $v_A = PL^3/3EI \qquad v_D = \frac{PL^3}{3EI} + \left(\frac{TL}{GJ}\right)R = \frac{PL^3}{3EI} + \frac{PR^2L}{2GI} = \frac{PL^3}{I}\left(\frac{1}{3E} + \frac{1}{2G}(R/L)^2\right)$ 9P.33

$$G = \frac{E}{2(1+\nu)} \Rightarrow v_D = \frac{PL^3}{EI}\left[1/3 + (1+\nu)(R/L)^2\right] = \frac{PL^3}{3EI}\left[1 + 3(1+\nu)(R/L)^2\right].$$

$$v_D/v_A = 1 + 3(1+\nu)(R/L)^2 \Rightarrow R/L = \left[(v_D/v_A - 1)/3(1+\nu)\right]^{1/2}$$

$$v_D/v_A = 1.0125,\ \nu = 0.25 \Rightarrow R/L = \left(\frac{1}{80}\cdot\frac{4}{15}\right)^{1/2} = \left(\frac{1}{300}\right)^{1/2} = \sqrt{3}/30 = 0.05774.$$

9.84 $EI(x)v^{IV}(x) = q(x) = q_0 \sin(3\pi x/L) \qquad$ B.C. $v(0) = v''(0) = v(L) = v''(L) = 0$

$$EI_0 \sin\frac{\pi x}{L}\, v^{IV}(x) = q_0 \sin(3\pi x/L) \Rightarrow EI_0 \sin\frac{\pi x}{L}\, v^{IV}(x) = q_0 \sin\frac{\pi x}{L}\left[2\cos\frac{2\pi x}{L} + 1\right].$$

$$\therefore EI_0 v^{IV}(x) = q_0\left(2\cos\frac{2\pi x}{L} + 1\right)$$

$$EI_0 v'''(x) = q_0\left[2(L/2\pi)\sin\frac{2\pi x}{L} + x\right] + A$$

$$EI_0 v''(x) = q_0\left[-2(L/2\pi)^2\cos\frac{2\pi x}{L} + x^2/2\right] + Ax + B$$

$$v''(0) = 0 \Rightarrow B = 2q_0(L/2\pi)^2;\ v''(L) = 0 \Rightarrow A = -q_0L/2$$

$$EI_0 v''(x) = q_0\left[-2(L/2\pi)^2\cos 2\pi x/L + x^2/2 - Lx/2 + 2(L/2\pi)^2\right]$$

$$EIv'(x) = q_0\left[-2(L/2\pi)^3\sin 2\pi x/L + x^3/6 - Lx^2/4 + 2(L/2\pi)^2 x\right] + C$$

$$EIv(x) = q_0\left[2(L/2\pi)^4\cos 2\pi x/L + x^4/24 - Lx^3/12 + (L/2\pi)^2x^2\right] + Cx + D$$

$$v(0) = 0 \Rightarrow D = -2(L/2\pi)^4 q_0$$

$$v(L) = 0 \Rightarrow C = L^3/24\pi^2(\pi^2 - 6)$$

$$\therefore v(x) = \frac{q_0L^4}{48\pi^4 EI}\left\{6\cos 2\pi x/L + 2\pi^4(x/L)^4 - 4\pi^4(x/L)^3 + 12\pi^2(x/L)^2 + 2\pi^2(\pi^2-6)(x/L) - 6\right\}$$

$$v(L/2) = \frac{q_0L^4}{384\pi^4 EI}\left[5\pi^4 - 24\pi^2 - 96\right] = 4.122\times10^{-3}\, q_0L^4/EI.$$

$$v'(x) = \frac{q_0L^4}{48\pi^4 EI}\left[-12\sin\frac{2\pi x}{L} + 8\pi^4(x/L)^3 - 12\pi^4(x/L)^2 + 24\pi^2(x/L) + 2\pi^2(\pi^2-6)\right]$$

$$v'(L/2) = 0$$

9.86

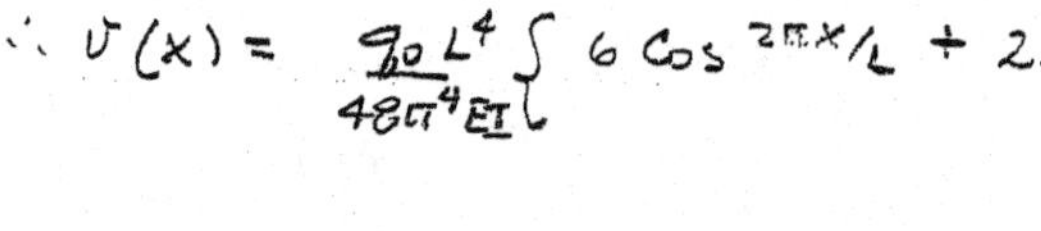

$$\Delta_c^P - \Delta_c^R = R/k \Rightarrow \frac{5PL^3}{48EI} - \frac{RL^3}{3EI} = R/k$$

$$R = \frac{5P}{16(1 + 3EI/kL^3)}$$

9P.34

9.87

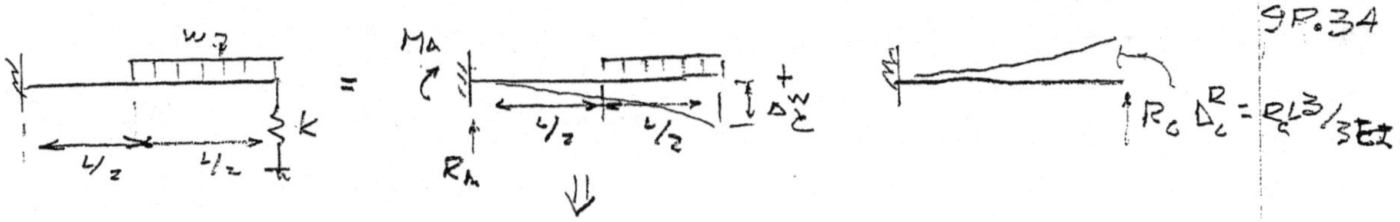

$\Downarrow$

$$M(x) = -3wL^2/8 + wLx/2 - \frac{w}{2}\langle x - L/2\rangle^2$$

$$EI\,v''(x) = 3wL^2/8 - wLx/2 + w/2\,\langle x - L/2\rangle^2 \;; \quad EI\,v'(x) = 3wL^2x/8 - wLx^2/4 + \frac{w}{6}\langle x - L/2\rangle^3 + \cancel{A}_{\to 0}$$

$$EI\,v(x) = 3wL^2x^2/16 - wLx^3/12 + w/24\,\langle x - L/2\rangle^4 + \cancel{B}_{\to 0}$$

$$v(L) = 41wL^4/384EI = \Delta_c^W$$

$$\Delta_c^W - \Delta_c^R = R/K \Rightarrow \frac{41wL^4}{384EI} - \frac{RL^3}{3EI} = R/K \Rightarrow R_c = \frac{41wL}{128}\;\frac{1}{3EI/KL^3 + 1}$$

$$R_A = wL/2 - R_c \;; \quad M_A = -3wL^2/8 + R_c L.$$

9.88

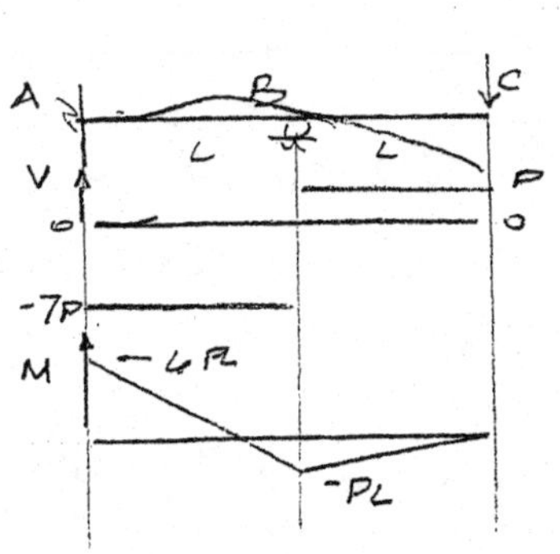

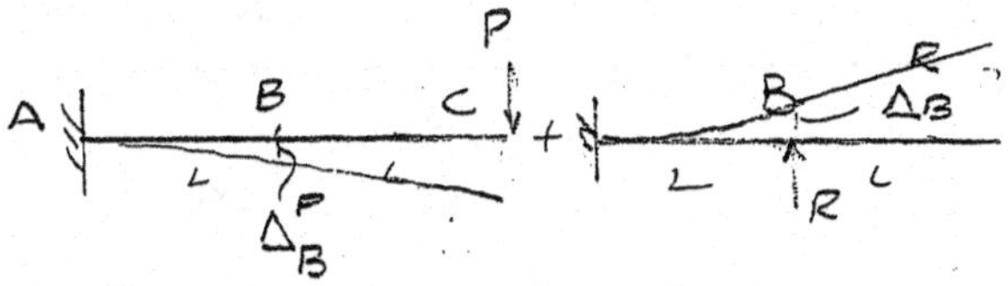

$$\Delta_B = \Delta_B^P - \Delta_B^R = 0 \Rightarrow \frac{P(2L)^3}{3EI} - \frac{RL^3}{3EI} = 0 \Rightarrow R_B = 8P \uparrow$$

$$\Sigma F_y = 0 \Rightarrow R_A = -7P\downarrow, \quad M_A = 6PL$$

9.89

$$M(x) = -P[L + x] \quad 0 \le x \le L.$$

$$\alpha EI_0 v(x) = -P[L + x], \quad 0 \le x \le L/2 \quad (a)$$

$$EI_0 v(x) = -P[L + x], \quad L/2 \le x \le L \quad (b)$$

$$v''(x) = \frac{P}{\alpha EI_0}\{L - x + (\alpha - 1)(L - x)\langle x - L/2\rangle^0\} \quad (c)$$

Note: Eq. (c) is equivalent to (a) & (b)

$$0 \le x \le L$$

$$v'(x) = P/\alpha EI_0\{Lx - x^2/2 + (\alpha-1)[(L-x)\langle x - L/2\rangle + \int\langle x - L/2\rangle dx]\}$$ ⇐ Integration by parts.

$$= P/\alpha EI_0\{Lx - x^2/2 + (\alpha-1)[(L-x)\langle x - L/2\rangle + \langle x - L/2\rangle^2/2]\} + \cancel{A}_{\to 0} \quad [v'(0) = 0]$$

$$v(x) = P/\alpha EI_0\{Lx^2/2 - x^3/6 + (\alpha-1)[(L-x)\langle x - L/2\rangle^2/2 + \int\tfrac{1}{2}\langle x - L/2\rangle^2 dx + \langle x - L/2\rangle^3/6]\} + B$$

$$= P/\alpha EI_0\{Lx^2/2 - x^3/6 + (\alpha-1)[(L-x)\langle x - L/2\rangle^2/2 + \langle x - L/2\rangle^3/3]\} + \cancel{B}_{\to 0} \quad [v(0) = 0]$$

$$v(x) = P/6\alpha EI_0\{x^2(3L - x) + (\alpha-1)\langle x - L/2\rangle^2[3(L - x) + 2\langle x - L/2\rangle]\}$$

$$v(L) = PL^3/24\alpha EI_0\,[8 + [\alpha - 1]]$$

$\alpha = 1 \Rightarrow v(L) = PL^3/3EI_0 \Rightarrow$ 

$\alpha \to \infty \quad v(L) = PL^3/24EI_0 \Rightarrow$

9.90 (i)

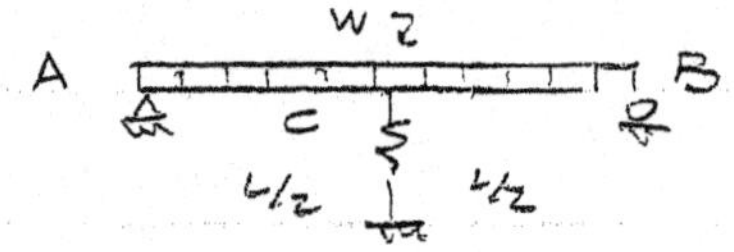

$M_C = R_A L/2 - wL^2/8 \Rightarrow R_A = 2M_C/L + wL/4$

By symmetry $R_A = R_B \Rightarrow R_C = wL - 2R_A$

$\therefore R_C = wL/2 - 4M_C/L$

By superposition:

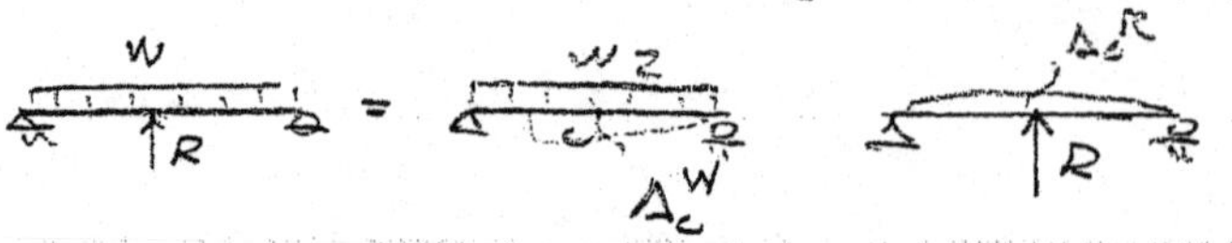

$$\Delta_C = \frac{5wL^4}{384EI} - \frac{R_C L^3}{48EI} + \frac{R_C}{k}$$

Substitute for R_C: $\Rightarrow \dfrac{wL^4}{384EI} + \dfrac{M_C L^2}{12EI} = \dfrac{wL/2 - 4M_C/L}{k}$

$$\therefore k = \frac{EI\,(wL/2 - 4M_C/L)}{wL^4/384 + M_C L^2/12}$$

(i) $M_C = 0 \Rightarrow k = 192\,EI/L^3$

(ii) $M_C = -wL^2/48 \Rightarrow k = \dfrac{EI}{L^3}\left(\dfrac{1/2 + 1/12}{1/384 - 1/12\cdot 48}\right) = 672\,EI/L^3$

9.91

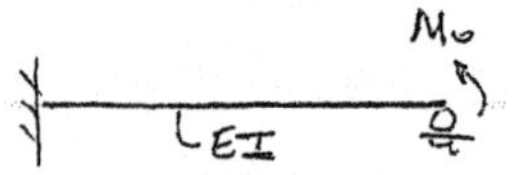

$q(x) = 0 \Rightarrow EIv^{iv}(x) = 0$ B.C. $v(0) = v'(0) = 0$, $v(L) = 0$, $v''(L) = -M_0/EI$

$EIv'''(x) = A$, $EIv''(x) = Ax + B$; $EIv'(x) = Ax^2/2 + Bx + C$ ($C \to 0$ [$v'(0) = 0$])

$EIv(x) = Ax^3/6 + Bx^2/2 + D$

$EIv''(L) = AL + B = -M_0$ (i); $v(0) = 0 \Rightarrow D = 0$, $v(L) = 0 \Rightarrow AL^3/6 + BL^2/2 = 0$ (ii)

Solving (i) & (ii): $A = -3M_0/2L$, $B = M_0/2$

$v(x) = \dfrac{M_0 x^2}{4EI}(x/L - 1)$; $V(0) = -EIv'''(0) = -A = +3M_0/2L = R_A$; $V(L) = 3M_0/2L = -R_B$

$M(0) = -EIv''(0) = -B = -M_0/2 = M_A$

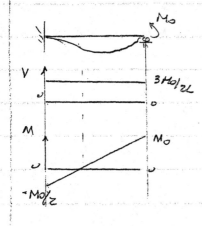

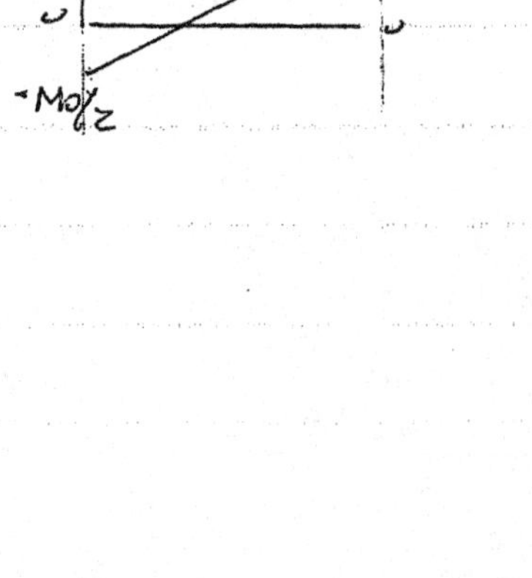

9.92

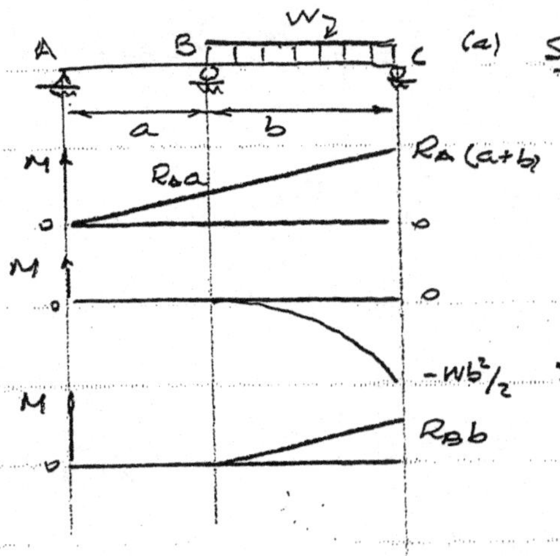

(a) **Segment A-B:** $v_A = v_B = 0$

$$a\theta_A = Q_B\Big|_A^B \Rightarrow \theta_A = \frac{Q_B}{a}\Big|_A^B$$

$$\theta_B = \theta_A - A\Big|_A^B = \left(\frac{Q_B}{a}\Big|_A^B - A\Big|_A^B\right)/EI$$

$$\theta_B = (R_A a^2/2)(a/3)/a - R_A a^2/2 = -R_A a^2/3EI$$

Segment B-C: $v_B = v_C = 0$

$$b\theta_B = Q_C\Big|_B^C \Rightarrow \theta_B = Q_C\Big|_B^C/bEI$$

$$\theta_B = [(-wb^2/2)(b/3)(b/4) + (R_B b^2/2)(b/3)]/bEI + [(R_A ab)(b/2) + (R_A ab/2)(b/3)]/bEI$$

$$\theta_B = \left(-wb^3/24 + \frac{R_A b(3a+b)}{6} + R_B b^2/6\right)/EI$$

$$(\theta_B)_{AB} = (\theta_B)_{BC} \Rightarrow 4[(3a+b)b + 2a^2]R_A + 4R_B b^2 = wb^3 \quad (i)$$

$$\Sigma M_C = 0 \Rightarrow (a+b)R_A + bR_B = wb^2/2 \quad (ii)$$

$$(i)\,\&\,(ii) \Rightarrow R_A = -\frac{wb^3}{8aL}, \quad R_B = \frac{wb}{8a}(4a+b)$$

$$\Sigma F_y = 0 \Rightarrow R_C = \frac{w}{8aL}(4aL + b^2 - bL) = \frac{wb}{8L}(4L-b)$$

(b) $M_B = R_A a = -wb^3/8L$

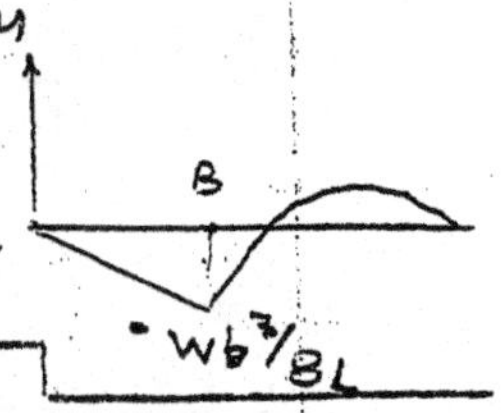

9.93

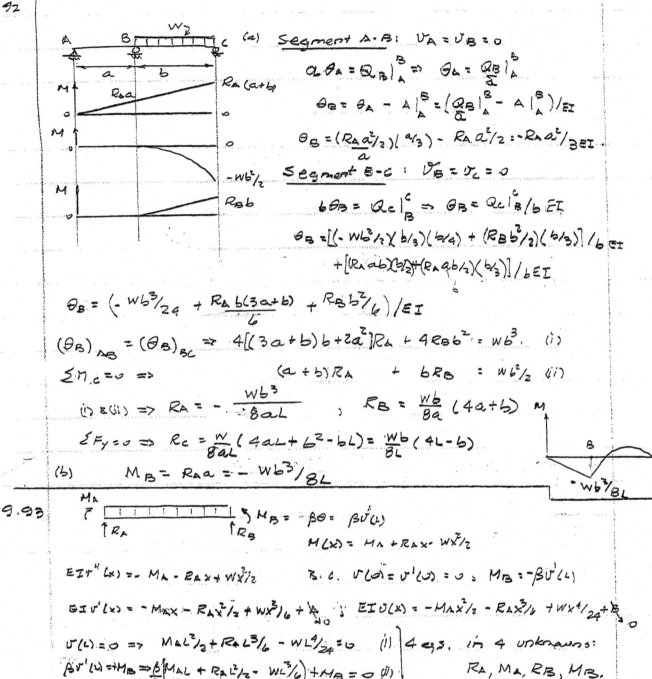

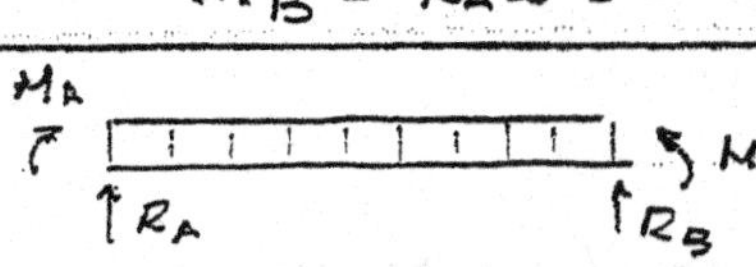

$M_B = -\beta\theta = \beta v'(L)$

$M(x) = M_A + R_A x - wx^2/2$

$EIv''(x) = -M_A - R_A x + wx^2/2$ B.C. $v(0) = v'(0) = 0$, $M_B = -\beta v'(L)$

$EIv'(x) = -M_A x - R_A x^2/2 + wx^3/6 + A_{=0}$; $EIv(x) = -M_A x^2/2 - R_A x^3/6 + wx^4/24 + B_{=0}$

$v(L) = 0 \Rightarrow M_A L^2/2 + R_A L^3/6 - wL^4/24 = 0$ (i)

$\beta v'(L) = +M_B \Rightarrow \frac{\beta}{EI}(M_A L + R_A L^2/2 - wL^3/6) + M_B = 0$ (ii)

STATICS:

$M_A + R_A L - wL^2/2 - M_B = 0$ (iii)

$R_A + R_B - wL = 0$ (iv)

4 eqs. in 4 unknowns: R_A, M_A, R_B, M_B.

Solving (i)-(iv) $\Rightarrow M_A = -\frac{wL^2}{12}\,\frac{6+\beta L/EI}{4+\beta L/EI}$, $R_A = \frac{wL}{2}\left(\frac{5+\beta L/EI}{4+\beta L/EI}\right)$

$\beta = 0 \Rightarrow M_A = -wL^2/8$, $R_A = 5wL/8$

$\beta \to \infty \Rightarrow M_A = -wL^2/12$, $R_A = wL/2$

9.94

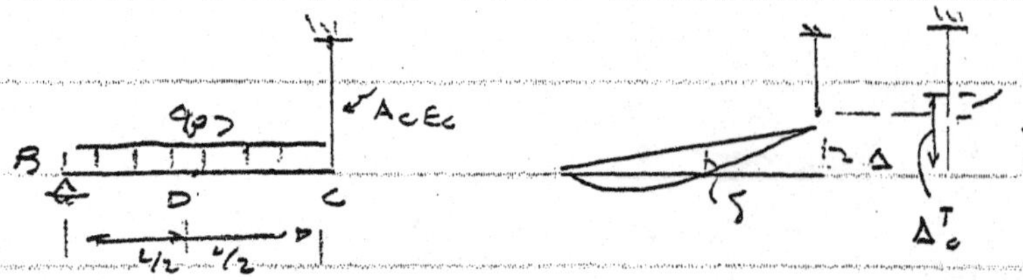

$\delta_D = 5q_0L^4/384EI$

$\delta = \Delta/2 \Rightarrow \Delta = 2\delta_D = 5q_0L^4/192EI$

$\Delta_C = \Delta_C^T - \Delta_C^R$, $R = q_0/2$

$\therefore \Delta_C = h\alpha(\Delta T) - (\frac{q_0L}{2})h/A_cE_c$

$$\frac{5q_0L^4}{192EI} = h\alpha(\Delta T) - \frac{q_0Lh}{2A_cE_c} \Rightarrow \Delta T = \frac{1}{\alpha}\left[\frac{5q_0L^4}{192EIh} + \frac{q_0L}{2A_cE_c}\right]$$

$$\Delta T = \frac{q_0L}{2\alpha}\left(\frac{5L^3}{96EIh} + \frac{1}{A_cE_c}\right) \quad \text{cools}$$

9.95

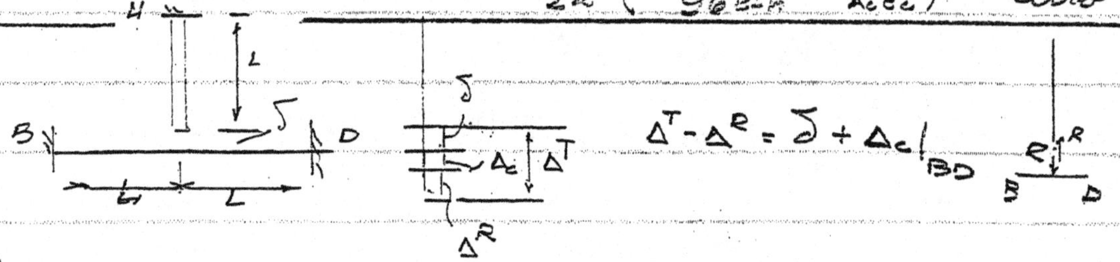

$\Delta^T - \Delta^R = \delta + \Delta_C|_{BD}$

Beam BD:

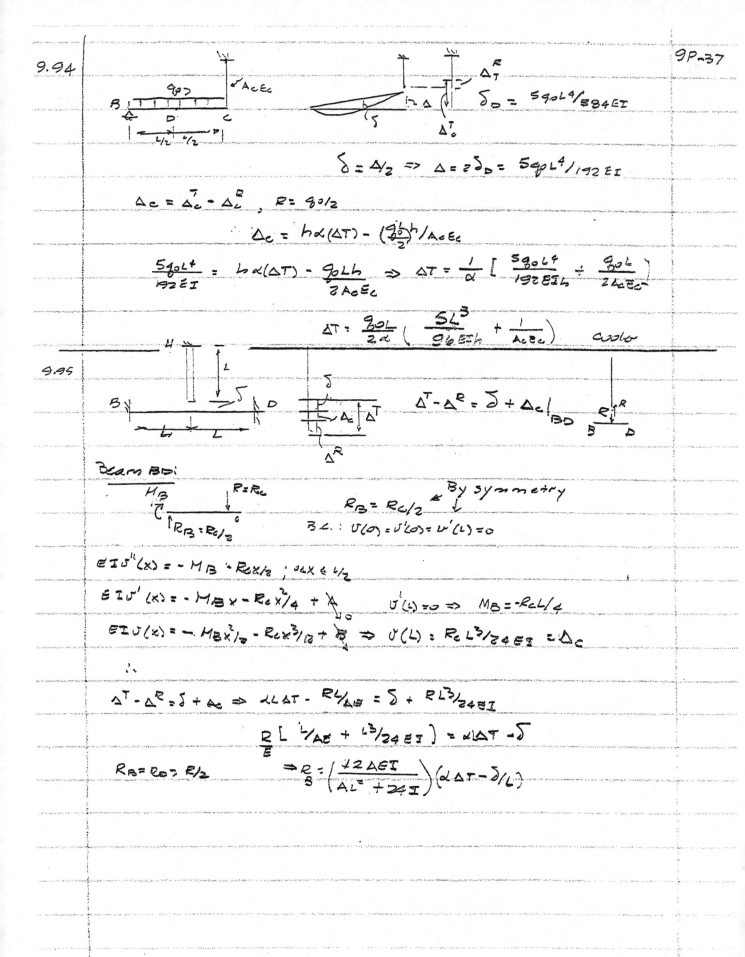

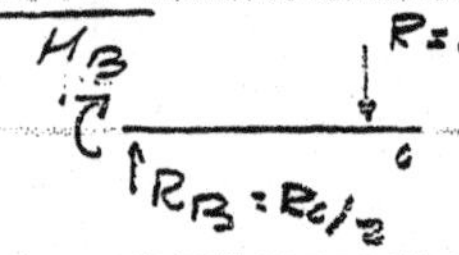

$R_B = R_C/2$ ← By symmetry

B.C.: $v(0) = v'(0) = v'(L) = 0$

$EIv''(x) = -M_B - R_Cx/2$; $0 \le x \le L/2$

$EIv'(x) = -M_Bx - R_Cx^2/4 + A_0$ $\quad v'(L) = 0 \Rightarrow M_B = -R_CL/4$

$EIv(x) = -M_Bx^2/2 - R_Cx^3/12 + B_0 \Rightarrow v(L) = R_CL^3/24EI = \Delta_C$

$\therefore$

$\Delta^T - \Delta^R = \delta + \Delta_C \Rightarrow \alpha L\Delta T - RL/AE = \delta + RL^3/24EI$

$$\frac{R}{E}\left[L/AE + L^3/24EI\right] = \alpha\Delta T - \delta$$

$R_B = R_D = R/2$ $$\Rightarrow \frac{R}{E} = \left(\frac{12AEI}{AL^2 + 24I}\right)(\alpha\Delta T - \delta/L)$$

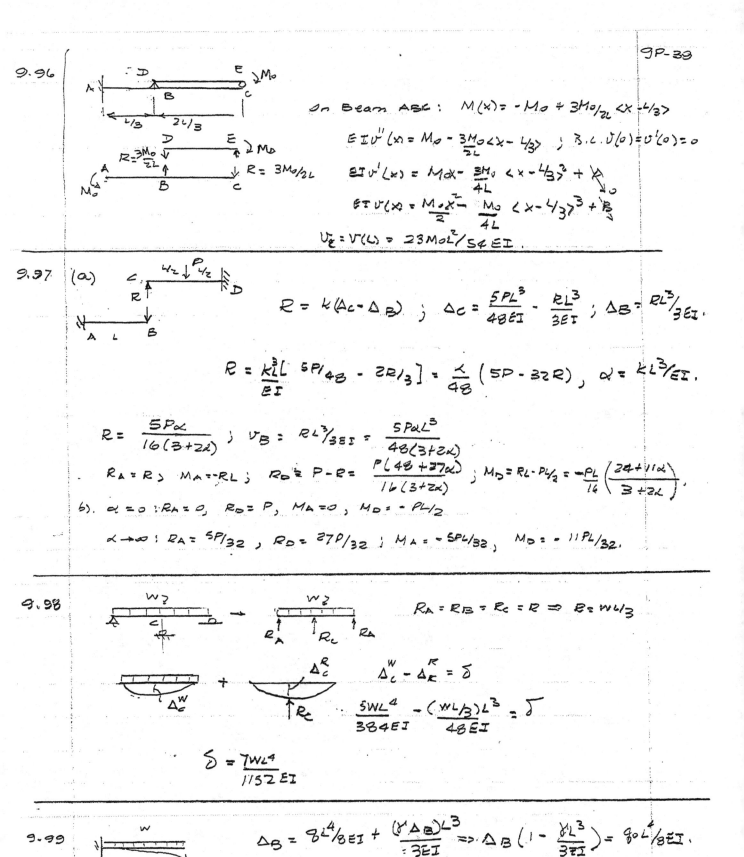

9.96

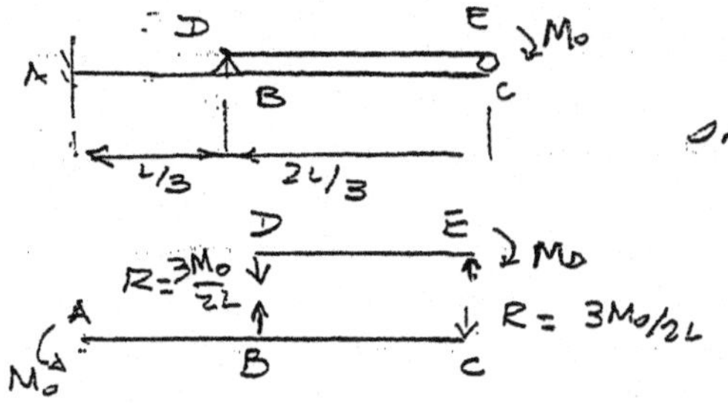

On Beam ABC: $M(x) = -M_0 + 3M_0/_{2L}\langle x - L/3\rangle$

$EIv''(x) = M_0 - \frac{3M_0}{2L}\langle x - L/3\rangle$; B.C. $v(0) = v'(0) = 0$

$EIv'(x) = M_0x - \frac{3M_0}{4L}\langle x - L/3\rangle^2 + A$ ($A \to 0$)

$EIv(x) = \frac{M_0x^2}{2} - \frac{M_0}{4L}\langle x - L/3\rangle^3 + B$ ($B \to 0$)

$v_E = v(L) = 23M_0L^2/54EI$.

9.97 (a)

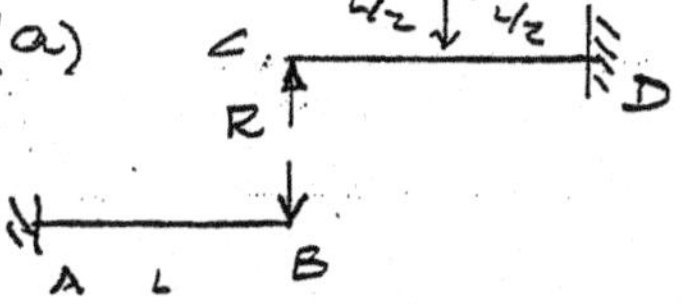

$R = k(\Delta_C - \Delta_B)$; $\Delta_C = \frac{5PL^3}{48EI} - \frac{RL^3}{3EI}$; $\Delta_B = RL^3/_{3EI}$.

$R = \frac{kL^3}{EI}[5P/48 - 2R/3] = \frac{\alpha}{48}(5P - 32R)$, $\alpha = kL^3/EI$.

$R = \frac{5P\alpha}{16(3+2\alpha)}$; $v_B = RL^3/_{3EI} = \frac{5P\alpha L^3}{48(3+2\alpha)}$

$R_A = R$, $M_A = -RL$; $R_D = P - R = \frac{P(48+27\alpha)}{16(3+2\alpha)}$; $M_D = RL - PL/2 = -\frac{PL}{16}\left(\frac{24+11\alpha}{3+2\alpha}\right)$.

b). $\alpha = 0$: $R_A = 0$, $R_D = P$, $M_A = 0$, $M_D = -PL/2$

$\alpha \to \infty$: $R_A = 5P/32$, $R_D = 27P/32$; $M_A = -5PL/32$, $M_D = -11PL/32$.

9.98

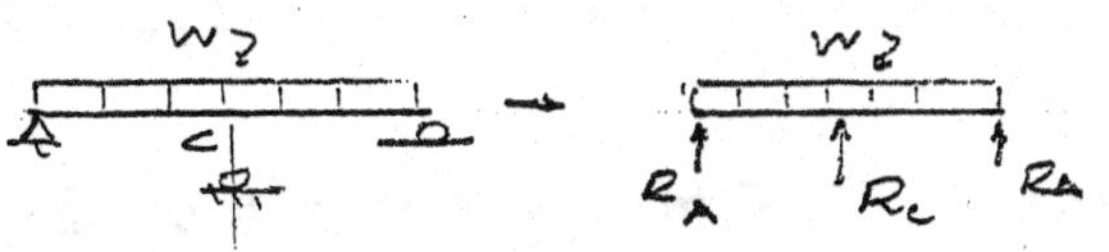

$R_A = R_B = R_C = R \Rightarrow R = wL/3$

$\Delta_C^W - \Delta_C^R = \delta$

$\frac{5wL^4}{384EI} - \frac{(wL/3)L^3}{48EI} = \delta$

$\delta = \frac{7wL^4}{1152EI}$

9.99

$\Delta_B = q_0L^4/_{8EI} + \frac{(\gamma\Delta_B)L^3}{3EI} \Rightarrow \Delta_B\left(1 - \frac{\gamma L^3}{3EI}\right) = q_0L^4/_{8EI}$.

$\Delta_B = \frac{q_0L^4}{8EI}\frac{1}{(1-\Gamma)}$, $\Gamma = \gamma L^3/_{3EI}$.

$\Delta_B \to \infty$ as $\Gamma \to 1$

$\Delta_B > 0$ (finite) $\Gamma < 1 \Rightarrow \gamma < \frac{3EI}{L^3}$

9.101.

a)
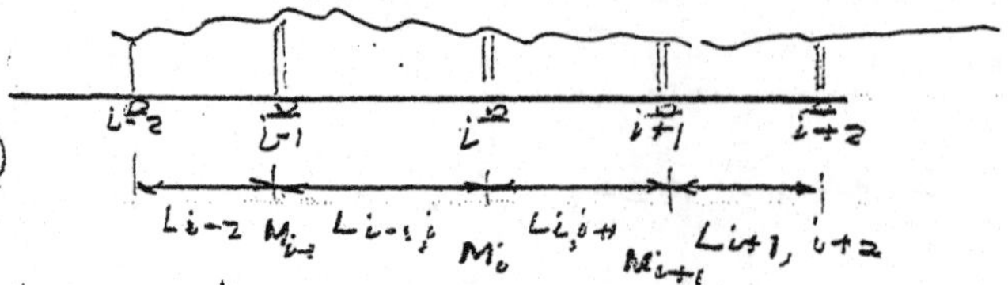

b)
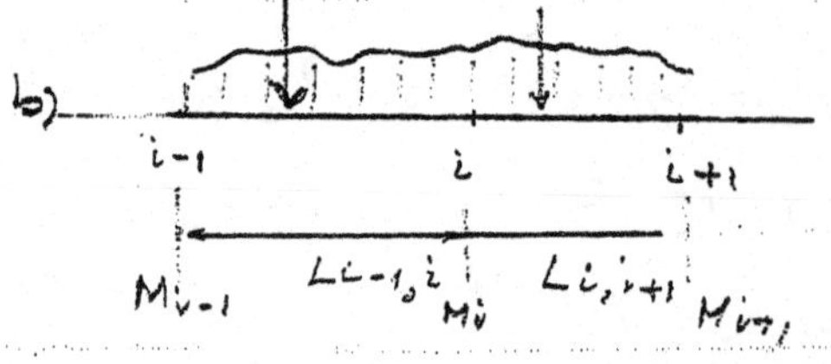

c)
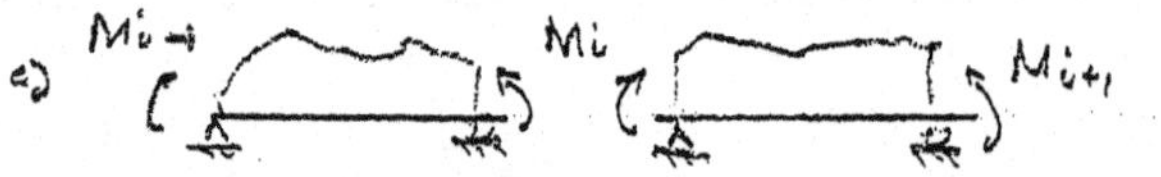

d)
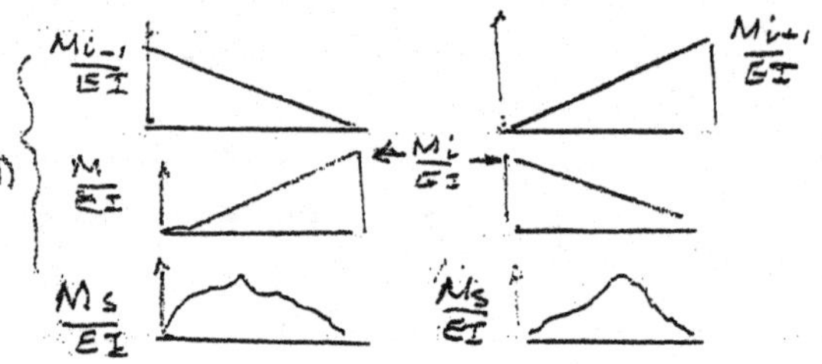

Span $i-1, i$: By eq. (9.13.13b) ⇒

$$\theta_i = -\frac{M_{i-1}(L_{i-1,i})^2}{6EI\,L_{i-1,i}} - \frac{M_i(L_{i-1,i})^2}{3EI(L_{i-1,i})} - \frac{\tilde{Q}_{i-1}\Big|_{i-1}^{i}}{L_{i-1,i}}$$

Span $i, i+1$ By eq (9.13.13a) ⇒

$$\theta_i = -\frac{M_i(L_{i,i+1})^2}{3EI(L_{i,i+1})} + \frac{M_{i+1}(L_{i,i+1})^2}{6EI(L_{i,i+1})} + \frac{\tilde{Q}_{i+1}\Big|_{i}^{i+1}}{L_{i,i+1}}$$

Setting the two θ_i's equal:

$$-\frac{M_{i-1}L_{i-1,i}}{6} - \frac{M_i L_{i-1,i}}{3} - \frac{EI\,\tilde{Q}_{i-1}\Big|_{i-1}^{i}}{L_{i-1,i}} = \frac{M_i L_{i,i+1}}{3} + \frac{M_{i+1}L_{i,i+1}}{6} + \frac{EI\,\tilde{Q}_{i+1}\Big|_{i}^{i+1}}{L_{i,i+1}}$$

$$M_{i-1}L_{i-1,i} + 2M_i(L_{i-1,i} + L_{i,i+1}) + M_{i+1}L_{i+1} = -6\left[\frac{\tilde{Q}_{i-1}\Big|_{i-1}^{i}}{L_{i-1,i}} + \frac{\tilde{Q}_{i+1}\Big|_{i}^{i+1}}{L_{i,i+1}}\right]$$

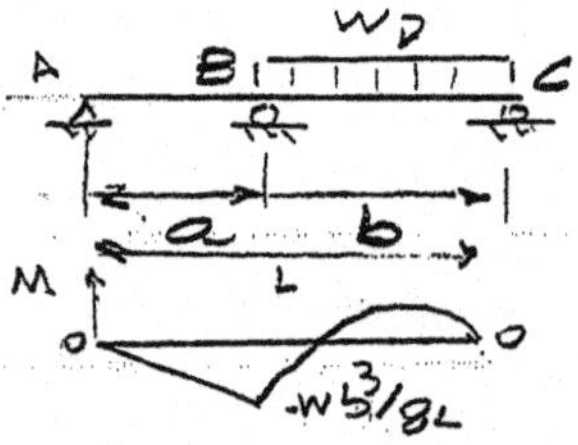

$$M_A = M_C = 0 \Rightarrow 2M_B(a+b) = -6\frac{\tilde{Q}_C\Big|_B^C}{b}$$

$$\tilde{Q}_C\Big|_B^C = \left(\frac{wb^2}{8}\right)(2b/3)(b/2) = wb^4/24$$

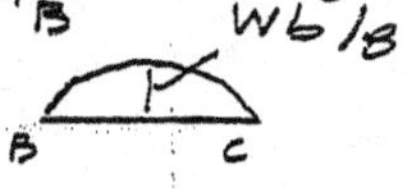

$$\therefore M_B = -\frac{wb^3}{8(a+b)} = -wb^3/8L$$

Span AB: $R_A \uparrow$ ⟶ $wb^3/8L$ ⇒ $R_A = wb^3/8aL$ ↓ Span BC: w_D, $wb^3/8L$, $\uparrow R_C$ ⇒ $R_C = \frac{wb}{8L}(4L-b)$

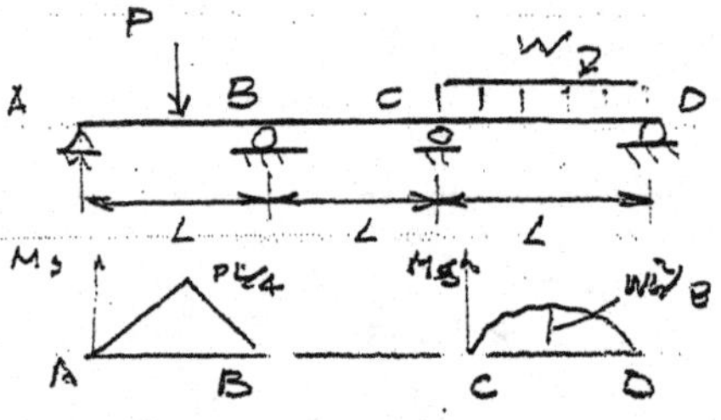

$M_A = M_D = 0$: Spans ABC: $4LM_B + LM_C = -\frac{6}{L}\tilde{Q}_A\Big|_A^B$

Spans BCD: $LM_B + 4LM_C = -\frac{6}{L}\tilde{Q}_D\Big|_C^D$

$$\tilde{Q}_A\Big|_A^B = (PL/4)(L/2)(L/2) = PL^3/16\ ;\ \tilde{Q}_D = wL^4/24$$

$$4M_B + M_C = -3PL/8$$

$$M_B + 4M_C = -wL^2/4$$

Solving for M_B and M_C:

$$M_B = \frac{wL^2}{60} - PL/10\,,\quad M_C = PL/40 - wL^2/15$$

10.1 $\sigma_x = pR/2t$, $\sigma_\theta = pR/t$ $D = 0.4\,m$

Normal $\sigma(\theta = 30^\circ) = \sigma_x \cos^2\theta + \sigma_\theta \sin^2\theta = 160$ MPa

$= (pR/t)\,[(1/2)(3/4) + (1/4)] = 160 \Rightarrow 5pR/8t = 150$ MPa

$p = 240\,t/R = 240(8\times10^{-3})/0.4 = 4.80$ N/m²

$p = 4.8$ MPa

$|\tau_{nt}|(\theta = 30^\circ) = |(\sigma_x - \sigma_\theta)\sin\theta\cos\theta| =$

$= (pR/2t)\,[(1/2)(\sqrt{3}/2)] = \sqrt{3}pR/8t = 90$ MPa

$p = (720/\sqrt{3})\,t/R = (240\sqrt{3})\,t/R = (240\sqrt{3})(8\times10^{-3}/0.4)$

$p = 8.3$ N/m² $= 8.3$ MPa $\qquad p_{all} = 4.8$ MPa

10.2 a) $\theta = 0, 90^\circ$: $\sigma_{max}\big|_{\theta=90^\circ} = \frac{pR}{t} = 120 \Rightarrow p = 120t/R = (120)(10^{-2})/0.8 = 1.5$ MPa

$\tau = 0$ in weld. (a) $p_{allow} = 1.5$ MPa

b) $\theta = \pm 45^\circ$: $\sigma_{max}\big|_{\theta=45^\circ} = \sigma_x\cos^2\theta + \sigma_\theta\sin^2\theta = \frac{pR}{t}[\cos^2\theta/2 + \sin^2\theta] = 120$

$0.75\,pR/t = 120 \Rightarrow p = 160t/R$

$p = 2.0$ MPa.

$|\tau_{max}|_{\theta=45^\circ} = |(\sigma_x - \sigma_\theta)\sin\theta\cos\theta| = \left(\frac{pR}{2t}\right)\left(\frac{1}{2}\right) = 80$

$p = 320\,t/R = (320)(.01)/0.8 = 4$ MPa.

(b) $p_{all} = 2.0$ MPa

10.3 a) $\varepsilon_x = \Delta/L$, $\Delta = \varepsilon_x L = \frac{L}{E}[\sigma_x - \nu\sigma_\theta] = \frac{pLR}{2Et}[1 - 2\nu]$

$\Rightarrow p = 2E\Delta t/(1-2\nu)RL$

b). $p = \dfrac{2(70\times10^9)(10^{-2})(.5\times10^{-2})}{(0.34)(0.4)(3.5)} = 14.7\times10^6$ N/m² $= 14.7$ MPa

---

10.4 $p = \rho h$ $\quad \gamma_{water} = \dfrac{1\,gram}{cm^3} = 10^6$ gram/m³ $= 10^3$ kg/m³ $= 9.807\times10^3$ N/m³ $= \rho$

$\sigma_\theta = p(x)R/t = \rho hR/t = (9.807\times10^3)(16)(2)/0.02 = 15.69$ N/m² $= 15.69$ MPa

---

10.5 $p = \rho h$ $\quad \gamma_{oil} = 0.9\gamma_{water} \Rightarrow \rho_{oil} = 8.8263\times10^3$ N/m3

$\sigma_{\theta\,all} = 20$ MPa ; $\rho hR/t = 80 \Rightarrow h = (\sigma_\theta)_{all}t/\rho R = 22.66$ m.

10.6

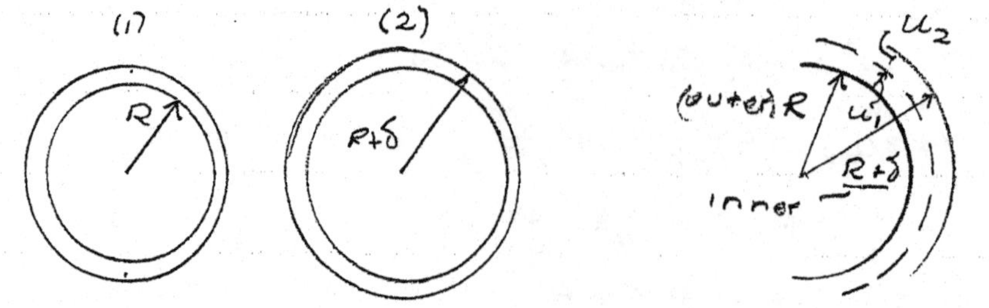

$u_1 + u_2 = \delta$; $u_1 = \varepsilon_\theta \cdot (R - t/2) = \varepsilon_\theta R(1 - t/2R) \doteq \varepsilon_\theta R \quad (t/R \ll 1)$

$\therefore u_1 = \sigma_\theta R/E = pR^2/Et$

$u_2 = \varepsilon_\theta \cdot [(R+\delta) - t/2] = \varepsilon_\theta \cdot R[1 + (\frac{\delta - t/2}{R})] \doteq \varepsilon_\theta R \quad (t/R \ll 1,\ \delta/R \ll 1)$

$= pR^2/Et$

$\therefore 2pR^2/Et = \delta \Rightarrow p = \delta Et/2R^2$

a) $\sigma_\theta^{(1)} = \frac{p(R - t/2)}{t} = \frac{pR}{t}(1 - t/2) \doteq pR/t = \frac{(\delta Et/2R^2)R}{t} = \delta E/2R$

$\sigma_\theta^{(2)} = -\delta E/2R$

b). $\sigma_{rr} = p = \delta Et/2R^2$, (c) $|p/\sigma_\theta| = t/R \ll 1$

10.7

a) 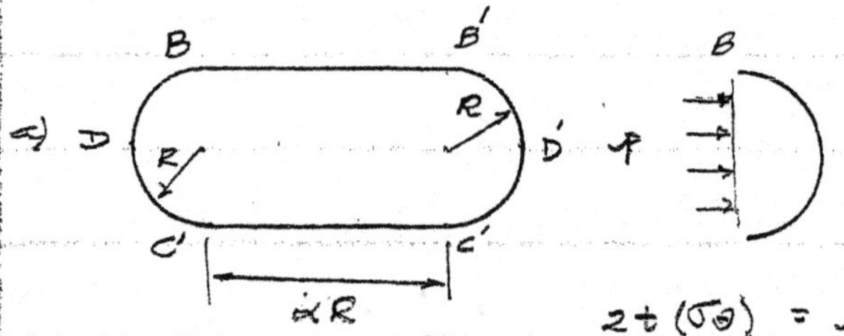

$(\sigma_\theta)_B = pR/t$

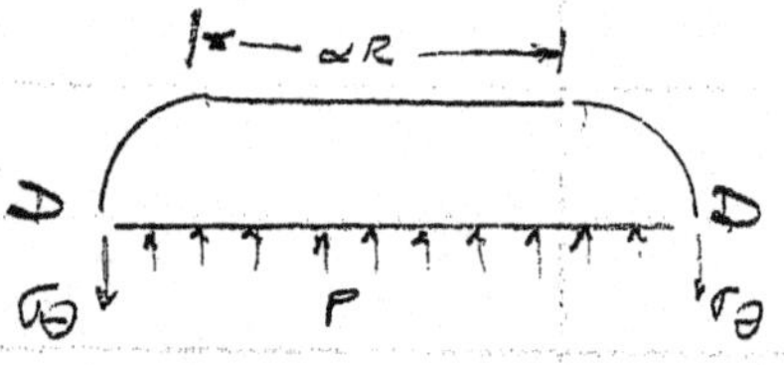

$2t(\sigma_\theta)_D = p[2R + 4R] = 6pR \Rightarrow (\sigma_\theta)_D = 6pR/2t = 3pR/t$
$\alpha = 4$

$\varepsilon_x = \text{const} \Rightarrow \sigma_x = \text{const}$.

$\sigma_x t[8R + 2\pi R] = p[(2R)(4R) + \pi R^2] = pR^2(8 + \pi)$

$\sigma_x = \frac{pR(8+\pi)}{2(4+\pi)t}$

b). $R = 0.25\,m$, $t = 2\,cm$. $\quad (\sigma_\theta)_{max} = 3(1.6\times10^6)(0.25)/0.02 = 60\,N/m^2 = 60\,MPa$

$(\sigma_x)_{max} = \frac{(1.6\times10^6)(0.25)(8+\pi)}{2(4+\pi)(0.02)} = 15.6\,MPa$

10.8

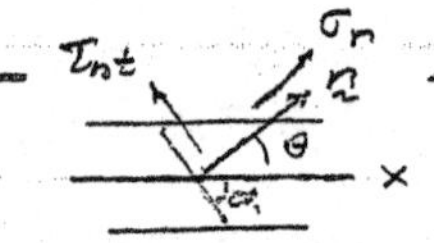

$\sigma_n = \sigma_x \cos^2\theta + \sigma_\theta \sin^2\theta = \frac{pR}{2t}[\cos^2\theta + 2\sin^2\theta]$

$\tau_{nt} = -(\sigma_x - \sigma_\theta)\sin\theta\cos\theta = -(pR/2t)\sin\theta\cos\theta$

$\sin\theta = \sin(\alpha - \pi/2) = \cos\alpha$;
$\cos\theta = \cos(\alpha - \pi/2) = +\sin\alpha$ $\quad \sigma_n = (pR/2t)(\sin^2\alpha + 2\cos^2\alpha) = 2\tau_0$

$|\tau_{nt}| = pR/2t \sin\alpha\cos\alpha = \tau_0$

$\sin^2\alpha + 2\cos^2\alpha = 1 + \cos^2\alpha > \sin\alpha\cos\alpha \Rightarrow \therefore$ tension governs

$p = \frac{4\tau_0 t}{R(1 - \cos^2\alpha)} = \frac{8\tau_0 t}{R(3 + \cos 2\alpha)} = \frac{4\sigma_0 t}{R(3 + \cos 2\alpha)}$

10.9 $u]_{press} = -\left[\frac{\sigma_x L}{E} - \frac{\nu \sigma_\theta L}{E}\right] = -\frac{L}{E}\left[\frac{(p\pi R^2)}{2\pi R t} - \frac{\nu p R}{t}\right] = -\frac{pRL}{2tE}(1-2\nu)$

$u^T = \alpha L \cdot \Delta T \qquad u^T - u_p < \delta \Rightarrow \alpha L \Delta T < \delta + pRL(1-2\nu)/2Et$

$$\Delta T|_{max} = \delta/\alpha L + \frac{pR(1-2\nu)}{2Et\alpha}$$

(a): See below.

10.10 $\tau_{x\theta} = 0$; $\sigma_x = pR/2t$, $\sigma_\theta = pR/t$; $\theta = 45°$

b) Weld: $\sigma_n = (\sigma_x + \sigma_\theta)/2 + (\sigma_x - \sigma_\theta)/2 \cos 2\theta = 3pR/4t \le (\sigma_n)_{all} = 90\text{ MPa}$

$$p = \frac{4(90)t}{3R} = 120\frac{t}{R}$$

$$p = \frac{(120)(6\times10^{-3})}{0.25} = 2.88\text{ MPa}$$

$|\tau_{nt}| = \frac{(\sigma_x - \sigma_\theta)}{2}\sin 2\theta = \frac{pR}{4t} \le \tau_{all} = 60\text{ MPa} \Rightarrow p = 240t/R = 5.76\text{ MPa}$

b). Aluminum: $\sigma_{max} = pR/t \Rightarrow p = \sigma_0 t/R = (100\times10^6)(6\times10^{-3})/0.25 = 2.4\text{ N/m}^2 = 2.4\text{ MPa}$

$\tau_{max} = pR/4t \Rightarrow p = 4\tau_0 t/R = 4(50\times10^6)(4\times10^{-3})/0.25 = 4.8\text{ N/m}^2 = 4.8\text{ MPa}$

$$p_{all} = 2.4\text{ MPa}$$

c). $V = \pi R^2 L \Rightarrow dV = \pi[2RL\,dR + R^2 dL] = \pi R(2L\,dR + R\,dL)$

$$= \pi R(2Lu + R\,dL)$$

$u = R\varepsilon_\theta = \frac{R}{E}[\sigma_\theta - \nu\sigma_x] = \frac{R}{E}[pR/t - \nu pR/2t] = \frac{pR^2}{2Et}(2-\nu)$

$dL = \varepsilon_x L = \frac{L}{E}[\sigma_x - \nu\sigma_\theta] = \frac{L}{E}[pR/2t - \nu pR/t] = \frac{pRL}{2Et}(1-2\nu)$

$\therefore dV = \pi R\left\{2L\frac{pR^2}{2Et}(2-\nu) + \frac{pR^2L}{2Et}(1-2\nu)\right\}$

$$= \frac{\pi p R^3 L}{2Et}[2(2-\nu) + (1-2\nu)] = \frac{\pi p R^3 L}{2Et}[5-4\nu]$$

$dV = \frac{\pi(2.4\times10^6)(0.25)^3(4)}{2(70\times10^9)(0.006)}[5 - 4(0.33)] = 2.064\times10^{-3}\text{ m}^3 = 2064\text{ cm}^3$

10.11 $\sigma_p = pR/2t$, $\sigma_w = \rho\frac{(4\pi R^3/3)}{2}\frac{1}{2\pi R t} = \rho R^2/3t$

$$\sigma = R/6t\,[3p + 2\rho R]$$

10.12 a)

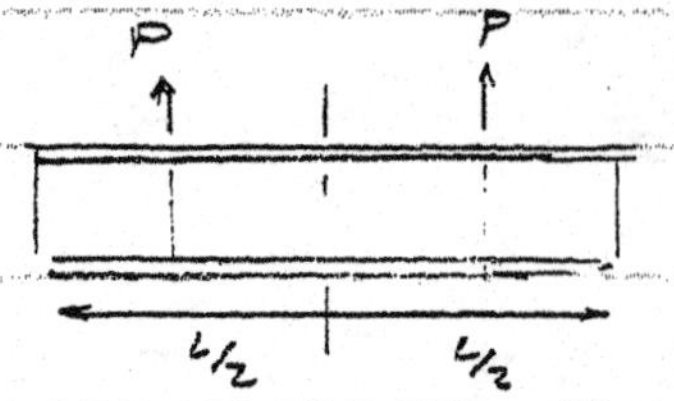

$W = \pi R^2 \rho L \quad P = W/2 \quad q = \rho \pi R^2 \ [N/m]$.

$M = P(L/2 - a) - q(L/2)^2/2 = P(L/2 - a) - qL^2/8$

$= \frac{\pi \rho R^2}{2} L[(L/2 - a) - L/4] = \frac{\pi \rho R^2 L}{8}(L - 4a)$

$\sigma_\theta|_p = pR/t, \quad \sigma_x|_p = pR/2t; \quad \sigma_x|_{W_A} = \frac{MR/2}{I} = \frac{-MR/2}{2\pi R^3 t} = -\frac{\rho L(L-a)}{32t}; \quad \sigma_x|_{W_B} = 0$

At A, C: $\sigma_{x_{A,C}} = \frac{pR}{2t} \mp \rho L(L-a)/32t = \frac{1}{2t}[pR \mp \rho L(L-a)/16]; \quad \sigma_\theta|_A = pR/t$

At B: $\sigma_x = pR/2t; \quad \sigma_\theta = pR/t$.

b)

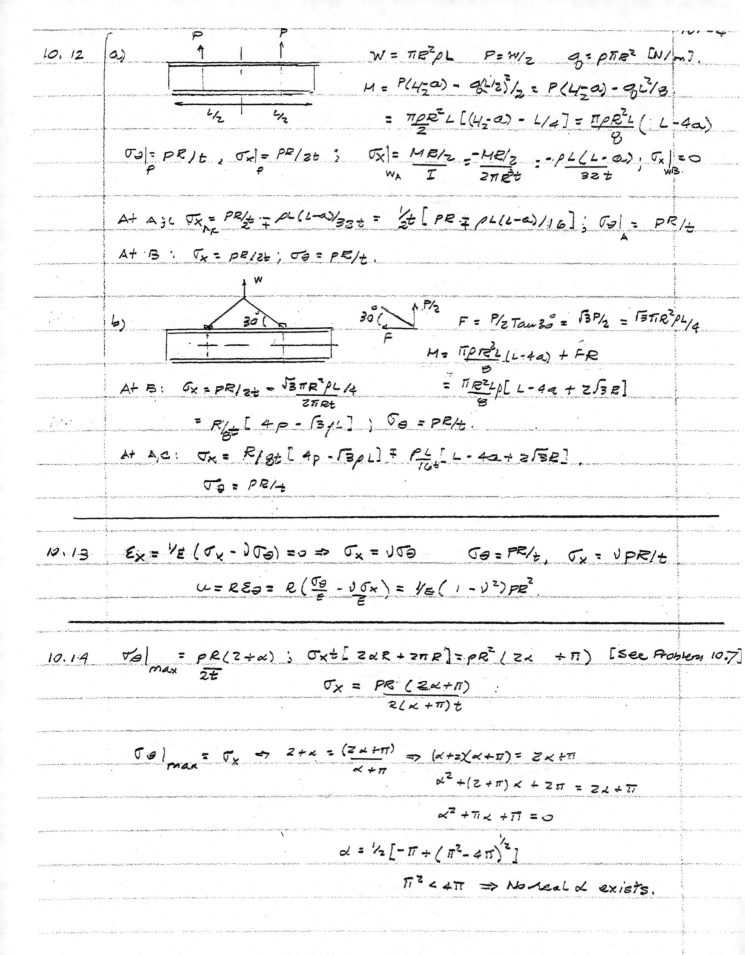

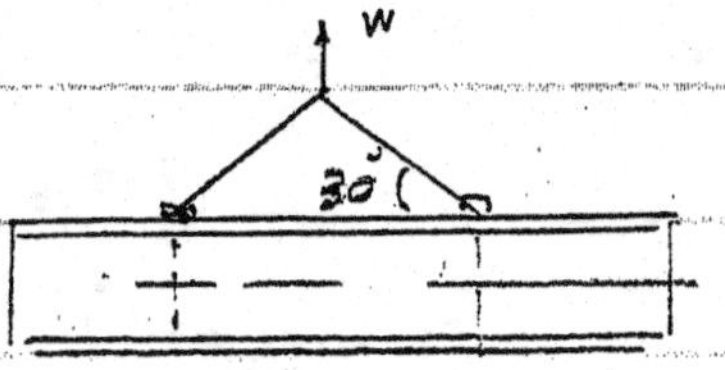

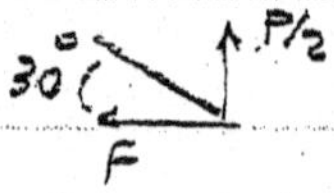

$F = P/2 \tan 30° = \sqrt{3}P/2 = \sqrt{3}\pi R^2 \rho L/4$

$M = \frac{\pi \rho R^2 L}{8}(L - 4a) + FR$

$= \frac{\pi R^2 L}{8}\rho[L - 4a + 2\sqrt{3}R]$

At B: $\sigma_x = pR/2t - \frac{\sqrt{3}\pi R^2 \rho L/4}{2\pi R t}$

$= R/8t[4p - \sqrt{3}\rho L]; \quad \sigma_\theta = pR/t$.

At A, C: $\sigma_x = R/8t[4p - \sqrt{3}\rho L] \mp \frac{\rho L}{16t}[L - 4a + 2\sqrt{3}R]$.

$\sigma_\theta = pR/t$

10.13 $\varepsilon_x = 1/E(\sigma_x - \nu\sigma_\theta) = 0 \Rightarrow \sigma_x = \nu\sigma_\theta \qquad \sigma_\theta = pR/t, \quad \sigma_x = \nu pR/t$

$u = R\varepsilon_\theta = R\left(\frac{\sigma_\theta}{E} - \frac{\nu\sigma_x}{E}\right) = 1/E(1 - \nu^2)pR^2$.

10.14 $\sigma_\theta|_{max} = \frac{pR(2+\alpha)}{2t}; \quad \sigma_x t[2\alpha R + 2\pi R] = pR^2(2\alpha + \pi)$ [See Problem 10.7]

$\sigma_x = \frac{pR(2\alpha + \pi)}{2(\alpha + \pi)t}$

$\sigma_\theta|_{max} = \sigma_x \Rightarrow 2 + \alpha = \frac{(2\alpha + \pi)}{\alpha + \pi} \Rightarrow (\alpha + 2)(\alpha + \pi) = 2\alpha + \pi$

$\alpha^2 + (2 + \pi)\alpha + 2\pi = 2\alpha + \pi$

$\alpha^2 + \pi\alpha + \pi = 0$

$\alpha = 1/2[-\pi + (\pi^2 - 4\pi)^{1/2}]$

$\pi^2 < 4\pi \Rightarrow$ No real α exists.

10.15

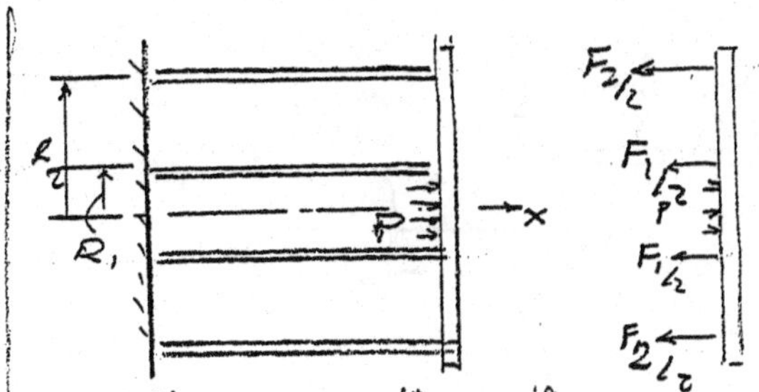

$$F_1 + F_2 = \pi R_1^2 p$$

$$F_2 = -F_1 + \pi R_1^2 p$$

$$\varepsilon_x^{(1)} = \frac{1}{E}\left(\sigma_x^{(1)} - \nu \sigma_\theta^{(1)}\right) = \frac{1}{E}\left[F_1/2\pi R_1 t - \nu p R_1/t\right]$$

$$\varepsilon_x^{(2)} = \frac{1}{E}\left(\sigma_x^{(2)} - \nu \sigma_\theta^{(2)}\right) = F_2/2\pi R_2 t$$

($\sigma_\theta^{(2)} \to 0$)

$$\Delta_x^{(1)} = \Delta_x^{(2)} \Rightarrow \varepsilon_x^{(1)} = \varepsilon_x^{(2)} \Rightarrow F_1/2\pi R_1 t - \nu p R_1/t = F_2/2\pi R_2 t$$

$$F_1/R_1 - F_2/R_2 = 2\pi \nu R_1 p \Rightarrow R_2 F_1 - R_1 F_2 = 2\pi \nu R_1^2 R_2 p$$

$$\therefore (R_2 + R_1) F_1 - \pi R_1^3 p = 2\pi \nu R_1^2 R_2 p$$

$$(R_2 + R_1) F_1 = \pi R_1^2 (2\nu R_2 + R_1) p$$

$$F_1 = \frac{\pi R_1^2 (2\nu R_2 + R_1) p}{R_1 + R_2}$$

$$\therefore \sigma_x^{(1)} = F_1/2\pi R_1 t = \frac{R_1 (R_1 + 2\nu R_2) p}{2t (R_1 + R_2)} \qquad \sigma_\theta^{(1)} = pR_1/t$$

$$\sigma_x^{(2)} = F_2/2\pi R_2 t = \frac{p}{2\pi R_2 t}\left[-\frac{\pi R_1^2 (2\nu R_2 + R_1)}{R_1 + R_2} + \pi R_1^2\right]$$

$$\sigma_x^{(2)} = \frac{p}{2R_2 t (R_1 + R_2)}\left[-2\nu R_1^2 R_2 + R_1^2 R_2\right] = \frac{R_1^2 (1 - 2\nu)}{2t (R_1 + R_2)} \; ; \; \sigma_\theta^{(2)} = p$$

$$\Delta x = \varepsilon_x L = L\sigma_x^{(2)}/E = \frac{R_1^2 (1 - 2\nu) L}{2t (R_1 + R_2) E} \cdot p$$

10.16

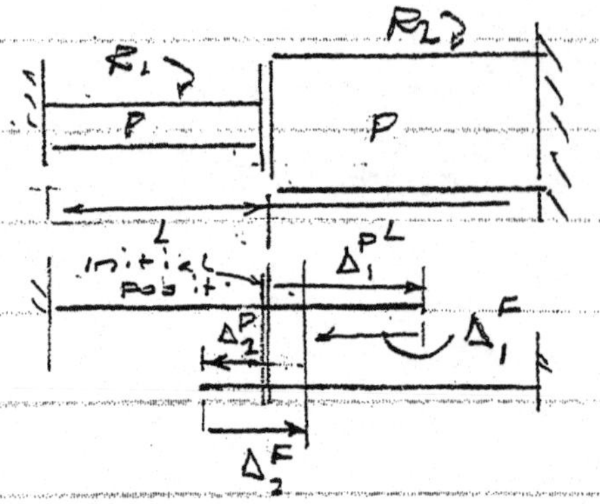

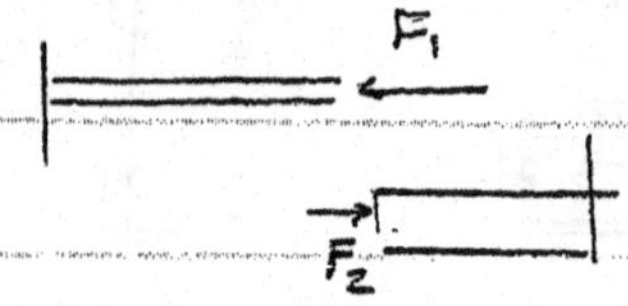

$$\Delta_1^F + \Delta_2^F = \Delta_1^P + \Delta_2^P$$

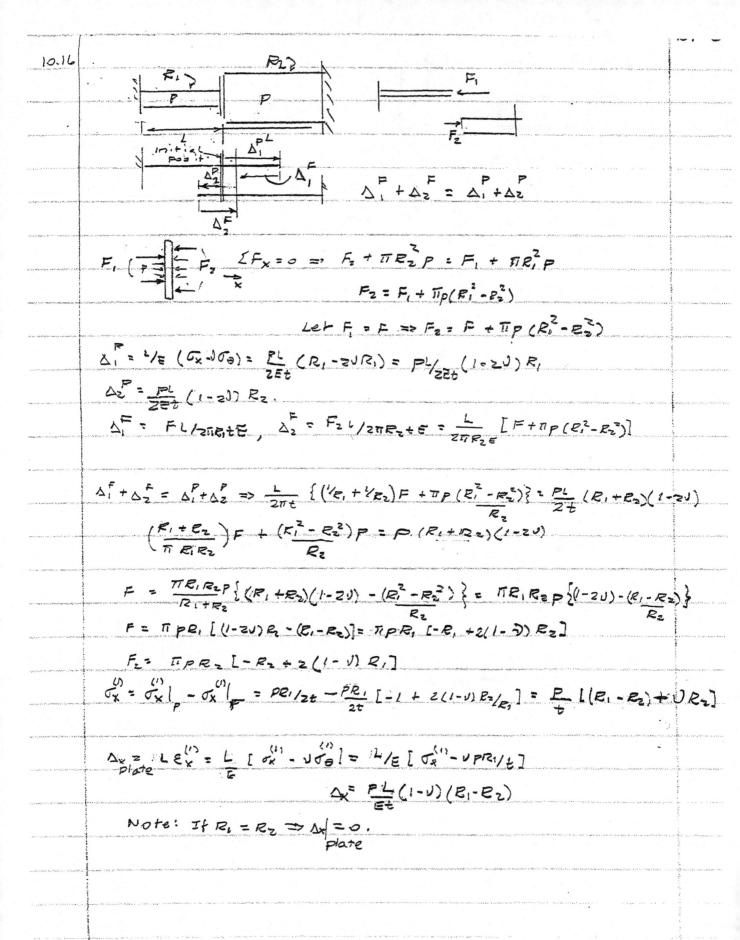

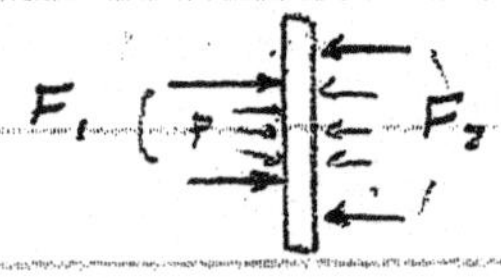

$$\Sigma F_x = 0 \Rightarrow F_2 + \pi R_2^2 p = F_1 + \pi R_1^2 p$$

$$F_2 = F_1 + \pi p (R_1^2 - R_2^2)$$

$$\text{Let } F_1 = F \Rightarrow F_2 = F + \pi p (R_1^2 - R_2^2)$$

$$\Delta_1^P = L/E\,(\sigma_x - \nu\sigma_\theta) = \frac{pL}{2Et}(R_1 - 2\nu R_1) = pL/_{2Et}(1-2\nu)R_1$$

$$\Delta_2^P = \frac{pL}{2Et}(1-2\nu)R_2.$$

$$\Delta_1^F = FL/2\pi R_1 tE, \quad \Delta_2^F = F_2 L/2\pi R_2 tE = \frac{L}{2\pi R_2 E}[F + \pi p (R_1^2 - R_2^2)]$$

$$\Delta_1^F + \Delta_2^F = \Delta_1^P + \Delta_2^P \Rightarrow \frac{L}{2\pi t}\left\{(1/R_1 + 1/R_2)F + \frac{\pi p (R_1^2 - R_2^2)}{R_2}\right\} = \frac{pL}{2t}(R_1 + R_2)(1-2\nu)$$

$$\left(\frac{R_1 + R_2}{\pi R_1 R_2}\right)F + \frac{(R_1^2 - R_2^2)}{R_2}p = p(R_1 + R_2)(1-2\nu)$$

$$F = \frac{\pi R_1 R_2 p}{R_1 + R_2}\left\{(R_1 + R_2)(1-2\nu) - \frac{(R_1^2 - R_2^2)}{R_2}\right\} = \pi R_1 R_2 p\left\{(1-2\nu) - \frac{(R_1 - R_2)}{R_2}\right\}$$

$$F = \pi p R_1 [(1-2\nu)R_2 - (R_1 - R_2)] = \pi p R_1 [-R_1 + 2(1-\nu)R_2]$$

$$F_2 = \pi p R_2 [-R_2 + 2(1-\nu)R_1]$$

$$\sigma_x^{(1)} = \sigma_x^{(1)}\Big|_p - \sigma_x^{(1)}\Big|_F = pR_1/2t - \frac{pR_1}{2t}[-1 + 2(1-\nu)R_2/R_1] = \frac{p}{t}[(R_1 - R_2) + \nu R_2]$$

$$\Delta_{x\,\text{plate}} = L\varepsilon_x^{(1)} = \frac{L}{E}[\sigma_x^{(1)} - \nu\sigma_\theta^{(1)}] = L/E\,[\sigma_x^{(1)} - \nu p R_1/t]$$

$$\Delta_x = \frac{pL}{Et}(1-\nu)(R_1 - R_2)$$

Note: If $R_1 = R_2 \Rightarrow \Delta_x\Big|_{\text{plate}} = 0.$

10.17 $\sigma_n = \sigma_x \cos^2\alpha + \sigma_\theta \sin^2\alpha = \frac{pR}{2t}(\cos^2\alpha + 2\sin^2\alpha) = \sigma_0 = n\tau_0$

$|\tau_{nt}| = |(\sigma_x - \sigma_\theta)| \sin\theta\cos\theta = pR/_{2t} \sin\theta\cos\theta = \tau_0$

$\sigma_n = \sigma_0$ & $|\tau_{nt}| = \tau_0 \Rightarrow \cos^2\alpha + 2\sin^2\alpha = n\sin\alpha\cos\alpha$

$1 + \sin^2\alpha = n\sin\alpha\cos\alpha = n\sin\alpha(1-\sin^2\alpha)^{1/2}$

$\therefore \sin^4\alpha + 2\sin^2\alpha + 1 = n^2\sin^2\alpha(1-\sin^2\alpha) = n^2(\sin^2\alpha - \sin^4\alpha)$

$(1+n^2)\sin^4\alpha + (2-n^2)\sin^2\alpha + 1 = 0$

Let $\xi = \sin^2\alpha \Rightarrow (1+n^2)\xi^2 + (2-n^2)\xi + 1 = 0$

$\xi = \frac{1}{2(1+n^2)}\{n^2 - 2 \pm n(n^2-8)^{1/2}\}$

For ξ real: $n^2 > 8 \Rightarrow n_{min} = \sqrt{8} = 2\sqrt{2}$

$\xi_{min} > 0 \Rightarrow \sin^2\alpha > 0$

For $n = n_{min} \Rightarrow \xi = 6/18 = 1/3 \Rightarrow \alpha = \sin^{-1}(1/3) = 35.26°$.

10.18

a).

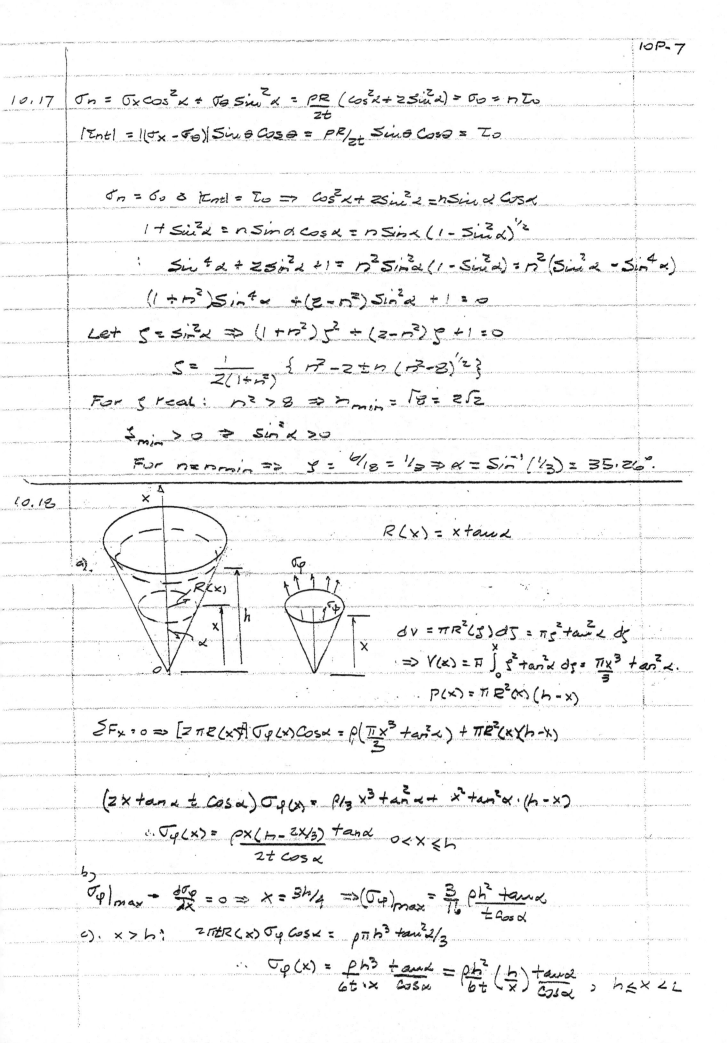

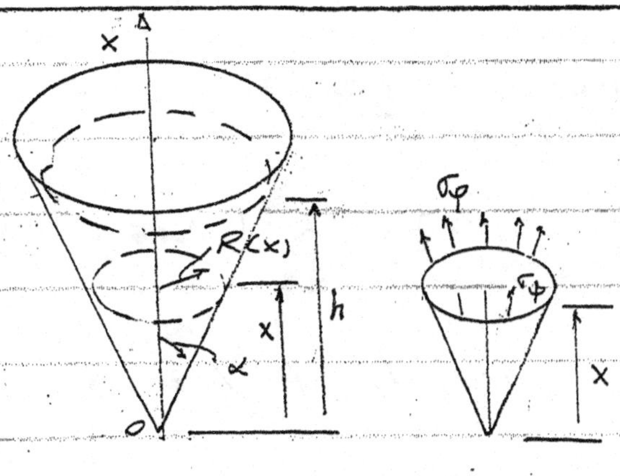

$R(x) = x\tan\alpha$

$dv = \pi R^2(\xi)d\xi = \pi\xi^2\tan^2\alpha\, d\xi$

$\Rightarrow V(x) = \pi\int_0^x \xi^2\tan^2\alpha\, d\xi = \frac{\pi x^3}{3}\tan^2\alpha.$

$\therefore P(x) = \pi R^2(x)(h-x)$

$\Sigma F_x = 0 \Rightarrow [2\pi R(x)t]\sigma_\varphi(x)\cos\alpha = \rho\left(\frac{\pi x^3}{3}\tan^2\alpha\right) + \pi R^2(x)(h-x)$

$(2x\tan\alpha\, t\cos\alpha)\sigma_\varphi(x) = \rho/3\, x^3\tan^2\alpha + x^2\tan^2\alpha\cdot(h-x)$

$\therefore \sigma_\varphi(x) = \frac{\rho x(h - 2x/3)\tan\alpha}{2t\cos\alpha} \quad 0 < x \leq h$

b)

$\sigma_\varphi|_{max} \rightarrow \frac{d\sigma_\varphi}{dx} = 0 \Rightarrow x = 3h/4 \Rightarrow (\sigma_\varphi)_{max} = \frac{3}{16}\frac{\rho h^2\tan\alpha}{t\cos\alpha}$

c). $x > h$: $\quad 2\pi t R(x)\sigma_\varphi\cos\alpha = \rho\pi h^3\tan^2\alpha/3$

$\therefore \sigma_\varphi(x) = \frac{\rho h^3}{6t\cdot x}\frac{\tan\alpha}{\cos\alpha} = \frac{\rho h^2}{6t}\left(\frac{h}{x}\right)\frac{\tan\alpha}{\cos\alpha}, \quad h \leq x < L$

10.19

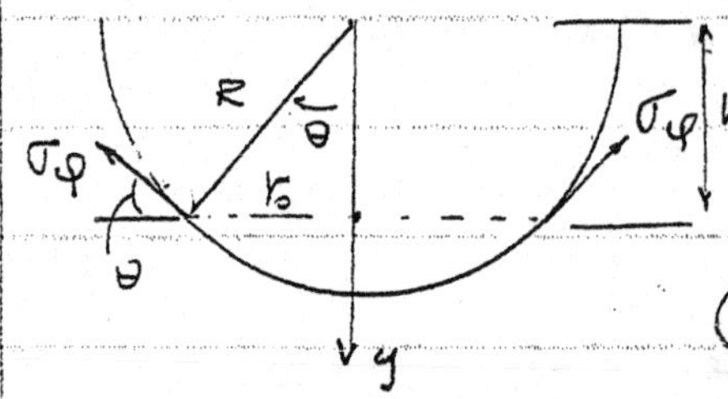

$\sum F_y = 0$

$(2\pi r_0 t \sigma_\varphi)\sin\theta = \rho h(\pi r_0^2) + W_{fluid}$

$r_0 = R\sin\theta \;,\; h = R\cos\theta$

$(2\pi R t \sin^2\theta)\,\sigma_\varphi = \rho(\pi R^2 \sin^2\theta) R\cos\theta + W_{fluid}$

$\therefore \sigma_\varphi = \dfrac{\rho R^2\cos\theta}{2t} + \dfrac{W_{fluid}}{2\pi R t \sin^2\theta}$

$dW_{fluid} = \rho[\pi r_0^2\, dy] = -\rho\pi R^2\sin^2\theta\; d[R\cos\theta]$

$= -\pi\rho R^3 \sin^2\theta \dfrac{d}{d\theta}(\cos\theta)\, d\theta$

$= \pi\rho R^3 \sin^3\theta\, d\theta$

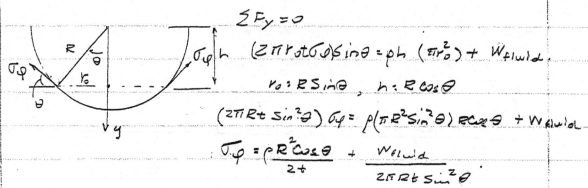

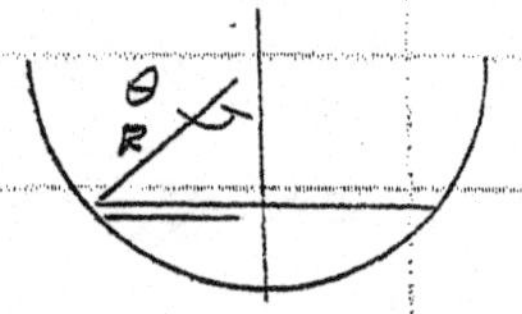

$W_{fluid} = \pi\rho R^3 \int_0^\theta \sin^3\eta\, d\eta = \pi\rho R^3 \int_0^\theta \sin\eta(1-\cos^2\eta)\,d\eta$

$= \pi\rho R^3\{-\cos\eta\,|_0^\theta + \dfrac{\cos^3}{3}\eta\,|_0^\theta\} = \pi\rho R^3[1-\cos\theta - \tfrac{1}{3}(1-\cos^3\theta)]$

$= \rho\pi R^3/3\,[\,2 + \cos\theta(\cos^2\theta - 3)] = \dfrac{\rho\pi R^3}{3}[2(1-\cos\theta) - \sin^2\theta\cos\theta]$

$\therefore \sigma_\varphi = \dfrac{\rho R^2\cos\theta}{2t} + \dfrac{\frac{\rho\pi R^3}{3}[2(1-\cos\theta) - \cos\theta\sin^2\theta]}{2\pi R t\sin^2\theta}$

$\sigma_\varphi = \dfrac{\rho R^2}{6t}\left[3\cos\theta + \dfrac{2(1-\cos\theta)}{\sin^2\theta} - \cos\theta\right] = \dfrac{\rho R^2}{3t}\left[\cos\theta + \dfrac{(1-\cos\theta)}{\sin^2\theta}\right]$

b)

$y = R/2 \Rightarrow \cos\theta = 0.5 \Rightarrow \theta = 60° \Rightarrow \sigma_\varphi = 7/18\, \rho R^2/t = 0.389 \quad \rho R^2/t$

11.1

a) 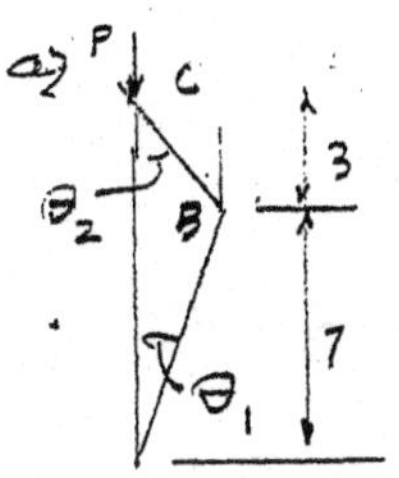

$7\theta_1 = 3\theta_2$

$\sum M_{ABC} = 0 \Rightarrow R_{cx} = 0$

$\therefore R_{Ax} = 0$

$pM = \beta(\theta_2 + \theta_1)$

BC:

$\curvearrowleft + (\sum M)_C = 0 \Rightarrow P(3\theta_2) - \beta(\theta_2 + \theta_1) = 0$

$$P = \frac{\beta(\theta_2+\theta_1)}{3\theta_2} = \frac{\beta(10\theta_2/7)}{3\theta_2} = 10\beta/21 = 200N$$

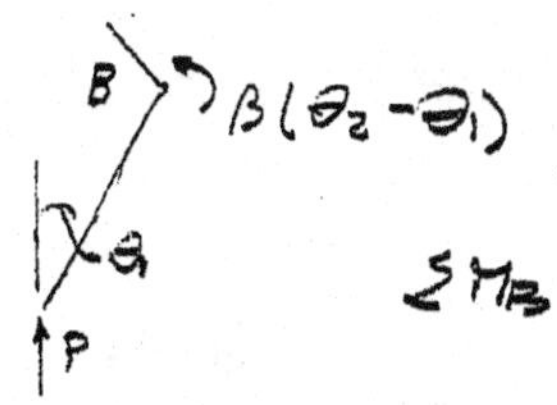

AB: (Check)

$\sum M_B = 0 \Rightarrow P(7\theta_1) - \beta(\theta_2 + \theta_1) = 0$

$$P = \beta(\theta_2+\theta_1)/7\theta_1 = \frac{\beta(10\theta_1/3)}{7\theta_1} = 10\beta/21 = 200\ N$$

b) 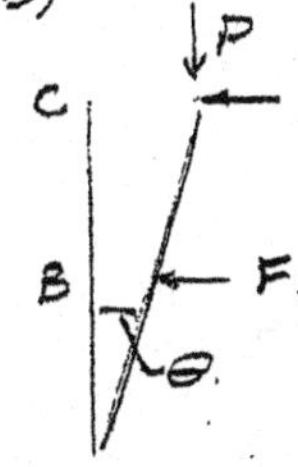

$F_c = k\Delta_c = 2kL\sin\theta \doteq 2kL\theta$

$F_B = k\Delta_B = kL\sin\theta \doteq kL\theta$

$\curvearrowleft + \sum M_A = 0 \Rightarrow 2PL\sin\theta - (2kL\theta)2L\cos\theta - (kL\theta)(L\cos\theta) = 0$

$\cos\theta \doteq 1 \qquad 2PL\theta - 5kL\theta = 0;\ \theta \neq 0 \Rightarrow P_{cr} = 5kL/2.$

11.2 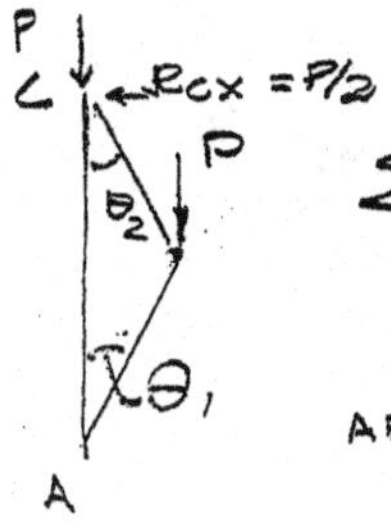

$\sum M_{ABC} = 0 \Rightarrow R_{cx} = PL\theta/2L = P\theta/2$

$\theta_1 = \theta_2 = \theta$

AB: $\sum M_B = 0 \Rightarrow 2\beta\theta + 2PL\theta - \frac{P}{2}\theta L = 0$

$(2\beta + 3PL/2)\theta = 0$

$\theta \neq 0 \Rightarrow P_{cr} = 4\beta/3L$

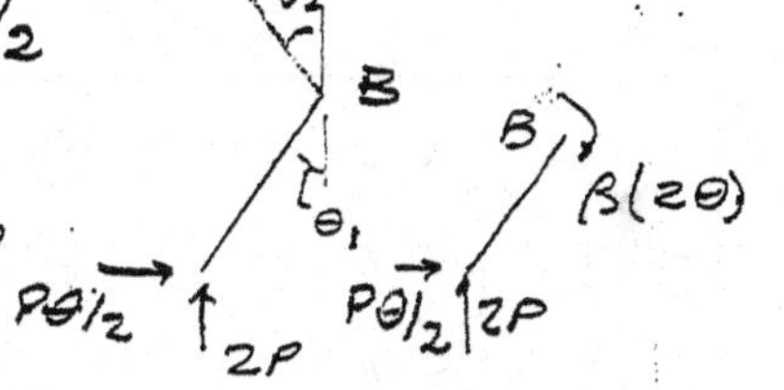

11.3 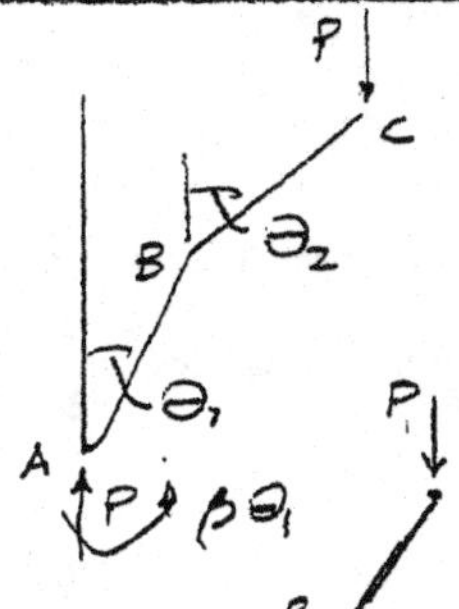

ABC: $(\sum M)_A = 0 \Rightarrow \beta\theta_1 - PL(\sin\theta_1 + \sin\theta_2) = 0$

$|\theta_i| \ll 1 \Rightarrow (\beta - PL)\theta_1 - PL\,\theta_2 = 0 \qquad (i)$

BC: $(\sum M)_B = 0 \Rightarrow \beta(\theta_2 - \theta_1) - PL\theta_2 = 0$

$-\beta\theta_1 + (\beta - PL)\theta_2 = 0 \qquad (ii)$

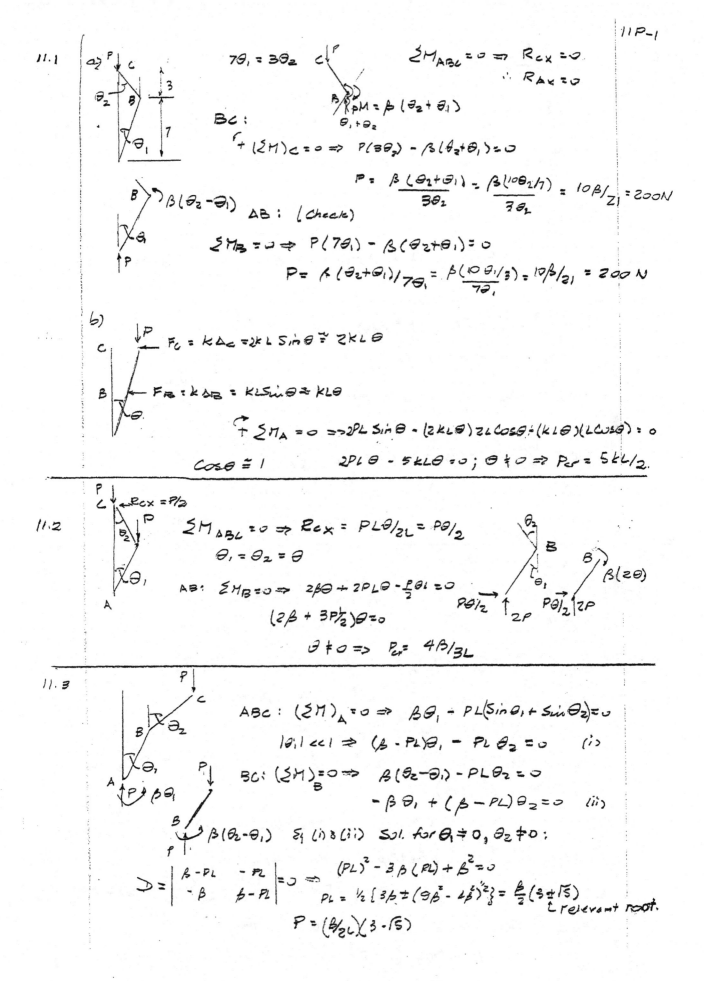

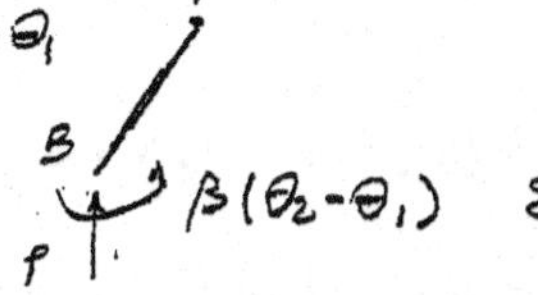

Eq (i) & (ii) Sol. for $\theta_1 \neq 0, \theta_2 \neq 0$:

$$D = \begin{vmatrix} \beta - PL & -PL \\ -\beta & \beta - PL \end{vmatrix} = 0 \Rightarrow (PL)^2 - 3\beta(PL) + \beta^2 = 0$$

$$PL = \tfrac{1}{2}\{3\beta \pm (9\beta^2 - 4\beta^2)^{1/2}\} = \frac{\beta}{2}(3 \pm \sqrt{5})$$ ← relevant root.

$$P = (\beta/2L)(3 - \sqrt{5})$$

11.4 a).

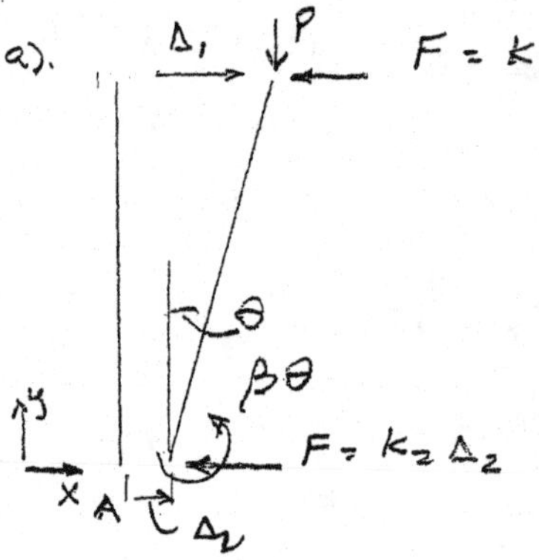

$F = k_1 \Delta_1$

$$\sum F_x = 0 \Rightarrow k_1 \Delta_1 + k_2 \Delta_2 = 0 \quad (i)$$

$$\curvearrowleft + \sum M_A = 0 \Rightarrow k_1 \Delta_1 (L\cos\theta) - PL\sin\theta + \beta\theta = 0 \quad (ii)$$

$$|\theta| << 1 \Rightarrow \cos\theta \approx 1, \ \sin\theta \approx \theta$$

$$\sin\theta = \frac{\Delta_1 - \Delta_2}{L} \Rightarrow \theta = (\Delta_1 - \Delta_2)/L.$$

(i) $\Rightarrow \Delta_2 = -k_1\Delta_1/k_2$; (ii) $k_1 L\Delta_1 + (\beta - PL)\theta = 0 \Rightarrow k_1 L \Delta_1 + (\beta - PL)\dfrac{\Delta_1 - \Delta_2}{L} = 0$

∴

$$k_1 L^2 \Delta_1 + (\beta - PL)(1 + k_1/k_2)\Delta_1 = 0$$

$$[k_1 k_2 L^2 + (\beta - PL)(k_1 + k_2)]\Delta_1 = 0$$

$$\Delta_1 \neq 0 \Rightarrow P_{cr} = \left(\frac{k_1 k_2}{k_1 + k_2}\right)L + \beta/L \qquad \text{Let } \gamma = \frac{k_1 k_2}{k_1 + k_2}$$

b).

$$dP_{cr}/dL = \frac{k_1 k_2}{k_1 + k_2} - \beta/L^2 = 0 \Rightarrow L = (\beta/\gamma)^{1/2}$$

$$(P_{cr})_{min} = \gamma(\beta/\gamma)^{1/2} + \beta(\gamma/\beta)^{1/2} = 2(\gamma\beta)^{1/2}$$

$$(P_{cr})_{min} = 2\left(\frac{k_1 k_2 \beta}{k_1 + k_2}\right)^{1/2}$$

11.5

a)

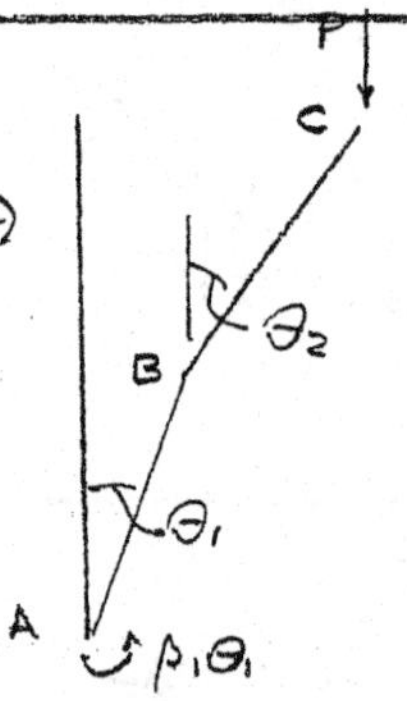

ABC:

$$\sum M_A = 0 \Rightarrow \beta_1\theta_1 - PL(\sin\theta_1 + \sin\theta_2) = 0$$

$$\Rightarrow (\beta_1 - PL)\theta_1 - PL\theta_2 = 0 \quad (i)$$

BC: $\left(\sum M_B\right)_{BC} = 0 \Rightarrow \beta_2(\theta_2 - \theta_1) - PL\theta_2 = 0$

$$\Rightarrow -\beta_2\theta_1 + (\beta_2 - PL)\theta_2 = 0 \quad (ii)$$

(i) & (ii) are 2 homog. eqs in θ_1, θ_2. Non-trivial sol. requires

$$D = \begin{vmatrix} \beta_1 - PL & -PL \\ -\beta_2 & (\beta_2 - PL) \end{vmatrix} = 0 \Rightarrow (PL)^2 - (\beta_1 + 2\beta_2)(PL) + \beta_1\beta_2 = 0$$

$$\therefore PL = \tfrac{1}{2}\left\{\beta_1 + 2\beta_2 \pm [(\beta_1 + 2\beta_2)^2 - 4\beta_1\beta_2]^{1/2}\right\}$$

$$P = \frac{1}{2L}\left\{\beta_1 + 2\beta_2 - [\beta_1^2 + 4\beta_2^2]^{1/2}\right\}.$$

b). Let $\gamma = \beta_2/\beta_1$ $\quad P = \frac{\beta_1}{2L}\left\{(1 + 2\gamma) - [1 + 4\gamma^2]^{1/2}\right\} = \frac{\beta_1}{2L}\left\{1 + 2\gamma - 2\gamma\left(1 + \frac{1}{4\gamma^2}\right)^{1/2}\right\}$

Let $\gamma >> 1$: $\qquad$ Binomial th.

$$P = \beta_1/2L\,[1 + 2\gamma - 2\gamma(1 + 1/8\gamma^2)] = \beta_1/2L\,[1 - 1/4\gamma] < \beta_1/2L$$

If $\gamma \to \infty \Rightarrow P = \beta_1/2L$.

11.6

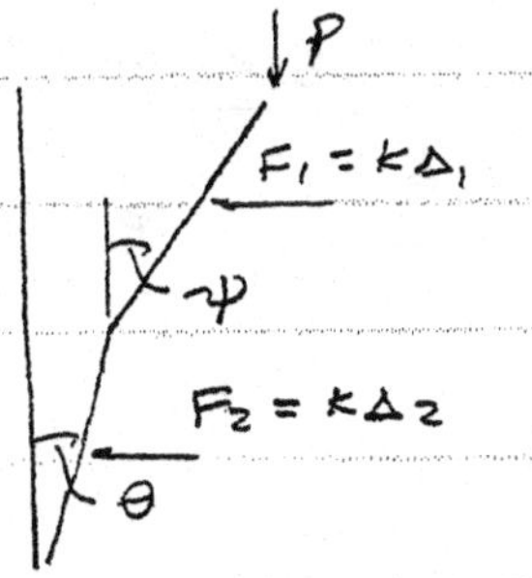

$\Delta_1 = 2L\theta + L\psi = L(2\theta + \psi)$

$\Delta_2 = L\theta$

CE: $\overset{\curvearrowleft}{+}\sum M_C = 0 \Rightarrow L\cos\psi\, F_1 - 2PL\psi = 0$

$\Rightarrow kL^2(2\theta + \psi) - 2PL\psi = 0$

$2kL\theta + (kL - 2P)\psi = 0 \qquad (i)$

ACE:

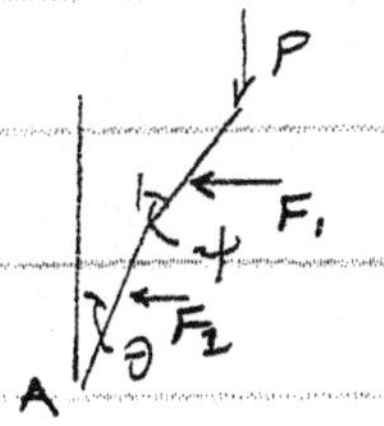

$\overset{\curvearrowleft}{+}\sum M_A = 0 \Rightarrow 3LF_1 + LF_2 - 2PL(\theta + \psi) = 0$

$3kL^2(2\theta + \psi) + kL^2\theta - 2P(\theta + \psi) = 0$

$(7kL - 2P)\theta + (3kL - 2P)\psi = 0 \qquad (ii)$

(a) & (b): $\begin{bmatrix} 2kL & kL - 2P \\ 7kL - 2P & 3kL - 2P \end{bmatrix} \begin{Bmatrix} \theta \\ \psi \end{Bmatrix} = 0$ Non-trivial sol

$D = \begin{vmatrix} 2kL & kL - 2P \\ 7kL - 2P & 3kL - 2P \end{vmatrix} = 0$

$\Rightarrow 4P^2 - 12kLP + k^2L^2 = 0 \Rightarrow P = kL(3/2 \pm \sqrt{2})$

$P_{cr} = kL/2\,(3 - 2\sqrt{2})$

11.7

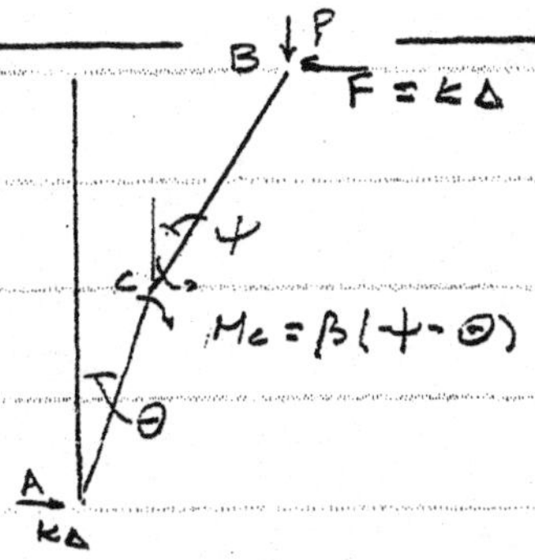

$\overset{\curvearrowleft}{+}\left(\sum M_B\right)_{AC} = 0 \Rightarrow PL\sin\theta - kL\Delta\cos\theta + \beta(\psi - \theta) = 0 \qquad (i)$

$\left(\sum M_B\right)_{BC} = 0 \Rightarrow PL\sin\psi - kL\Delta\cos\psi - \beta(\psi - \theta) = 0 \qquad (ii)$

$\Delta = L(\sin\theta + \sin\psi). \qquad (iii)$

Using $|\theta| \ll 1$, $|\psi| \ll 1 \Rightarrow \sin\theta \doteq \theta$, $\sin\psi \doteq \psi$, $\cos\theta, \psi \doteq 1$, and (iii)

(i) $\Rightarrow (PL - kL^2 - \beta)\theta + (\beta - kL^2)\psi = 0$

(ii) $\Rightarrow (\beta - kL^2)\theta + (PL - kL^2 - \beta)\psi = 0$

$D \equiv |\quad| = 0 \Rightarrow (PL - kL^2 - \beta)^2 - (\beta - kL^2)^2 = 0$

$\Rightarrow P = 1/L\,[kL^2 + \beta \pm (\beta - kL^2)]$

$P_1 = 2\beta/L$, $P_2 = 2kL$.

b). $k = 5$ N/cm, $\beta = 500$ N-cm/rad. (i) $L = 12$ cm $\Rightarrow P_{cr} = 2\beta/L = 83.3$ N

(ii) $L = 8$ cm $\Rightarrow P_{cr} = 2kL = 80$ N.

c).

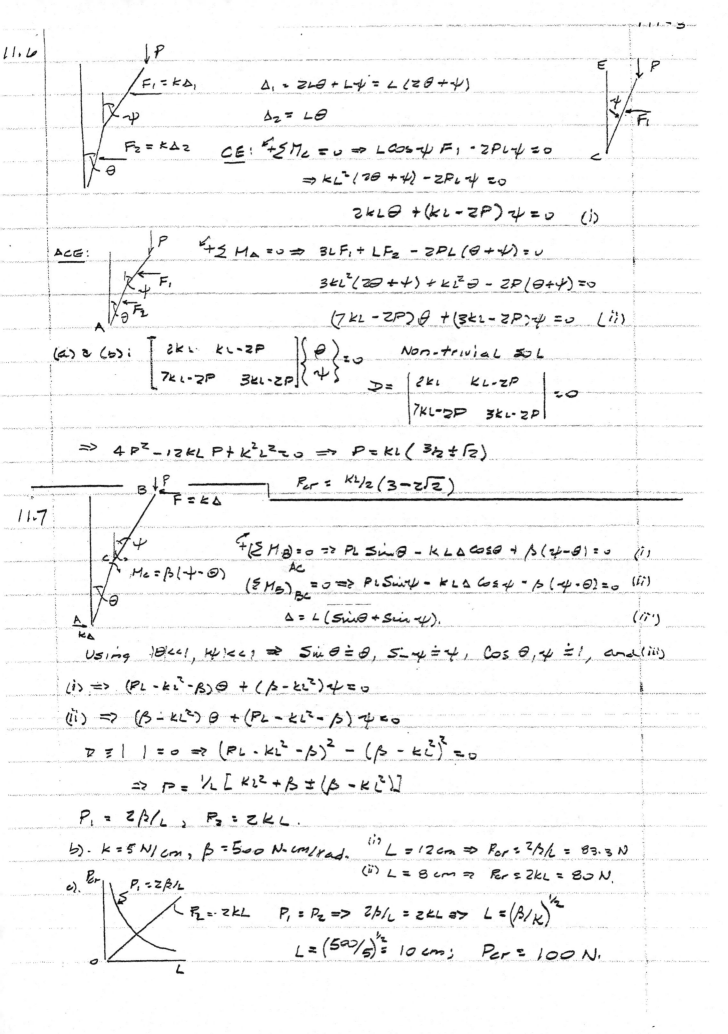

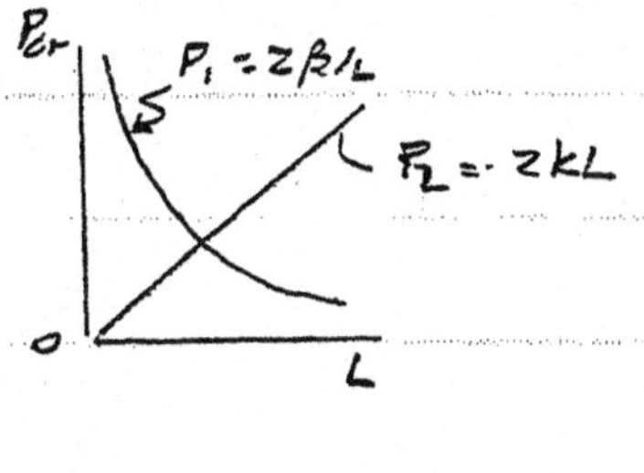

$P_1 = P_2 \Rightarrow 2\beta/L = 2kL \Rightarrow L = (\beta/k)^{1/2}$

$L = (500/5)^{1/2} = 10$ cm; $P_{cr} = 100$ N.

11.8

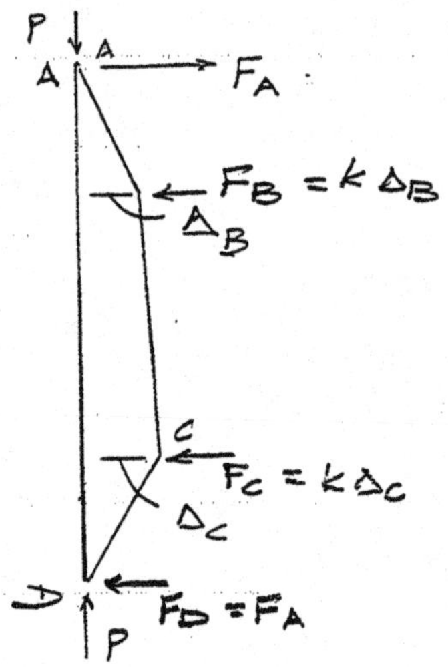

$$\sum M_D = 0 \Rightarrow F_A = \frac{k[\Delta_C a + \Delta_B(a+b)]}{2a+b} \quad \text{(i)}$$

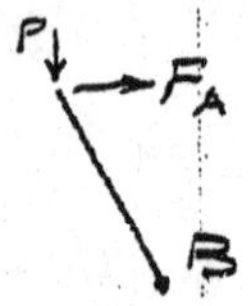

AB:

$$\curvearrowleft + \sum M_B = 0 \Rightarrow P\Delta_B - F_A a = 0 \quad \text{(ii)}$$

C-D:

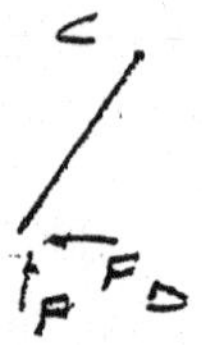

$$\curvearrowleft + \sum M_C = 0 \Rightarrow P\Delta_C + F_D a = 0$$

$$F_D = F_A \Rightarrow P\Delta_C + F_A a = 0 \quad \text{(iii)}$$

(i) & (ii) $\Rightarrow$ $$P\Delta_B - \frac{ka}{2a+b}[\Delta_C a + \Delta_B(a+b)] = 0$$

$$[(2a+b)P - ka(a+b)]\Delta_B - ka^2\Delta_C = 0$$

(i) & (iii) $\Rightarrow$ $$P\Delta_C + \frac{ka}{2a+b}[\Delta_C a + \Delta_B(a+b)] = 0$$

$$ka(a+b)\Delta_B + [(2a+b)P + ka^2]\Delta_C = 0$$

Non-trivial sol. for Δ_B & Δ_C:

$$\begin{vmatrix} (2a+b)P - ka(a+b) & -ka^2 \\ ka(a+b) & (2a+b)P + ka^2 \end{vmatrix} = 0$$

Expanding determinant:

$$(2a+b)P[(2a+b)P - kab] = 0$$

$$P \neq 0,\ 2a+b \neq 0 \Rightarrow P = \frac{kab}{2a+b}$$

b). $$P = \frac{ka(L-2a)}{L}$$

$$dP/da = \frac{k}{L}[L-4a] = 0 \Rightarrow a = L/4$$

$$(P_{cr})_{max} = \frac{k}{L}(L/4)(L/2) = kL/8 \qquad b/a = 2$$

11.9 a). $P_{cr} = \pi^2 EI/L^2 \quad I_{yy} = I_{min} = (16)(10)^3/12 = 1333\ mm^4$

$$P_{cr} = \frac{\pi^2(10\times10^9)(1.333\times10^{-9})}{4} = 32.9\ N$$

b). In x-z plane: $P_{cr} = 4(32.9) = 131.6\ N$

In x-y plane: $P_{cr} = \frac{\pi^2 E I_{zz}}{L^2} \qquad I_{zz} = \frac{(10)(16)^3}{12} = 3,413\ mm^4$

$$P_{cr} = \frac{\pi^2(10\times10^9)(3.413\times10^{-9})}{4} = 84.2\ N$$

$(P_{cr}) = 84.2\,N$ in x-y plane.

11.10 a). $\angle\ 76\times76\times12.7 \qquad r = 14.8\ mm,\ A = 1775\ mm^2$

$$P_{cr} = \pi^2 EI/L^2 = \frac{\pi^2 EAr^2}{L^2} = \frac{\pi^2(200\times10^3)(1775)(14.8)^2}{(3\times10^3)^2} = 85.28\ kN$$

b). $2\cdot(\angle\ 51\times51\times9.5)$ For each angle: $I_{zz} = I_{yy} = 0.199\times10^6\ mm^4$

$\bar{y} = \bar{z} = 20.5\ mm,\ A = 1450\ mm^2$

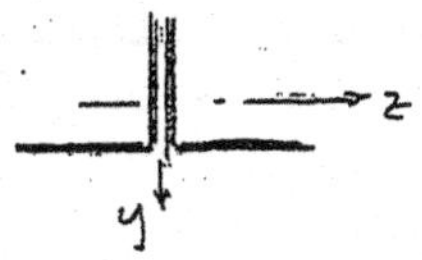

$I_{zz} = 2(0.199\times10^6) = 0.398\times10^6\ mm^4$

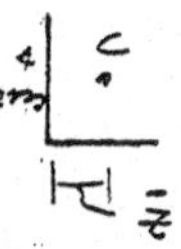

$I_{yy} = 2[(0.199\times10^6) + 1450(20.5)^2] = 1.617\times10^6\ mm^4$

$I_{min} = I_{zz} < I_{yy} \qquad I_{min} = 0.398\times10^6\ mm^4 \qquad P_{cr} = \frac{\pi^2(200\times10^3)(0.398\times10^6)}{(3\times10^3)^2} = 87.29\ kN$

11.11

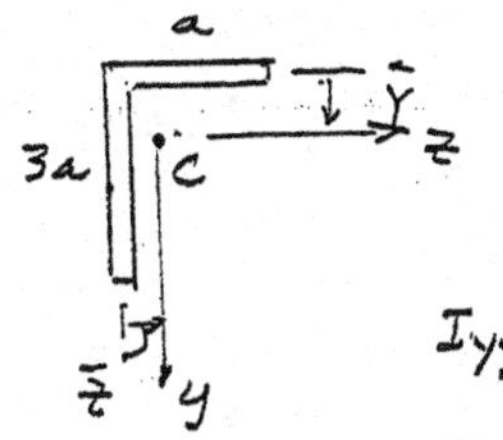

$$\bar{Y} = \frac{(3at)(3a/2)}{4at} = 9a/8, \quad \bar{z} = \frac{(at)(a/2)}{4at} = a/8$$

For $t \ll a$:

$$I_{yy} = [a^3t/12 + (at)(\tfrac{a}{2} - \tfrac{a}{8})^2] + (3at)(a/8)^2 = 13a^3t/48$$

$$I_{zz} = (at)(9a/8)^2 + [\tfrac{27a^3t}{12} + (3at)(\tfrac{3a}{2} - \tfrac{9a}{8})^2] = 63a^3t/16$$

$$I_{yz} = (at)[(-9a/8)(\tfrac{a}{2} - \tfrac{a}{8})] + (3at)[(\tfrac{3a}{2} - 9a/8)(-a/8)] = -9a^3t/16$$

$$I_{min} = \frac{I_{yy}+I_{zz}}{2} - \left[\left(\frac{I_{yy}-I_{zz}}{2}\right)^2 + I_{yz}^2\right]^{1/2} = 0.18648\ a^3t$$

$$P_{cr} = \pi^2 EI/L^2 = \pi^2 E(0.18648\,a^3t)/4L^2 = 0.4601\,Ea^3t/L^2$$

11.12

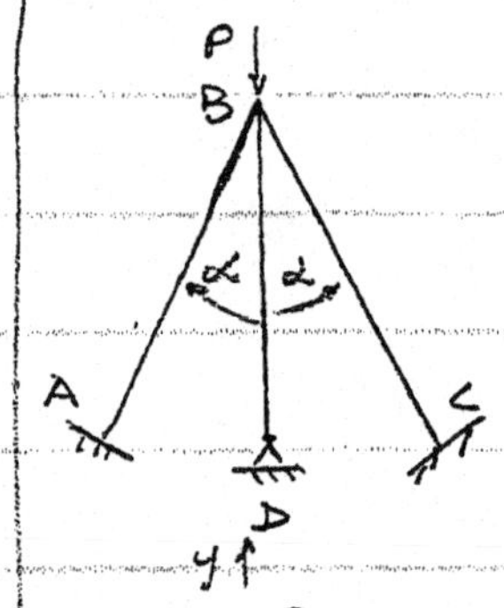

NOTE: Buckling can only occur with AB, BC and BD buckle simultaneously.

$$(P_{cr})_{BD} = \pi^2 EI/L^2$$

$$(P_{cr})_{AB} = (P_{cr})_{BC} = \frac{\pi^2 EI}{(.7L/\cos\alpha)^2} = \frac{\pi^2 EI\cos^2\alpha}{(.7L)^2}$$

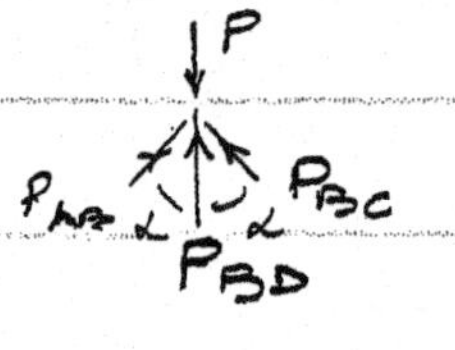

$$\Sigma F_y = 0 \Rightarrow P = P_{BD} + 2P_{AB}\cos\alpha$$

$$\alpha = 20^\circ$$

$$\therefore P_{cr} = \frac{\pi^2 EI}{L^2}\left[1 + \frac{2\cos^3\alpha}{(.7)^2}\right] = \frac{\pi^2 EI}{L^2}(43.848) = 43.3\ EI/L^2$$

11.13

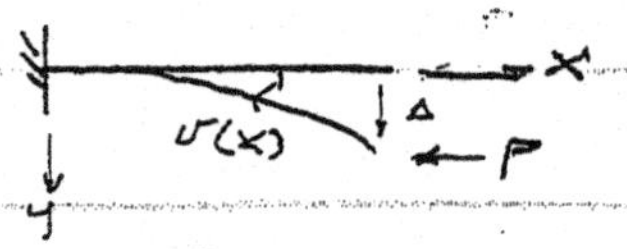

$$M(x) = P(\Delta - v(x)).$$

$$EIv''(x) = -M(x) \Rightarrow EIv''(x) + Pv(x) = P\Delta$$

a) $v''(x) + \lambda^2 v(x) = \lambda^2\Delta$, $\lambda^2 = P/EI$

B.C. $v(0) = v'(0) = 0$.

$v(x) = A\sin\lambda x + B\cos\lambda x + \Delta$ $\quad v(0) = 0 \Rightarrow B = -\Delta$

$v(x) = A\sin\lambda x + \Delta(1-\cos\lambda x)$; $v'(x) = A\lambda\cos\lambda x + \lambda\Delta\sin\lambda x$; $v'(0) = 0 \Rightarrow A = 0$

$\therefore v(x) = \Delta(1-\cos\lambda x)$

$v(L) = \Delta \Rightarrow \cos\lambda L = 0 \Rightarrow \lambda = n\pi/2L$, $n = 1, 3, 5 \cdots$

$v(x) = \Delta(1 - \cos n\pi x/2L)$, $P = n^2\pi^2 EI/4L^2$, n odd

b) $P_{cr} = \pi^2 EI/4L^2$; $v(x) = \Delta(1 - \cos \pi x/2L)$

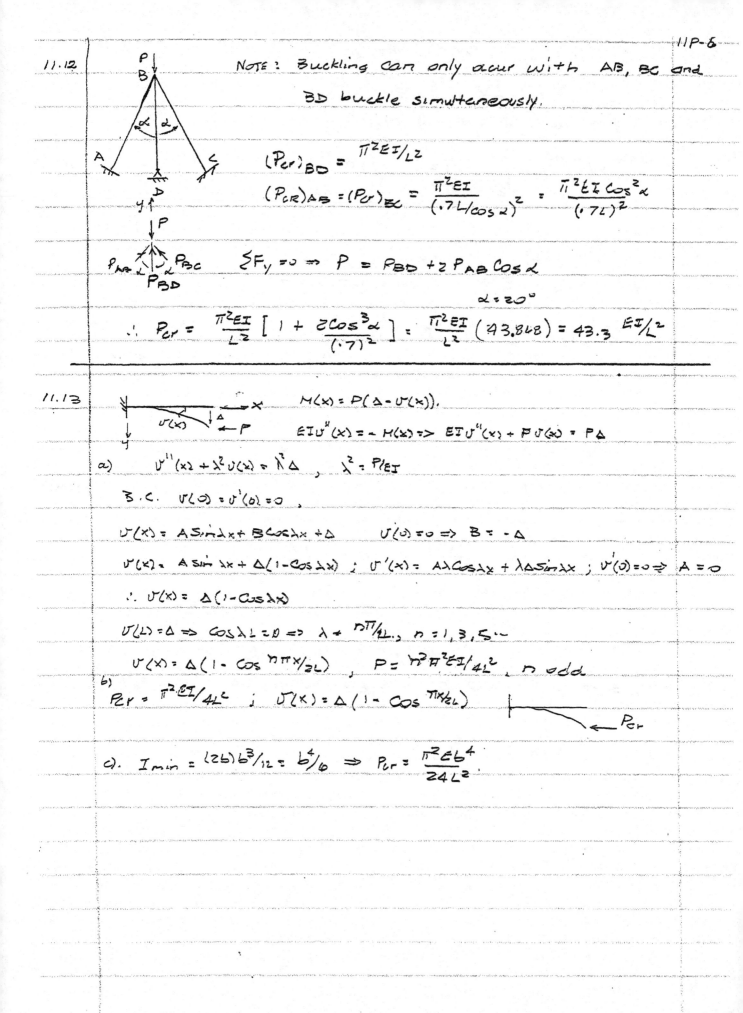

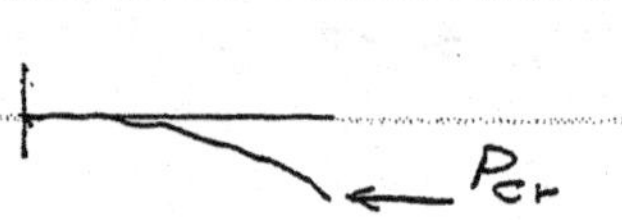

c). $I_{min} = (2b)b^3/12 = b^4/6 \Rightarrow P_{cr} = \dfrac{\pi^2 E b^4}{24L^2}$

11.14

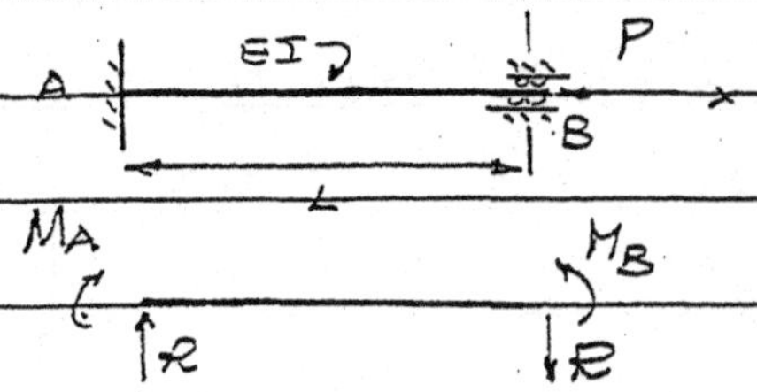

a).

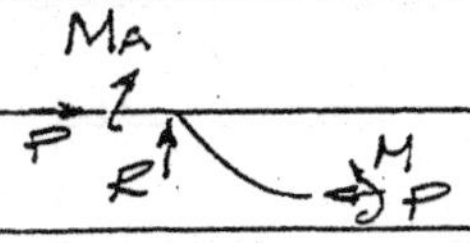

$$M(x) = M_A + Rx + Pv(x)$$

$$EIv''(x) = -M(x)$$

$$EIv''(x) + Pv(x) = -(M_A + RX)$$

$$v''(x) + \lambda^2 v(x) = -M_A/EI - \frac{RX}{EI}, \quad \lambda^2 = P/EI$$

$$v(X) = A\sin\lambda x + B\cos\lambda x - M_A/P - RX/P$$

B.C. $v(0) = v'(0) = v(L) = v'(L) = 0$

$$v'(x) = A\lambda\cos\lambda x - B\lambda\sin\lambda x - R/P$$

$$v(0) = 0 \Rightarrow B - M_A/P = 0 \Rightarrow B = M_A/P$$

$$v'(0) = 0 \Rightarrow A\lambda - R/P = 0 \Rightarrow A = R/P\lambda$$

$$\Rightarrow v(x) = R/P\left(\frac{\sin\lambda x}{\lambda} - x\right) + M_A/P(\cos\lambda x - 1)$$

$$v'(x) = R/P(\cos\lambda x - 1) - M_A\lambda/P\,\sin\lambda x$$

$$v(L) = 0 \Rightarrow R\left(\frac{\sin\lambda L}{\lambda} - L\right) + M_A(\cos\lambda L - 1) = 0$$

$$v'(L) = 0 \Rightarrow R(\cos\lambda L - 1) - M_A\lambda\sin\lambda L = 0$$

$$(\sin\lambda L - \lambda L)R + \lambda(\cos\lambda L - 1)M_A = 0 \qquad (a)$$

$$(\cos\lambda L - 1)R - \lambda\sin\lambda L\, M_A = 0 \qquad (b)$$

$$|\ | = 0 \Rightarrow \lambda\sin\lambda L(\sin\lambda L - \lambda L) + \lambda(\cos\lambda L - 1)^2 = 0$$

$$\sin^2\lambda L - \lambda L\sin\lambda L + \cos^2\lambda L - 2\cos\lambda L + 1 = 0$$

$$\underline{\underline{2 - \lambda L\sin\lambda - 2\cos\lambda L = 0}}$$

or

$$2(1 - \cos\lambda L) = \lambda L\sin\lambda L$$

11.14 (continued) $2(1-\cos\lambda L) = \lambda L \sin\lambda L$ $\qquad \cos^2 x = \frac{1}{2}(1+\cos 2x)$

$\sin^2 x = \frac{1}{2}(1-\cos 2x)$

$\therefore\ 4\sin^2 \lambda L/2 = 2\lambda L \sin\lambda L/2 \cos\lambda L/2$

$\therefore\ \sin\lambda L/2\,(2\sin\lambda L/2 - \lambda L\cos\lambda L/2) = 0$

or $\sin\lambda L/2\,[\tan\lambda L/2 - \lambda L/2] = 0$ $\Rightarrow$ Two possibilities

b) $\sin\lambda L/2 = 0 \Rightarrow \lambda L/2 = n\pi \Rightarrow \lambda = 2n\pi/L$ $\qquad n = 1, 2, 3 \ldots$

$P = 4\pi^2 EI/L^2 \quad (n=1)$

$$R = \frac{\lambda(\cos\lambda L - 1)}{\lambda L - \sin\lambda L}\cdot M_A = 0$$

$$\therefore\ v(x) = \frac{M_A}{P}(\cos\lambda x - 1) = \frac{M_A}{P}\left(\cos\frac{2\pi x}{L} - 1\right)$$

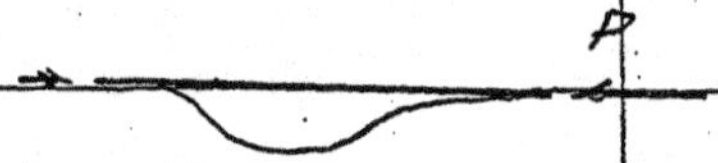

Bucking mode

c). $\tan\lambda L/2 = \lambda L/2$

$\Rightarrow P = \pi^2 EI/(1.7L/2)^2$

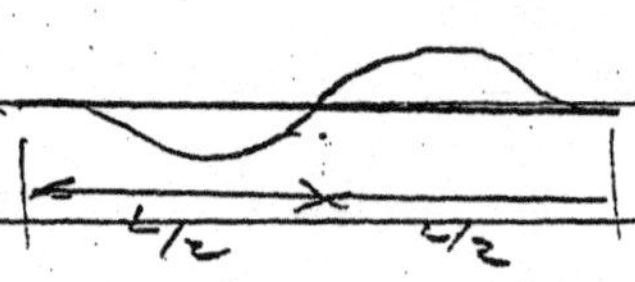

Note: $-2\sin^2 x\, R - \lambda\sin 2x \cdot M_A$ $\qquad x = \lambda L/2$

$2\sin^2 x\, R + 2\lambda\sin x\cos x \cdot M_A = 0$

$\sin x\,(\sin x \cdot R + \lambda\cos x \cdot M_A) = 0$

$\sin x \neq 0 \Rightarrow \quad M_A = -\frac{1}{\lambda}\tan x \cdot R = -\tan\lambda L/2 \cdot \frac{R}{\lambda} = -\frac{\lambda L}{2}\frac{R}{\lambda}$

$\therefore\ M_A = -RL/2$

Note: M_A & R of opposite sign.

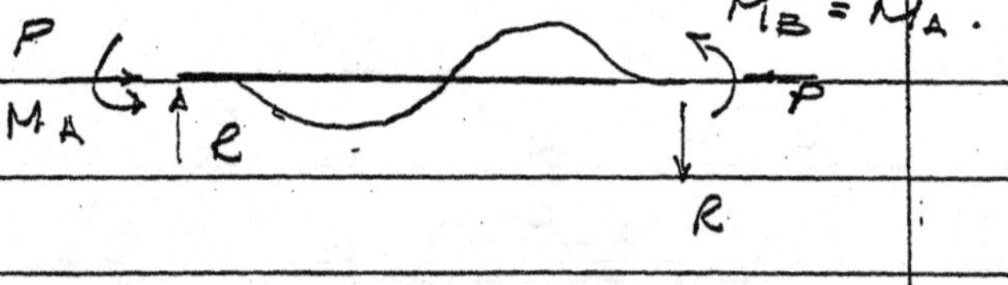

11.15

$$M(x) = -P[\Delta - v(x)] \quad ; \quad M_A = P\Delta$$

$$EIv''(x) = -M(x) \Rightarrow EIv''(x) + Pv(x) = P\Delta$$

a).

$$v''(x) + \lambda^2 v(x) = \lambda^2 \Delta \ , \quad \lambda^2 = P/EI$$

$$\text{B.C.} \quad v(0) = 0, \quad M_A = P\Delta \Rightarrow v'(0) = P\Delta/\beta$$

$$v(x) = A\sin\lambda x + B\cos\lambda x + \Delta \ ; \quad v(0) = 0 \Rightarrow B = -\Delta$$

$$v(x) = A\sin\lambda x + \Delta(1 - \cos\lambda x) \ ; \quad v'(x) = A\lambda\cos\lambda x + \Delta\lambda\sin\lambda x$$

$$v'(0) = P\Delta/\beta \Rightarrow A = P\Delta/\beta\lambda$$

$$v(x) = \Delta + \Delta\left[(P/\beta\lambda)\sin\lambda x - \cos\lambda x\right] ; \quad v(x) = \Delta\left[1 + \frac{EI\lambda}{\beta}\sin\lambda x - \cos\lambda x\right]$$

$$v(L) = \Delta \Rightarrow \frac{P\tan\lambda L}{\beta\lambda} = 1 \Rightarrow \tan\lambda L = \beta\lambda/P = \frac{\beta\lambda}{\lambda^2 EI} = \frac{\beta}{\lambda EI} = \alpha/\lambda L \ ; \quad \alpha = \beta L/EI$$

b) $\beta \to \infty \Rightarrow \tan\lambda L \to \infty \Rightarrow \lambda = \pi/2L \Rightarrow P = \frac{\pi^2 EI}{4L^2}$; $EI \to \infty$ Rigid Body

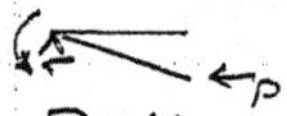

$P = \beta/L$

11.16 a)

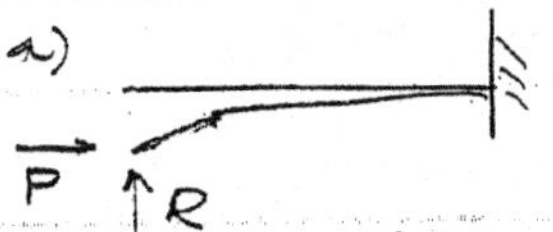

$$M(x) = -P[\Delta - v(x)] + Rx \quad ; \quad R = k\Delta$$

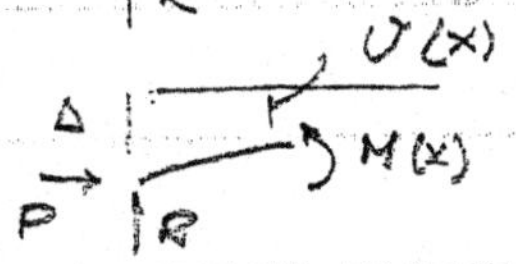

$$EIv''(x) = -M(x) \Rightarrow EIv''(x) + Pv(x) = P\Delta - Rx$$

$$v''(x) + \lambda^2 v(x) = \lambda^2\Delta - \frac{k\cdot\Delta}{EI}x \ , \quad \lambda^2 = P/EI$$

$$\text{B.C.} \quad v(L) = v'(L) = 0 \ ; \quad \text{Also } v(0) = \Delta$$

$$v(x) = A\sin\lambda x + B\cos\lambda x + \Delta - k\Delta x/P$$

$$v(0) = \Delta \Rightarrow B = 0$$

$$v(x) = A\sin\lambda x + \Delta(1 - kx/P) \ ; \quad v(L) = 0 \Rightarrow (\sin\lambda L)A + (1 - kL/P)\Delta = 0 \quad (a)$$

$$v'(x) = A\lambda\cos\lambda x - k\Delta/P \quad ; \quad v'(L) = 0 \Rightarrow (\lambda\cos\lambda L)A - (k/P)\Delta = 0 \quad (b)$$

Non-trivial sol. $(\Delta \neq 0, A \neq 0) \Rightarrow \begin{vmatrix} \sin\lambda L & (1 - kL/P) \\ \lambda\cos\lambda L & -k/P \end{vmatrix} = 0$

$$\frac{k}{P}\sin\lambda L + \lambda(1 - kL/P)\cos\lambda L = 0$$

$$\sin\lambda L + \lambda(P/k - L)\cos\lambda L = 0$$

$$\tan\lambda L = \lambda(L - P/k) = \lambda(L - EI\lambda^2/k)$$

$$\tan(\lambda L) = \lambda L\left[1 - \frac{(\lambda L)^2 EI}{kL^3}\right]$$

$$\eta \equiv \lambda L$$

$$\tan\eta = \eta\left[1 - \eta^2/\alpha\right] \ , \quad \alpha = kL^3/EI$$

b). $k \to \infty \Rightarrow \tan\eta = \eta \qquad P_{cr} = \pi^2 EI/(0.7L)^2.$

$$\eta = \lambda L = 4.4934$$

11.17

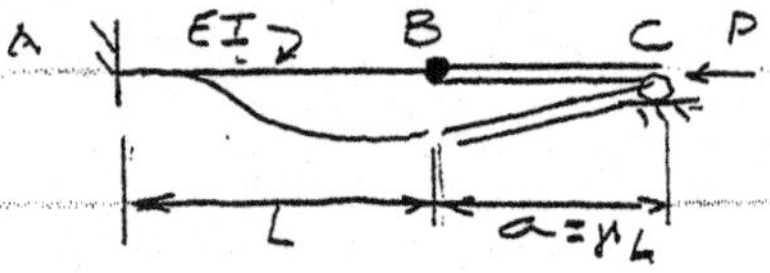

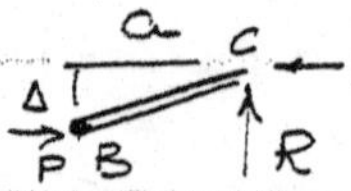

$(\Sigma M_B)_{BC} = 0 \Rightarrow R = -P\Delta/a$

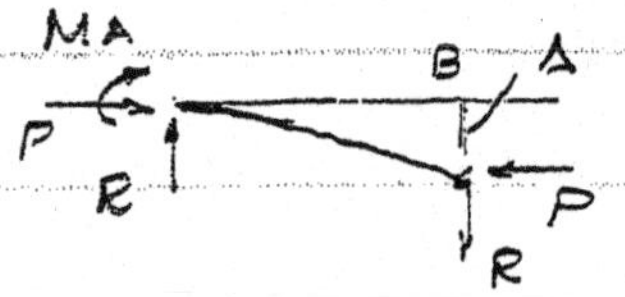

$M_A = -P\Delta - RL$

$M(x) = M_A + Pv(x) + Rx = -P\Delta + Pv(x) + R(x-L)$

$M(x) = Pv(x) - \frac{P\Delta}{a}(a+L-x)$

$EIv''(x) = -M(x) \Rightarrow EIv''(x) + Pv(x) = \frac{P\Delta}{a}(a+L-x)$

$v''(x) + \lambda^2 v(x) = \frac{\lambda^2 \Delta}{a}(a+L-x)$, $\lambda^2 = P/EI$. B.C. $v(0) = v'(0) = 0$ $[v(L) = \Delta]$

$v(x) = A\sin\lambda x + B\cos\lambda x + \frac{\Delta}{a}(a+L-x)$.

$v(0) = 0 \Rightarrow B = -\frac{\Delta}{a}(a+L)$

$v'(x) = A\lambda\cos\lambda x - B\lambda\sin\lambda x - \Delta/a$; $v'(0) = 0 \Rightarrow A = \Delta/a\lambda$

$\therefore v(x) = \frac{\Delta}{a}\left[\frac{\sin\lambda x}{\lambda} - (a+L)\cos\lambda x + (a+L-x)\right]$

$v(L) = \Delta \Rightarrow \frac{\Delta}{\lambda a}[\sin\lambda L - \lambda(a+L)\cos\lambda L + \lambda a] = \Delta$

$\therefore \frac{\sin\lambda L - \lambda(a+L)\cos\lambda L}{\lambda a} = 0$

$\Rightarrow \tan\lambda L = \lambda(L+a)$.

$\tan\lambda L = \lambda L(1+\gamma)$

11.18

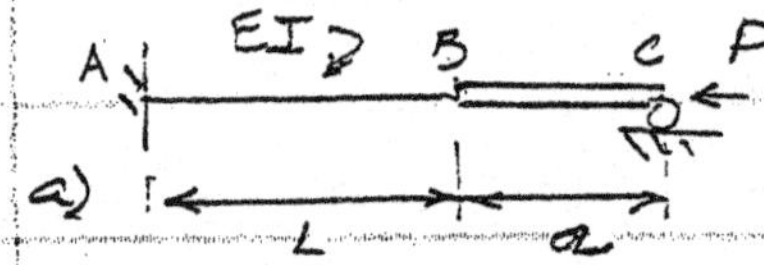

a) $M_A = R(L+a)$ $M(x) = M_A + Pv(x) - Rx$

$= Pv(x) + R(L+a-x)$

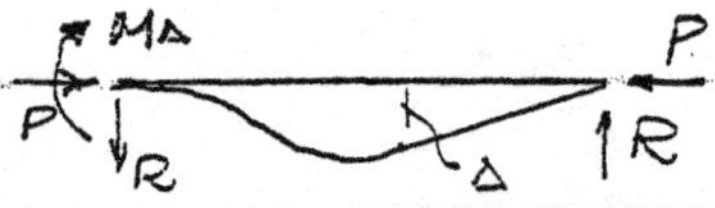

$EIv''(x) = -M(x) \Rightarrow EIv''(x) + Pv(x) = R[x-(L+a)]$

$v''(x) + \lambda^2 v(x) = \frac{R}{EI}[x-(L+a)]$; $(0 < x < L)$, $\lambda^2 = P/EI$

$v(x) = A\sin\lambda x + B\cos\lambda x + R/P\,[x-(L+a)]$; B.C. $v(0) = v'(0) = 0$; $v'(L) = -\Delta/a$
$v(L) = \Delta$

$v(0) = 0 \Rightarrow B = \frac{R}{P}(L+a)$; $v'(x) = A\lambda\cos\lambda x - B\lambda\sin\lambda x + R/P$.

$v'(0) = 0 \Rightarrow A = -R/P\lambda$; $v(x) = R/P\left[-\frac{\sin\lambda x}{\lambda} + (L+a)(\cos\lambda x - 1) + x\right]$

$v'(x) = R/P[-\cos\lambda x - (L+a)\lambda\sin\lambda x + 1]$

$\Delta \equiv v(L) = -av'(L) \Rightarrow -\frac{\sin\lambda L}{\lambda} + (L+a)\cos\lambda L - a = a\cos\lambda L + a(L+a)\lambda\sin\lambda L - a$

$\Rightarrow [1 + a(L+a)\lambda^2]\sin\lambda L = \lambda L\cos\lambda L \Rightarrow [1 + a(L+a)\lambda^2]\tan\lambda L = \lambda L$

b) $a \to \infty \Rightarrow \sin\lambda L = 0 \Rightarrow P_{cr} = \pi^2 EI/L^2$

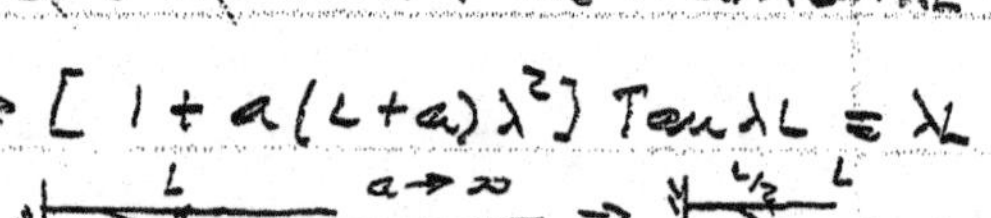

$P_{cr} = \frac{\pi^2 EI}{2(L/2)^2} = \pi^2 EI/L^2$

c) $a \to 0 \Rightarrow \tan\lambda L = \lambda L \Rightarrow P_{cr} = \pi^2 EI/(.7L)^2$

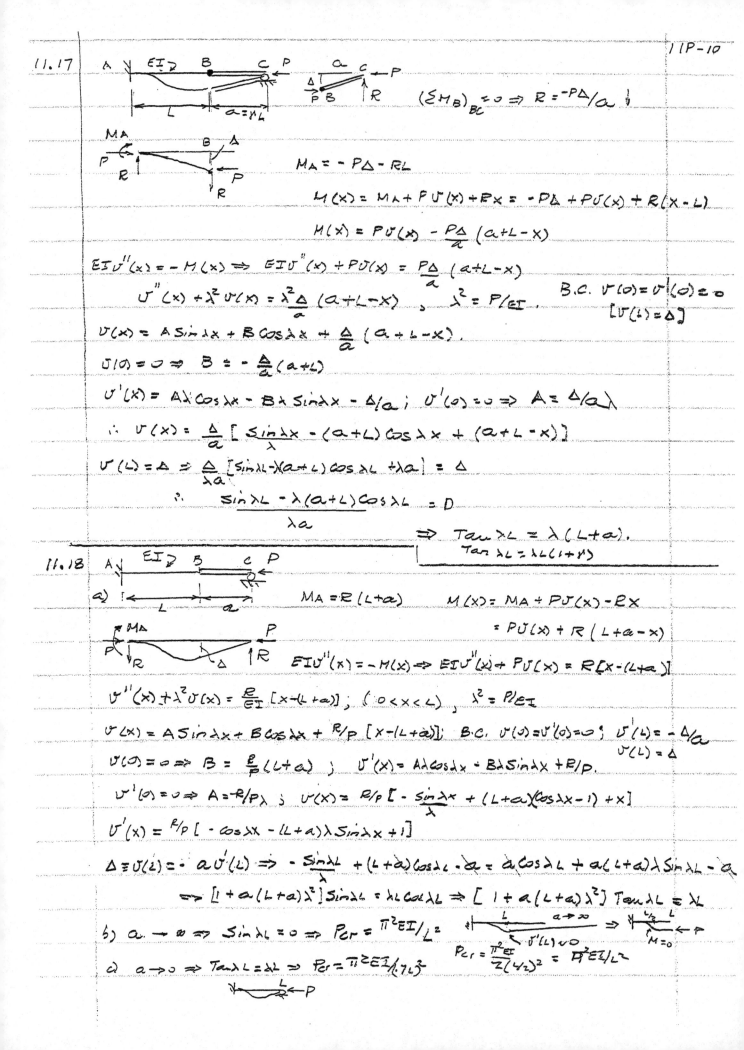

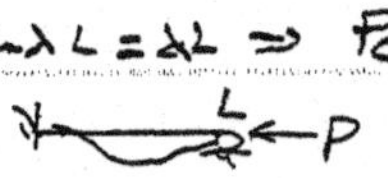

11.19

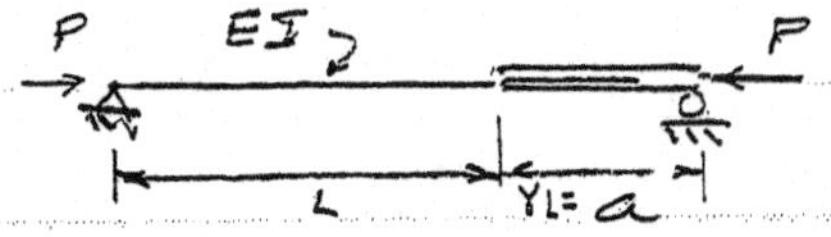

$M(x) = P\,v(x)$

$EI\,v''(x) = -M(x) \Rightarrow EI\,v''(x) + P\,v(x) = 0$

a)

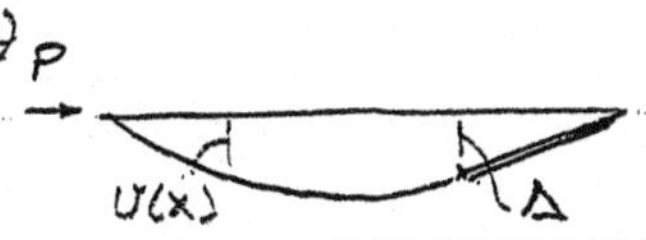

$v''(x) + \lambda^2 v(x) = 0 \;,\; \lambda^2 = P/EI \;;\; 0 < x < L.$

B.C. $v(0) = 0 \;,\; v(L) = \Delta \;,\; v'(L) = -\Delta/a$

$v(x) = A\sin\lambda x + B\cos\lambda x$

$v(0) = 0 \Rightarrow B = 0 \qquad v(x) = A\sin\lambda x \;;\; v'(x) = \lambda A\cos\lambda x.$

$v'(L) = -\Delta/a \Rightarrow \lambda A\cos\lambda L = -\Delta/a$

$v(L) = \Delta \Rightarrow A\sin\lambda L = \Delta \quad \Rightarrow \lambda\Delta\dfrac{\cos\lambda L}{\sin\lambda L} = -\Delta/a$

$$\tan\lambda L = -\lambda a \Rightarrow \tan\lambda L = -\gamma\lambda L \;,\; \gamma = a/L$$

b). $a \to 0 \Rightarrow \sin\lambda L = 0 \Rightarrow \lambda = \pi/L \Rightarrow P_{cr} = \pi^2 EI/L^2$

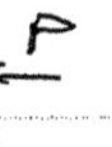

c). $a \to \infty \Rightarrow \tan\lambda L \to \infty \Rightarrow \lambda = \pi/2L \quad P_{cr} = \pi^2 EI/4L^2$

11-20

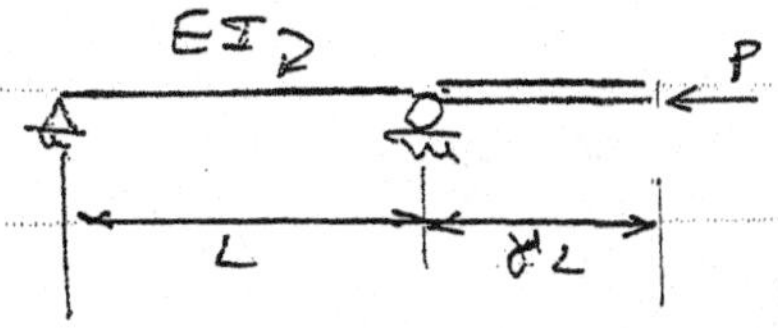

$R = P\Delta/L = -P\,v'(L)\,a/L \qquad (a = \alpha L)$

$M(x) = P\,v(x) + Rx$

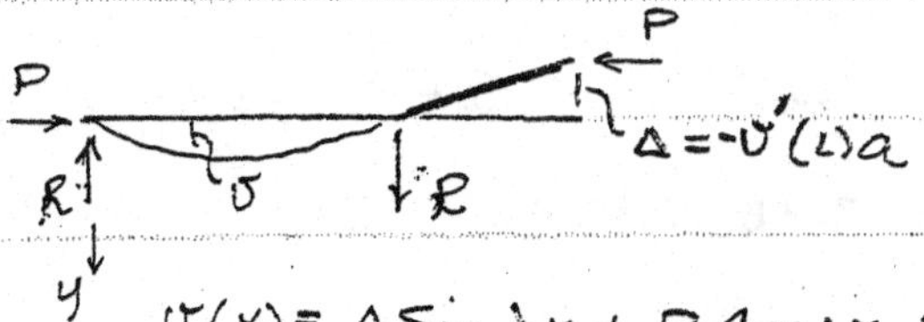

$EI\,v''(x) = -M(x) \Rightarrow EI\,v''(x) + P\,v(x) = -Rx$

$v''(x) + \lambda^2 v(x) = \dfrac{P\,v'(L)\,a x}{EIL} \;;\; \lambda^2 = P/EI \;,\; 0 < x < L.$

$v(x) = A\sin\lambda x + B\cos\lambda x + \dfrac{a\,v'(L)}{L}\cdot x \qquad$ B.C. $v(0) = v(L) = 0$

$v(0) = 0 \Rightarrow B = 0 \;;\; v(x) = A\sin\lambda x + \dfrac{a\,v'(L)}{L}x \;;\; v'(x) = A\lambda\cos\lambda x + a\,v'(L)/L$

$v'(L) = A\lambda\cos\lambda L + a\,v'(L)/L$

$v(L) = 0 \Rightarrow A\sin\lambda L + a\,v'(L) = 0 \Rightarrow v'(L) = -\dfrac{A\sin\lambda L}{a}$

$\therefore\; A\lambda\cos\lambda L - \dfrac{A}{L}\sin\lambda L = -\dfrac{A\sin\lambda L}{a} \Rightarrow A[\lambda\cos\lambda L + (1/a - 1/L)\sin\lambda L] = 0$

$\dfrac{A}{aL}[\lambda a L\cos\lambda L + (L - a)\sin\lambda L] = 0$

$A \neq 0 \Rightarrow \tan\lambda L = \dfrac{a\lambda L}{a - L} \Rightarrow \tan\lambda L = \dfrac{\gamma\lambda L}{\gamma - 1}$

b) $\gamma \to 0 \Rightarrow \tan\lambda L = 0 \Rightarrow \sin\lambda L = 0 \Rightarrow \lambda = \pi/L \Rightarrow P_{cr} = \dfrac{\pi^2 EI}{L^2}$

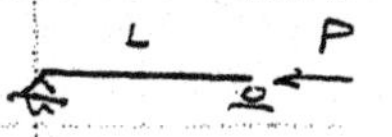

c). $\gamma \to \infty \Rightarrow \tan\lambda L = \lambda L \Rightarrow P_{cr} = \dfrac{\pi^2 EI}{(.7L)^2}$

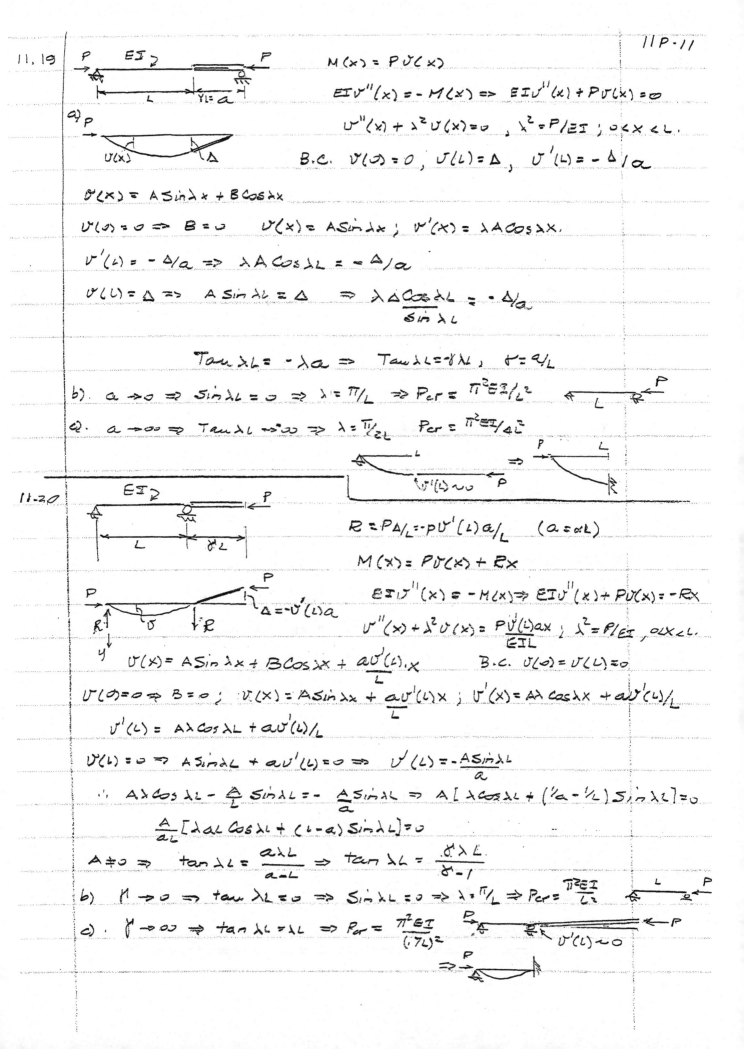

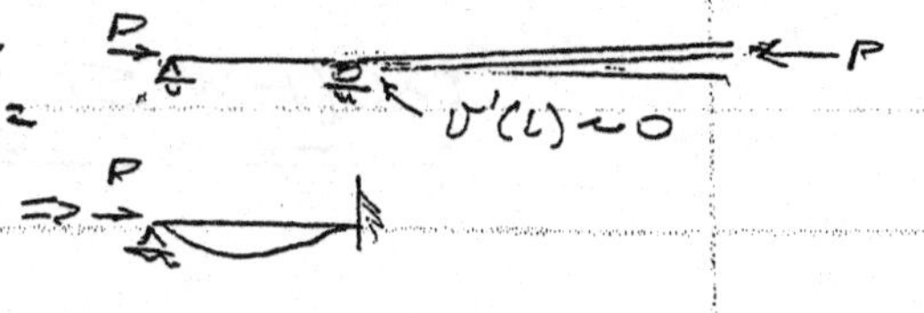

11.21 a)

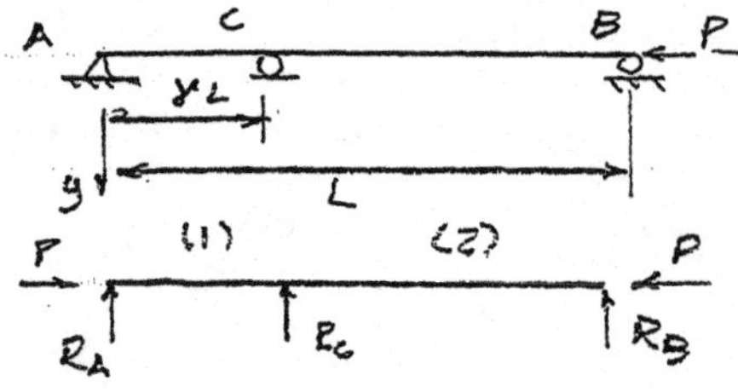

$$\sum M_C = 0 \Rightarrow R_A \gamma = R_C(1-\gamma)L$$

$$M(x) = R_A x + P v_1(x), \quad 0 < x < \gamma L$$

$$EI v''(x) = -M(x) \qquad M(x) = R_C(L-x) + P v_2(x), \quad \gamma L < x < L$$

$0 < x < \gamma L$ $\qquad\qquad$ $\gamma L < x < L$

$$EI v_1''(x) + P v_1(x) = -R_A x \qquad EI v_2''(x) + P v_2(x) = -R_C(L-x)$$

$$v_1''(x) + \lambda^2 v_1(x) = -R_A x / EI \qquad v_2''(x) + \lambda^2 v_2(x) = -\frac{R_C}{EI}(L-x), \quad \lambda^2 = P/EI$$

B.C. $v_1(0) = 0$, $v_1(\gamma L) = 0$ $\qquad$ $v_2(L) = 0$, $v_2(\gamma L) = 0$

$$v_1'(\gamma L) = v_2'(\gamma L)$$

$$v_1(x) = A_1 \sin\lambda x + B_1 \cos\lambda x - \frac{R_A x}{P} \qquad v_2(x) = A_2 \sin\lambda x + B_2 \cos\lambda x - \frac{R_C}{P}(L-x)$$

$$v_1'(x) = A_1\lambda \cos\lambda x - B_1 \lambda \sin\lambda x - R_A/P \qquad v_2'(x) = A_2\lambda \cos\lambda x - B_2\lambda \sin\lambda x + R_C/P$$

$$v_1(0) = 0 \Rightarrow B_1 = 0$$

$$v_1(\gamma L) = 0 \Rightarrow A_1 \lambda \sin\gamma\lambda L - R_A \gamma L/P = 0 \qquad (i)$$

$$v_2(L) = 0 \Rightarrow A_2 \sin\lambda L + B_2 \cos\lambda L = 0 \qquad (ii)$$

$$v_2(\gamma L) = 0 \Rightarrow A_2 \sin\gamma\lambda L + B_2 \cos\gamma\lambda L - \frac{R_C}{P}(1-\gamma)L = 0 \qquad (iii)$$

$$v_1'(\gamma L) = v_2'(\gamma L) \Rightarrow A_1\lambda \cos\gamma\lambda L - A_2\lambda\cos\gamma\lambda L + B_2\lambda \sin\gamma\lambda L - (R_A + R_C)/P = 0 \qquad (iv)$$

Substitute $R_A = R_C(1-\gamma)L/\gamma$ in (i) $\Rightarrow$ $R_A + R_C = R_C/\gamma$; $\sin\gamma\lambda L = (1-\gamma)LR_C/P$

$$\begin{bmatrix} 0 & \sin\lambda L & \cos\lambda L \\ -\sin\gamma\lambda L & \sin\gamma\lambda L & \cos\gamma\lambda L \\ \left(\lambda\cos\gamma\lambda L - \frac{\sin\gamma\lambda L}{(1-\gamma)\gamma L}\right) & -\lambda\cos\gamma\lambda L & \lambda\sin\gamma\lambda L \end{bmatrix} \begin{Bmatrix} A_1 \\ A_2 \\ B_2 \end{Bmatrix} = \begin{Bmatrix} 0 \\ 0 \\ 0 \end{Bmatrix}$$

Non-trivial sol: $A_1 \neq 0$, $A_2 \neq 0$, $B_2 \neq 0$:

$$D = |\ \ | = 0 \Rightarrow \cos\lambda L \sin^2\gamma\lambda L + \sin\lambda L\,[(1-\gamma)\gamma\lambda L - \tfrac{1}{2}\sin(2\gamma\lambda L)] = 0$$

$$\Rightarrow \cos\lambda L \sin^2\gamma\lambda L + \sin\lambda L\,[(1-\gamma)\gamma\lambda L - \cos\gamma\lambda L \sin\gamma\lambda L] = 0$$

$$\Rightarrow \sin\gamma\lambda L\,[\sin\gamma\lambda L \cos\lambda L - \cos\gamma\lambda L \sin\lambda L] + (1-\gamma)\gamma\lambda L \sin\lambda L = 0$$

$$\sin\gamma\lambda L \sin[(\gamma-1)\lambda L] + (1-\gamma)\gamma\lambda L \sin\lambda L = 0$$

$$\underline{\underline{\gamma\lambda L(1-\gamma)\sin\lambda L - \sin\gamma\lambda L \sin[(1-\gamma)\lambda L] = 0}}$$

11.21 (cont'd)

b) $\gamma = 0.5$

$$\frac{\lambda L}{4}\sin\lambda L - \sin^2\frac{\lambda L}{2} = 0 \Rightarrow \frac{\lambda L}{4}\left(2\sin\frac{\lambda L}{2}\cos\frac{\lambda L}{2}\right) - \sin^2\frac{\lambda L}{2} = 0$$

$$\sin\frac{\lambda L}{2}\left[\frac{\lambda L}{2}\cos\frac{\lambda L}{2} - \sin\frac{\lambda L}{2}\right] = 0$$

$$\sin\frac{\lambda L}{2}\left[\frac{\lambda L}{2} - \tan\frac{\lambda L}{2}\right] = 0$$

$\sin\lambda L/2 = 0$ yields lowest root: $\lambda = 2\pi/L \Rightarrow P_{cr} = \frac{4\pi^2 EI}{L^2}$.

c) $\gamma \to 0 \Rightarrow \gamma\lambda L(1-\gamma)\sin\lambda L - \sin\gamma\lambda L[\sin\lambda L\cos\gamma\lambda L - \sin\gamma\lambda L\cos\lambda L] = 0$

$|\gamma| << 1 \Rightarrow \sin\gamma\lambda L \sim \gamma\lambda L,\ \cos\gamma\lambda L \sim 1$

$$\therefore \gamma\lambda L(1-\gamma)\sin\lambda L - \gamma\lambda L[\sin\lambda L - \gamma\lambda L\cos\lambda L] = 0$$

$$\gamma^2[-\lambda L\sin\lambda L + (\lambda L)^2\cos\lambda L] = 0 \Rightarrow \tan\lambda L = \lambda L.$$

11.22 a) 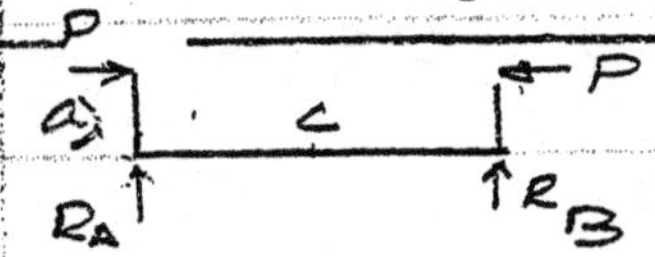

$\Sigma M_A = 0 \Rightarrow R_B = 0 \Rightarrow R_A = 0,\ V_c = 0.$

b). $V(x) = -EIv'''(x) \Rightarrow V(x) = -EI\lambda^3 e\left[-\tan\frac{\lambda L}{2}\cos\lambda x + \sin\lambda x\right]$

$V_A = V(0) = EI\lambda^3 e\tan\frac{\lambda L}{2}$ $\quad V_B = V(L) = -EI\lambda^3 e\,[$

$$V_B = V(L) = -EI\lambda^3 e\left[-\frac{\sin\frac{\lambda L}{2}}{\cos\lambda L/2}\cos\lambda L + \sin\lambda L\right] = -EI\lambda^3 e\,\frac{\left[-\sin\lambda L/2\cos\lambda L + \sin\lambda L\cos\frac{\lambda L}{2}\right]}{\cos\lambda L/2}$$

$$= -EI\lambda^3 e\cdot\frac{\sin\lambda L/2}{\cos\lambda L/2} = -EI\lambda^3 e\tan\lambda L/2$$

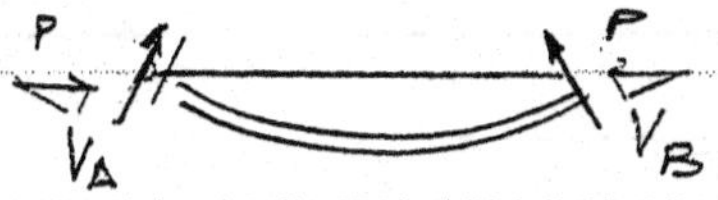

$V_C = V(L/2) = 0$ since $v'(L/2) = 0$.

11.23

$$EI v^{iv} + P v''(x) = q(x) = 0$$

$$v^{iv}(x) + \lambda^2 v''(x) = 0 \quad , \quad \lambda^2 = P/EI$$

B.C. $v(0) = v'(0) = v''(L) = 0, \quad EI v'''(L) = -P v'(L) \Rightarrow v'''(x) = -\lambda^2 v'(L)$

$$EI v''(0) = P\Delta \Rightarrow v''(0) = \lambda^2 \Delta$$

$$v(x) = A \sin\lambda x + B\cos\lambda x + Cx + D$$

$$v(0) = 0 \Rightarrow B = -D \ ; \quad v'(x) = A\lambda\cos\lambda x - B\lambda\sin\lambda x + C$$

$$v''(x) = -A\lambda^2 \sin\lambda x - B\lambda^2\cos\lambda x$$

$$v'''(x) = -A\lambda^3\cos\lambda x + B\lambda^3 \sin\lambda x .$$

$$v'(0) = 0 \Rightarrow A = -C/\lambda \ ; \quad v''(L) = 0 \Rightarrow \quad B = -A\tan\lambda L$$

$$v'''(L) = -\lambda^2 v'(L) \Rightarrow \lambda^3(-A\cos\lambda L + B\sin\lambda L) = -\lambda^3 (A\cos\lambda L - B\sin\lambda L + C)$$

$$\Rightarrow C = 0$$

$$v''(0) = \lambda^2 \Delta \Rightarrow -B = \Delta \Rightarrow A = \Delta/\tan\lambda L$$

$$\therefore \ v(x) = \frac{\Delta}{\tan\lambda L}\sin\lambda x - \Delta(1-\cos\lambda x)$$

$$v(x) = \Delta \Rightarrow \quad \Delta\left[\frac{\sin\lambda L}{\tan\lambda L} - 1 + \cos\lambda L\right] = \Delta$$

$$\therefore \ 2\cos\lambda L - 1 = 1 \Rightarrow \underline{\cos\lambda L = 1}$$

$$\Rightarrow \lambda = \frac{n\pi}{2L}, \ n = 1, 3 \ldots \quad P_{cr} = EI\lambda^2\Big|_{n=1} = \pi^2 EI/4L^2$$

Note: $\tan\lambda L \to \infty \Rightarrow \quad v(x) = \Delta\,(\cos \pi x/2L - 1)$

11.24

$v^{iv}(x) + \lambda^2 v''(x) = 0$; B.C. $v(0) = 0, \quad EI v''(0) = \beta v'(0)$

$$v''(L) = 0, \quad v'''(L) = -\lambda^2 v'(L)$$

$$v(x) = A\sin\lambda x + B\cos\lambda x + Cx + D \quad ; \quad v(0) = 0 \Rightarrow B = -D \quad (a)$$

$$v'(x) = A\lambda\cos\lambda x - B\lambda\sin\lambda x + C$$

$$v''(x) = -A\lambda^2\sin\lambda x - B\lambda^2\cos\lambda x \ ; \quad v''(L) = 0 \Rightarrow B = -A\tan\lambda L \ (b)$$

$$v'''(x) = -A\lambda^3\cos\lambda x + B\lambda^3\sin\lambda x \ ; \ v'''(L) = -\lambda^2 v'(L) \Rightarrow \lambda^3[-A\cos\lambda L + B\sin\lambda L] = -\lambda^3[A\cos\lambda L - B\sin\lambda L + C]$$

$$\therefore \Rightarrow C = 0 \quad (c)$$

$$EI v''(0) = \beta v'(0) \Rightarrow -EI B\lambda^2 = \beta A\lambda \Rightarrow B = -\frac{A\beta}{EI\lambda} \ (d)$$

From (b) & (d) $\Rightarrow \underline{\tan\lambda L = \beta/EI\lambda}$. From (a) - (d): $B = \Delta, \ A = \dfrac{\Delta}{\tan\lambda L}, \ C = 0, \ D = -\Delta$

$$\Rightarrow \quad v(x) = \Delta\left[\frac{\sin\lambda x}{\tan\lambda L} + \cos\lambda x - 1\right]$$

Note: For $\beta \to \infty \quad \lambda \to n\pi/2L \Rightarrow v(x) = \Delta\,(\cos n\pi x/2L - 1)$, n odd

11.25

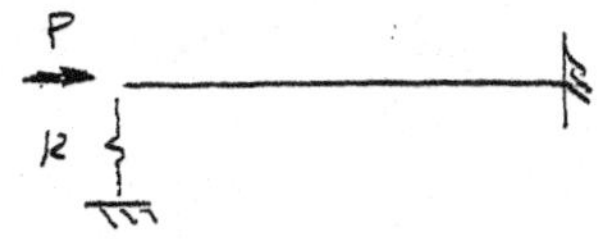

$$v^{IV}(x) + \lambda^2 v''(x) = 0 \quad , \quad \lambda^2 = P/EI.$$

B.C. $v(L) = v'(L) = 0$

$v''(0) = 0, \quad EIv'''(0) = -[Pv'(0) + kv(0)]$

$$v(x) = A\sin\lambda x + B\cos\lambda x + Cx + D$$

$$v'(x) = A\lambda\cos\lambda x - B\lambda\sin\lambda x + C$$

$$v''(x) = -A\lambda^2\sin\lambda x - B\lambda^2\cos\lambda x$$

$$v'''(x) = -A\lambda^3\cos\lambda x + B\lambda^3\sin\lambda x$$

$v''(0) = 0 \Rightarrow B = 0 \; ; \quad v'(L) = 0 \Rightarrow C = -A\lambda\cos\lambda L \quad (a)$

$v'''(0) = -[\lambda^2 v'(0) + \frac{k}{EI}v(0)] \Rightarrow -A\lambda^3 = -[\lambda^3 A + \lambda^2 C + \frac{k}{EI}D] \Rightarrow D = -\left(\frac{\lambda^2 EI}{k}\right)C \; (b)$

$v(L) = A\sin\lambda L + CL + D = 0$

From (a) & (b): $C = -A\lambda\cos\lambda L, \; D = \frac{\lambda^2 EI}{k}C = \left(\frac{\lambda^3 EI}{k}\cos\lambda L\right)A$

$\therefore \; A[\sin\lambda L - \lambda L\cos\lambda L + \frac{\lambda^3 EI}{k}\cos\lambda L] = 0$

$A \neq 0 \Rightarrow \tan\lambda L = \lambda L\left[1 - \frac{(\lambda L)^2 EI}{kL^3}\right]$

In non-dimensional terms: Let $\eta = \lambda L \Rightarrow \tan\eta = \eta(1 - \eta^2/\kappa) \; ; \; \kappa = kL^3/EI$

$$v(x) = A[\sin\lambda x + (\frac{\lambda^3 EI}{k} - \lambda x)\cos\lambda L] = A[\sin\eta x/L + \eta(\eta^2/\kappa - x/L)\cos\eta]$$

11.26

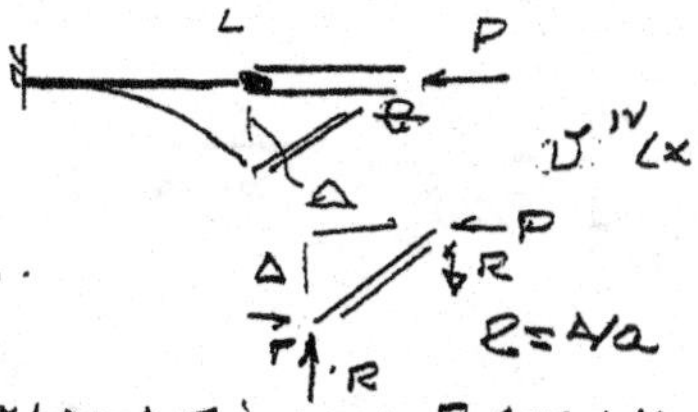

$v^{IV}(x) + \lambda^2 v''(x) = 0 \qquad$ B.C. $v(0) = v'(0) = 0 \; , \; v''(L) = 0,$

$EIv'''(L) = -[R + Pv'(L)] = -P[\Delta/a + v'(L)]$

$\Rightarrow v'''(L) = -\lambda^2[\Delta/a + v'(L)]$

$v(x) = A\sin\lambda x + B\cos\lambda x + Cx + D \quad ; \quad v(0) = 0 \Rightarrow B = -D \quad (a)$

$$v'(x) = A\lambda\cos\lambda x - B\lambda\sin\lambda x + C$$

$$v''(x) = -A\lambda^2\sin\lambda x - B\lambda^2\cos\lambda x$$

$$v'''(x) = -A\lambda^3\cos\lambda x + B\lambda^3\sin\lambda x$$

$v'(0) = 0 \Rightarrow A = -C/\lambda \; (b) \; ; \quad v''(L) = 0 \Rightarrow B = -A\tan\lambda L \quad (c)$

$v'''(L) = -\lambda^2[\Delta/a + v'(L)] \Rightarrow \lambda^3[-A\cos\lambda L + B\sin\lambda L] = -\lambda^2[\Delta/a + A\lambda\cos\lambda L - B\lambda\sin\lambda L + C]$

$\Rightarrow C = -\Delta/a \; (d) \Rightarrow (b)\&(d) \Rightarrow A = \Delta/a\lambda$

From (a)-(d)

$A = \Delta/a\lambda, \; B = -\frac{\Delta}{a\lambda}\tan\lambda L, \; C = -\Delta/a, \; D = \frac{\Delta}{a\lambda}\tan\lambda L$

$$v(x) = \frac{\Delta}{a\lambda}[\sin\lambda x + \tan\lambda L(1 - \cos\lambda x) - \lambda x]$$

$v(L) = \Delta \Rightarrow \tan\lambda L - \lambda L = a\lambda \Rightarrow \underline{\underline{\tan\lambda L = (a+L)\lambda}}$

11.27

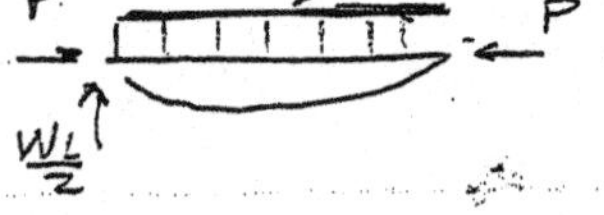

$EI v''(x) = -M(x)$; $M(x) = \frac{WL}{2} x - Wx^2/2 + P v(x)$

$EI v''(x) + P v(x) = -WLx/2 + Wx^2/2$ B.C. $v(0) = v(L) = 0$

$v''(x) + \lambda^2 v(x) = -WLx/2EI + Wx^2/2EI$.

Sol. $v_c(x) = A \sin \lambda x + B \cos \lambda x$; $v_p = C + Dx + Fx^2$; $v_p' = D + 2Fx$; $v_p'' = 2F$

$\therefore 2F + \lambda^2 (C + Dx + Fx^2) = -WLx/2EI + Wx^2/2EI \Rightarrow D = -WL/2\lambda^2 EI = -WL/2P$

$F = W/2P$, $C = -\frac{2F}{\lambda^2} = -W/P\lambda^2$

$\therefore v_p = -\frac{W}{P\lambda^2} - \frac{WLx}{2P} + \frac{Wx^2}{2P} \Rightarrow v(x) = A \sin \lambda x + B \cos \lambda x - W/P\lambda^2 - WLx/2P + Wx^2/2P$

$v(0) = 0 \Rightarrow B = W/P\lambda^2$ $v(L) = 0 \Rightarrow A \sin \lambda L + B \cos \lambda L - W/P\lambda^2 = 0$

$A = \frac{1}{\sin \lambda L}\left[W/P\lambda^2 - \frac{W}{P\lambda^2}\cos \lambda L\right] = \frac{W}{P\lambda^2 \sin \lambda L}(1 - \cos \lambda L)$

$\therefore v(x) = \frac{W}{P}\left[\frac{1}{\lambda^2}\left(\frac{1 - \cos \lambda L}{\sin \lambda L} \sin \lambda x + \cos \lambda x - 1\right) - Lx/2 + x^2/2\right]$

$v(L/2) = \frac{W}{P}\left[\frac{1}{\lambda^2 \sin \lambda L}\left(\overbrace{\sin \lambda L/2 - \cos \lambda L \sin \lambda L/2 + \cos \lambda L/2 \sin \lambda L}^{\sin \lambda L/2} - \sin \lambda L\right) - L^2/8\right]$

$= \frac{W}{P\lambda^2}\left[\frac{2 \sin \lambda L/2}{\sin \lambda L} - 1 - \lambda^2 L^2/8\right] = \frac{W}{P\lambda^2}\left[\frac{1}{\cos \lambda L/2} - 1 - \lambda^2 L^2/8\right]$

$\mu = \lambda L/2 \Rightarrow \mu = \frac{\pi}{2}\sqrt{P/P_E}$

$W/P\lambda^2 = \frac{W}{(P/\lambda^2)\lambda^4} = \frac{WL^4}{EI(\lambda L)^4} \Rightarrow v(L/2) = \frac{WL^4}{EI(2\mu)^4}\left[\frac{1}{\cos \mu} - 1 - \mu^2/2\right]$

$v(L/2) = \frac{WL^4}{16 EI \mu^4}\left(1/\cos \mu - 1 - \mu^2/2\right) = \delta\left[\frac{24}{5}\left(1/\cos \mu - 1 - \mu^2/2\right)\right]$

$\delta = 5WL^4/384EI$

a).

$P \Rightarrow P_E \Rightarrow \mu \to \pi/2 \Rightarrow v(L/2) \to \infty$ instability

b) $\mu \to 0 \Rightarrow v(L/2) = \delta \left\{\frac{24}{5\mu^4}\left[\left(1 + \mu^2/2 + \frac{5\mu^4}{24} + \dots\right) - 1 - \mu^2/2\right]\right\}$

$= \delta\{1 + O(\mu^6)\} \to \delta$

Similarly

$M(x) = -EI v''(x) = \frac{EIW}{P}\left[\left(\frac{1 - \cos \lambda L}{\sin \lambda L}\right) \sin \lambda x + \cos \lambda x - 1\right]$

$\Rightarrow M(L/2) = \frac{WEI}{P}\left[\frac{1}{\cos \lambda L/2} - 1\right] = \frac{WL^2}{4\mu^2}\left[\frac{1}{\cos \lambda L/2} - 1\right]$

$P \to P_E \Rightarrow \mu \to \pi/2 \to M(L/2) \to \infty$

$\mu \to 0 \Rightarrow M(L/2) = \frac{WL^2}{4}\left[\frac{1}{\mu^2}\left(1 + \mu^2/2 + \frac{5\mu^4}{24} + \dots\right) - 1\right] = WL^2/8$ as $\mu \to 0$.

Similarly: $V(x) = \frac{WL}{2\mu}\left[\frac{1 - \cos \lambda L}{\sin \lambda L} \cos \lambda x - \sin \lambda x\right] \Rightarrow V(L/2) = 0$

$V(0) = \frac{WL}{2\mu} \tan \frac{\lambda L}{2}$

$P \to P_E \to V(0) \to \infty$; $\mu \to 0$ $V(0) \to WL/2$

11.28

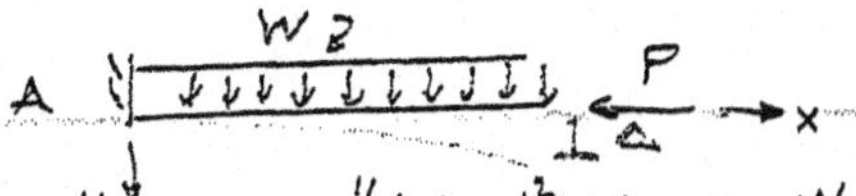

$M(x) = -WL^2/2 + WLx - Wx^2/2 - P[\Delta - v(x)]$

$v''(x) + \lambda^2 v(x) = \frac{Wx^2}{2EI} - \frac{WLx}{EI} + WL^2/2EI + \lambda^2\Delta, \quad \lambda^2 = P/EI.$

B.C. $v(0) = v'(0) = 0$

$v(x) = A\sin\lambda x + B\cos\lambda x + Wx^2/2P - \frac{WLx}{P} + \frac{WL^2}{2P} - \frac{W}{P\lambda^2} + \Delta$

$v(0) = 0 \Rightarrow \quad B = W/P\lambda^2 - WL^2/2P - \Delta$

$v'(x) = A\lambda\cos\lambda x - B\lambda\sin\lambda x + Wx/P - WL/P$

$v'(0) = 0 \Rightarrow \quad A\lambda = WL/P$

$v(x) = \frac{WL}{P\lambda}\sin\lambda x + \frac{W}{P}(1/\lambda^2 - L^2/2)\cos\lambda x + Wx^2/2P - WLx/P + WL^2/2P - W/P\lambda^2 + \Delta(1-\cos\lambda x)$

$v(L) = \frac{WL}{P\lambda}\sin\lambda L + \frac{W}{P\lambda^2}(1 - \lambda^2L^2/2)\cos\lambda x - W/P\lambda^2 + \Delta(1-\cos\lambda x)$

$= \frac{WL^4}{EI(\lambda L)^3}\sin\lambda L + \frac{WL^4}{EI(\lambda L)^4}(\cos\lambda L - 1) - \frac{WL^4}{2EI(\lambda L)^2}\cos\lambda L + \Delta(1-\cos\lambda L)$

$v(L) = \frac{WL^4}{EI}\left\{\frac{1}{(\lambda L)^3}\sin\lambda L + \frac{1}{(\lambda L)^4}(\cos\lambda L - 1) - \frac{1}{2(\lambda L)^2}\cos\lambda L\right\} + \Delta(1-\cos\lambda L)$

$\tilde{\mu} = \lambda L = \frac{\pi}{2}\sqrt{P/P_{cr}}, \; P_{cr} = \pi^2 EI/(2L)^2$

$v(L) = \Delta \Rightarrow \quad \Delta \equiv v(L/2) = \frac{WL^4}{EI\mu^2}\left[\frac{1}{\tilde{\mu}}\sin\tilde{\mu} + \frac{1}{\tilde{\mu}^2}(\cos\tilde{\mu} - 1) - \frac{1}{2}\cos\tilde{\mu}\right] + \Delta(1-\cos\tilde{\mu})$

$\Delta = \frac{WL^4}{\mu^2 EI\cos\tilde{\mu}}\left[\frac{1}{\tilde{\mu}}\sin\tilde{\mu} + 1/\mu^2(\cos\tilde{\mu} - 1) - \frac{1}{2}\cos\tilde{\mu}\right]$

b).

$\mu \to \pi/2 \Rightarrow \cos\mu \to 0 \Rightarrow \Delta \to \infty \;;\;$ As $P \to P_{cr} = \frac{\pi^2 EI}{(2L)^2}$.

$\tilde{\mu} \to 0$

$\Delta = \frac{WL^4}{EI\cos\tilde{\mu}(\tilde{\mu}^2)}\left\{\frac{1}{\tilde{\mu}}\sin\tilde{\mu} + \left[\frac{1}{\tilde{\mu}^2} - 1/2\right]\cos\tilde{\mu} - \frac{1}{\tilde{\mu}^2}\right\}$

$= \frac{WL^4}{EI\cos\tilde{\mu}\,(\tilde{\mu})^2}\left\{\frac{1}{\tilde{\mu}}\left(\tilde{\mu} - \frac{\tilde{\mu}^3}{6}\right) + \left(\frac{1}{\tilde{\mu}^2} - 1/2\right)\left(1 - \frac{\tilde{\mu}^2}{2} + \frac{\tilde{\mu}^4}{24}\ldots\right) - \frac{1}{\tilde{\mu}^2}\right\}$

$= \frac{WL^4}{\mu^2\cos\tilde{\mu}\,EI}\left(\tilde{\mu}^2/8\right) = \frac{WL^4}{8EI}\left[1 + \tilde{\mu}^2/2 + \ldots\right] \to WL^4/8EI = \delta$

11.29

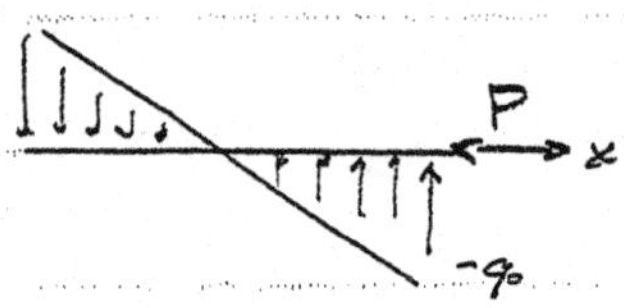

$$EI\,v^{IV}(x) + P\,v''(x) = q(x) = q_0(L-2x)/L$$

$$v^{IV}(x) + \lambda^2 v''(x) = \frac{q_0}{EIL}(L-2x)\;; \qquad \lambda^2 = P/EI$$

a) B.C. $v(0) = v(L) = 0\;;\; v''(0) = v''(L) = 0$

$$v(x) = A\sin\lambda x + B\cos\lambda x + Cx + D + \frac{q_0}{}(3L-2x)x^2/6PL$$

$v(0) = 0 \Rightarrow B + D = 0 \Rightarrow B = -D$

$$v''(x) = -A\lambda^2\sin\lambda x - B\lambda^2\cos\lambda x + \frac{q_0}{6PL}(6L - 12x)$$

$v''(0) = 0 \Rightarrow B = q_0/P\lambda^2$

$$v''(L) = 0 \Rightarrow A = -\frac{q_0}{P\lambda^2\sin\lambda L}(\cos\lambda L + 1)$$

$$v(L) = 0 \Rightarrow A\sin\lambda L + B\cos\lambda L + CL + D + q_0L^2/6P = 0$$

$$B = \frac{q_0}{P}\left(2/\lambda^2 L - L/6\right)$$

$$\therefore\; v(x) = \frac{q_0}{P}\left\{\tfrac{1}{\lambda^2}(\cos\lambda x - 1) + \left(\tfrac{2}{\lambda^2 L} - \tfrac{L}{6}\right)x - \tfrac{1}{\lambda^2}\left(\cot\lambda L + \frac{1}{\sin\lambda L}\right)\sin\lambda x + (3L-2x)x^2/6L\right\}$$

$$v''(x) = \frac{q_0}{P}\left\{-\cos\lambda x + \left(\cot\lambda L + \frac{1}{\sin\lambda L}\right)\sin\lambda x + 1 - 2x/L\right\}$$

b).

$$M(x) = -EI\,v''(x) = -\frac{q_0L^2}{(\lambda L^2)}\left[-\cos\lambda x + \left(\cot\lambda L + \frac{1}{\sin\lambda L}\right)\sin\lambda x + 1 - 2x/L\right]$$

$$V(x) = \frac{dM}{dx} = -\frac{q_0L}{(\lambda L)}\left[\sin\lambda x + \left(\cot\lambda L + \frac{1}{\sin\lambda L}\right)\cos\lambda x - 2/\lambda L\right]$$

$$V(0) = -\frac{q_0}{\lambda}\left(\cot\lambda L + \frac{1}{\sin\lambda L} - 2/\lambda L\right)$$

c)

Note:

$$v'(x) = \frac{q_0}{P}\left[-\tfrac{1}{\lambda}\sin\lambda x - L/6 + 2/L\lambda^2 - \tfrac{1}{\lambda}\left(\cot\lambda L + \frac{1}{\sin\lambda L}\right)\cos\lambda x + (x - x^2/L)\right\}$$

$$\Rightarrow v'(0) = q_0/P\left\{-L/6 + \tfrac{1}{\lambda}\left[2/\lambda L - \cot\lambda L - \frac{1}{\sin\lambda L}\right]\right\}$$

$$\therefore \Rightarrow q_0/\lambda\,(2/\lambda L - \cot\lambda L - 1/\sin\lambda L) = q_0L/6 + P\,v'(0)$$

$$\therefore\; V(0) = q_0L/6 + P\,v'(0)$$

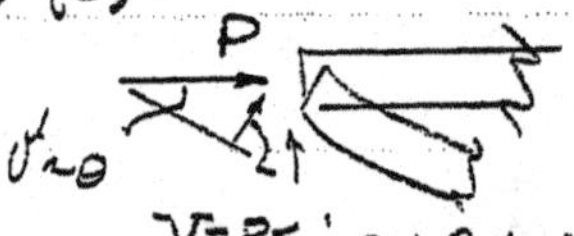

$$V = P\sin\theta + \frac{q_0L}{6}\cos\theta \sim P\,v'(0) + q_0L/6$$

11.30

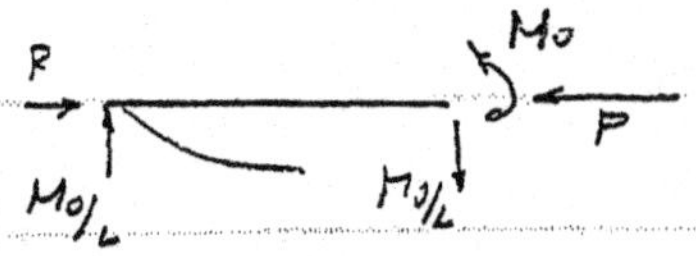

a) $M(x) = Pv(x) - M_o x/L$

$EIv''(x) = -M(x) \Rightarrow v''(x) + \lambda^2 v(x) = -M_o x/EIL$; $\lambda^2 = P/EI$.

BC: $v(0) = v(L) = 0$

$v(x) = A\sin\lambda x + B\cos\lambda x - M_o x/PL$.

$v(0) = 0 \Rightarrow B = 0$; $v(L) = 0 \Rightarrow A = M_o/P\sin\lambda L$.

$$v(x) = \frac{M_o}{P}\left(\frac{\sin\lambda x}{\sin\lambda L} - \frac{x}{L}\right);$$

b)

$$M_o = Pe\ ;\quad v(x) = e\left(\frac{\sin\lambda x}{\sin\lambda L} - \frac{x}{L}\right).$$

c).

$$\Delta_c = a\theta = -av'(L)\ ;\quad v'(x) = e\left(\frac{\lambda\cos\lambda x}{\sin\lambda L} - 1/L\right).$$

$$v'(L) = e/L\,(\lambda L\cot\lambda L - 1)$$

$$\Delta_c = \frac{ae}{L}(1 - \lambda L\cot\lambda L).$$

d).

$EIv''(x) = -M(x) = -Pex/L$; $v(0) = v(L) = 0$

$EIv'(x) = -Pex^2/2L + A$; $EIv(x) = -Pex^3/6L + Ax + B = 0$

$v(0) = 0 \Rightarrow B = 0$

$v(L) = 0 \Rightarrow A = PeL/6$

$\therefore EIv'(L) = -PeL/2 + PeL/6 = -PeL/3$

$$\Delta_c = -av'(L) = PeaL/3EI. \Rightarrow \Delta_c/L = \frac{PL^2}{EI}\left(\frac{ea}{3L^2}\right).$$

e).

$$\cot\lambda L \cong \frac{1}{\lambda L} - \frac{\lambda L}{3} - \frac{(\lambda L)^3}{45} \Rightarrow 1 - \lambda L\cot\lambda L \cong (\lambda L)^2/3 + (\lambda L)^4/45$$

From (c) $\Rightarrow$

$$\Delta_c/L = \frac{ae}{L^2}(1 - \lambda L\cot\lambda L) = \frac{ea\lambda^2}{3}(1 + \lambda^2L^2/15)$$

$\therefore$

$$\Delta_c/L = \frac{Pea}{3EI}(1 + PL^2/15EI \cdots)$$

11.31

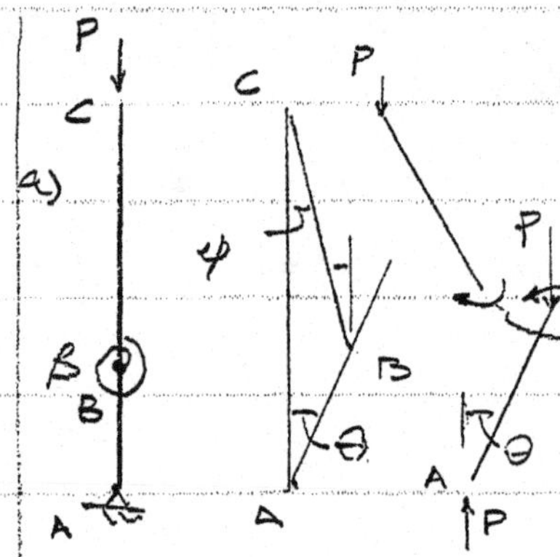

$\theta a = -\psi b \Rightarrow \psi = \theta a / b$.

$\circlearrowleft + (\sum M_A)_{AB} = 0 \Rightarrow \beta(\theta+\psi) - Pa\theta = 0$

$\circlearrowleft + (\sum M_C)_{BC} = 0 \Rightarrow \beta(\theta+\psi) - Pb\psi = 0$

$$\begin{aligned}(\beta - Pa)\theta + \beta\psi &= 0 \\ \beta\theta + (\beta - Pb)\psi &= 0\end{aligned} \Rightarrow D = |\ \ | = 0 \Rightarrow (\beta - Pa)(\beta - Pb) - \beta^2 = 0$$

$(ab)P^2 - [(a+b)\beta]P = 0$

$P[abP - (a+b)\beta] = 0;\ P \neq 0 \Rightarrow P_{cr} = \left(\frac{a+b}{ab}\right)\beta = L\beta/ab$

$b = L - a = L - cb \Rightarrow (1+c)b = L \Rightarrow b = L(1+c)^{-1}$

$$P_{cr} = \frac{\beta L}{ab} = \frac{\beta L}{cb^2} = \frac{\beta}{cL(1+c)^{-2}} = \frac{\beta}{L} c^{-1}(1+c)^2$$

b) $\frac{dP_{cr}(c)}{dc} = 0 \Rightarrow -c^{-2}(1+c)^2 + 2(1+c)c^{-1} = 0$

$2(1+c)c - (1+c)^2 = 0 \Rightarrow (1+c)(c-1) = 0 \Rightarrow c = 1 \Rightarrow a/b = 1$

$$(P_{cr})_{min} = 4\beta/L$$

11.32

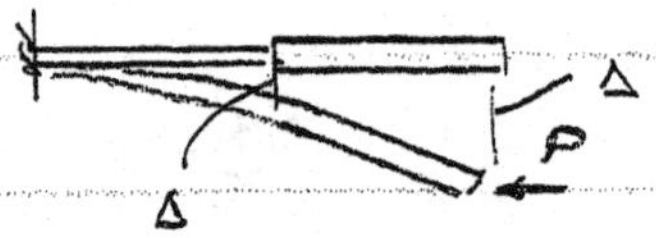

$M(x) = -P[\Delta - v(x)]$

$EIv''(x) = -M(x) \Rightarrow v''(x) + \lambda^2 v(x) = P\Delta/EI,\quad \lambda^2 = P/EI.$

B.C. $v(0) = 0,\ v'(0) = 0$

$v(x) = A\sin\lambda x + B\cos\lambda x + \Delta;\quad v(0) = 0 \Rightarrow B = -\Delta;\quad v'(x) = A\lambda\cos\lambda x - B\lambda\sin\lambda x$

$v'(0) = 0 \Rightarrow A = 0$

$v(x) = \Delta(1 - \cos\lambda x)$

$v(L) = \Delta(1 - \cos\lambda L);\quad \Delta = v(L) + a v'(L) = \Delta(1 - \cos\lambda L + a\lambda\sin\lambda L)$

$\therefore -\cos\lambda L + a\lambda\sin\lambda L = 0 \Rightarrow \underline{\tan\lambda L = 1/a\lambda}.$

b). $a \to 0 \Rightarrow \tan\lambda L \to \infty \Rightarrow \lambda \to \pi/2L \Rightarrow P_{cr} = \frac{\pi^2 EI}{4L^2}$

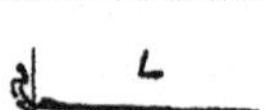

$a \to \infty \Rightarrow \tan\lambda L \to 0 \Rightarrow P_{cr} = 0$

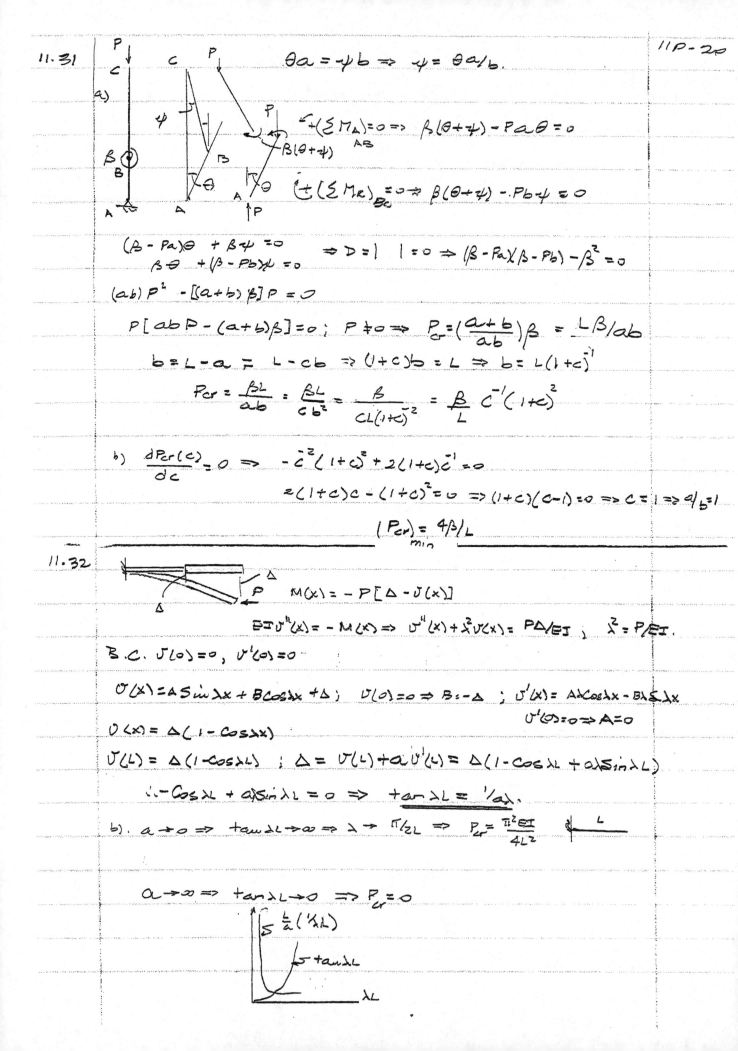

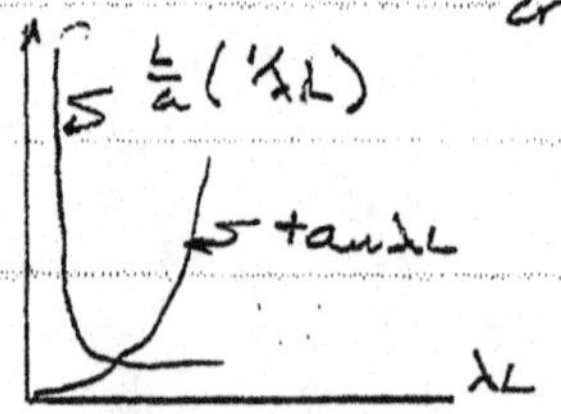

11.33 a)

$$EI\upsilon'''(x) = -\frac{F\lambda^2 \sin\lambda L/2}{P\,\sin\lambda L}\cos\lambda x \Rightarrow V(0) = -EI\upsilon'''(0) = \frac{EIF\lambda^2\sin\lambda L/2}{2P\sin\lambda L/2\,\cos\lambda L/2}$$

$$V(0) = \frac{F}{2}\,\frac{1}{\cos\lambda L/2}$$

$$\upsilon'(0) = \frac{F}{2P\cos\lambda L/2} - \frac{F}{2P} \;;\quad F/2 + P\upsilon'(0) = \frac{F}{2\cos\frac{\lambda L}{2}} = V(0)$$

b).

$$V(0) = \frac{F}{2} + P\upsilon'(0) = \frac{F}{2}\left[1 + \left(\frac{1}{\cos\lambda L/2} - 1\right)\right]$$

Let $\mu = \lambda L/2 = \pi/2\sqrt{P/P_E}$.

$$V(0) = F/2\,[1 + (1 + \mu^2/2 + \ldots) - 1] = \frac{F}{2}(1 + \mu^2/2)$$

$$V(0) = \frac{F}{2}\left[1 + \frac{\pi^2}{8}(P/P_E)\right]$$

11.34

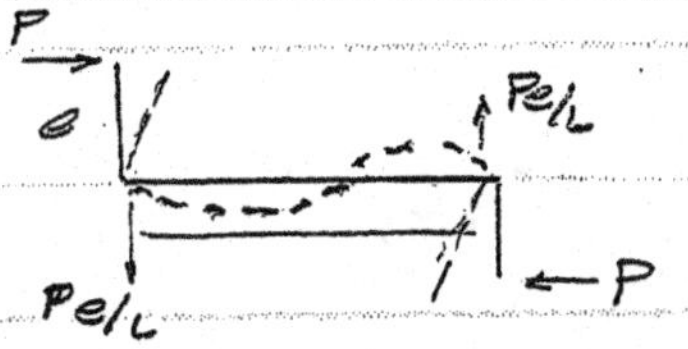

$$M(x) = Pe - 2Pex/L + P\upsilon(x)$$

$$EI\upsilon''(x) = -M(x) \Rightarrow EI\upsilon''(x) + P\upsilon(x) = -Pe + 2Pex/L$$

$$\upsilon''(x) + \lambda^2\upsilon(x) = -e\lambda^2 + 2\lambda^2 ex/L \;;\quad \lambda^2 = P/EI$$

B.C. $\upsilon(0) = \upsilon(L) = 0$

$$\upsilon(x) = A\sin\lambda x + B\cos\lambda x + e\,(2x/L - 1)$$

B.C. $\upsilon(0) = 0 \Rightarrow B = e$; $\upsilon(L) = 0 \Rightarrow A = -e\dfrac{(\cos\lambda L + 1)}{\sin\lambda L}$

$$\upsilon(x) = e\left\{\cos\lambda x - (\cos\lambda L + 1)\frac{\sin\lambda x}{\sin\lambda L} + (2x/L - 1)\right\}$$

11.35

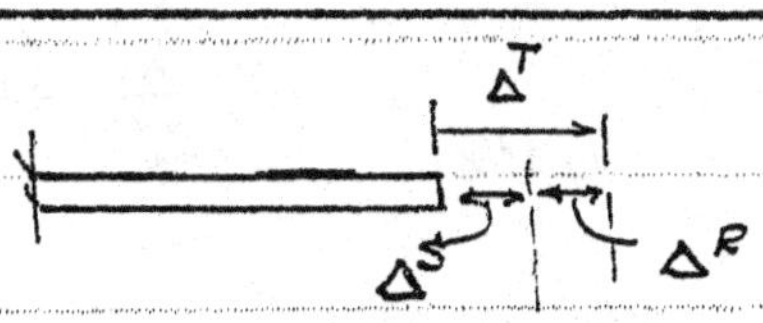

$$\Delta^{(S)} + \Delta^{(R)} = \Delta^{(T)} \;;\quad \Delta^{(T)} = L(\alpha\Delta T) \;;\quad \Delta^{(R)} = P_{cr}L/AE \;;\quad \Delta^{(S)} = P_{cr}/k$$

$$P_{cr}[L/AE + 1/k] = L(\alpha\Delta T)$$

$$\Delta T = \frac{P_{cr}}{\alpha L}\left(L/AE + 1/k\right) = \frac{P_{cr}}{\alpha\Delta E k L}(kL + AE)$$

$P_{cr} = \pi^2 EI/4L^2$; $A = \pi R^2$, $I = \pi R^4/4$

$$\text{a) } \Delta T_{cr} = \frac{\pi^2 E(\pi R^4/4)}{4\alpha(\pi R^2)EkL^3}[kL + \pi R^2 E] = \frac{\pi^2 R^2}{16\alpha kL^3}(kL + \pi R^2 E)$$

b) $e = \dfrac{P_{cr}}{k} = \dfrac{\pi^3 ER^4}{16L^2k}$; c) $T_{cr}\Big|_{k\to\infty} = \dfrac{\pi^2 R^2}{16\alpha L^2}$

11.37

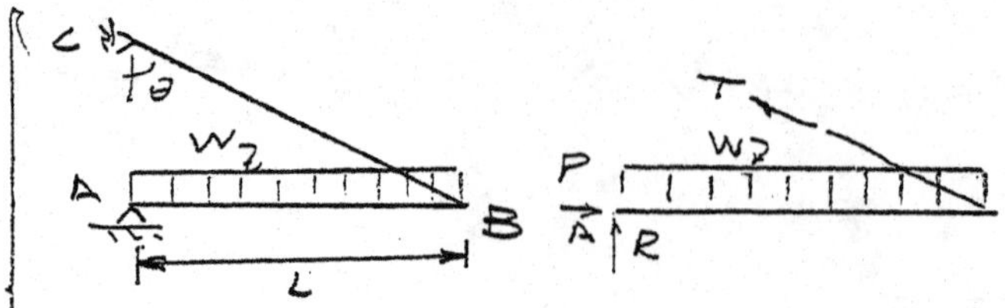

$\sum M_A = (T\cos\theta)L - wL^2/2 = 0$

$T = wL/2\cos\theta;\ R = wL/2$

$P = T\sin\theta = wL\tan\theta/2.$

$M(x) = \frac{w}{2}(Lx - x^2) + P\,v(x).$

$EI\,v''(x) = -M(x) \Rightarrow EI\,v''(x) + P\,v(x) = -\frac{w}{2}(Lx - x^2)$

$v''(x) + \lambda^2 v(x) = -\frac{w}{EI}(Lx - x^2),\quad \lambda^2 = P/EI.$

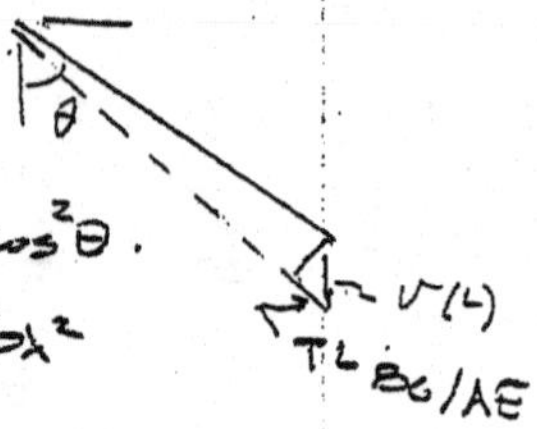

B.C. $v(0) = 0,\ v(L) = \left(\frac{TL/\sin\theta}{AE\cos\theta}\right) = \frac{TL}{AE}\frac{1}{\sin\theta\cos\theta} = \frac{wL^2}{2AE\sin\theta\cos^2\theta}.$

$v(x) = A\sin\lambda x + B\cos\lambda x + wx^2/2P - wLx/2P - w/P\lambda^2$

$v(0) = 0 \Rightarrow B = w/P\lambda^2$

$v(L) = \frac{wL^2}{2AE\sin\theta\cos^2\theta} \Rightarrow A\sin\lambda L + B\cos\lambda L - wL^2/2P - w/P\lambda^2 = v(L)$

$$A = \frac{1}{\sin\lambda L}\left\{v(L) - \frac{w}{P\lambda^2}(\cos\lambda L - 1) + wL^2/2P\right\}$$

$$= \frac{wL^4}{\sin\lambda L}\left\{\frac{1}{2AEL^2\sin\theta\cos^2\theta} - \frac{\cos\lambda L - 1}{EI(\lambda L)^4} + \frac{1}{2EI(\lambda L)^2}\right\}$$

$$A = \frac{wL^4}{EI\sin\lambda L}\left[\left(\frac{I}{A}\right)\cdot\frac{1}{2L^2\sin\theta\cos^2\theta} - \frac{\cos\lambda L - 1}{(\lambda L)^4} + \frac{1}{(\lambda L)^2}\right]$$

$$v(x) = \frac{wL^4}{EI}\left\{\frac{1}{\sin\lambda L}\left[\left(\frac{I}{AL^2}\right)\cdot\frac{1}{2\sin\theta\cos^2\theta} - \frac{\cos\lambda L - 1}{(\lambda L)^4} + \frac{1}{(\lambda L)^2}\right]\sin\lambda x + \frac{(\cos\lambda x - 1)}{(\lambda L)^4} + \frac{1}{2(\lambda L)^2}\left(\frac{x^2}{L^2} - \frac{x}{L}\right)\right\}$$

$$v(L/2) = \frac{wL^4}{EI}\left\{\frac{1}{\sin\lambda L}\left[\left(\frac{I}{AL^2}\right)\frac{1}{2\sin\theta\cos^2\theta} + \frac{(1 - \cos\lambda L) - (\lambda L)^2}{(\lambda L)^4}\right]\sin\frac{\lambda L}{2} + \frac{\cos\lambda L/2 - 1}{(\lambda L)^4} - \frac{1}{8(\lambda L)^2}\right\}$$

where

$$\lambda = [P/EI]^{1/2} = \left(\frac{wL\tan\theta}{2EI}\right)^{1/2}$$

11.38

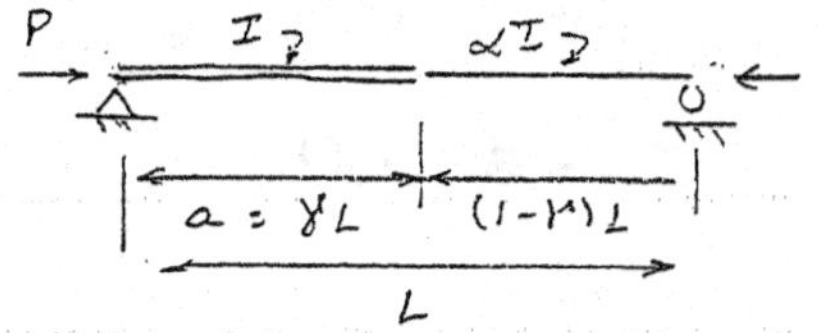

$0 \le x \le a$ $\qquad\qquad$ $a \le x \le L$

$EI\, v_1''(x) + P v_1(x) = 0 \qquad \alpha EI\, v_2''(x) + P v(x) = 0 \quad ; \quad$ B.C. $v_1(0) = v_2(L) = 0;\ v_1(a) = v_2(a)$

$v_1'(a) = v_2'(a)$.

$v_1''(x) + \lambda^2 v_1(x) = 0 \qquad v_2''(x) + \dfrac{\lambda^2}{\alpha} v(x) = 0 \quad , \lambda^2 = P/EI$

$v_2''(x) = \dfrac{\lambda^2}{c^2} v(x) = 0 \quad ; \quad c^2 = \alpha$

Solution: $v_1(x) = A_1 \sin \lambda x + B_1 \cos \lambda x$; $\qquad v_2(x) = A_2 \sin \lambda x/c + B_2 \cos \lambda x/c$.

$v_1(0) = 0 \Rightarrow B_1 = 0 \qquad\qquad v_2(L) = 0 \Rightarrow B_2 = -A_2 \tan \lambda L/c$

$v_1(a) = v_2(a) \Rightarrow \quad A_1 \sin \lambda a = A_2 (\sin \lambda a/c - \tan \lambda L/c \cos \lambda a/c)$

$v_1'(a) = v_2'(a) \Rightarrow \lambda A_1 \cos \lambda a = \dfrac{\lambda}{c} A_2 [\cos \lambda a/c + \tan \lambda L/c \sin \lambda a/c]$

$\therefore \dfrac{\lambda \cos \lambda a}{\sin \lambda a} [\sin \lambda a/c - \tan \lambda L/c \cos \lambda a/c] = \dfrac{\lambda}{c} [\cos \dfrac{\lambda a}{c} + \tan \dfrac{\lambda L}{c} \sin \dfrac{\lambda a}{c}]$

$\therefore c \cos \lambda a (\sin \lambda a/c - \tan \lambda L/c \cos \lambda a/c) = \sin \lambda a (\cos \dfrac{\lambda a}{c} + \tan \lambda L/c \sin \lambda a/c]$

In terms of $\gamma = a/L$,

$$c \cdot \cos(\gamma \lambda L) [\sin(\frac{\gamma \lambda L}{c}) - \tan(\frac{\lambda L}{c}) \cos \frac{\lambda \gamma L}{c}] = \sin(\lambda \gamma L) [\cos \frac{\lambda \gamma L}{c} + \tan \frac{\lambda L}{c} \sin \frac{\lambda \gamma L}{c}] \qquad (I)$$

$$c \left\{ \frac{\sin(\gamma \lambda L/c) - \tan(\lambda L/c) \cos(\lambda \gamma L/c)}{\cos(\gamma \lambda L/c) + \tan(\lambda L/c) \sin(\lambda \gamma L/c)} \right\} = \tan(\gamma \lambda L)$$

$$c \left\{ \frac{\tan(\gamma \lambda L/c) - \tan(\lambda L/c)}{1 + \tan(\lambda L/c) \tan(\gamma \lambda L/c)} \right\} = \tan(\gamma \lambda L)$$

Note: 1) $\gamma = 1 \Rightarrow \tan(\lambda L) = 0 \Rightarrow \sin \lambda L = 0 \Rightarrow \lambda = n\pi/L \Rightarrow P_{cr} = \pi^2 EI/L^2$

2) $\gamma = 0 \Rightarrow \tan \lambda L/c = 0 \Rightarrow \sin \lambda L/c = 0 \Rightarrow \lambda = c\pi/L \Rightarrow P_{cr} = \dfrac{\pi^2 c^2 EI}{L^2} = \dfrac{\pi^2 \alpha EI}{L^2}$

11.39

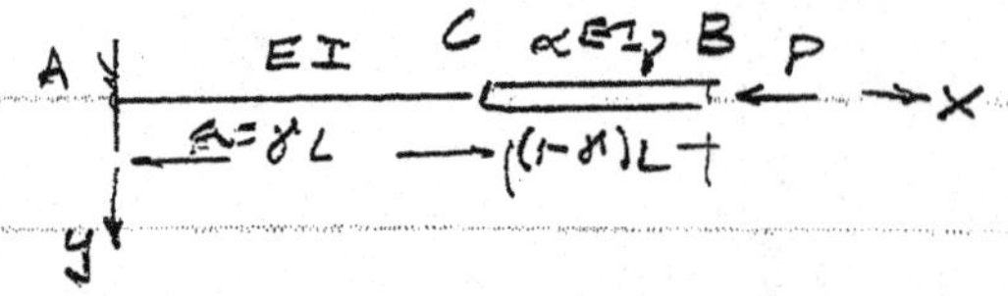

$M(x) = -P[\Delta_B - v(x)]$

$0 \le x \le a$ $\qquad\qquad$ $a \le x \le L$

$EI\,v_1''(x) + P v_1(x) = P\Delta_B$ $\qquad$ $\alpha EI\,v_2''(x) + P v_2(x) = P\Delta_B$ $\qquad$ $\lambda^2 = P/EI$

$v_1''(x) + \lambda^2 v_1(x) = \lambda^2 \Delta_B$ $\qquad$ $v_2''(x) + \dfrac{\lambda^2}{c^2} v_2(x) = \dfrac{\lambda^2}{c^2}\Delta_B$, $\quad \alpha = c^2$

B.C. $v_1(0) = 0$, $v_1'(0) = 0$ $\qquad$ $v_2(a) = v_1(a)$; $v_2'(a) = v_1'(a)$

$v_1(x) = A_1 \sin\lambda x + B_1 \cos\lambda x + \Delta_B$ $\qquad$ $v_2(x) = A_2 \sin\dfrac{\lambda x}{c} + B_2 \cos\dfrac{\lambda x}{c} + \Delta_B$

$v_1(0) = 0 \Rightarrow B_1 = -\Delta_B$

$v_1'(0) = 0 \Rightarrow A_1 = 0 \Rightarrow v_1(x) = \Delta_B(1 - \cos\lambda x)$, $\quad v_1'(x) = \lambda\Delta_B \sin\lambda x$.

$v_1(a) = v_2(a) \Rightarrow \Delta_B(1 - \cos\lambda a) = A_2 \sin\dfrac{\lambda a}{c} + B_2 \cos\dfrac{\lambda a}{c} + \Delta_B \qquad (i)$

$v_1'(a) = v_2'(a) \Rightarrow \lambda\Delta_B \sin\lambda a = A_2\lambda \cos\dfrac{\lambda a}{c} - B_2 \lambda \sin\dfrac{\lambda a}{c} \qquad (ii)$

$v_2(L) = \Delta_B \Rightarrow \quad A_2 \sin\dfrac{\lambda L}{c} + B_2 \cos\dfrac{\lambda L}{c} + \Delta_B = \Delta_B$

$\therefore \quad B_2 = -A_2 \tan\dfrac{\lambda L}{c}$.

$(i) \Rightarrow \quad \Delta_B(1 - \cos\lambda a) = A_2\left(\sin\dfrac{\lambda a}{c} - \cos\dfrac{\lambda a}{c}\tan\dfrac{\lambda L}{c}\right) + \Delta_B$

$$\Delta_B = \frac{A_2}{\cos\lambda a}\left[\cos\frac{\lambda a}{c}\tan\frac{\lambda L}{c} - \sin\frac{\lambda a}{c}\right] = A_2\left(\frac{\cos\lambda a/c\,\sin\frac{\lambda L}{c}}{\cos\lambda a\,\cos\lambda L/c} - \frac{\sin\lambda a/c}{\cos\lambda a}\right)$$

$$(ii) \Rightarrow \frac{\sin\lambda a}{\cos\lambda a}\left(\frac{\cos\frac{\lambda a}{c}\sin\frac{\lambda L}{c}}{\cos\lambda L/c} - \sin\frac{\lambda a}{c}\right) = \left(\cos\frac{\lambda a}{c} + \tan\frac{\lambda L}{c}\sin\frac{\lambda a}{c}\right)A_2$$

$$\Rightarrow \tan\lambda a = \frac{\cos\frac{\lambda a}{c} + \dfrac{\sin\lambda L/c\,\sin\lambda a/c}{\cos\lambda L/c}}{\dfrac{\sin\frac{\lambda L}{c}\cos\lambda a}{\cos\lambda L/c} - \sin\frac{\lambda a}{c}} = \frac{\cos\lambda a/c\,\cos\lambda L/c + \sin\lambda L/c\,\sin\lambda a/c}{\sin\frac{\lambda L}{c}\cos\lambda a - \sin\lambda a/c\,\cos\lambda L/c}$$

$$\tan\lambda a = \frac{\cos[\lambda(L-a)/c]}{\sin[\lambda(L-a)/c]} = \cot[\lambda(L-a)/c]$$

$$\tan\gamma\lambda L = \cot\left[\frac{(1-\gamma)\lambda L}{c}\right], \quad c = \sqrt{\alpha}$$

$\gamma \to 1 \Rightarrow \tan\lambda L = \cot(0) \to \infty \Rightarrow \cos\lambda L = 0 \Rightarrow \lambda = n\pi/2L$, $\; n = 1, 3, \ldots$

$\gamma \to 0 \Rightarrow \cot(\lambda L/c) = 0 \Rightarrow \lambda = nc\pi/2L \Rightarrow P_{cr} = \pi^2 c^2 EI/(2L)^2 = \pi^2 \alpha EI/(2L)^2$

11.40

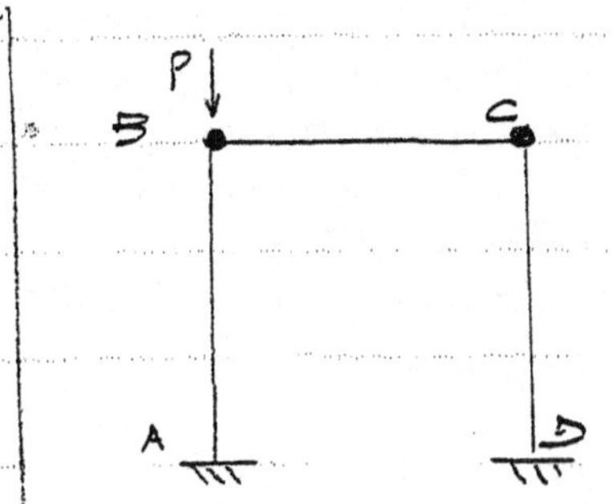

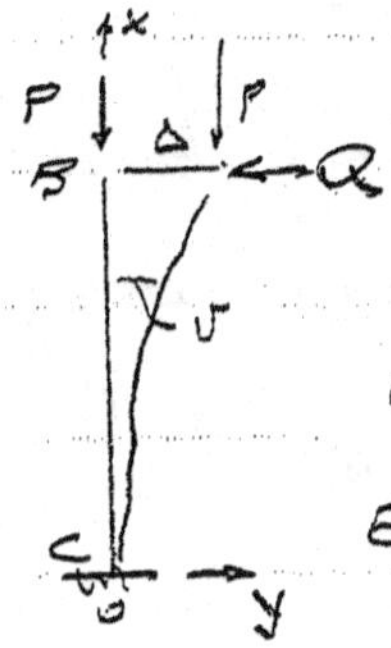

$$M(x) = -P[\Delta - v(x)] + Q(L-x)$$

$$EI\,v''(x) = -M(x) \Rightarrow$$

$$EI\,v''(x) + P\,v(x) = P\Delta - Q(L-x)$$

$$v''(x) + \lambda^2 v(x) = \lambda^2\Delta - Q(L-x)/EI \quad , \quad \lambda^2 = P/EI$$

B.C. $v(0) = v'(0) = 0$

$$v(x) = A\sin\lambda x + B\cos\lambda x + \Delta - Q(L-x)/P$$

$v(0) = 0 \Rightarrow B = -\Delta + QL/P$

$v'(x) = A\lambda\cos\lambda x - B\lambda\sin\lambda x - Q/P$; $v'(0) = 0 \Rightarrow A = Q/P\lambda$.

$$v(x) = -\frac{Q}{P}\left[\frac{1}{\lambda}\sin\lambda x - L\cos\lambda x + (L-x)\right] + \Delta(1-\cos\lambda x)$$

$$v(L) = -Q/P\left[1/\lambda\,\sin\lambda L - L\cos\lambda L\right] + \Delta(1-\cos\lambda L) \qquad (I)$$

$v(L) = \Delta$, $Q \neq 0$:

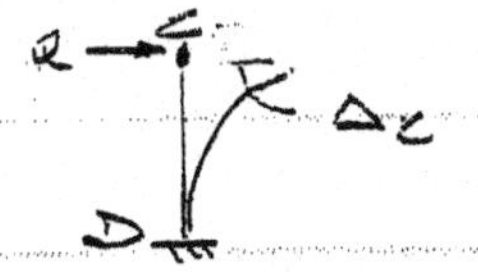

$\Delta_C = \Delta_B \equiv \Delta \Rightarrow \Delta_C = QL^3/3EI \equiv \Delta \Rightarrow Q = 3EI\Delta/L^3$.

$$Q/P = 3\Delta/\lambda^2L^3$$

$$\therefore (I) \Rightarrow \Delta\left[-\frac{3}{\lambda^2L^3}\left(1/\lambda\,\sin\lambda L - L\cos\lambda L\right)\right] + \Delta(1-\cos\lambda L) = \Delta$$

$$-\frac{3}{\lambda^3L^3}(\sin\lambda L - \lambda L\cos\lambda L) - \cos\lambda L = 0$$

$$\sin\lambda L - \lambda L\cos\lambda L + \frac{\lambda^3L^3}{3}\cos\lambda L = 0$$

$$\underline{\tan\lambda L = \lambda L - \lambda^3L^3/3}$$

11.41

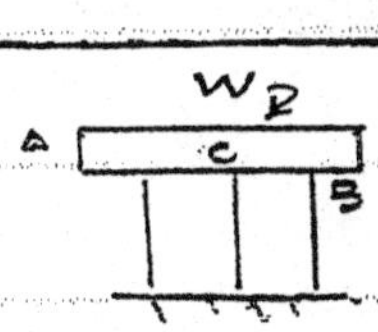

Note:
Due to symmetry, ACB displaces vertically ⇒ buckling occurs simultaneously in the 3 rods and each carries the same load, P.

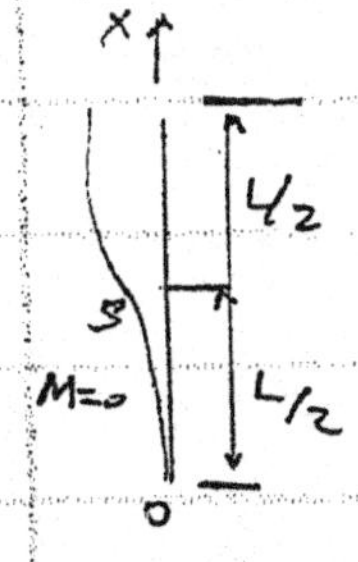

Solution A:
$K = 0$ at $x = L/2 \Rightarrow M(L/2) = 0$. $\Rightarrow P_{cr} = \dfrac{\pi^2EI}{[4(L/2)^2]} = \pi^2EI/L^2$

$$W_{cr} = 3P_{cr} = 3\pi^2EI/L^2$$

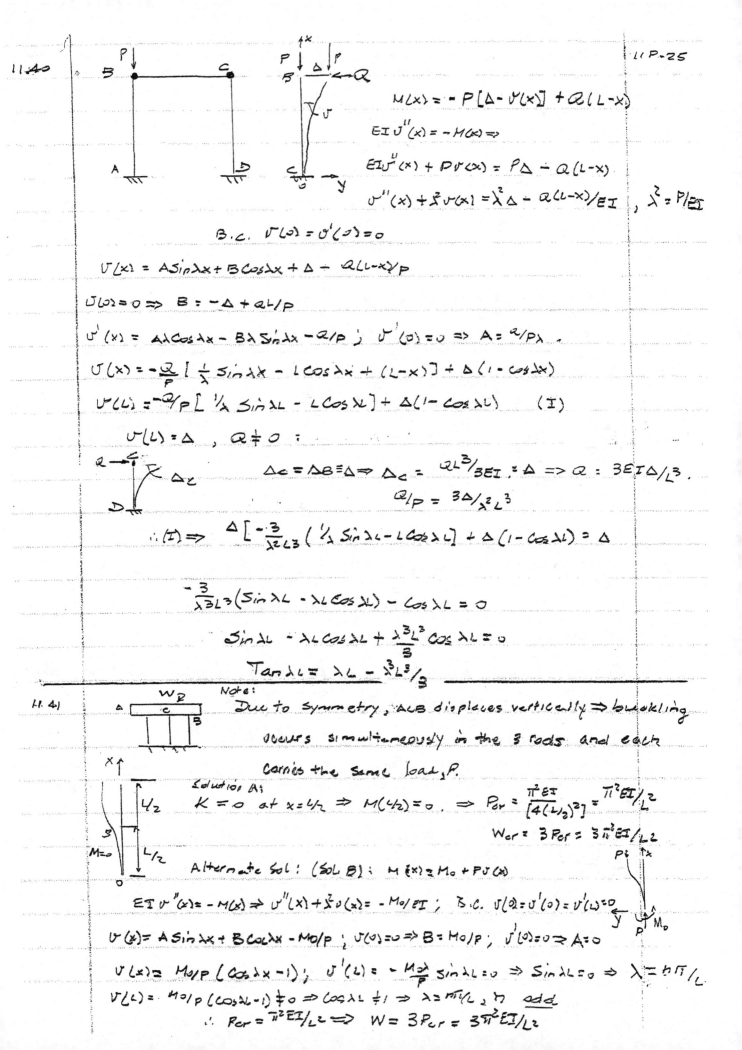

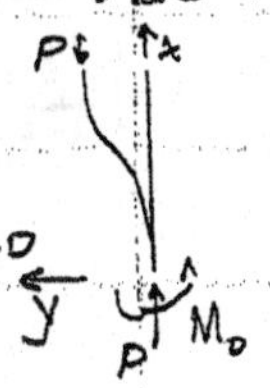

Alternate Sol: (Sol. B): $M(x) = M_0 + Pv(x)$

$EI\,v''(x) = -M(x) \Rightarrow v''(x) + \lambda^2 v(x) = -M_0/EI$; B.C. $v(0) = v'(0) = v'(L) = 0$

$v(x) = A\sin\lambda x + B\cos\lambda x - M_0/P$; $v(0) = 0 \Rightarrow B = M_0/P$; $v'(0) = 0 \Rightarrow A = 0$

$v(x) = M_0/P(\cos\lambda x - 1)$; $v'(L) = -\dfrac{M_0\lambda}{P}\sin\lambda L = 0 \Rightarrow \sin\lambda L = 0 \Rightarrow \lambda = n\pi/L$

$v(L) = M_0/P(\cos\lambda L - 1) \neq 0 \Rightarrow \cos\lambda L \neq 1 \Rightarrow \lambda = n\pi/L$, n odd

$$\therefore P_{cr} = \pi^2EI/L^2 \Rightarrow W = 3P_{cr} = 3\pi^2EI/L^2$$

11.42

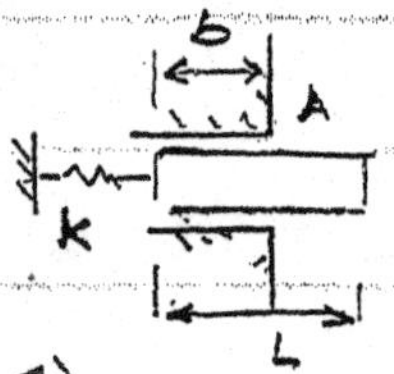

Note: Rod cannot buckle within the slot.

a). External portion, L-b, is fixed at A

a)

$$P_{cr} = \frac{\pi^2 EI}{4(L-b)^2} \Rightarrow (L-b)^2 = \frac{\pi^2 EI}{4P_{cr}}$$

$$b = L - \frac{\pi}{2}\left(EI/P_{cr}\right)^{1/2}$$

$$P_{cr} = bk$$

$$\therefore k = \frac{P_{cr}}{L-(\pi/2)\left(\frac{EI}{P_{cr}}\right)^{1/2}} = \frac{P_{cr}}{L-\frac{\pi R^2}{4}\left(E\pi/P_{cr}\right)^{1/2}}$$

b).

$P_{cr} = 50\times10^3$ N, $L = 2$m, $E = 200$GPa, $R = 2$cm.

$$k = \frac{50\times10^3}{2-\frac{\pi}{4}(2\times10^{-2})^2\left(\frac{200\times10^9}{50\times10^3}\pi\right)^{1/2}} = 56,412 \text{ N/m}$$

11.43

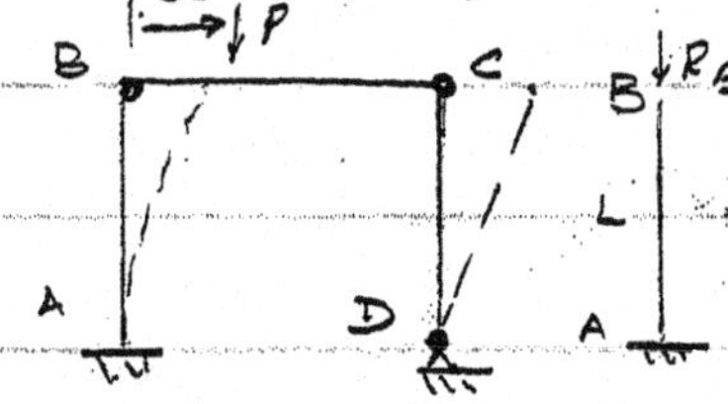

Rod AB:

$$(R_B)_{cr} = \frac{\pi^2 EI}{(2L)^2}$$

Note: Pt. B can move horizontally with no restraint if $|\Delta_B| \ll l$.

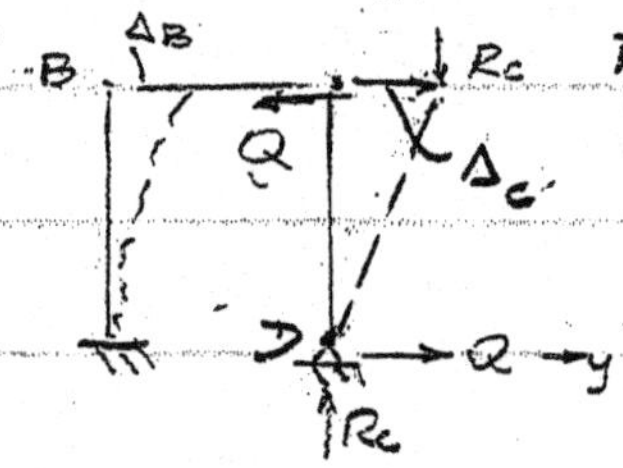

Rod CD:

Since $\Delta_c = \Delta_B$, rod CD is restrained by Q, i.e. by Rod AB when R_c acts.

Note: CD is a two-force Body ⇒ Forces act axially ⇒ $M(x) = 0$, $0 < x < L$.

Equilibrum: $\sum M_D = 0 \Rightarrow Q = R_c\Delta_c/L$

But $\Delta_c = \Delta_B$, Δ_B due to $Q \Rightarrow \Delta_c = \Delta_B = QL^3/3EI \Rightarrow R_c = \frac{QL}{\Delta_c} = 3EI/L^2$

$$\therefore (R_c)_{cr} = QL/\Delta_c = 3EI/L^2 \quad (\Delta_c \neq 0)$$

From BC: $(R_c)_{cr} = P_{cr}a/L$, $(R_B)_{cr} = P_{cr}(L-a)/L$

$$(P_{cr})_{AB} = \left(\frac{L}{L-a}\right)(R_B)_{cr} = \frac{\pi^2 EI}{4L(L-a)} = \frac{\pi^2 EI}{4L^2(1-a/L)};\quad (P_{cr})_{CD} = \frac{(R_c)_{cr}L}{a} = \frac{3EI}{(a/L)L^2}$$

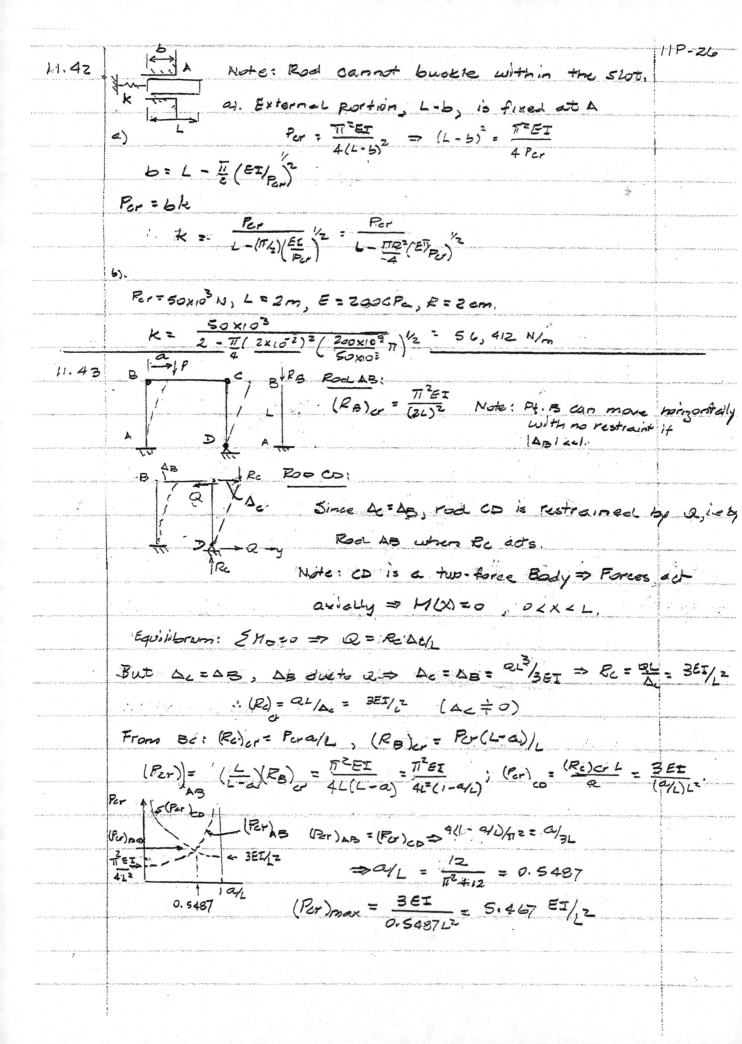

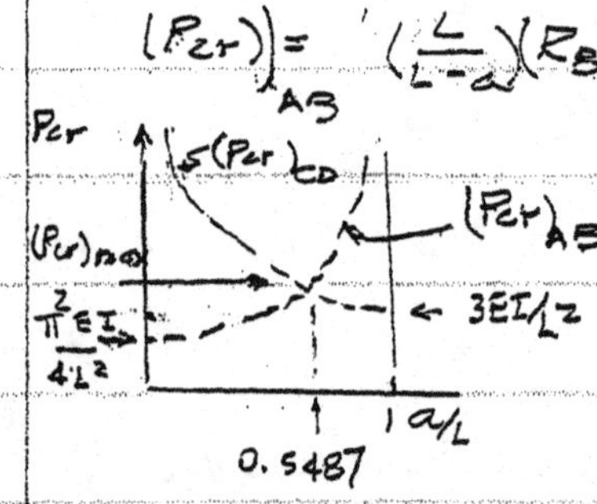

$$(P_{cr})_{AB} = (P_{cr})_{CD} \Rightarrow 4(1-a/L)/\pi^2 = a/3L$$

$$\Rightarrow a/L = \frac{12}{\pi^2+12} = 0.5487$$

$$(P_{cr})_{max} = \frac{3EI}{0.5487L^2} = 5.467\ EI/L^2$$

11.44

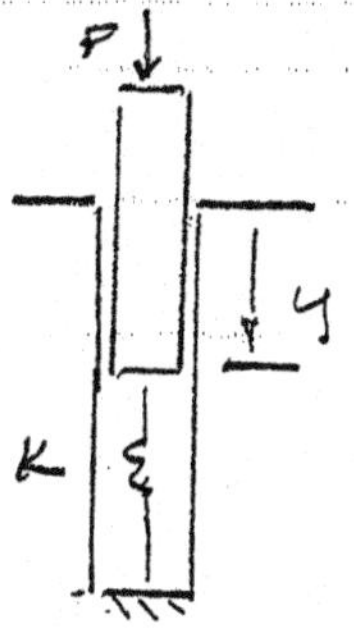

$$P_{cr}(y) = \pi^2 EI / 4(L-y)^2$$

CRITERION: $F_{sp} < P_{cr} \Rightarrow ky < \dfrac{\pi^2 EI}{4(L-y)^2}$

$$\therefore k < \frac{\pi^2 EI}{4(L-y)^2 y} \qquad \text{Let } \Gamma(y) = A y^{-1}(L-y)^{-2};\ A = \frac{\pi^2 EI}{4}$$

$$\Gamma(y)|_{min} \Rightarrow \frac{d\Gamma(y)}{dy} = 0 \Rightarrow -Ay^{-2}(L-y)^{-2} + 2Ay^{-1}(L-y)^{-3} = 0$$

$$\frac{2}{y(L-y)^3} - \frac{1}{y^2(L-y)^2} = 0 \Rightarrow 2y - (L-y) = 0 \Rightarrow y = L/3$$

$$k < \Gamma(y)|_{min} \Rightarrow k < \frac{A}{L/3}\left(\frac{1}{2L/3}\right)^2 = \frac{27A}{4L^3} = \frac{27\pi^2 EI}{16L^3}.$$

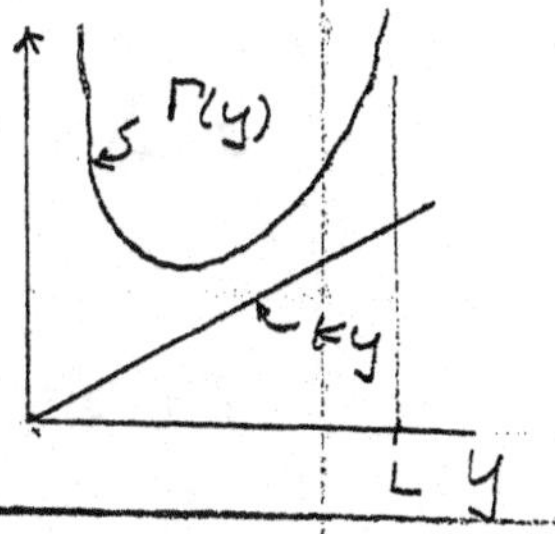

11.45

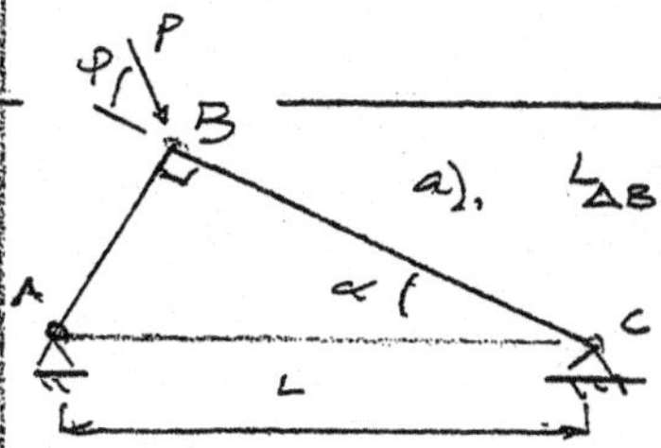

a). $L_{AB} = L\sin\alpha,\ L_{BC} = L\cos\alpha$

$$F_{AB} = P\sin\varphi,\quad F_{BC} = P\cos\varphi;\quad (F_{cr})_{AB} = \frac{\pi^2 EI}{L^2\sin^2\alpha},\quad (F_{cr})_{BC} = \frac{\pi^2 EI}{L^2\cos^2\alpha}$$

Note: AB & BC can buckle independently since AB⊥BC.

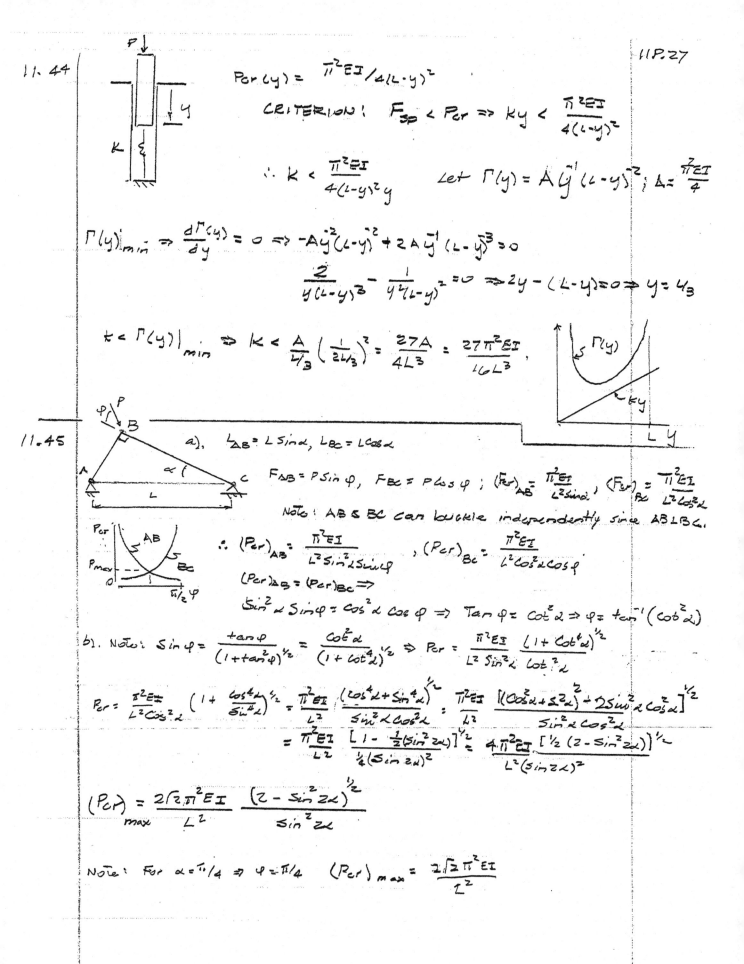

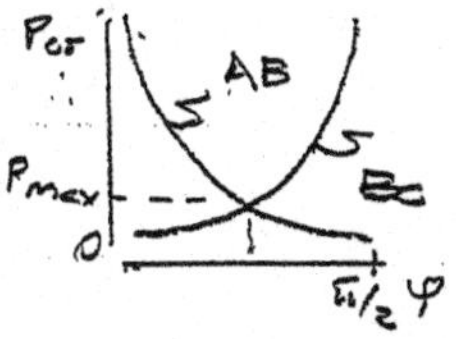

$$\therefore (P_{cr})_{AB} = \frac{\pi^2 EI}{L^2\sin^2\alpha\sin\varphi},\quad (P_{cr})_{BC} = \frac{\pi^2 EI}{L^2\cos^2\alpha\cos\varphi}$$

$(P_{cr})_{AB} = (P_{cr})_{BC} \Rightarrow$

$$\sin^2\alpha\sin\varphi = \cos^2\alpha\cos\varphi \Rightarrow \tan\varphi = \cot^2\alpha \Rightarrow \varphi = \tan^{-1}(\cot^2\alpha)$$

b). Note: $\sin\varphi = \dfrac{\tan\varphi}{(1+\tan^2\varphi)^{1/2}} = \dfrac{\cot^2\alpha}{(1+\cot^4\alpha)^{1/2}} \Rightarrow P_{cr} = \dfrac{\pi^2 EI}{L^2\sin^2\alpha}\dfrac{(1+\cot^4\alpha)^{1/2}}{\cot^2\alpha}$

$$P_{cr} = \frac{\pi^2 EI}{L^2\cos^2\alpha}\left(1+\frac{\cos^4\alpha}{\sin^4\alpha}\right)^{1/2} = \frac{\pi^2 EI}{L^2}\frac{(\cos^4\alpha+\sin^4\alpha)^{1/2}}{\sin^2\alpha\cos^2\alpha} = \frac{\pi^2 EI}{L^2}\frac{[(\cos^2\alpha+\sin^2\alpha)^2 - 2\sin^2\alpha\cos^2\alpha]^{1/2}}{\sin^2\alpha\cos^2\alpha}$$

$$= \frac{\pi^2 EI}{L^2}\frac{[1-\frac{1}{2}(\sin^2 2\alpha)]^{1/2}}{\frac{1}{4}(\sin 2\alpha)^2} = \frac{4\pi^2 EI}{L^2(\sin 2\alpha)^2}[\tfrac{1}{2}(2-\sin^2 2\alpha)]^{1/2}$$

$$(P_{cr})_{max} = \frac{2\sqrt{2}\pi^2 EI}{L^2}\frac{(2-\sin^2 2\alpha)^{1/2}}{\sin^2 2\alpha}$$

Note: For $\alpha = \pi/4 \Rightarrow \varphi = \pi/4$ $\quad (P_{cr})_{max} = \dfrac{2\sqrt{2}\pi^2 EI}{L^2}$

11.46

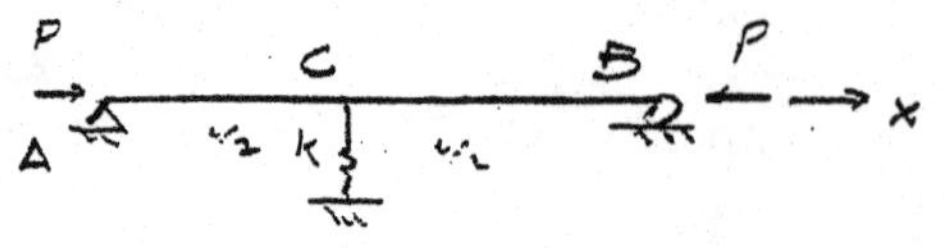

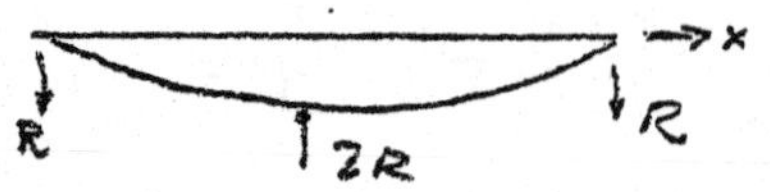

$$M(x) = -Rx + Pv(x)$$

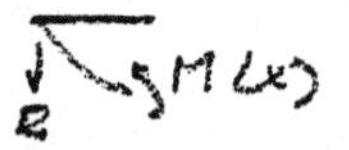

$$EIv''(x) = -M(x) \Rightarrow EIv''(x) + Pv(x) = Rx$$

Symmetric Mode:

$$v''(x) + \lambda^2 v(x) = Rx/EI \;;\; \lambda^2 = P/EI \,, \quad B.C.\; v(0)=0,\; v'(L/2)=0$$

$$v(x) = A\sin\lambda x + B\cos\lambda x + Rx/P \qquad v(L/2) = 2R/k$$

$$v'(x) = A\lambda\cos\lambda x - B\lambda\sin\lambda x + R/P.$$

$$v(0)=0 \Rightarrow B=0$$

$$v'(L/2)=0 \Rightarrow A\lambda\cos\lambda L/2 + R/P = 0 \qquad (a)$$

$$v(L/2) = 2R/k \Rightarrow A\sin\lambda L/2 + RL/2P = 2R/k \qquad (b)$$

$$(a) \to (b) \Rightarrow A\sin\lambda L/2 - \frac{A\lambda L}{2}\cos\lambda L/2 = -\frac{2A\lambda P}{k}\cos\lambda L/2$$

$$A\left[\sin\lambda L/2 + \frac{\lambda L}{2}\left(\frac{4P}{kL} - 1\right)\cos\lambda L/2\right] = 0$$

$$A\left[\tan\lambda L/2 + \lambda L/2\,(4P/kL - 1)\right] = 0$$

$$P = \pi^2 EI/(.7L/2)^2 = 4\pi^2 EI/(.7L)^2$$

$$k \to \infty \Rightarrow \tan\lambda L/2 = \lambda L/2$$

$$k \to 0 \Rightarrow \tan\lambda L/2 = -\infty \Rightarrow \lambda L/2 = n\pi/2,\; n = \text{odd};\; \lambda = \frac{n\pi}{L} \qquad P_{cr} = n^2\pi^2 EI/L^2,\; n\ \text{odd}$$

$$\text{For } 0<k<\infty \Rightarrow \tan\lambda L/2 - \lambda L/2 + 2P\lambda/k = 0 \qquad 2\frac{P\lambda}{k} = \frac{2P}{EI}\cdot\lambda(EI/k)$$

$$\text{or } \tan\lambda L/2 - \lambda L/2 + \lambda^3(2EI/k) = 0$$

$$\text{Let } \alpha \equiv \frac{kL^3}{16EI}, \quad \zeta = \lambda L/2$$

$$\tan\zeta - \zeta + \zeta^3/\alpha = 0 \quad \text{solve for } \zeta = \zeta(\alpha). \quad (I)$$

11.46 (cont'd)

$v(x) = A(\sin\lambda x - \cos\lambda L/2 \cdot (\lambda x))$

$= A[\sin 2\xi x/L - \cos\xi\,(2\xi x/L)]$, $\xi = \lambda L/2$.

For $v(L/2) = 0 \Rightarrow \sin\xi - \xi\cos\xi = 0$.

$\tan\xi = \xi \Rightarrow$ From (I) $\rightarrow$ Require $\alpha = 0 \Rightarrow k \to \infty$.

$v(L/2) = 0$ for symmetric buckling only if $k \to \infty$.

For Anti-symmetric buckling: $P_{cr} = 4\pi^2 EI/L^2 \Rightarrow (\lambda L)^2 = 4\pi^2$

$\therefore \xi^2 = \pi^2 \Rightarrow \xi = \pi$.

Setting $\xi = \pi$ in (I):

$-\xi + \dfrac{\xi^3}{\alpha} = 0 \Rightarrow \xi(\xi^2 - \alpha) = 0 \Rightarrow \alpha = \xi^2 = \pi^2$

$\therefore k = \dfrac{16\pi^2 EI}{L^3}$.

For $\xi = \pi$ $(P_{cr})_{symm} = (P_{cr})_{anti\text{-}sym}$.

For $\xi \le \pi$ $(P_{cr})_{symm} \le (P_{cr})_{anti\text{-}sym}$.

$\therefore$ $k > 16\pi^2 EI/L^2$ for buckling with $v_L = 0$.

1.54

From 11.46: $\tan\zeta - \zeta + \frac{\zeta^3}{\alpha} = 0$, $\alpha = kL^3/16EI$

$\zeta = \lambda L/2$

$\therefore \tan\zeta = \zeta - \zeta^3/\alpha$.

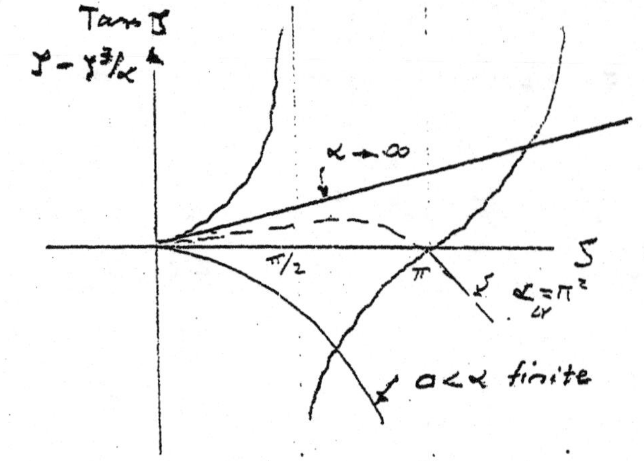

$$\alpha = \frac{\zeta^3}{\zeta - \tan\zeta}$$

Note: $\zeta = \pi/2 \Rightarrow \alpha \to 0 \Rightarrow P/P_E = 1$

Note: $P/P_E = \frac{\lambda^2 EI}{\pi^2 EI/L^2} = \lambda^2 L^2/\pi^2 = 4\zeta^2/\pi^2 \Rightarrow \zeta = \frac{\pi}{2}\sqrt{P/P_E} \equiv \frac{\pi}{2}\sqrt{R}$

Note: $\zeta = \pi \Rightarrow \alpha_{cr} = \pi^2$

$\alpha = \alpha[\zeta(R)]$

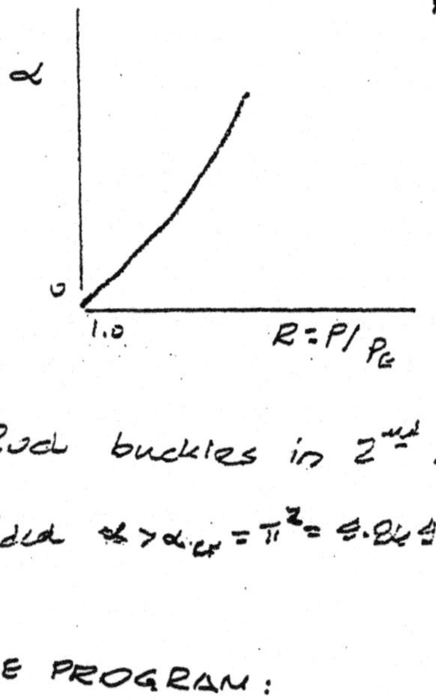

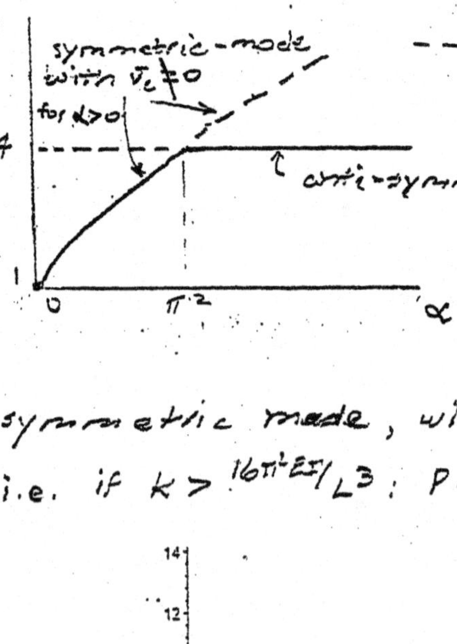

Rod buckles in 2nd anti-symmetric mode, with $v(L/2) = 0$, provided $\alpha > \alpha_{cr} = \pi^2 = 9.8696$, i.e. if $k > 16\pi^2 EI/L^3$: $P = 4P_E$.

MAPLE PROGRAM:

```
>
> zeta:= (Pi*sqrt(r))/2;
```

$$\zeta := \frac{1}{2}\pi\sqrt{r}$$

```
> α := zeta**3/(zeta-tan(zeta));
```

$$:= \frac{1}{8}\frac{\pi^3 r^{(3/2)}}{\frac{1}{2}\pi\sqrt{r} - \tan\left(\frac{1}{2}\pi\sqrt{r}\right)}$$

```
> plot ( α ,r=1.0..5.0,labels=['R',' α ']);
>
```

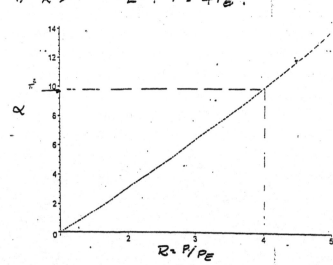

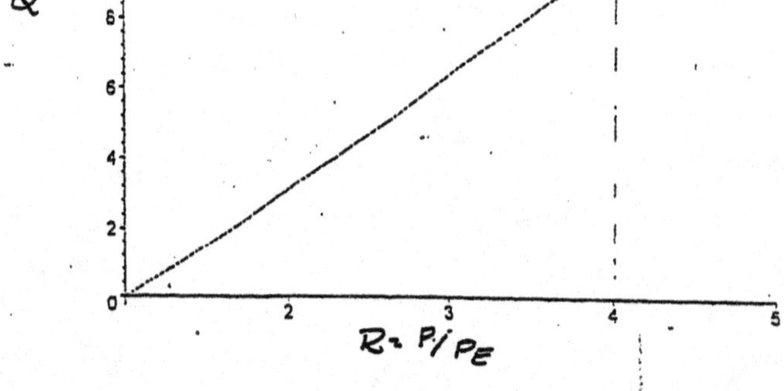

12.1 $\tau_{xy} = \frac{\partial \phi}{\partial z}, \quad \tau_{xz} = -\frac{\partial \phi}{\partial y}$

$V_y = \iint_A \tau_{xy}\, dA = \iint_A \frac{\partial \varphi}{\partial z} dA = -\oint_C \varphi\, dy = k_0 \oint_c dy$

† By Green's Th.

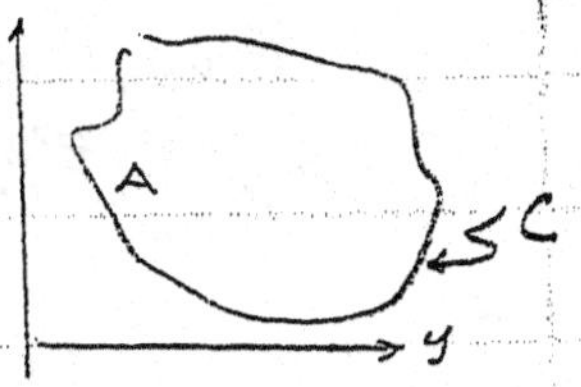

$\therefore V_y = 0$ since $\oint dy = 0$; $V_z = \iint_A \tau_{xz}\, dA = -\iint_A \frac{\partial \phi}{\partial y} dA = -\oint_c \phi\, dz = -k_0 \oint_c dz = 0$

12.2

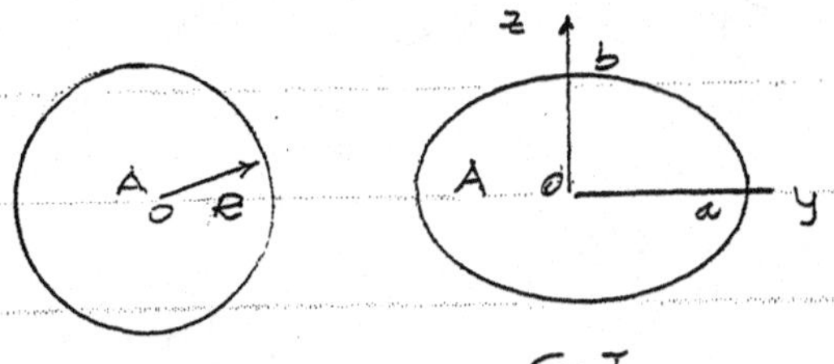

C_0, J_0 $\qquad$ C, J

$a/b = k$

$A_0 = \pi R^2$; $A_e = \pi a b = \pi k b^2$

$\therefore R = b\sqrt{k}$.

$J = \frac{\pi}{4}(a^3 b + a b^3) = \frac{\pi a b}{4}(a^2 + b^2) = \frac{\pi k b^2}{4}(k^2 b^2 + b^2) = \frac{\pi k b^4}{4}(1+k^2)$

$J_0 = \pi R^4/2 = \pi b^4 k^2/2$

$J/J_0 = (1+k^2)/2k.$

$C_0 = J_0 = \pi b^4 k^2/2$; $C = \pi\left(\frac{a^3 b^3}{a^2 + b^2}\right) = \pi \frac{k^3 b^6}{(1+k^2) b^2} = \frac{\pi k^3 b^4}{1+k^2}$

$C/C_0 = \frac{2k}{1+k^2} = J_0/J$

12.3 a) Assume: $\phi(y,z) = k\left[(z - \sqrt{3}y/3)(z + \sqrt{3}y/3)(y - \frac{\sqrt{3}a}{2})\right]$

$= k\left[(z^2 - y^2/3)(y - \sqrt{3}a/2)\right]$

$= k\left[z^2 y - \frac{\sqrt{3}a}{2} z^2 - y^3/3 + \frac{\sqrt{3}a}{6} y^2\right]$

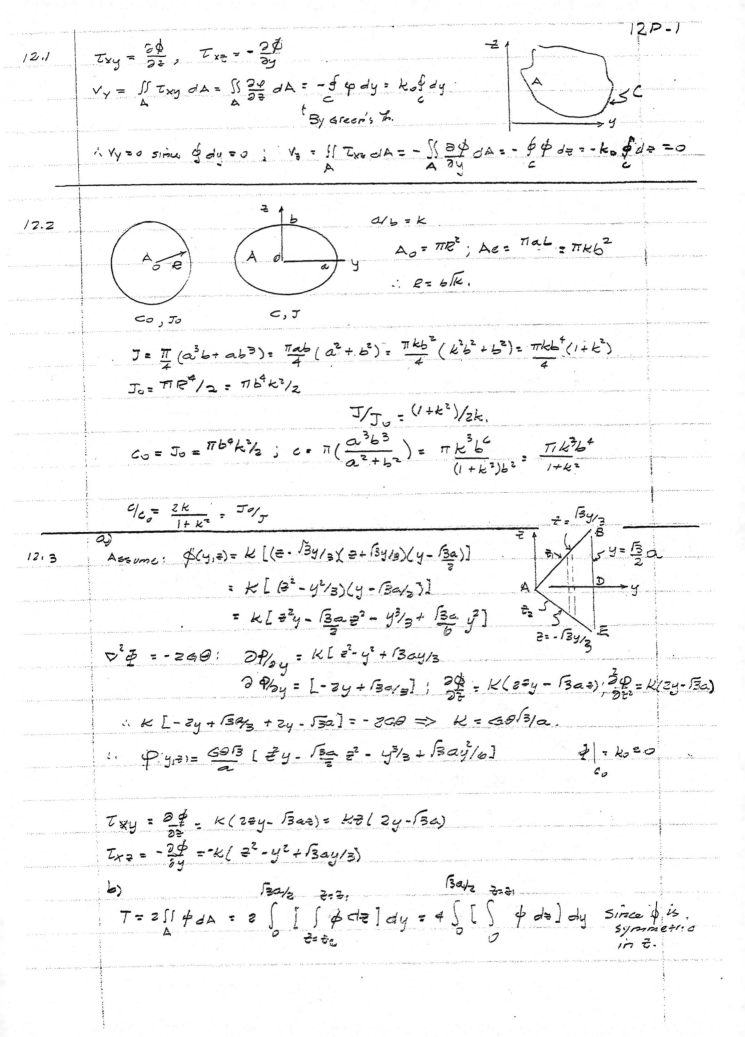

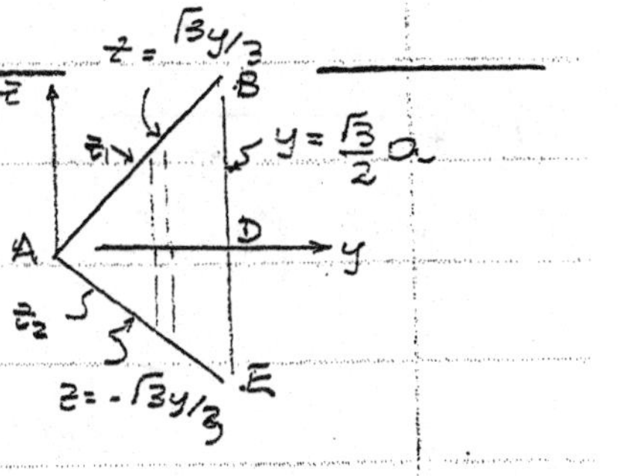

$\nabla^2 \Phi = -2G\theta$: $\partial \phi/\partial y = k\left[z^2 - y^2 + \sqrt{3}ay/3\right]$

$\partial^2 \phi/\partial y^2 = \left[-2y + \sqrt{3}a/3\right]$; $\frac{\partial \phi}{\partial z} = k(2zy - \sqrt{3}az)$, $\frac{\partial^2 \phi}{\partial z^2} = k(2y - \sqrt{3}a)$

$\therefore k\left[-2y + \sqrt{3}a/3 + 2y - \sqrt{3}a\right] = -2G\theta \Rightarrow k = G\theta\sqrt{3}/a.$

$\therefore \phi(y,z) = \frac{G\theta\sqrt{3}}{a}\left[z^2 y - \frac{\sqrt{3}a}{2} z^2 - y^3/3 + \sqrt{3}ay^2/6\right]$ $\qquad \phi\Big|_{C_0} = k_0 = 0$

$\tau_{xy} = \frac{\partial \phi}{\partial z} = k(2zy - \sqrt{3}az) = kz(2y - \sqrt{3}a)$

$\tau_{xz} = -\frac{\partial \phi}{\partial y} = -k(z^2 - y^2 + \sqrt{3}ay/3)$

b)

$T = 2\iint_A \phi\, dA = 2\int_0^{\sqrt{3}a/2}\left[\int_{z=z_2}^{z=z_1} \phi\, dz\right] dy = 4\int_0^{\sqrt{3}a/2}\left[\int_0^{z=z_1} \phi\, dz\right] dy$ since ϕ is symmetric in z.

12.3 (cont'd)

$$T = 4k\int_0^{\sqrt{3}a/2}\left\{\int_0^{\sqrt{3}y/3}\left[z^2 y - \frac{\sqrt{3}a}{3}z^2 - y^3/3 + \sqrt{3}ay^2/6\right]dz\right\}dy$$

$$= 4k\int_0^{\sqrt{3}a/2}\left\{ z^3y/3 - \sqrt{3}az^3/6 - y^3z/3 + \frac{\sqrt{3}ay^2z}{6}\Big]_0^{\sqrt{3}y/3}\right\}dy$$

$$= 4k\int_0^{\sqrt{3}a/2}\left\{\sqrt{3}y^4/27 - ay^3/18 - \sqrt{3}y^4/9 + ay^3/6\right\}dy$$

$$= 4k\int_0^{\sqrt{3}a/2}\left\{-2\sqrt{3}y^4/27 + y^3a/9\right\}dy = 4k\left[-2\sqrt{3}y^5/135 + ay^4/36\right]_0^{\sqrt{3}a/2}$$

$$T = \frac{3ka^5}{8}\cdot\frac{1}{30} = ka^5/80 \qquad \text{But } k = G\theta\sqrt{3}/a$$

$$\Rightarrow T = G\sqrt{3}a^4\theta/80 \Rightarrow C = T/\theta = \sqrt{3}a^4/80.$$

STRESSES: $\tau_{xy} = kz(2y - \sqrt{3}a)$; $\tau_{xz} = k(z^2 - y^2 + \sqrt{3}ay/3)$

Along BE: $\tau_{xy} = \tau_{xy}(\sqrt{3}/2, z) = 0$

$$\tau_{xz} = \tau_{xz}(\sqrt{3}/2, z) = -k(z^2 - 3a^2/4 + a^2/2)$$

$(\tau_{xz})_{max}$: $\partial\tau_{xz}/\partial z = 0 \Rightarrow z = 0 \Rightarrow \tau_{max}$ occur at point D.

$$\tau_{xz}\Big|_{max} = ka^2/4$$

$$k = G\theta\sqrt{3}/a = \frac{T}{Ca} = 80T/a^5. \qquad \therefore \tau_{xz}\Big|_D = 20T/a^3$$

12.4

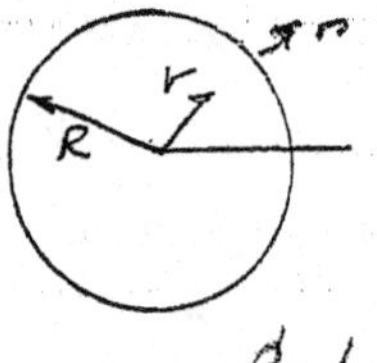

$$\nabla^2\phi = -2G\theta;\quad \phi = \phi(r) \Rightarrow \frac{1}{r}\frac{d}{dr}\left(r\frac{d\phi}{dr}\right)\Big] = -2G\theta$$

$$\text{B.C. } \phi(r=R) = k_0 = 0$$

$$\frac{d}{dr}\left(r\frac{d\phi}{dr}\right) = -2G\theta r \Rightarrow r\frac{d\phi}{dr} = -G\theta r^2 + A_1$$

$(r \sim n)$

$$\frac{d\phi}{dr} = -G\theta r + A/r$$

Note: $\tau_{x\theta} = -\frac{d\phi}{dr}$; $\tau_{x\theta}(r \to 0)$ is finite $\Rightarrow \therefore A = 0$

$$\phi(r) = -G\theta r^2/2 + B \qquad \phi(R) = 0 \Rightarrow B = G\theta R^2/2.$$

$$\therefore \phi(r) = G\theta/2\,(R^2 - r^2)$$

$$T = 2\iint_A \phi\, dA = G\theta\iint(R^2 - r^2)dA = 2\pi G\theta\int_0^R (R^2 - r^2)\,r\,dr = 2\pi G\theta\left(R^2r^2/2 - r^4/4\right)_0^R$$

$$T = 2\pi G\theta(R^4/4) = G\theta(\pi R^4/2) = G\theta J$$

$$\therefore \theta = T/GJ$$

$$\tau_{x\theta} = -\frac{d\Phi(r)}{dr} = G\theta r = Tr/J.$$

12.5 Ellipse: $\tau_{xy} = -\frac{2T}{\pi a b^3} z = -\frac{2T}{\pi ab}(z/b^2) = -\alpha\left(\frac{z}{b^2}\right)$

$\tau_{xz} = \frac{2T}{\pi a^3 b} y = \frac{2T}{\pi ab}(y/a^2) = \alpha(y/a^2)$

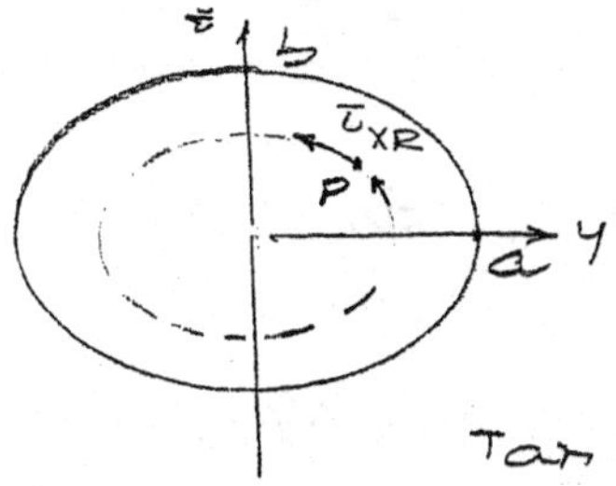

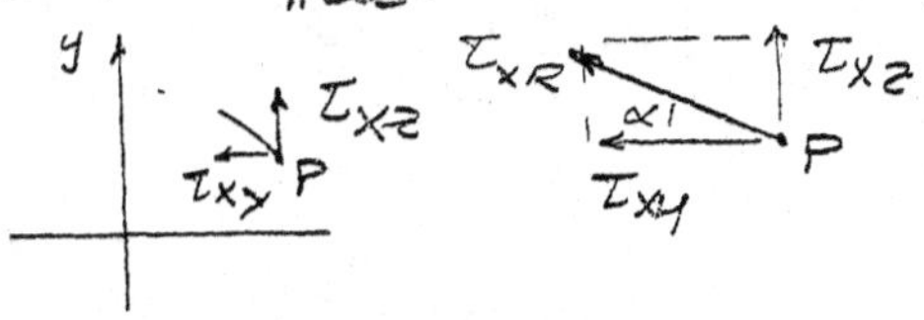

$$\text{Tan}\,\alpha = \tau_{xz}/\tau_{xy} = -\frac{y}{z}\left(b/a\right)^2$$

τ_{zR} acts at $P(y,z)$ in direction with slope $\text{Tan}\,\alpha = \frac{dz}{dy} = -\frac{y}{z}\left(\frac{b}{a}\right)^2$

For ellipse: $y^2/a^2 + z^2/b^2 = 1$

$$\frac{d}{dy}\left(y^2/a^2 + z^2/b^2\right) = 0 \Rightarrow 2y/a^2 + \frac{2z}{b^2}\left(dz/dy\right) = 0$$

$\therefore dz/dy = -\left(\frac{b}{a}\right)^2 (y/z)$, i.e. slope is the

same as for τ_{xR}. $\Rightarrow \therefore \tau_{zR}$ acts along ellipse

---

12.6

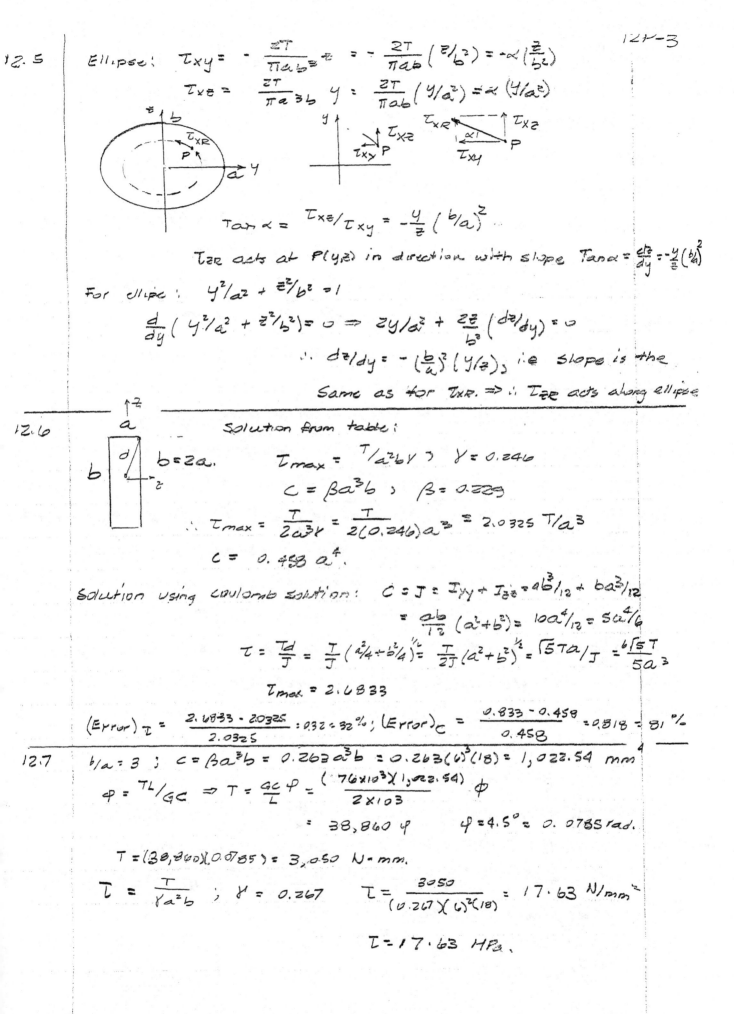

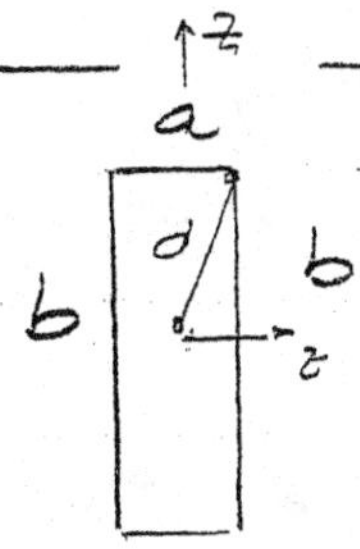

$b = 2a$.

Solution from table:

$\tau_{max} = T/a^2 b\gamma$; $\gamma = 0.246$

$C = \beta a^3 b$; $\beta = 0.229$

$\therefore \tau_{max} = \frac{T}{2a^3\gamma} = \frac{T}{2(0.246)a^3} = 2.0325\ T/a^3$

$C = 0.458\ a^4$.

Solution using coulomb solution: $C = J = I_{yy} + I_{zz} = ab^3/12 + ba^3/12$

$= \frac{ab}{12}(a^2+b^2) = 10a^4/12 = 5a^4/6$

$$\tau = \frac{Td}{J} = \frac{T}{J}\left(a^2/4 + b^2/4\right)^{1/2} = \frac{T}{2J}\left(a^2+b^2\right)^{1/2} = \sqrt{5}\,Ta/J = \frac{6\sqrt{5}\,T}{5a^3}$$

$\tau_{max} = 2.6833$

$$(\text{Error})_\tau = \frac{2.6833 - 2.0325}{2.0325} = 0.32 = 32\%;\ (\text{Error})_C = \frac{0.833 - 0.458}{0.458} = 0.818 = 81\%$$

12.7 $b/a = 3$; $C = \beta a^3 b = 0.263 a^3 b = 0.263(6)^3(18) = 1{,}022.54\ mm^4$

$\varphi = TL/GC \Rightarrow T = \frac{GC}{L}\varphi = \frac{(76\times10^3)(1{,}022.54)}{2\times10^3}\varphi$

$= 38{,}860\ \varphi$ $\qquad \varphi = 4.5° = 0.0785$ rad.

$T = (38{,}860)(0.0785) = 3{,}050$ N·mm.

$\tau = \frac{T}{\gamma a^2 b}$; $\gamma = 0.267$ $\qquad \tau = \frac{3050}{(0.267)(6)^2(18)} = 17.63\ N/mm^2$

$\tau = 17.63$ MPa.

12.8

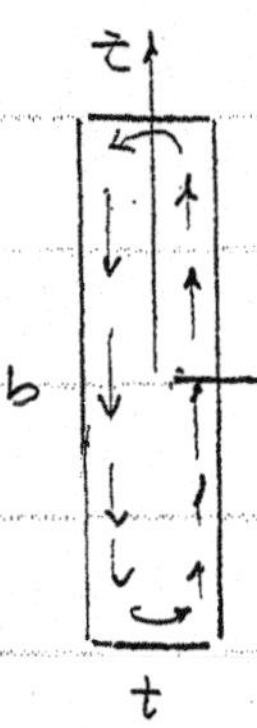

$t \ll b$

$\tau = \tau_{xz} = Ty/C$, $\tau_{xy} \sim 0$. $C = bt^3/3$

Due to τ_{xz}: $T = \iint_A y\,\tau_{xz}(y)\,dA = \frac{T}{C}\iint_A y^2\,dA$

$$T = \frac{T}{C}\cdot I_{zz} = \frac{T}{C}\left(bt^3/12\right) = \frac{T\left(bt^3/12\right)}{bt^3/3} = T/4.$$

τ_{xz} stresses produce a torsional moment $T/4$ instead of T!

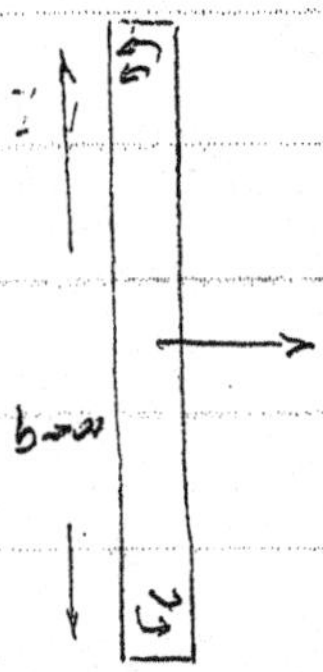

Discrepancy: τ_{zy} stresses are neglected near ends

The τ_{zy} acting near ends $z \sim b/2$, over a width t contribute $3T/4$ since "lever arm", $b \to \infty$

12.9

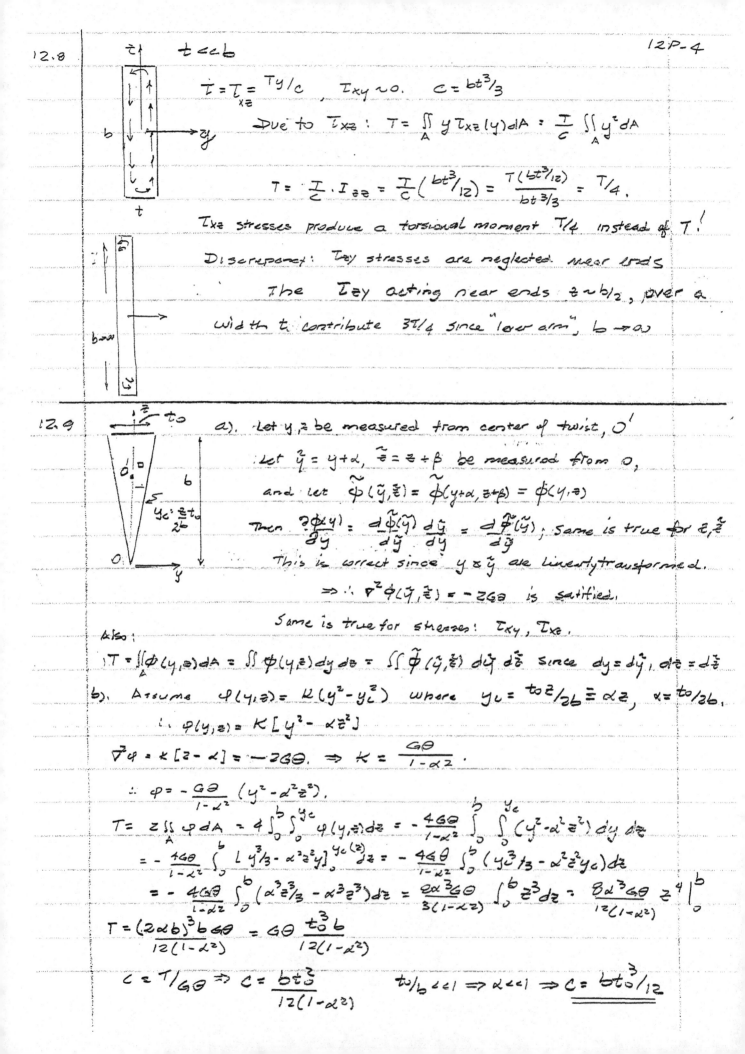

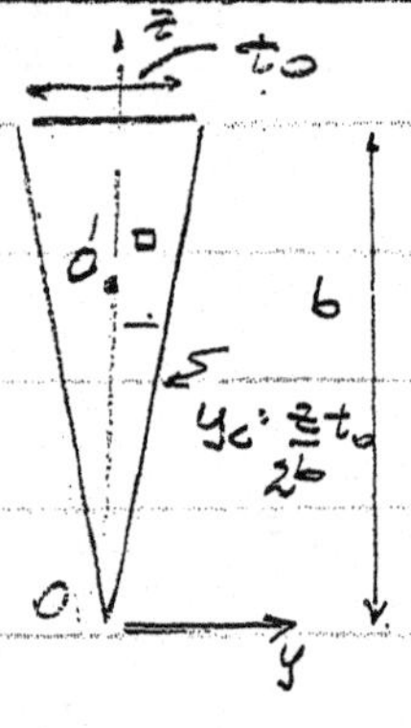

a). Let y, z be measured from center of twist, O'

Let $\tilde{y} = y + \alpha$, $\tilde{z} = z + \beta$ be measured from O,

and let $\tilde{\phi}(\tilde{y}, \tilde{z}) = \tilde{\phi}(y+\alpha, z+\beta) = \phi(y, z)$

Then $\frac{\partial \phi(y)}{\partial y} = \frac{d\tilde{\phi}(\tilde{y})}{d\tilde{y}}\frac{d\tilde{y}}{dy} = \frac{d\tilde{\phi}(\tilde{y})}{d\tilde{y}}$; Same is true for $z, \tilde{z}$

This is correct since y & $\tilde{y}$ are linearly transformed.

$\Rightarrow \therefore \nabla^2 \tilde{\phi}(\tilde{y}, \tilde{z}) = -2G\theta$ is satified.

Same is true for stresses: τ_{xy}, τ_{xz}.

Also:

$T = \iint_A \phi(y,z)\,dA = \iint \phi(y,z)\,dy\,dz = \iint \tilde{\phi}(\tilde{y},\tilde{z})\,d\tilde{y}\,d\tilde{z}$ since $dy = d\tilde{y}$, $dz = d\tilde{z}$

b). Assume $\varphi(y,z) = K(y^2 - y_c^2)$ where $y_c = t_0 z/2b \equiv \alpha z$, $\alpha = t_0/2b$.

$\therefore \varphi(y,z) = K[y^2 - \alpha z^2]$

$\nabla^2 \varphi = K[2 - \alpha] = -2G\theta. \Rightarrow K = \frac{G\theta}{1-\alpha 2}.$

$\therefore \varphi = -\frac{G\theta}{1-\alpha^2}(y^2 - \alpha^2 z^2).$

$$T = 2\iint_A \varphi\,dA = 4\int_0^b\int_0^{y_c}\varphi(y,z)\,dz = -\frac{4G\theta}{1-\alpha^2}\int_0^b\int_0^{y_c}(y^2 - \alpha^2 z^2)\,dy\,dz$$

$$= -\frac{4G\theta}{1-\alpha^2}\int_0^b\left[y^3/3 - \alpha^2 z^2 y\right]_0^{y_c(z)}dz = -\frac{4G\theta}{1-\alpha^2}\int_0^b\left(y_c^3/3 - \alpha^2 z^2 y_c\right)dz$$

$$= -\frac{4G\theta}{1-\alpha 2}\int_0^b\left(\alpha^3 z^3/3 - \alpha^3 z^3\right)dz = \frac{8\alpha^3 G\theta}{3(1-\alpha 2)}\int_0^b z^3\,dz = \frac{8\alpha^3 G\theta}{12(1-\alpha^2)}\,z^4\Big|_0^b$$

$$T = \frac{(2\alpha b)^3 b\,G\theta}{12(1-\alpha^2)} = G\theta\,\frac{t_0^3 b}{12(1-\alpha^2)}$$

$C = T/G\theta \Rightarrow C = \frac{b t_0^3}{12(1-\alpha^2)}$ $\quad t_0/b \ll 1 \Rightarrow \alpha \ll 1 \Rightarrow$ $\underline{\underline{C = b t_0^3/12}}$

12.10

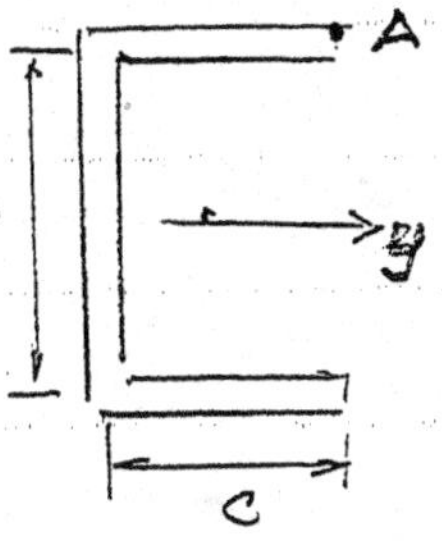

$$C = \sum \frac{b_i t_i^3}{3} = \frac{t^3}{3}[2c + h]$$

$$\alpha = \theta L = \frac{TL}{GC}$$

a) Since center of twist lies on y-axis (as axis of symmetry), $\Delta_y = -\frac{h}{2}\alpha = -TLh/2GC$

$$\Delta y = \frac{-3ThL}{2Gt^3(2c+h)} = \frac{-3TL}{2Gt^3(1+2c/h)}$$

b) Δz cannot be determined since location of center of twist on the z-axis is unknown.

c) $c \to 0 \Rightarrow$ channel becomes narrow rectangle. Center of twist is then at centroid.

$$\Delta y = -\frac{3Tc}{2Gt^3}, \qquad \Delta z = 0.$$

12.11

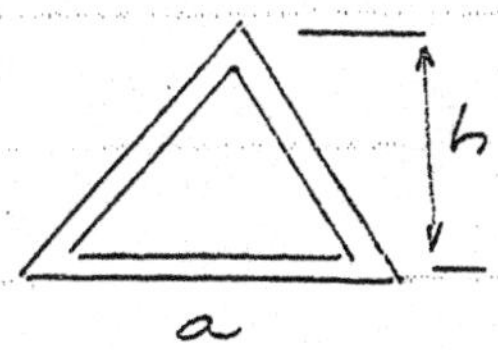

a) $\tau = T/2At$ $\qquad A = ha/2 \,;\; h = a/2\tan 30^\circ = \frac{\sqrt{3}}{2}a$

$$A = \sqrt{3}a^2/4$$

$$\tau = \frac{T}{(\sqrt{3}a^2/2)t} = \frac{2\sqrt{3}\,T}{3a^2t}$$

b) $$C = \frac{4A^2G}{S/t} = \frac{4(3a^4/16)G}{3a/t} = \frac{a^3t}{4}G.$$

12.12

$$\tau_{ex} = \frac{TR}{J} = \frac{TR}{\frac{\pi}{2}(R_o^4 - R_i^4)}$$

$$(R_o^4 - R_i^4) = (R_o^2 + R_i^2)(R_o^2 - R_i^2) = [(R+t/2)^2 + (R-t/2)^2](R_o^2 - R_i^2)$$

$$= [2R^2 + 2(t/2)^2](R_o + R_i)(R_o - R_i)$$

$$= 2R^2[1 + (t/2R)^2](2R)t = 4R^3t[1 + (t/2R)^2]$$

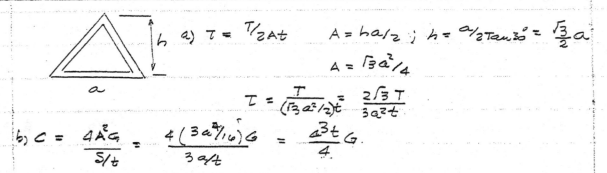

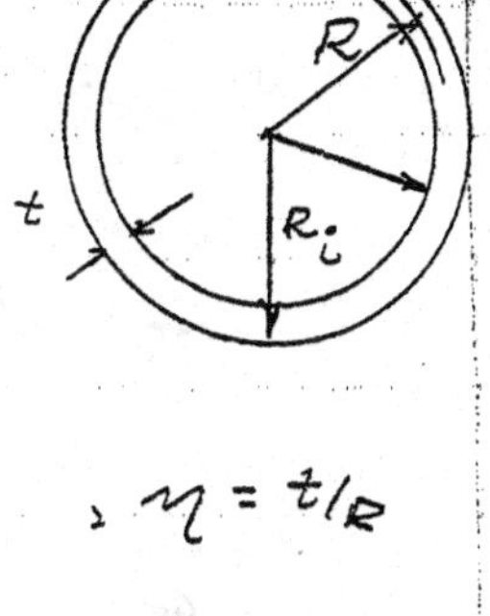

$$\tau_{ex} = \frac{TR}{2\pi R^3 t[1 + (t/2R)^2]}$$

$$\frac{\tau_{ma} - \tau_{ex}}{\tau_{ex}} = \frac{\dfrac{T}{2\pi R^2 t}}{\dfrac{T}{2\pi R^2 t[1 + (t/2R)^2]}} - 1 = (1 + t/2R)^2) - 1 = \frac{\eta^2}{4}, \quad \eta = t/R$$

12.13

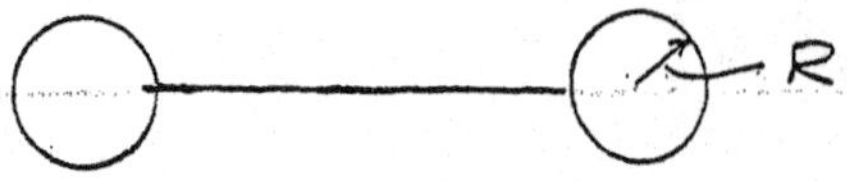

$C_{plate} = bt^3/3$

$C_o = 2\left[\frac{4A^2}{S/t}\right] = \frac{8t(\pi R^2)^2}{2\pi R} = 4\pi R^3 t$

$C = \sum C = \frac{bt^3}{3} + 4\pi R^3 t$ $\qquad b = \alpha t \qquad b = 48t$

$C/C_{pl} = 1 + \frac{4\pi R^3 t}{bt^3/3} = 10 \Rightarrow \frac{12\pi R^3}{bt^2} = 9 \Rightarrow 12\alpha \frac{\pi R^3}{b^3} = 9$

$b/R = \left(\frac{4\cdot 48}{3}\right)^{1/2} = 64^{1/3} = 4$

12.14 a). No slit: $\tau = \frac{T}{2\pi R^2 t}$

$C_a = \frac{4A^2}{S/t} = \frac{4\pi^2 R^4 t}{2\pi R} = 2\pi R^3 t$

slit: $\tau_b = Tt/C_b = \frac{Tt}{2\pi R t^3/3} = \frac{3T}{2\pi R t^2}$

$C_b = 2\pi R t^3/3$

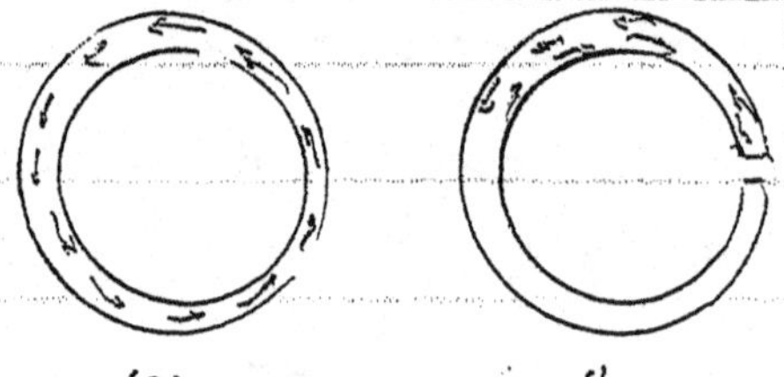

(a) (b)

a). $\tau_a/\tau_b = t/3R$ $\qquad$ b) $C_a/C_b = 3R^2/t^2 \Rightarrow \frac{\theta_a}{\theta_b} = \frac{C_b}{C_a} = t^2/3R^2$

12.15 CIRCLE: $\qquad$ TRIANGLE

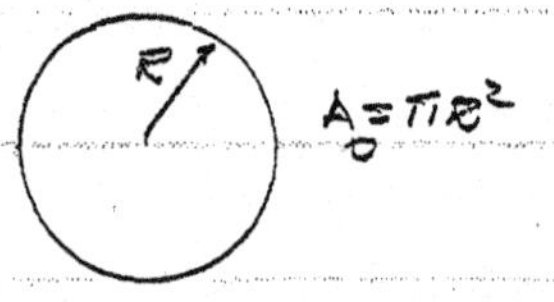

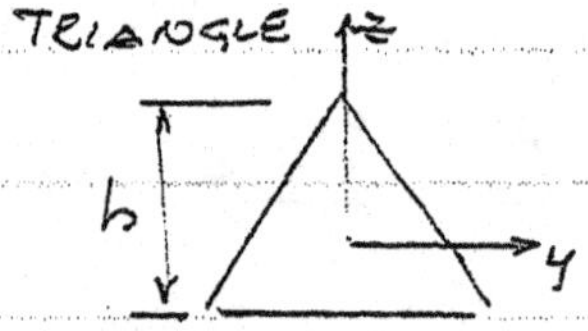

$h = \sqrt{3}a/2 \qquad A_T = \sqrt{3}a^2/4$

$A_O = A_T \Rightarrow \sqrt{3}a^2/4 = \pi R^2 \Rightarrow a^2 = \frac{4\sqrt{3}\pi R^2}{3} \Rightarrow R^4 = 3a^4/16\pi^2.$

$J_O = \pi R^4/2 = 3\pi a^4/32\pi^2 = 3a^4/32\pi = C_O$

$J_T = I_{yy} + I_{zz} \qquad I_{yy} = ah^3/36 \ ; \ I_{zz} = 2h\frac{(a/2)^3}{12} = ha^3/48$

$J_T = \frac{ha}{12}\left(\frac{h^2}{3} + \frac{a^2}{4}\right) = \frac{ha}{144}(4h^2 + 3a^2) = \frac{6ha^3}{144} = \sqrt{3}a^4/48$

$J_T/J_O = \frac{\sqrt{3}a^4/48}{3a^4/32\pi} = 1.209$

$C_T/C_O = \frac{\sqrt{3}a^4/80}{3a^4/32\pi} = \frac{32\pi\sqrt{3}}{80\cdot 3} = \frac{2\sqrt{3}}{15}\pi = 0.726$

$J_T/J_O > 1 \Rightarrow C_T/C_O < 1$

12.16 (a). A

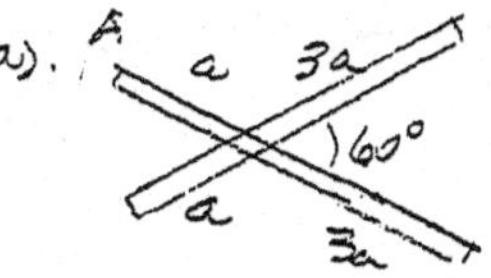

$$C_A = \sum \frac{b_i t_i^3}{3} = \frac{t^3}{3}\left[2a + 2(3a)\right] = 8at^3/3.$$

B

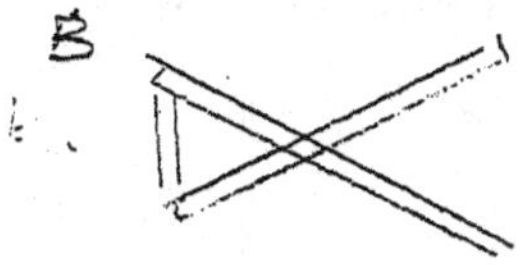

$C_{open} = 6at^3/3 \qquad C_{closed} = 4A^2t/S \qquad A = \sqrt{3}a^2/4,\ S = 3a$

$$C_{closed} = \frac{4(3a^4/16)t}{3a} = a^3t/4$$

$$C_B = 6at^3/3 + a^3t/4 = \frac{at}{12}(24t^2 + 3a^2)$$

(b) $a = 30\,mm, \quad t = 4\,mm.$

$$C_B/C_A = \frac{1}{32t^2}(24t^2 + 3a^2) = \frac{3}{4} + \frac{3}{32}(a/t)^2 = 38.25$$

c). A: $C = 13{,}653\ mm^4$

$$\tau_A = Tt/C = \frac{(150\times10^3)(4)}{13{,}653}$$

$\tau_A = 43.9\ N/mm^2 = 43.9\,MPa$

$$\theta_A = \frac{T}{GC} = \frac{150\times10^3}{(200\times10^3)(13{,}653)}$$

$= 0.549\times10^{-4}/mm = 0.0549\ rad/m$

$\theta_A = 3.147°/m$

B: $C_{closed} = 512\times10^3\ mm^4$, $C_{open} = 10{,}240\ mm^4$

$C = 522{,}240\ mm^4 \qquad A = 2771\ mm^2$

$$\frac{T_c}{T} = \frac{C_c}{C} \Rightarrow T_c = (C_c/C)\,T = 147{,}059\ N\text{-}mm.$$

$T_o = 2{,}941$

$$\tau_{open} = Tt/C = (2{,}941)(4)/522{,}240 = 0.023\ MPa.$$

$$\tau_{cl} = \frac{T}{2At} = \frac{147{,}059}{2(2771)(4)} = 6.63\ N/mm^2 = 6.63\,MPa$$

$$\theta_B = \frac{T}{GC} = \frac{150{,}000}{(200\times10^3)(522{,}240)} = 1.436\times10^{-6}\ rad/mm$$

$= 1.4361\times10^{-3}\ rad/m = 0.0823°/m.$

12.17 a). A:

$$C_A = 4A^2t/S + 8at^3/3 = a^3t\left[1 + \tfrac{8}{3}(t/a)^2\right] \approx a^3t$$

B:

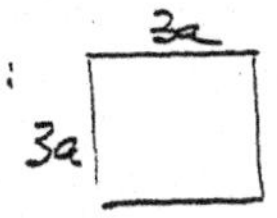

$$C_B = 4A^2t/S = \frac{4(9a^2)^2t}{12a} = 27a^3t$$

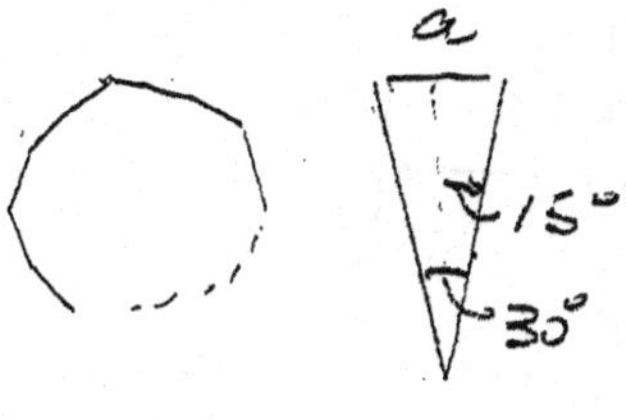

$$A_0 = \tfrac{1}{2}\left\{\tfrac{a}{2}\left[\frac{a}{2\,\mathrm{Tan}\,15°}\right]\right\} = a^2/8\,\mathrm{Tan}\,15°.$$

$A = 24A_0 = 3a^2/\mathrm{Tan}\,15°$

$$C = 4A^2t/S \qquad S = 12a \Rightarrow C = \frac{4\cdot 9a^4t}{12a\,\mathrm{Tan}^2\theta} = 3a^3t/\mathrm{Tan}^2 15°$$

$C_c = 41.785\,a^3t.$

b)

$$\tau_A = \tau_{cl} \approx \frac{T}{2At} = \frac{T}{2a^2t}\ ; \quad \tau_B = \frac{T}{18at}\ ; \quad \tau_c = \frac{T\cdot\tan 15°}{6a^2t} = 0.04466\ T/a^2t$$

c). As $n \to \infty \Rightarrow$ section becomes a circle; circum $= 12a \Rightarrow R = 6a/\pi$

$$C = J = (2\pi Rt)R^2 = 2\pi R^3t = 2\pi t(6a/\pi)^3 = 432a^3/\pi^2 = 43.77\,a^3t.$$

2.18

$$T = 2\iint_{A_{net}} \Phi\, dA + 2k_1 A_1 = 2\iint_{A_0} \Phi\, dA - 2\iint_{A_1} \Phi\, dA + 2k_1 A_1$$

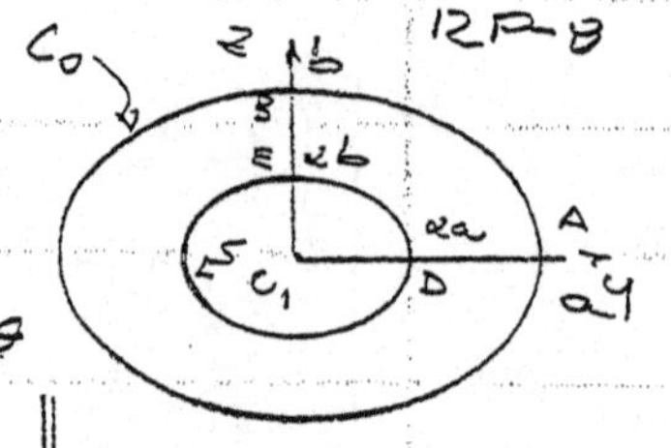

Let $\Phi = K(1 - y^2/a^2 - z^2/b^2)$; $\nabla^2\Phi = -2G\theta \Rightarrow K = -\dfrac{a^2b^2}{a^2+b^2}G\theta$

$$T = 2K\left[(A_0 - A_1) - \frac{1}{a^2}(I_{zz_0} - I_{zz_1}) - \frac{1}{b^2}(I_{yy_0} - I_{yy_1})\right] + 2k_1A_1$$

$$= 2K\left\{\pi ab(1-\alpha^2) - \frac{\pi a^3 b}{4a^2}(1-\alpha^4) - \frac{\pi ab^3}{4b^2}(1-\alpha^4)\right\} + 2k_1A_1$$

$$= 2K\pi ab\left[(1-\alpha^2) - \tfrac{1}{2}(1-\alpha^4)\right] + 2k_1A_1$$

A_0: Area within C_0
A_1: Area within C_1
A_{net}: Area between C_1 and C_0.

On C_1: $y^2/\alpha^2a^2 + z^2/\alpha^2b^2 = 1 \Rightarrow y^2/a^2 + z^2/b^2 = \alpha^2$

$$\Phi\Big|_{C_1} = K(1-\alpha^2) = \text{const.} \Rightarrow k_1 = K(1-\alpha^2)$$

$$A_1 = \pi ab\alpha^2$$

$$T = 2\pi abK\left[(1-\alpha^2) - \tfrac{1}{2}(1-\alpha^4) + \alpha^2 - \alpha^4\right]$$

$$= 2\pi abK(1-\alpha^4) = \frac{\pi a^3b^3}{a^2+b^2}(1-\alpha^4)G\theta$$

$$C = \frac{T}{G\theta} = \frac{\pi a^3b^3}{a^2+b^2}(1-\alpha^4) = C_s(1-\alpha^4) \quad \text{where } C_s \text{ is } C \text{ of solid ellipse}$$

$$K = \frac{T}{\pi ab(1-\alpha^4)} \qquad \tau_{xy} = \frac{\partial\Phi}{\partial z} = -2Kz/b^2 \Rightarrow \tau_{xy}\Big|_B = -2K/b,\ \tau_{xy}\Big|_E = -2K\alpha/b$$

$$\tau_{xz} = -\frac{\partial\Phi}{\partial y} = -2Ky/a^2 \Rightarrow \tau_{xz}\Big|_A = -2K/a,\ \tau_{xz}\Big|_D = -2K\alpha/a$$

$$\tau_{max} = \tau\Big|_B = \frac{2T}{\pi ab^2(1-\alpha^4)}$$

2.19

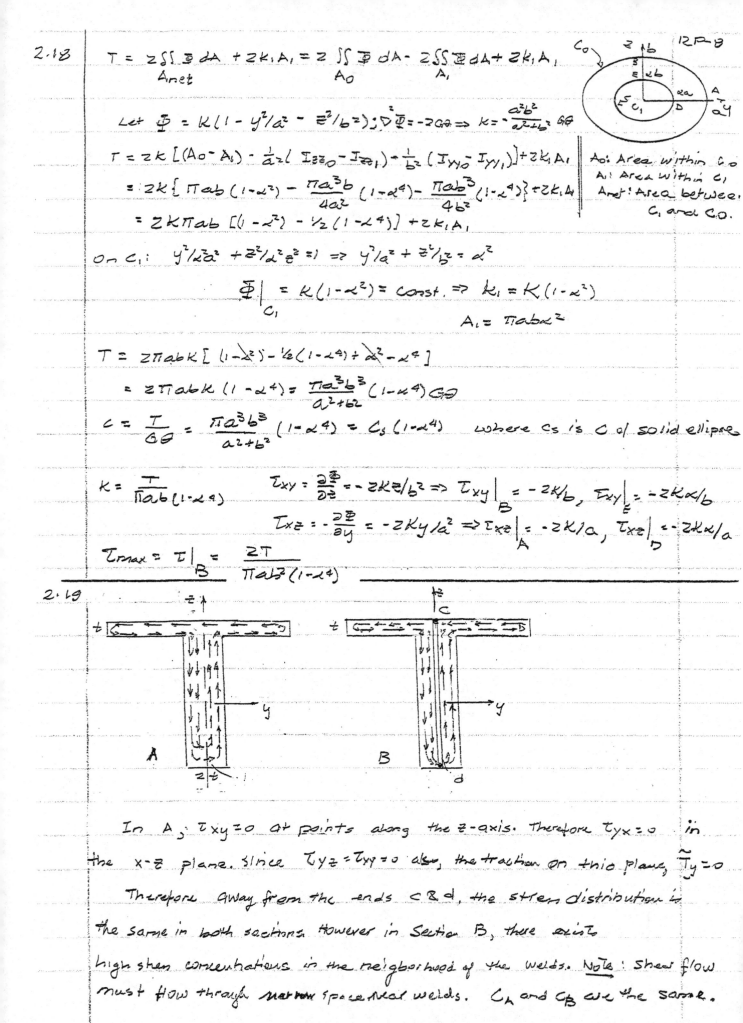

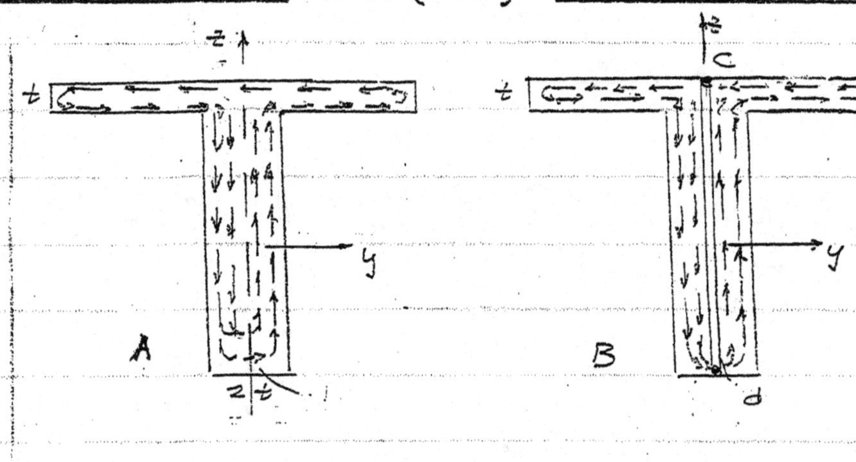

In A, $\tau_{xy} = 0$ at points along the z-axis. Therefore $\tau_{yx} = 0$ in the x-z plane. Since $\tau_{yz} = \tau_{xy} = 0$ also, the traction on this plane, $\tilde{T}_y = 0$

Therefore away from the ends c & d, the stress distribution is the same in both sections. However in Section B, there exists high stress concentrations in the neighborhood of the welds. Note: Shear flow must flow through ~~narrow~~ space near welds. C_A and C_B are the same.

12.20 a)

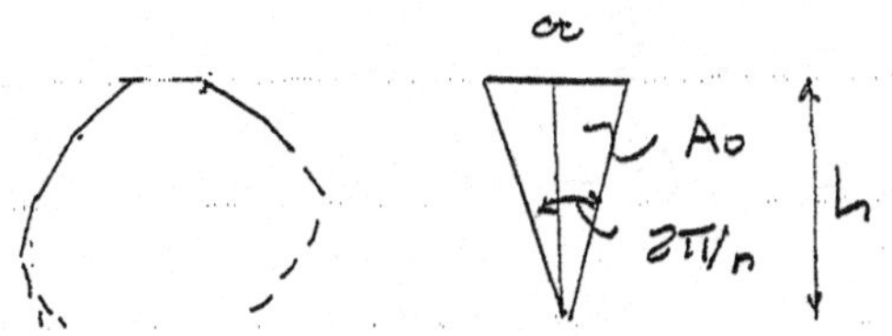

$h = \frac{a}{2\tan(\pi/n)}$, $A_0 = ah/2 = \frac{na^2}{4\tan(\pi/n)}$

$$C = 4A^2 t/s = \frac{4n^2a^4}{16\tan^2(\pi/n)\cdot na} \Rightarrow C = \frac{na^3 t \cot^2(\pi/n)}{4}$$

b). Δ : $(n=3)$ $\quad C = \frac{3a^3t}{4}\cot^2(60°) = a^3t/4$

$\square$: $(n=4)$ $\quad C = a^3 t \cot^2(\pi/4) = a^3 t$

Hexagon: $\quad C = \frac{3a^3t}{2}\cot^2(30°) = \frac{9}{2}a^3t$

c). $\tau = \frac{T}{2At} = \frac{2T\cdot\tan(\pi/n)}{na^2t}$

d). Δ : $\tau = 2\sqrt{3}T/3a^2t$; $\square$: $\tau = \frac{T}{2a^2t}$; Hexagon : $\tau = \sqrt{3}a^2t/9$

e) $$C = \lim_{\substack{n\to\infty \\ a\to 0 \\ na = \text{const} = 2\pi R}} \frac{na^3t}{4}\cot^2(\pi/n) = \lim_{\substack{n\to\infty \\ a\to 0 \\ na \to \text{const} = 2\pi R}} \frac{na^3t}{4}\left[n/\pi - \pi/3n - \frac{\pi^3}{45n^3}\cdots\right]^2$$

$$C \cong \lim_{\substack{n\to\infty \\ a\to 0}} \frac{na^3t}{4}(n/\pi)^2 = \lim \frac{(na)^3t}{4\pi^2} = \frac{8\pi^3R^3t}{4\pi^2} = 2\pi R^3 t = J \text{ (thin-wall circle)}$$

$$\tau = \lim_{\substack{n\to\infty \\ a\to 0}} \frac{2T\tan(\pi/n)}{na^2t} = \lim_{\substack{n\to\infty \\ a\to 0 \\ na\to 2\pi R}} \frac{2T}{na^2t}\left[\pi/n - (\pi/n)^3 \cdots\right]$$

$$= \lim_{\substack{n\to\infty \\ a\to 0 \\ na\to 2\pi R}} \frac{2T\pi}{(na)^2t} = \frac{2T\pi}{(2\pi R)^2t} = \frac{T}{2\pi R^2 t} = \frac{T}{2At}$$

12.23 $\quad C = \beta a^3 b$, $\quad \tau = T/a^2b\gamma$

$A_s = a_s^2$, $A_r = ba_r = \alpha a_r^2$; $A_s = A_r \Rightarrow a_r^2 = a_s^2/\alpha$

$C_s = \beta_s a_s^4 = 0.141\, a_s^4$

$C_r = \beta_r a_r^3 b_r = \beta_r \alpha a_r^4 = \beta_r a_s^2/\alpha$

$\Rightarrow \frac{C_r}{C_s} = \frac{\beta_r/\alpha}{.141} = \frac{7.0922\,\beta_r}{\alpha}$

$\tau_s = T/\gamma_s a_s^3 = 4.8077\, a_s^3$

$\tau_r = T/\gamma_r a_r^2 b_r = T/\gamma_r \alpha a_r^3 = T/\gamma_r \alpha a_s^3/\alpha^{3/2}$

$\Rightarrow \frac{\tau_r}{\tau_s} = \frac{0.208\,\alpha^{1/2}}{\gamma_r}$

13.1 a).

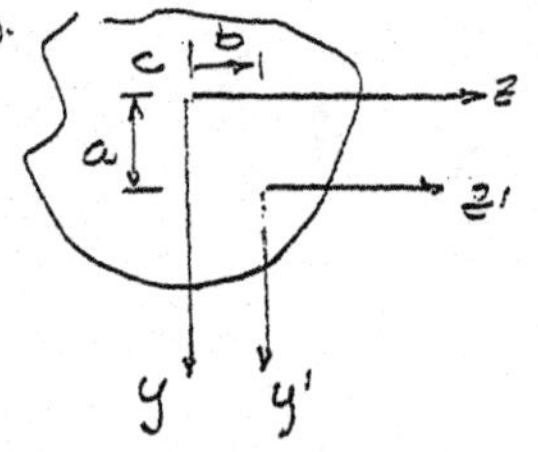

$y' = y - a,\ z' = z - b$

$I_{y'z'} = \iint_A y'z'\,dA = \iint_A [(y-a)(z-b)]\,dA$

$= I_{yz} - b\iint y\,dA - a\iint z\,dA + abA = I_{yz} + abA$

if C is centroid. $ab > 0$ or < 0.

b). $a = 0 \Rightarrow I_{y'z'} = I_{yz}$.

APPENDIX E:

13.2. $\angle 89 \times 76 \times 12.7 \Rightarrow I_{zz} = 0.97 \times 10^6\ mm^4,\ I_{yy} = 1.44 \times 10^6\ mm^4$

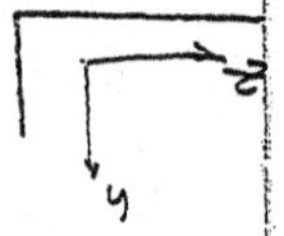

$I_{min} = (15.8)^2(1935) = 0.48305 \times 10^6\ mm^4.$

$I_{min} = \frac{I_{yy} + I_{zz}}{2} - \left[\left(\frac{I_{yy} - I_{zz}}{2}\right)^2 + I_{yz}^2\right]^{1/2} \Rightarrow I_{yz} = -0.6826 \times 10^6\ mm^4.$

Note: $I_{yy} I_{zz} - I_{yz}^2 = 9.3082 \times 10^{11}\ mm^8$.

$\sigma_x = \frac{(I_{yy} y - I_{yz} z) M_z}{I_{yy} I_{zz} - I_{yz}^2}$. $M_z = 2500\ N\text{-}m = 2.5 \times 10^6\ N\text{-}mm;\ M_y = 0$

$\sigma_x = 3.876\,y + 1.833\,z.$

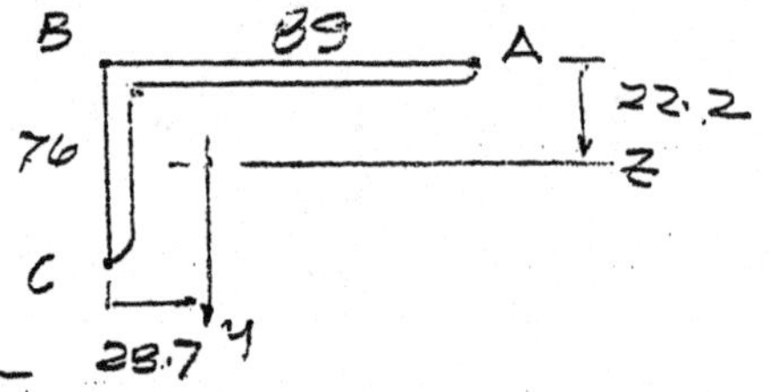

A: $y = -22.2,\ z = 60.3\ mm$

$\sigma_A = 24.5\ N/mm^2 = 24.5\ MPa$

B: $y = -22.2,\ z = -28.7 \Rightarrow \sigma_B = -138.7\ MPa$

C: $y = 53.8,\ z = -28.7 \Rightarrow \sigma_C = 155.9\ MPa.$

13.3

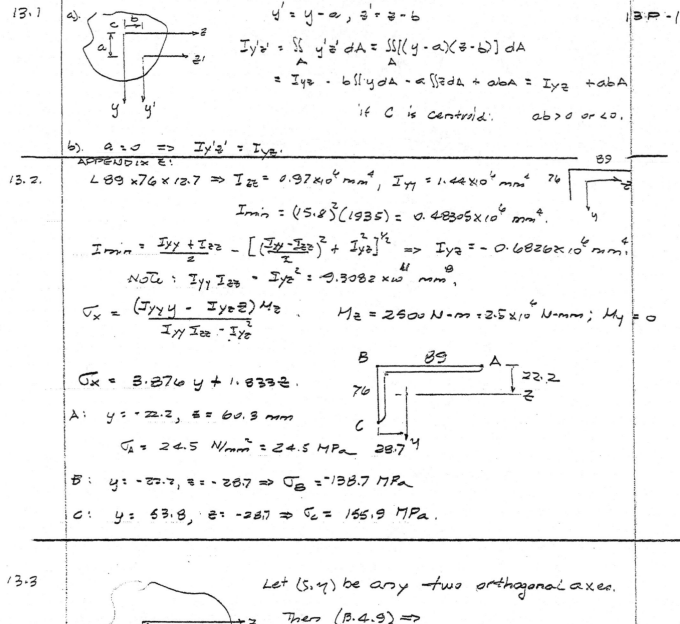

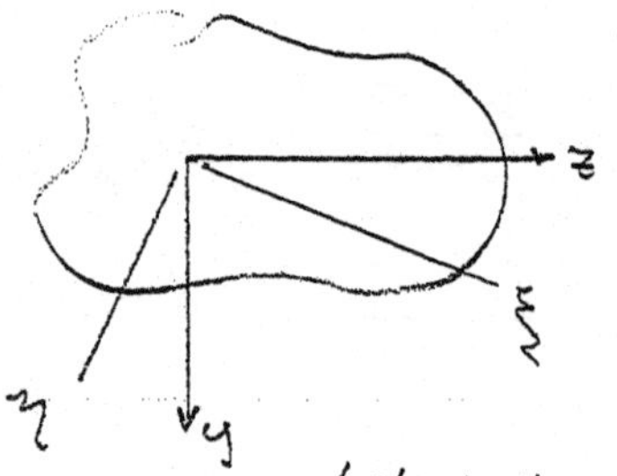

Let (ξ, η) be any two orthogonal axes.

Then (B.4.9) $\Rightarrow$

$$\sigma_x = \frac{(M_\xi I_{\eta\eta} - M_\eta I_{\eta\xi})\eta + (M_\eta I_{\xi\xi} - M_\xi I_{\eta\xi})\xi}{I_{\eta\eta} I_{\xi\xi} - I_{\eta\xi}^2}$$

Let ξ be a Neutral axis $\Rightarrow \sigma_x(\eta = 0) = 0$

$\therefore M_\eta I_{\xi\xi} - M_\xi I_{\eta\xi} = 0 \Rightarrow M_\eta = M_\xi (I_{\eta\xi}/I_{\xi\xi}).$

$\therefore$

$$\sigma_x = \frac{[I_{\eta\eta} - I_{\eta\xi}(I_{\eta\xi}/I_{\xi\xi})] M_\xi \eta}{I_{\eta\eta} I_{\xi\xi} - I_{\eta\xi}^2} = \frac{M_\xi \eta}{I_{\xi\xi}} \equiv \frac{M_n \eta}{I_n}$$

since ξ axis is Neutral axis.

13.4

$$\iint_A (y-\lambda z)^2 dA > 0 \Rightarrow \iint_A [y^2 - 2\lambda yz + \lambda^2 z^2]\, dA > 0$$

$$F(\lambda) \equiv I_{yy}\lambda^2 - 2I_{yz}\lambda + I_{zz} > 0$$

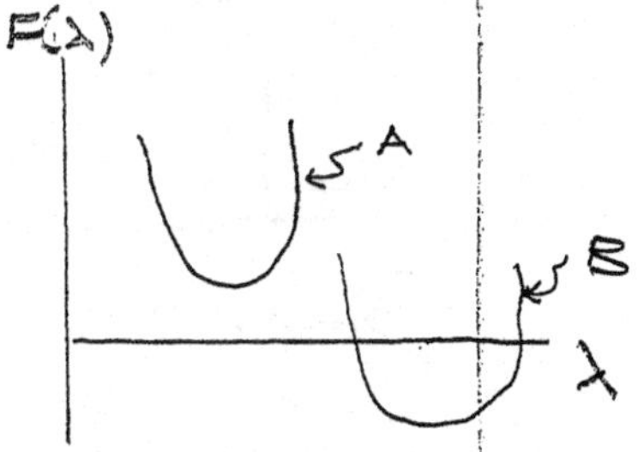

Since $I_{yy} > 0$, $F(\lambda)$ has shape shown:

Note: Only "A" is possible $\Rightarrow F(\lambda) = 0$ cannot exist $\Rightarrow F(\lambda) = 0$ has no real roots.

$$F(\lambda) = 0 \Rightarrow \lambda = \frac{1}{2I_{yy}}\left\{2I_{yz} \pm [4I_{yz}^2 - 4I_{yy}I_{zz}]^{1/2}\right\}$$

$\therefore$ Require: $I_{yy}I_{zz} - I_{yz}^2 > 0$

13.5

a) $\tan\beta = \left(\frac{I_{zz}}{I_{yy}}\right)\tan\theta$; $\theta = 30°$; $I_{zz}/I_{yy} = \frac{(15)(20)^3}{(20)(15)^3} = (\frac{3}{4})(4/3)^3 = (4/3)^2$

$\tan\beta = (4/3)^2 \sqrt{3}/3 = 1.0264 \Rightarrow \beta = 45.746°$

b) $M_z = (1.25)(3500)(\sqrt{3}/2) = 3789$ N-m $\qquad I_{yy} = (20)(15)^3/12 = 5625$ cm^4

$M_y = -(1.25)(3500)(\frac{1}{2}) = -2187.5$ N-m. $\qquad I_{zz} = (15)(20)^3/12 = 10^4$ cm^4

$$(13.4.9) \Rightarrow \sigma_x = \frac{M_z y}{I_{zz}} + \frac{M_y z}{I_{yy}} \Rightarrow \sigma_x = (3.789\times10^7)\,y - (3.889\times10^7)\,z$$

$(\sigma_x)_A = \sigma_x(y=-10, z=7.5\text{ cm}) = -6.71\times10^5$ N/m^2 $= -6.71$ MPa

$(\sigma_x)_B = \sigma_x(y=10, z=7.5\text{ cm}) = 0.872\times10^6$ N/m^2 $=$ 0.872 MPa

$(\sigma_x)_C = \sigma_x(y=10, z=-7.5\text{ cm}) = 6.71\times10^5$ N/m^2 $= 6.71$ MPa

$(\sigma_x)_D = \sigma_x(y=-10, z=-7.5\text{ cm}) = -0.872$ MPa

c).

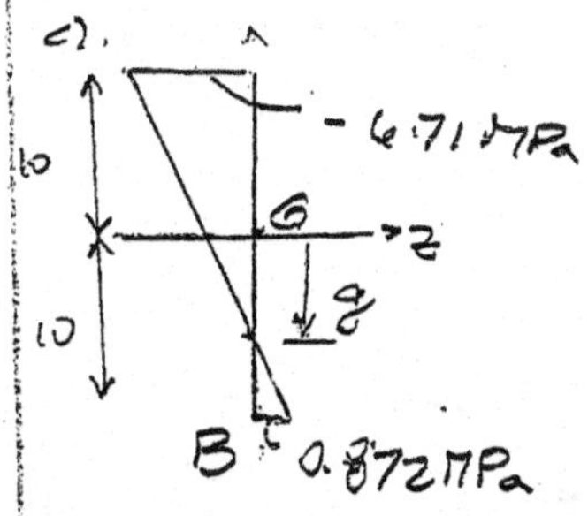

d).

$$\frac{10+g}{6.71} = \frac{10-g}{0.872} \Rightarrow 8.72 + 0.872g = 67.1 - 6.71g$$

$7.582\,g = 58.38 \Rightarrow g = 7.7$

$g = 7.5\tan\beta = 7.7$ ✓

e). $(\sigma_x)_G = \sigma_x(y=0, z=7.5) = -2.92$ MPa

$M_n = (1.25)(3500)\cos[(45.75-30)°] = 4211$ N-m

$\beta = 45.75°$

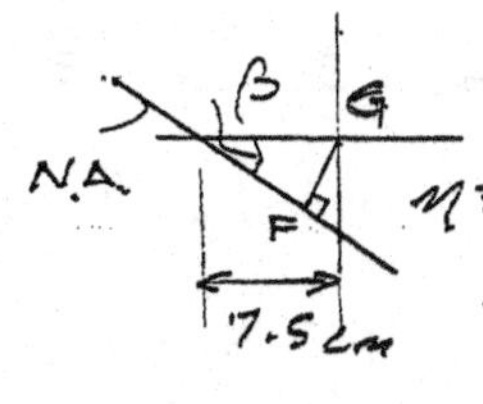

$\eta = \overline{FG} = -7.5\sin\beta = -5.37$ cm.

$I_{NA} = I_{zz}\cos^2\beta + I_{yy}\sin^2\beta = 7755$ cm^4 $\qquad \sigma_x = -\frac{M_n\eta}{I_{NA}} = \frac{(4211)(0.0537)}{7.755\times10^{-8}}$

$(\sigma_x)_G = -2.92$ N/m^2 $= -2.92$ MPa

13.6

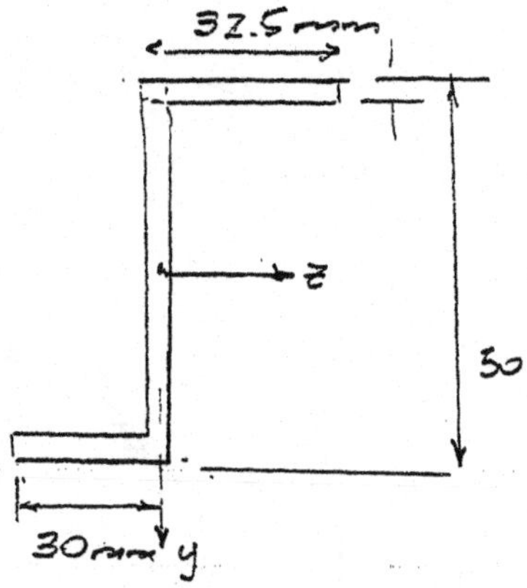

$$I_{zz} = \frac{(5)(50)^3}{12} + 2\left[\frac{(27.5)(5)^3}{12} + (27.5)(5)(22.5)^2\right]$$

$$I_{zz} = 1.919\times10^5\ mm^4 = 19.2\ cm^4$$

$$I_{yy} = (40)(5)^3/12 + 2\left[\frac{(5)(32.5)^3}{12} + (32.5)(5)(13.75)^2\right]$$

$$= 9.047\times10^4\ mm^4 = 9.05\ cm^4.$$

$$I_{yz} = -2[(32.5\times5)(13.75)(22.5)] = -1.006\times10^5\ mm^4$$

$$I_{yz} = -10.05\ cm^4.$$

$$I_{1,2} = \frac{I_{yy}+I_{zz}}{2} \pm \left[\left(\frac{I_{yy}-I_{zz}}{2}\right)^2 + I_{yz}^2\right]^{1/2} = 25.4,\ 2.9\ cm^4.$$

13.7

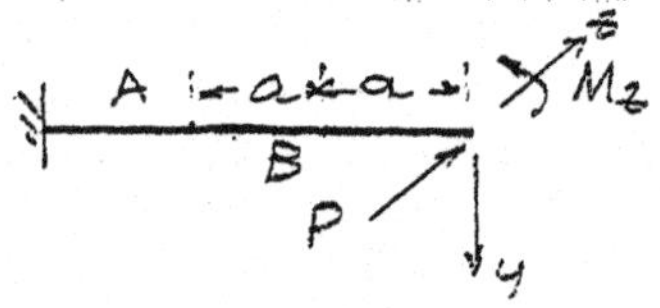

At A: M_z, $M_y = -2Pa$

$$I_{yy}I_{zz} - I_{yz}^2 = 7I_0^2.$$

$$\sigma_x\Big|_D = \frac{1}{7I_0^2}\{[(4I_0)M_z + I_0(-2Pa)](-2b) + [(2I_0)(-2Pa) + I_0M_z](b)\} = \sigma_0$$

$$\Rightarrow (4M_z - 2Pa)(-2b) + (-4Pa + M_z)b = 7I_0\sigma_0 \Rightarrow M_z = -\sigma_0 I_0/b$$

At B: M_z, $M_y = -Pa$

$$\sigma_x = \frac{1}{7I_0^2}\{(4I_0M_z - I_0Pa)y + (-2I_0Pa + I_0M_z)z\} = 0$$

$$\Rightarrow \left(-4\frac{\sigma_0 I_0^2}{b} - I_0Pa\right)y + (-2I_0Pa - \sigma_0 I_0^2/b)z = 0$$

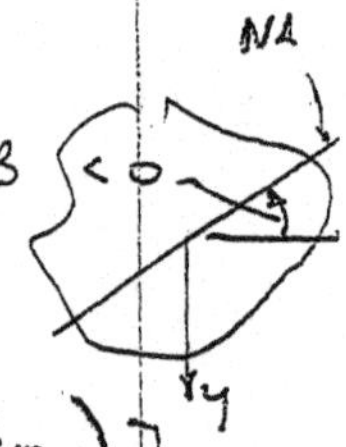

$$\tan\beta = y/z = -\frac{(2Pa + \sigma_0 I_0/b)}{Pa + 4\sigma_0 I_0/b}$$

$$\beta = \tan^{-1}\left[-\left(\frac{2Pab + \sigma_0 I_0}{Pab + 4\sigma_0 I_0}\right)\right]$$

13.8

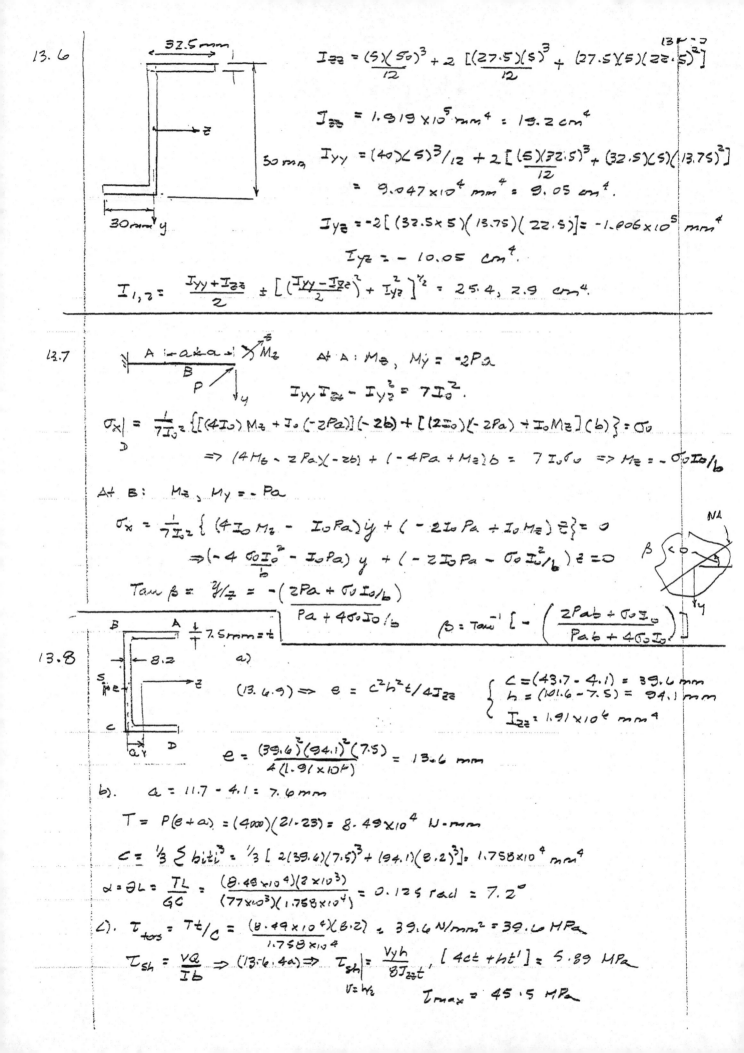

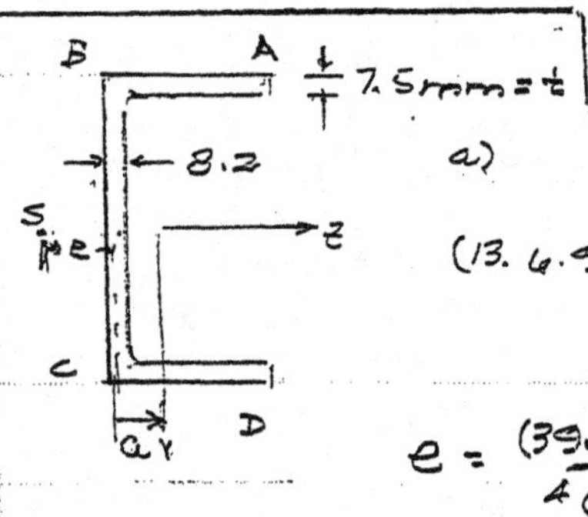

a)

$$(13.6.9) \Rightarrow e = c^2h^2t/4I_{zz}$$

$$\begin{cases} c = (43.7 - 4.1) = 39.6\ mm \\ h = (101.6 - 7.5) = 94.1\ mm \\ I_{zz} = 1.91\times10^6\ mm^4 \end{cases}$$

$$e = \frac{(39.6)^2(94.1)^2(7.5)}{4(1.91\times10^6)} = 13.6\ mm$$

b). $a = 11.7 - 4.1 = 7.6\ mm$

$$T = P(e+a) = (4000)(21.23) = 8.49\times10^4\ N\cdot mm$$

$$C = \tfrac{1}{3}\sum b_i t_i^3 = \tfrac{1}{3}[2(39.6)(7.5)^3 + (94.1)(8.2)^3] = 1.758\times10^4\ mm^4$$

$$\alpha = \theta L = \frac{TL}{GC} = \frac{(8.49\times10^4)(2\times10^3)}{(77\times10^3)(1.758\times10^4)} = 0.125\ rad = 7.2^\circ$$

c). $$\tau_{tors} = Tt/C = \frac{(8.49\times10^4)(8.2)}{1.758\times10^4} = 39.6\ N/mm^2 = 39.6\ MPa$$

$$\tau_{sh} = \frac{VQ}{Ib} \Rightarrow (13.6.4a) \Rightarrow \tau_{sh}\Big|_{V=V_y} = \frac{V_y h}{8I_{zz}t'}[4ct + ht'] = 5.89\ MPa$$

$$\tau_{max} = 45.5\ MPa$$

13.9

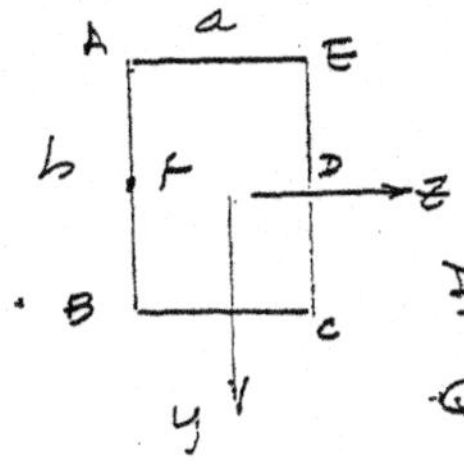

$\tau = V_y Q_z / I_{zz} t$; $I_{zz} = 2th^3/12 + 2at(h/2)^2 = \frac{h^2 t}{6}(h+3a)$

DE:

$Q|_{DE} = tv^2/2$, $\tau = V_y v^2 / 2I_{zz}$

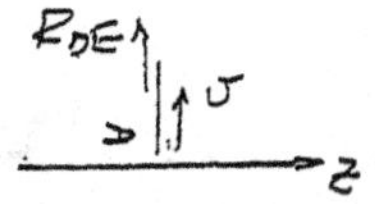

$$R_{DE} = t\int \tau(v)\,dv = \frac{V_y t}{2I_{zz}}\int_0^{h/2} v^2 dv = \frac{V_y t h^3}{48 I_{zz}} \uparrow$$

E-A:

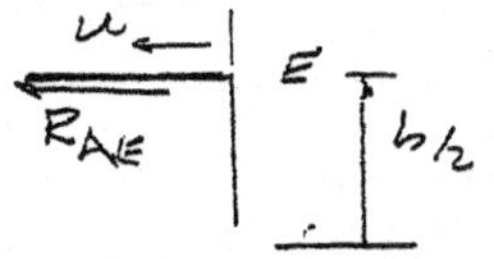

$Q|_{EA} = th^2/8 + thu/2$

$$R_{AE} = t\int_0^a \tau(u)\,du = \frac{V_y}{I_{zz} t}\left(\frac{t^2 a h^2}{8} + \frac{t^2 a^2 h}{4}\right) = \frac{V_y a h t}{8 I_{zz}}(h+2a)$$

$\sum M_F = 0$

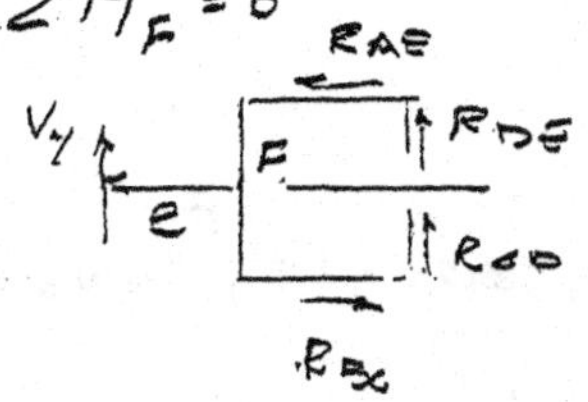

$$V_y e = 2R_{DE}a + R_{AE}h$$
$$= \frac{V_y t h^3 a}{24 I_{zz}} + \frac{V_y a h^2 t}{8 I_{zz}}(h+2a)$$
$$\Rightarrow e = \frac{a}{2}\left(\frac{2h+3a}{h+3a}\right).$$

13.10

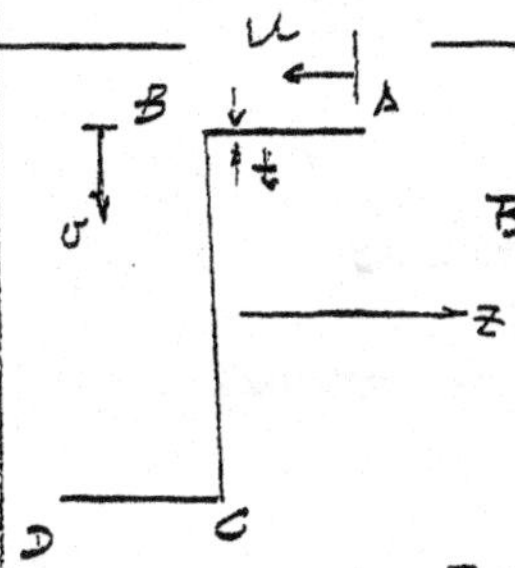

$B = I_{yy}I_{zz} - I_{yz}^2 = 72.7\ cm^4$, $b = 5\ mm = 0.5\ cm$.

$$\tau = \frac{(I_{yy}Q_z - I_{yz}Q_y)}{I_{yy}I_{zz} - I_{yz}^2}\cdot\frac{V_y}{b} = \frac{(9.1Q_z + 10.1Q_y)}{36.35}\cdot V_y$$

$$\tau = (0.250 Q_z + 0.278 Q_y)V_y \times 10^{-3} \quad Q\ [mm^3]$$

A-B: $Q_z = -22.5(5u) = -112.5u$; $Q_y = ut(c - u/2)$; $c = 30\ mm$, $t = 5\ mm$ mm. $Q_y = 5u(30 - u/2)$ mm.

B-C $Q_z = -3375 - 2.5v(45 - v)$; $Q_y = 2250$

(A-B) : $\tau/V_y = -0.028u + 0.0014u(30 - u/2) = 0.014u - 0.0007u^2$

(B-C) : $\tau/V_y = -0.844 - 6.25\times10^{-4}v(45 - v) + 0.6255 = -0.218 - 0.028v + 6.25\times10^{-4}v^2$

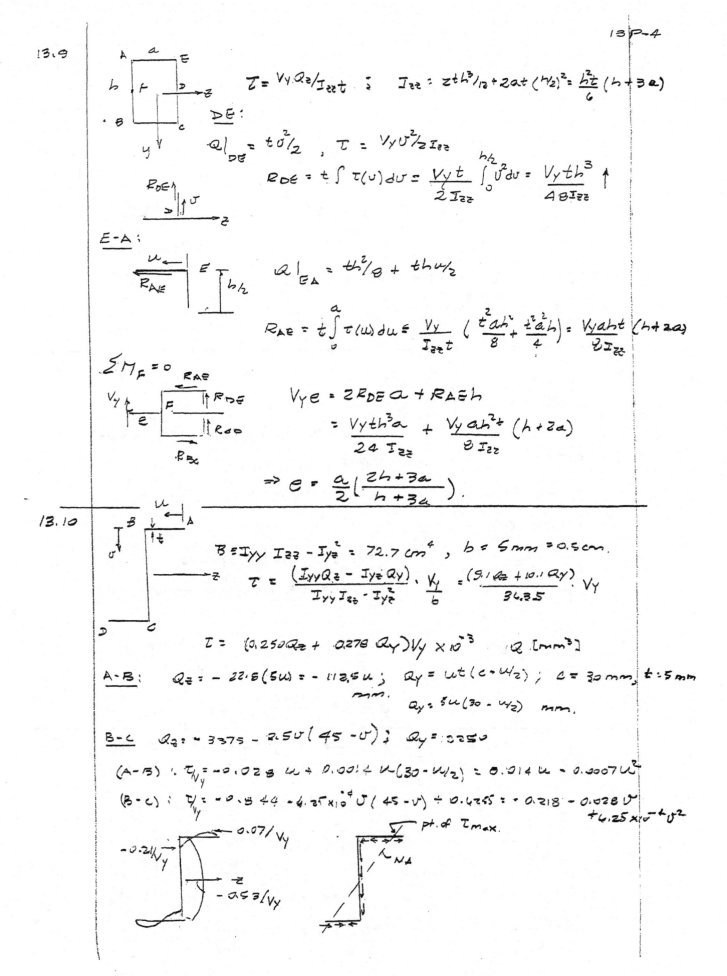

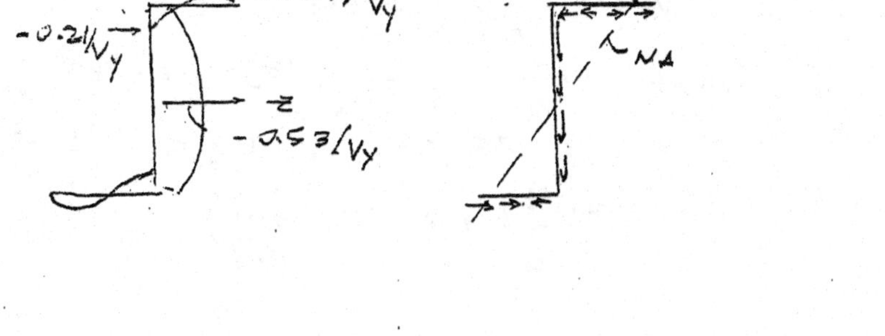

13.11

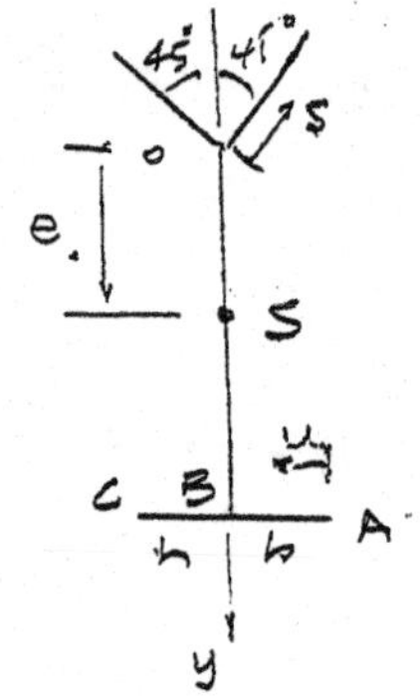

$$\tau = \frac{V_z Q_y}{I_{yy} t} \qquad Q_y = (tu)(h - u/2)$$

$$\tau = \frac{V_z (h - u/2)\, u}{I_{yy}}$$

$$R_{AB} = t\int_0^h \tau(u)\, du = \frac{tV_z}{I_{yy}} \int_0^h (h - u/2)\, u\, du = \frac{V_z t h^3}{3I_{yy}} \longrightarrow$$

$$R_{BC} = R_{AB}.$$

$$\sum M_o = 0 \Rightarrow V_z e = 2R_{AB}\, d \Rightarrow e = \frac{2}{3}\frac{th^3}{I_{yy}} \cdot d$$

$$I_{yy} = t(2h)^3/12 + 2\int_0^h (s\sqrt{2}/2)^2 t\, ds = 2th^3/3 + t\int_0^h s^2 ds = th^3.$$

$$\therefore e = 2d/3.$$

13.12

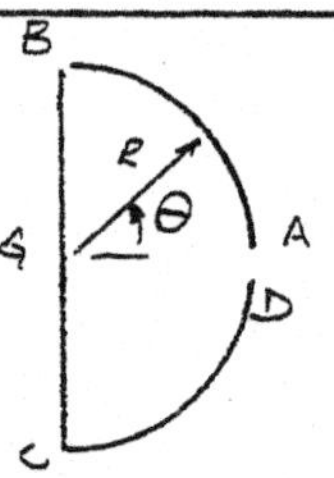

$$\tau = V_y Q_z / I_{zz} t$$

$$Q_z = -t\int_0^\theta (R\sin\varphi) R\, d\varphi = -tR^2 \int_0^\theta \sin\varphi\, d\varphi$$

$$Q_z(\theta) = tR^2(\cos\theta - 1)$$

$$I_{zz} = t(2R)^3/12 + 2tR^3\int_0^{\pi/2} \sin^2\varphi\, d\varphi = 2tR^3/3 + tR^3\pi/2 = (tR^3/6)(4 + 3\pi)$$

$$\tau(\theta) = \frac{V_y R^2(\cos\theta - 1)}{I_{zz}}$$

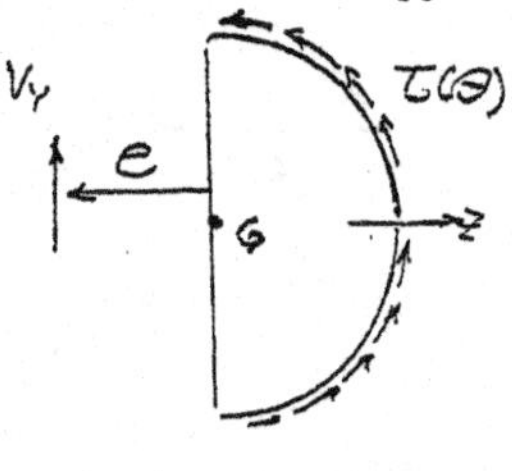

$$\sum M_G = V e - 2R\int_0^{\pi/2} \tau(\theta)\, tR\, d\theta = 0$$

$$V_y e = \frac{2R^4 V_y t}{I_{zz}} \int_0^{\pi/2} (1 - \cos\theta)\, d\theta$$

$$e = \frac{2R^4 t}{I_{zz}} \left[\theta - \sin\theta\right]_0^{\pi/2} = \frac{2R^4 t}{I_{zz}} (\pi/2 - 1)$$

$$\therefore e = 12R\left(\frac{\pi/2 - 1}{4 + 3\pi}\right) = 6R\left(\frac{\pi - 2}{3\pi + 4}\right).$$

13.13

(a) $\tau = V_Y Q_z / I_{zz} t$; $Q_z(\varphi) = \int_0^\varphi (-R\cos\theta) t\, d\theta = -R^2 t \sin\theta \Big|_0^\varphi = -R^2 t \sin\varphi$

$$I_{zz} = \int_0^\pi (R^2\cos^2\theta) tR\, d\theta = R^3 t/2 \int_0^\pi (1 - \cos 2\theta)\, d\theta = R^3 t\pi/2.$$

$\therefore \tau = -2V_y \sin\varphi / R\pi t$. (b) $\sum M_o = 0 \Rightarrow V_Y e + \int_0^\pi R\tau(\varphi) \cdot tR\, d\varphi = 0$

$$\Rightarrow V_Y e = \frac{2R^2 V_Y}{R\pi} \int_0^\pi \sin\varphi\, d\varphi \Rightarrow e = 4R/\pi$$

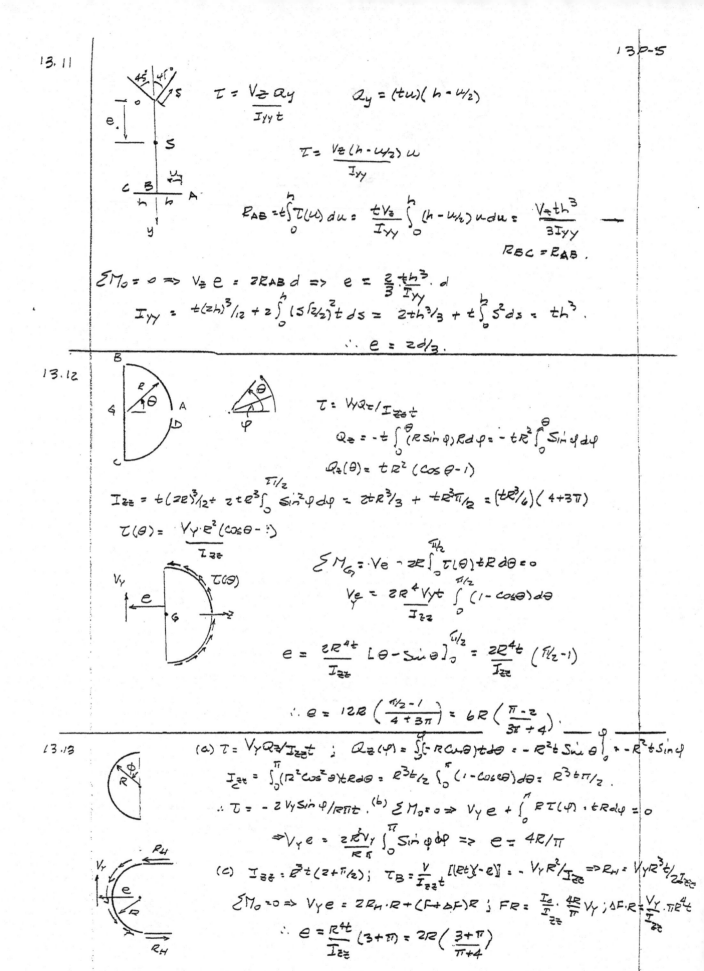

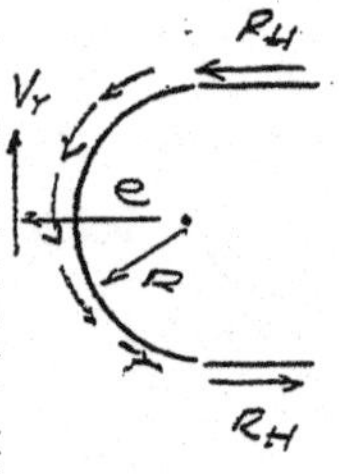

(c) $I_{zz} = R^3 t(2 + \pi/2)$; $\tau_B = \frac{V}{I_{zz} t}[(Rt)(-e)] = -V_Y R^2 / I_{zz} \Rightarrow R_H = V_Y R^3 t / 2I_{zz}$

$\sum M_o = 0 \Rightarrow V_Y e = 2R_H \cdot R + (F + \Delta F)R$; $FR = \frac{I_0}{I_{zz}} \cdot \frac{4R}{\pi} V_Y$; $\Delta F \cdot R = \frac{V_Y}{I_{zz}} \cdot \frac{\pi R^4 t}{2}$

$$\therefore e = \frac{R^4 t}{I_{zz}}(3 + \pi) = 2R\left(\frac{3 + \pi}{\pi + 4}\right)$$

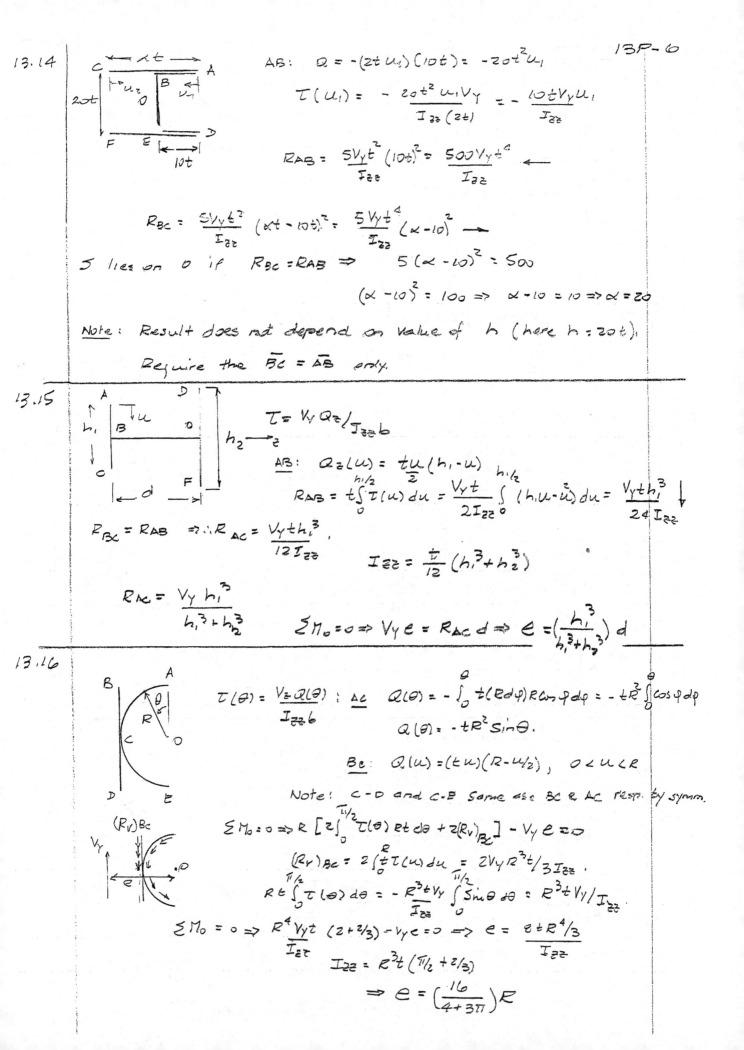

13.14

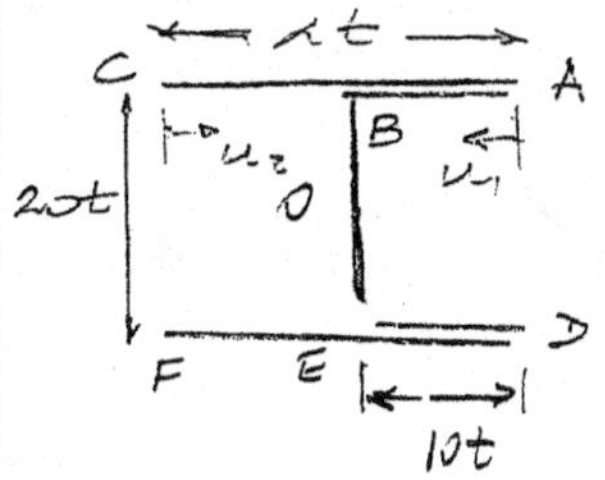

AB: $Q = -(2t\,u_1)(10t) = -20t^2 u_1$

$$\tau(u_1) = -\frac{20t^2 u_1 V_Y}{I_{zz}(2t)} = -\frac{10tV_Y u_1}{I_{zz}}$$

$$R_{AB} = \frac{5V_Y t^2}{I_{zz}}(10t)^2 = \frac{500V_Y t^4}{I_{zz}} \leftarrow$$

$$R_{BC} = \frac{5V_Y t^2}{I_{zz}}(\alpha t - 10t)^2 = \frac{5V_Y t^4}{I_{zz}}(\alpha - 10)^2 \rightarrow$$

S lies on O if $R_{BC} = R_{AB} \Rightarrow 5(\alpha - 10)^2 = 500$

$(\alpha - 10)^2 = 100 \Rightarrow \alpha - 10 = 10 \Rightarrow \alpha = 20$

Note: Result does not depend on value of h (here $h = 20t$).

Require the $\overline{BC} = \overline{AB}$ only.

13.15

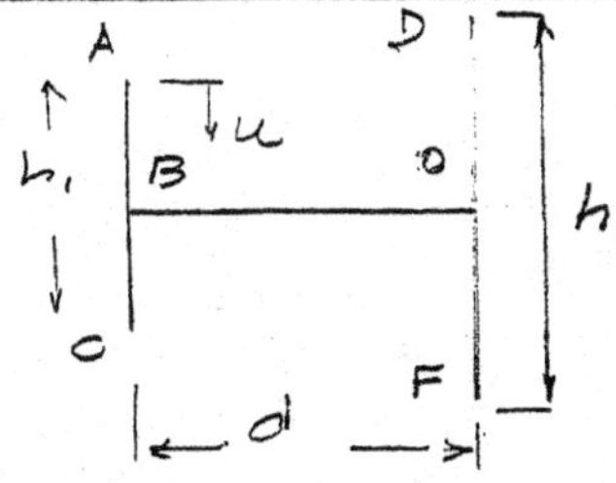

$\tau = V_Y Q_z / I_{zz} b$

AB: $Q_z(u) = \frac{tu}{2}(h_1 - u)$

$$R_{AB} = t\int_0^{h_1/2} \tau(u)\,du = \frac{V_Y t}{2I_{zz}}\int_0^{h_1/2}(h_1 u - u^2)\,du = \frac{V_Y t h_1^3}{24 I_{zz}} \downarrow$$

$R_{BC} = R_{AB} \Rightarrow \therefore R_{AC} = \frac{V_Y t h_1^3}{12 I_{zz}}$

$$I_{zz} = \frac{t}{12}(h_1^3 + h_2^3)$$

$$R_{AC} = \frac{V_Y h_1^3}{h_1^3 + h_2^3}$$

$$\Sigma M_O = 0 \Rightarrow V_Y e = R_{AC}\, d \Rightarrow e = \left(\frac{h_1^3}{h_1^3 + h_2^3}\right) d$$

13.16

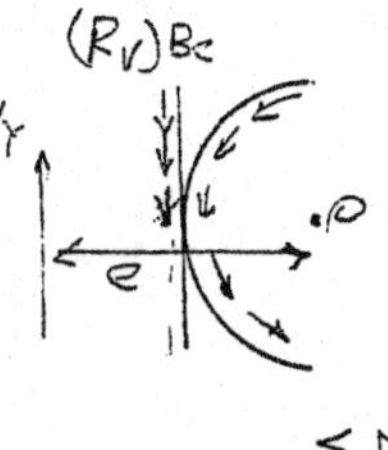

$$\tau(\theta) = \frac{V_z Q(\theta)}{I_{zz} b} \; : \; \underline{AC} \quad Q(\theta) = -\int_0^\theta t(R\,d\varphi)R\cos\varphi\,d\varphi = -tR^2\int_0^\theta \cos\varphi\,d\varphi$$

$$Q(\theta) = -tR^2\sin\theta.$$

$\underline{BC}$: $Q(u) = (tu)(R - u/2)$, $0 < u < R$

Note: C-D and C-E same as BC & AC resp. by symm.

$$\Sigma M_O = 0 \Rightarrow R\left[2\int_0^{\pi/2}\tau(\theta)\,Rt\,d\theta + 2(R_V)_{BC}\right] - V_Y e = 0$$

$$(R_V)_{BC} = 2\int_0^R t\,\tau(u)\,du = 2V_Y R^3 t / 3I_{zz}.$$

$$Rt\int_0^{\pi/2}\tau(\theta)\,d\theta = -\frac{R^3 t V_Y}{I_{zz}}\int_0^{\pi/2}\sin\theta\,d\theta = R^3 t V_Y / I_{zz}$$

$$\Sigma M_O = 0 \Rightarrow \frac{R^4 V_Y t}{I_{zz}}(2 + 2/3) - V_Y e = 0 \Rightarrow e = \frac{8tR^4/3}{I_{zz}}$$

$$I_{zz} = R^3 t(\pi/2 + 2/3)$$

$$\Rightarrow e = \left(\frac{16}{4 + 3\pi}\right)R$$

3.17

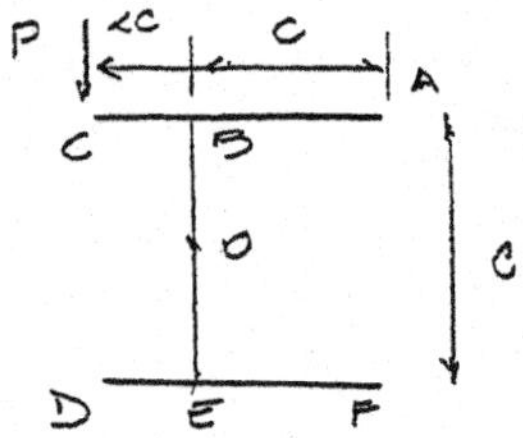

$\tau = V_y Q_z / I_z b$

A − B: $Q_z = -(t\omega)(c/2) = -tc\omega/2$

$\tau(\omega) = -V_y c\omega/2I_{zz}$; $R_{AB} = t\int_0^c \tau(\omega)\,d\omega = -V_y c^3 t/4I_{zz}$ ←

Similarly: $R_{BC} = -V_y(\alpha c)^3 t/4I_{zz}$ ⟶

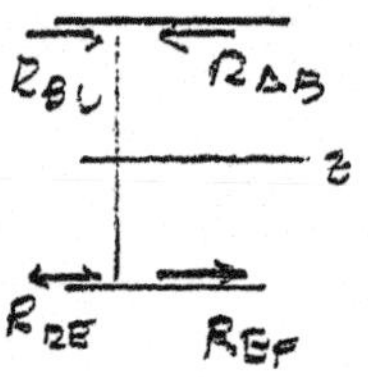

← By symmetry

$\sum M_0 = 0 \Rightarrow V_y e = (R_{AB} - R_{BC})c \Rightarrow e = \frac{h^4 t(1-\alpha^2)}{4I_{zz}}$

Require: $e = \alpha c$

$\therefore 4\alpha c I_{zz} = c^4 t(1-\alpha^2)$

$I_{zz} = tc^3/12 + 2[(1+\alpha)c](c/2)^2 = tc^3/12\,[1+6(1+c)]$

$\therefore 9\alpha^2 + 7\alpha - 3 = 0 \Rightarrow \alpha = 1/18[-7 \pm \sqrt{157}] = \frac{1}{18}(-7+12.58)$; $\alpha = 0.307$

3.18 $\tau(\varphi) = V_y Q_z(\varphi)/I_{zz}t(\varphi)$

$Q_z(\varphi) = \iint_A y\,dA = -R\int_0^\varphi \cos\theta \cdot t(\theta) R\,d\theta = -R^2 t_0 \int_0^\varphi \cos\theta \sin\theta\,d\theta$

$Q_z(\varphi) = R^2 t_0/2\,(\cos^2\varphi - 1)$. $\Rightarrow \tau(\varphi) = \frac{V_y R^2 t_0(\cos^2\varphi - 1)}{2I_{zz}(t_0 \sin\varphi)} = -\frac{V_y R^2}{2I_{zz}}\sin\varphi$.

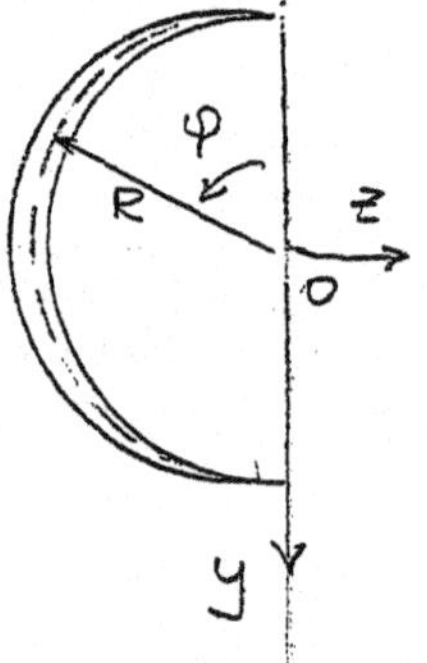

$\sum M_0 = 0 \Rightarrow V_y e = R\iint_A R\,\tau(\varphi)\,dA = R\int_0^\pi \tau(\varphi)t(\varphi)R\,d\varphi$

$\Rightarrow V_y e = R^2 t_0 \int_0^\pi \tau(\varphi)\sin\varphi\,d\varphi = \frac{V_y R^4 t_0}{2I_{zz}}\int_0^\pi \sin^2\varphi\,d\varphi = \frac{V_y R^4 t_0 \pi}{4I_{zz}}$

$I_{zz} = \iint_A y^2\,dA = R^2\int_0^\pi t(\theta)\cos^2\theta\,R\,d\theta = R^3 t_0 \int_0^\pi \cos^2\theta \sin\theta\,d\theta = 2R^3 t_0/3$

$\therefore e = \frac{\pi R^4 t_0}{8R^3 t_0/3} = 3\pi R/8$

3.19 If $\Delta \parallel P$, then $P \perp NA$ since $\Delta \perp NA$. $\Rightarrow \theta = \beta$

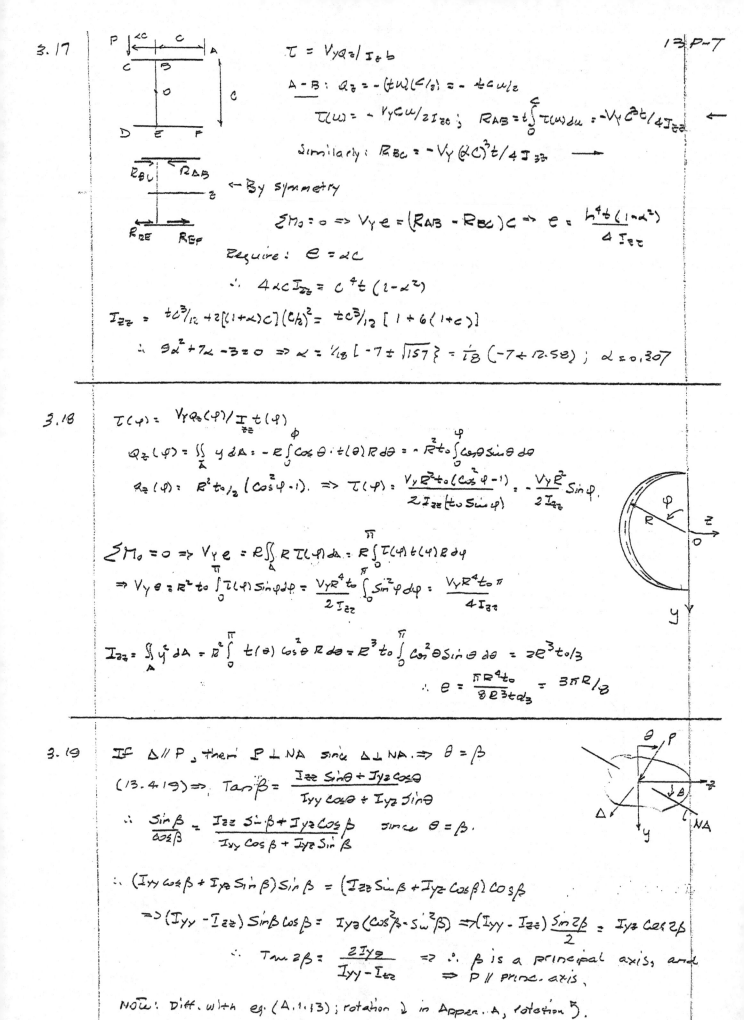

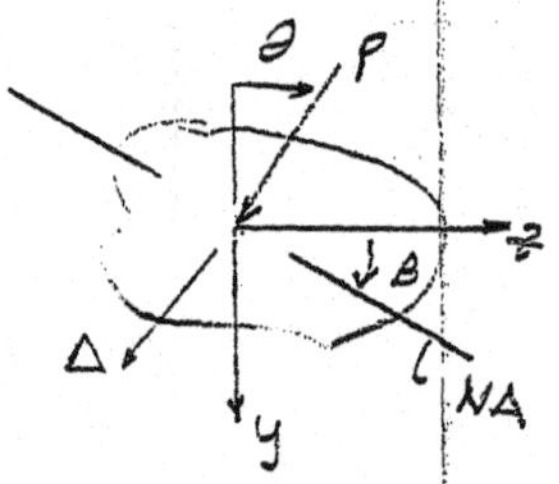

$(13.4.19) \Rightarrow \tan\beta = \frac{I_{zz}\sin\theta + I_{yz}\cos\theta}{I_{yy}\cos\theta + I_{yz}\sin\theta}$

$\therefore \frac{\sin\beta}{\cos\beta} = \frac{I_{zz}\sin\beta + I_{yz}\cos\beta}{I_{yy}\cos\beta + I_{yz}\sin\beta}$ since $\theta = \beta$.

$\therefore (I_{yy}\cos\beta + I_{yz}\sin\beta)\sin\beta = (I_{zz}\sin\beta + I_{yz}\cos\beta)\cos\beta$

$\Rightarrow (I_{yy} - I_{zz})\sin\beta\cos\beta = I_{yz}(\cos^2\beta - \sin^2\beta) \Rightarrow (I_{yy} - I_{zz})\frac{\sin 2\beta}{2} = I_{yz}\cos 2\beta$

$\therefore \tan 2\beta = \frac{2I_{yz}}{I_{yy} - I_{zz}}$ ⇒ ∴ β is a principal axis, and ⇒ P ∥ princ. axis.

NOTE: Diff. with eq. (A.1.13); rotation ↺ in Appen. A, rotation ↻.

13.20 $\Delta \parallel P \Rightarrow P$ acts in principal direction

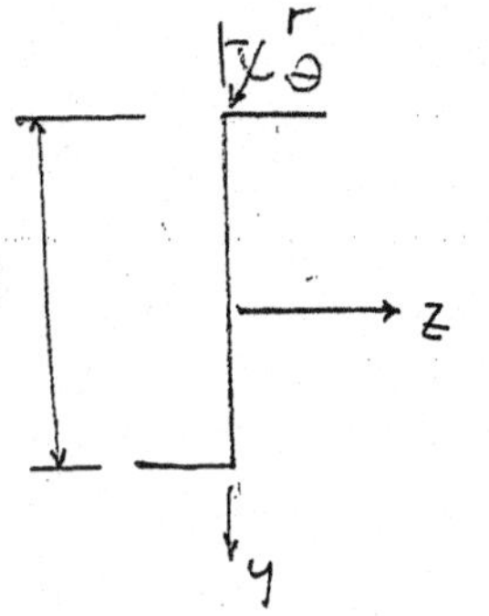

$$\tan 2\theta = \frac{2I_{yz}}{I_{zz} - I_{yy}}$$

$$I_{zz} = td^3/12 + 2[(at)\, d^2/4] = t[d^3/12 + ad^2/2]$$

$$= \frac{td^2}{12}[d + 6a].$$

$$I_{yy} = 2a^3t/3 \quad ; \quad I_{yz} = -2(at)\left(\frac{a}{2}\cdot\frac{d}{2}\right) = -a^2dt/2$$

$$d = 5a \Rightarrow I_{zz} = 275ta^3/12,\; I_{yy} = 2a^3t/3,\; I_{yz} = -5a^3t/2.$$

$$\tan 2\theta = \frac{-5a^3t}{\frac{275}{12} - \frac{2}{3}} = -0.2247 \Rightarrow \theta = -6.33°,\ 83.67°.$$

13.21

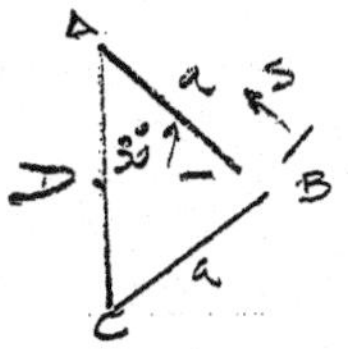

$$Q(s) = -st\left(\frac{s}{2}\sin 30°\right) = -s^2t/4 \qquad \tau(s) = -V_y s^2/4I_{zz}$$

$$R_{AB} = t\int_0^a \tau(s)\,ds = -V_y a^3/12I_{zz} \;;\; R_{BC} = R_{AB}.$$

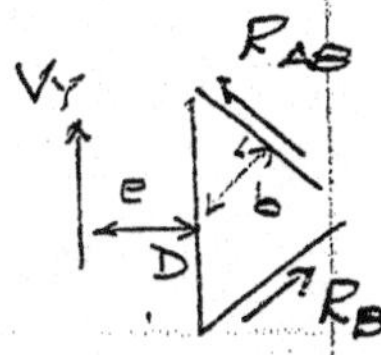

$$\Sigma M_o = 0 \Rightarrow V_y e = 2b R_{AB}, \quad b = \frac{a}{2}\cos 30° = a\sqrt{3}/4.$$

$$\therefore e = \frac{a\sqrt{3}}{2}\cdot\frac{ta^3}{12I_{zz}}$$

$$I_{zz} = ta^3/12 + 2\left[t\int_0^a (s^2\sin^2 30)\,ds\right]$$

$$= ta^3/12 + \frac{t}{2}\cdot\frac{a^3}{3} = ta^3/4$$

$$\therefore e = \frac{a\sqrt{3}}{6}$$

13.22

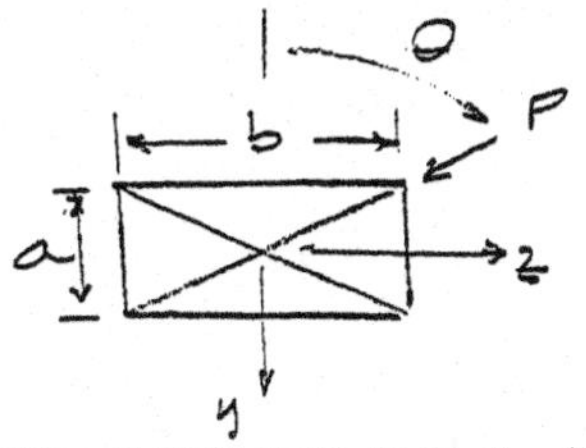

$$M_z = \pm M\cos\theta = \pm Ma/d,\quad M_y = \mp M\sin\theta = \mp M\,b/d;\quad \tan\theta = \mp b/a$$

$$\sigma_x = (I_{yy}M_z)y + (I_{zz}M_y)z = 0$$

$$\tan\beta = y/z = -\left(\frac{I_{zz}}{I_{yy}}\right)\left(\frac{M_y}{M_z}\right) = -\frac{I_{zz}}{I_{yy}}(-b/a)$$

$$I_{zz}/I_{yy} = ba^3/ab^3 = (a/b)^2$$

$$\therefore \tan\beta = a/b$$

13.23

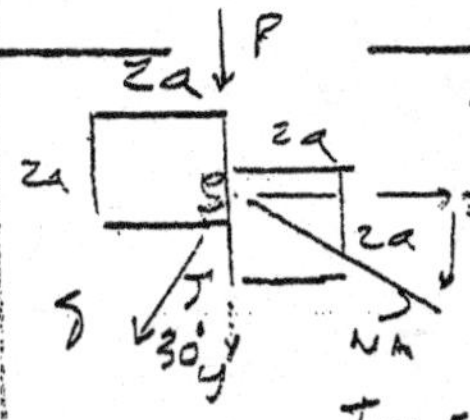

a) $M_y = 0 \Rightarrow \sigma_x = (I_{yy}\,y - I_{yz}\,z)M_z = 0$

$$\tan\beta = y/z = \frac{I_{yz}}{I_{yy}}$$

$$I_{yy} = 2\left(\frac{16a^4}{3}\right) = 32a^4/3\;;\quad I_{yz} = 2[(4a^2)(a-e)] = 8a^3(a-e)$$

$$\therefore \tan\beta = \frac{8a^3(a-e)}{32a^4/3} = \sqrt{3}/3 \Rightarrow \frac{3}{4}(a-e) = \sqrt{3}a/3 \Rightarrow \frac{e}{a} = 1 - \frac{4\sqrt{3}}{9} = 0.2302$$

b). $e = 0 \Rightarrow I_{yz} = 8a^4 \Rightarrow \beta = \tan^{-1}(3/4) = 36.87°$

13.24

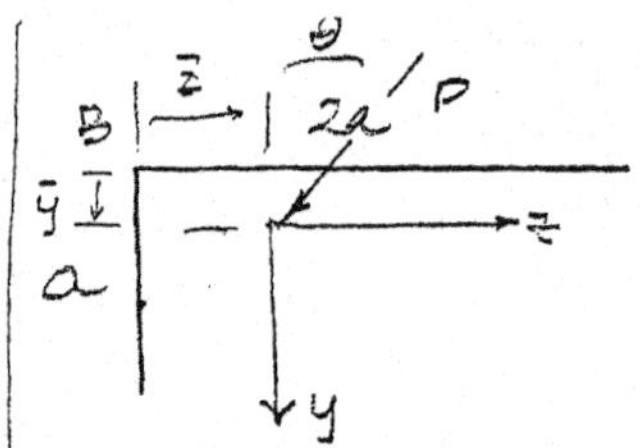

a) $\delta \parallel z \Rightarrow NA \parallel y \Rightarrow \beta = 90^\circ \Rightarrow T$

$\tan\beta \to \infty$ –

$\therefore$ (13.4.19) $\Rightarrow I_{yy}\cos\theta + I_{yz}\sin\theta = 0$

$\tan\theta = -I_{yy}/I_{yz}$

Centroid: $\bar{z} = 2a^2t/3at = 2a/3$; $\bar{y} = \dfrac{a^2t/2}{3at} = a/6$

$I_{yy} = (at)(2a/3)^2 = 4a^3t/9$;

$I_{yz} = at[(a/2 - a/6)(-a/3)] + 2at[(a - a/3)(-a/6)] = -a^3t/3$.

$\theta = \tan^{-1}(4/3) = 53.13^\circ$. Note: $\cos\theta = 0.6$, $\sin\theta = 0.8$

b)

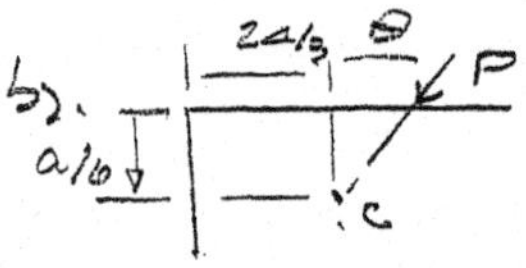

$\Delta_y = (\theta L)(2a/3) = \dfrac{2TLa}{3GC}$.

$T = (P\cos\theta)(2a/3) + (P\sin\theta)(a/6) = \dfrac{Pa}{6}(4\cos\theta + \sin\theta) = 8Pa/15$

$C = 3at^3/3 = at^3$

$\Delta_y = \dfrac{(16Pa/15)La}{3Gat^3} = \dfrac{16}{45}\cdot\dfrac{PLa}{Gt^3}$,

---

13.25

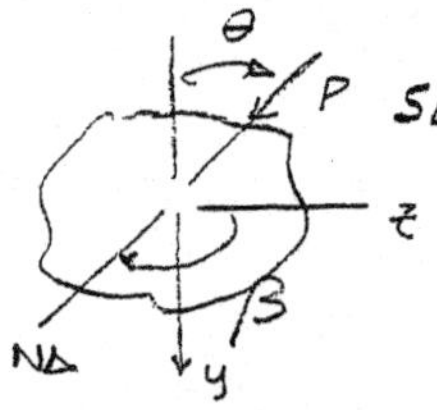

Proof by contradiction:

Suppose $\delta \perp P \Rightarrow NA \parallel P \Rightarrow \beta = \theta + 90^\circ$

(13.4.9) $\Rightarrow \tan\beta = \tan(\theta + 90^\circ) = \dfrac{I_{zz}\sin\theta + I_{yz}\cos\theta}{I_{yy}\cos\theta + I_{yz}\sin\theta}$

$\tan(\theta + 90^\circ) = -\dfrac{\cos\theta}{\sin\theta} = \dfrac{I_{zz}\sin\theta + I_{yz}\cos\theta}{I_{yy}\cos\theta + I_{yz}\sin\theta}$

$\therefore I_{yy}\cos^2\theta + I_{zz}\sin^2\theta = -2I_{yz}\sin\theta\cos\theta$

$\Rightarrow I_{yy}^2\cos^4\theta + 2I_{yy}I_{zz}\sin^2\theta\cos^2\theta + I_{zz}^2\sin^4\theta = 4I_{yz}^2\sin^2\theta\cos^2\theta$

$\Rightarrow I_{yy}^2\cos^4\theta - 2I_{yy}I_{zz}\sin^2\theta\cos^2\theta + I_{zz}^2\sin^4\theta = 4(I_{yz}^2 - I_{yy}I_{zz})\sin^2\theta\cos^2\theta$

or $(I_{yy}\cos^2\theta - I_{zz}\sin^2\theta)^2 = 4(I_{yz}^2 - I_{yy}I_{zz})\sin^2\theta\cos^2\theta$

Note: Left hand side is always positive

Right hand side is always negative since $I_{yy}I_{zz} - I_{yz}^2 > 0$, (see Prob. 13.4). $\Rightarrow$ Contradiction $\Rightarrow \therefore \delta \perp P$ is impossible.

13.26

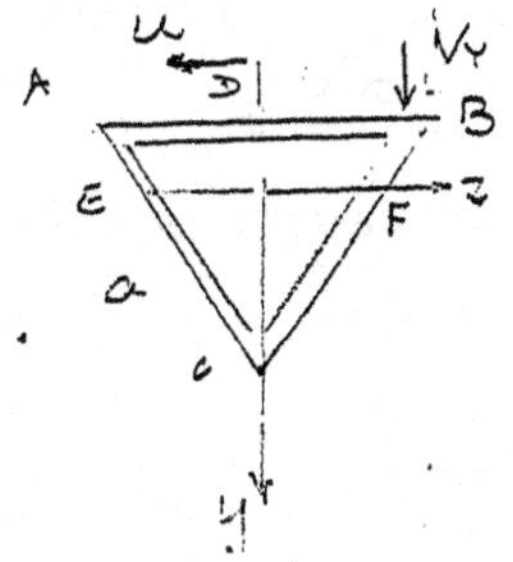

$\tau_f\big|_D = 0 \qquad Q(u) = -tu\,a\sqrt{3}/6 = -20u\sqrt{3}/3$

$(a = 40\text{ cm},\ t = 1\text{cm})$

$\tau_f\big|_{DA} = -\dfrac{20\sqrt{3}/3\,u V_y}{I_{zz}} \leftarrow \qquad \tau_f\big|_{DB} = -\dfrac{20\sqrt{3}/3\,u V_y}{I_{zz}} \rightarrow$

A-C: $\quad Q_E = Q_A - t\dfrac{(a\sqrt{3}/4)^2}{2} = -\dfrac{400\sqrt{3}}{3} - \dfrac{400}{6} = -\dfrac{200}{3}(1+2\sqrt{3})$

$\tau_{fE} = -\dfrac{200(1+2\sqrt{3})}{3}\dfrac{V_y}{I_{zz}} \qquad I_{zz} = I_{AB} + 2I_{AC}$

$= (at)a^2/12 + 2\left[\dfrac{ta^3}{12}\cos^2 30 + at\left(\dfrac{5\sqrt{3}a}{6}\right)^2\right]$

$= a^3t/12 + 103a^3t/24 = 35a^3t/8 = 7500\text{ cm}^4$

$V_f = 10^3\text{ N} \Rightarrow \quad (\tau_f)_A = -\dfrac{400\sqrt{3}/3}{7500}\cdot 2\times10^5 = 6158\text{ N/cm}^2 = 61.58\text{ MPa}$

$(\tau_f)_E = -7936\text{ N/cm}^2 = 79.36\text{ MPa}.$

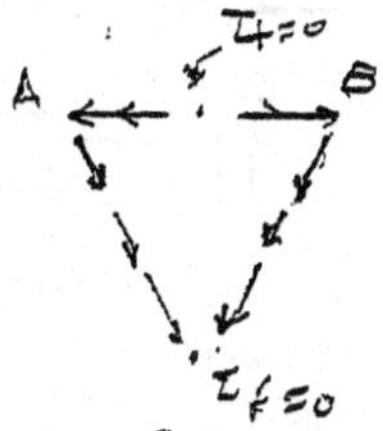

Note: By symmetry: $(\tau_f)_B = (\tau_f)_A$, $(\tau_f)_F = (\tau_f)_E$

$T_f = 0 \Rightarrow T = V_y\cdot 10 = 2\times10^6\text{ N-cm}, \quad A = a^2\sqrt{3}/4 = 693\text{ cm}^2$

$\tau_c = \dfrac{T}{2At} = \dfrac{2\times10^6}{2(693)(1)} = 1443\text{ N/cm}^2 = 14.43\text{ MPa}$

$\tau_D = \tau_c = 14.43\text{ MPa}, \quad \tau_{A,B} = 61.58 \mp 14.43 = 47.15, 76\text{ MPa}$

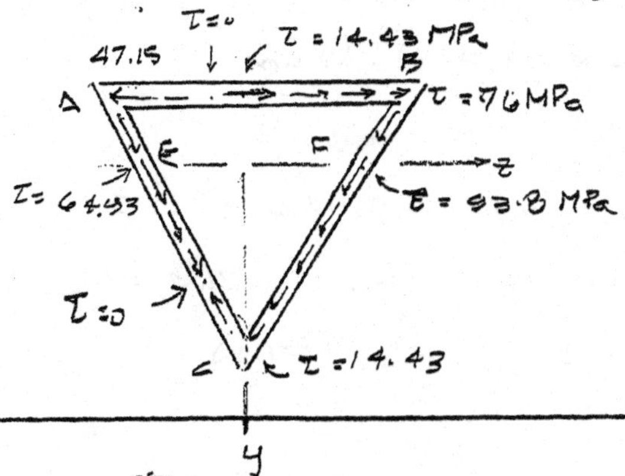

$\tau_E = 79.36 - 14.43 = 64.93\text{ MPa}$

$\tau_F = 79.36 + 14.43 = 93.8\text{ MPa}$

13.27

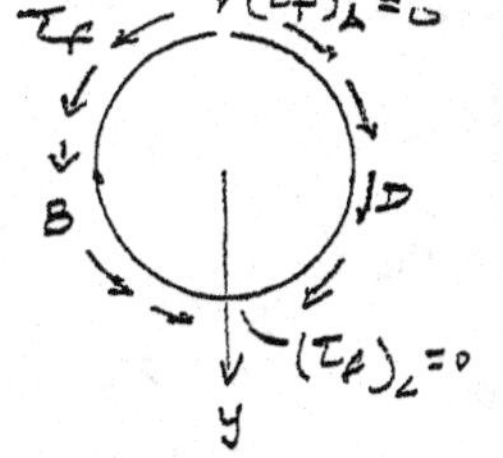

Due to W (acting along y-axis)

$Q_B = -t\int_0^{\pi/2} R\cos\theta\, R\,d\theta = -R^2t = Q_D$

$(\tau_f)_B = -\dfrac{WR^2t}{(\pi R^3 t)t} = -\dfrac{W}{\pi Rt} \downarrow = (\tau_f)_D$

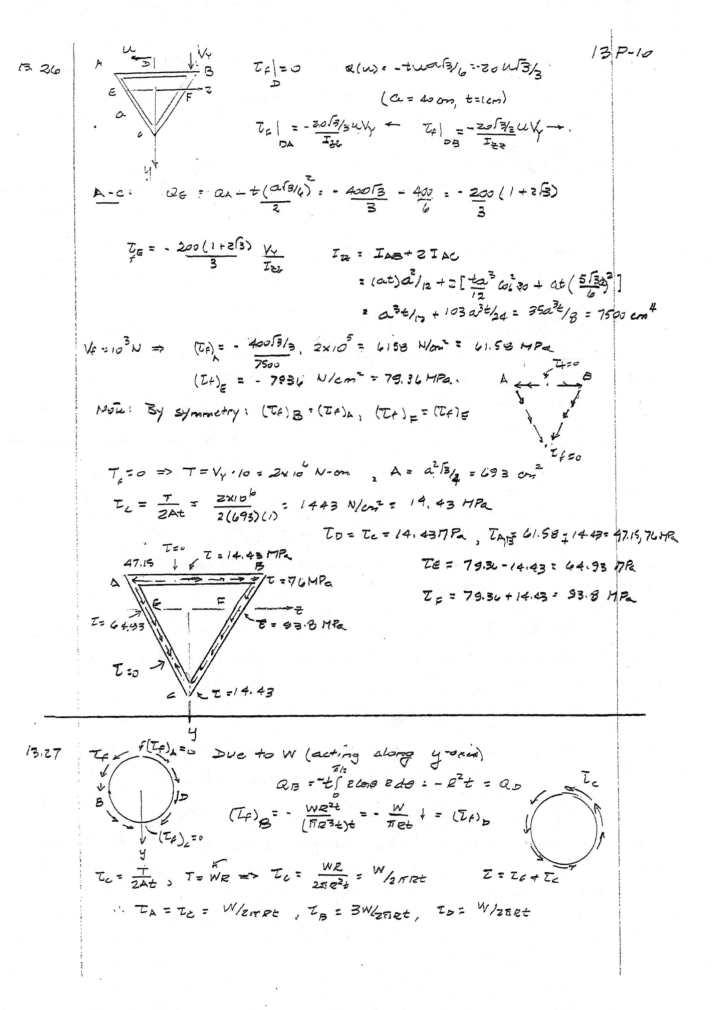

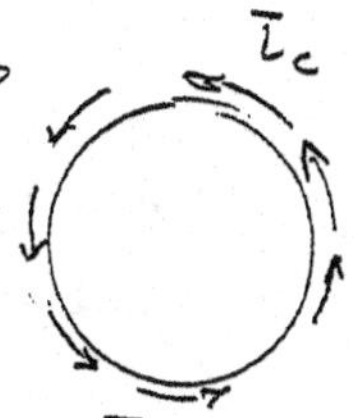

$\tau_c = \dfrac{T}{2At}, \quad T = WR \Rightarrow \tau_c = \dfrac{WR}{2\pi R^2 t} = W/2\pi Rt \qquad \tau = \tau_f + \tau_c$

$\therefore\ \tau_A = \tau_c = W/2\pi Rt, \quad \tau_B = 3W/2\pi Rt, \quad \tau_D = W/2\pi Rt$

13.29

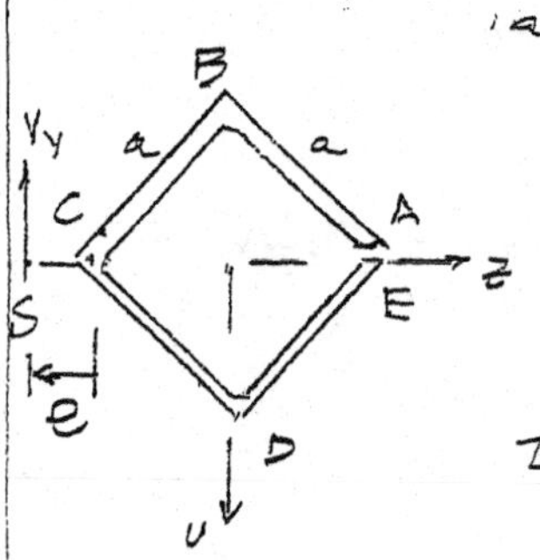

(a) $I = 2\left[\frac{ta^3}{12} + (at)(a/2)^2\right] = 2a^3t/3$

Note: I is same about all centroidal axes.

A-B:

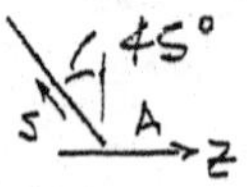

$Q(s) = (st)\frac{s}{2}\cos 45° = \frac{\sqrt{2}}{4}s^2t$

$\tau(s) = \frac{(\sqrt{2}s^2t/4)V_y}{(2a^3t/3)t} = \frac{3\sqrt{2}s^2}{8a^3t}V_y$

$R_{AB} = \int_0^a t\,\tau(s)\,ds = \frac{3\sqrt{2}V_y}{8a^3}\int_0^a s^2\,ds = \frac{\sqrt{2}V_y}{8}$; $R_{DE} = R_{AB}$

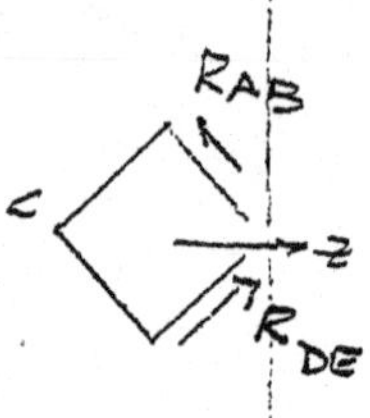

$\sum M_G = 0 \Rightarrow V_y e = \frac{2\sqrt{2}V_y}{8}.a \Rightarrow e = \sqrt{2}a/4$

b) TRUE! SINCE ALL AXES ARE PRINCIPAL AXES.

13.30

a).

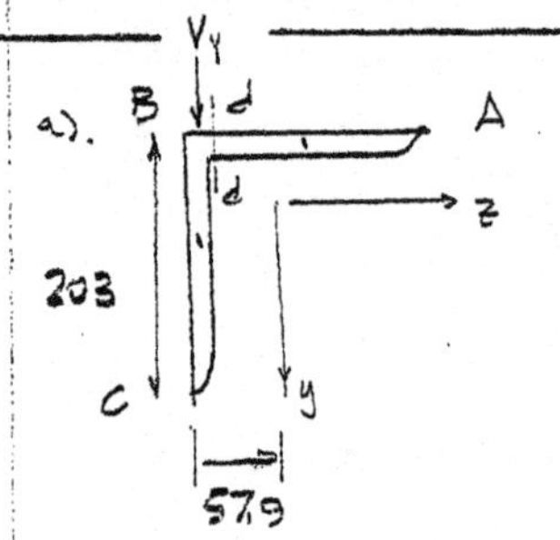

$V_y = 1,400$ N L203×203×19.1

$I_{yy} = I_{zz} = 29.0\times10^6$ mm^4, $A = 7355$ mm^2

$r_{min} = 40.1$ mm $\Rightarrow I_{min} = 11.83\times10^6$ mm^4

$I_{min} = 29.0\times10^6 - \sqrt{I_{yz}^2} = 11.83\times10^6 \Rightarrow I_{yz} = -17.17\times10^6$ mm^4 ← NOTE SIGN!

$I_{yy}I_{zz} - I_{yz}^2 = 5.46\times10^{14}$ mm^8

$M_z = PL = (1.4\times10^3)(250) = 3.5\times10^5$ N-cm. $= 3.5\times10^7$ N-mm ; $M_y = 0$

$\sigma_x = \frac{(I_{yy}y - I_{yz}z)M_z}{I_{yy}I_{zz} - I_{yz}^2} = (1.86\,y + 1.10\,z)$ N/mm^2

$(\sigma_x)_A = \sigma_x(y = -57.9, z = 145.1) = 111.6$ N/mm^2 = 51.9 N/mm^2 = 51.9 MPa (Tens.)

$(\sigma_x)_B = \sigma_x(y = -57.9, z = -57.9) = -171.4$ N/mm^2 = -171.4 MPa (comp.)

$(\sigma_x)_C = \sigma_x(y = 145.1, z = -57.9) = 204.2$ MPa (TENS.)

$\tau = \left(\frac{I_{yy}Q_z - I_{yz}Q_y}{I_{yy}I_{zz} - I_{yz}^2}\right)V_y$

$\tau|_{d-d}$: $Q_z|_{d-d} = (184.9\times19.1)(-48.35) = -170.8\times10^3$ mm^3

$Q_y|_{d-d} = (184.9\times19.1)(203 - 57.9) = 509.9\times10^3$ mm^3

$\tau_{d-d} = \frac{(29\times10^6)(-170.8\times10^3) - (-17.17\times10^6)(509.9\times10^3)}{5.46\times10^{14}} = 0.016$ MPa.

b). $\tan\beta = I_{yz}/I_{yy} = -17.17/29 = 0.5921$ $\beta = 30.63°$

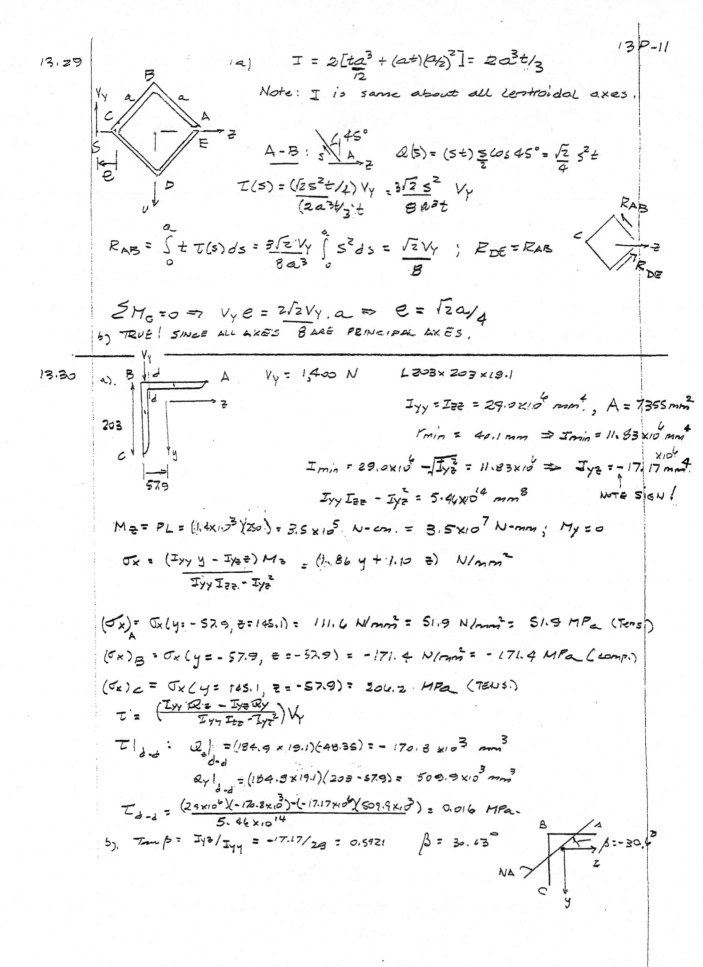

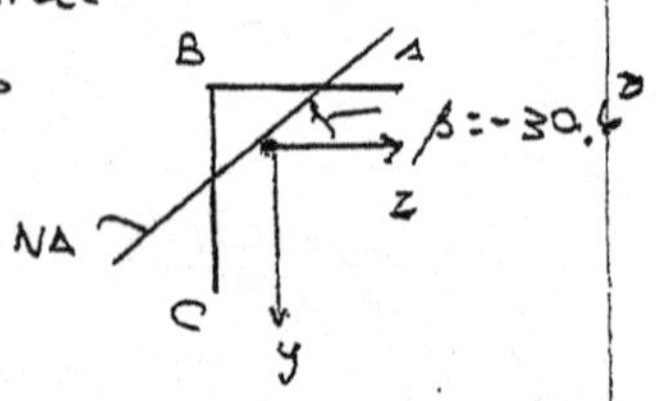

13.31 $\delta \parallel y \Rightarrow$ NA lies on z-axis.

a) $\therefore \; I_{zz} M_y - I_{yz} M_z = 0 \qquad M_z = \pm M \cos\alpha, \; M_y = \mp M \sin\alpha$

$$I_{zz}\sin\alpha + I_{yz}\cos\alpha = 0 \Rightarrow \tan\alpha = -I_{yz}/I_{zz}$$

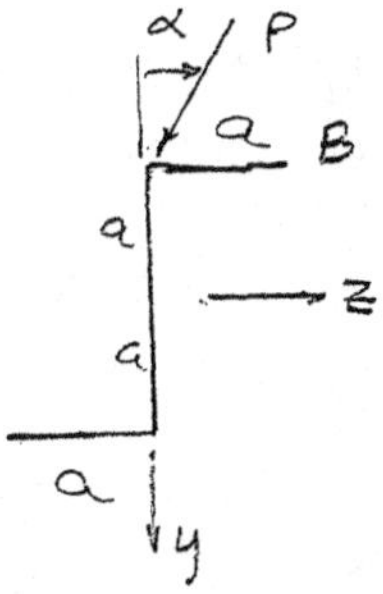

$$I_{zz} = t(2a)^3/12 + 2[(ta)a^2] = \frac{8a^3t}{3}$$

$$I_{yz} = 2[(at)(-a)(a/2)] = -a^3t.$$

$$\tan\alpha = +3/8 \qquad \alpha = 20.56^\circ$$

b). $(\Delta_z)_B = a\varphi = a\frac{TL}{GC} \qquad T = aP\sin\alpha\; ; \; \sin\alpha = \frac{\tan\alpha}{(1+\tan^2\alpha)^{1/2}} = \frac{3/8}{(1+9/64)^{1/2}}$

$$\sin\alpha = \frac{3}{\sqrt{73}}$$

$\therefore \; C = 4at^3 \Rightarrow (\Delta_z)_B = \frac{(3Pa/\sqrt{73})aL}{G4at^3} = \frac{3PaL}{4\sqrt{73}Gt^3} = \frac{3\sqrt{73}PaL}{292Gt^3}.$

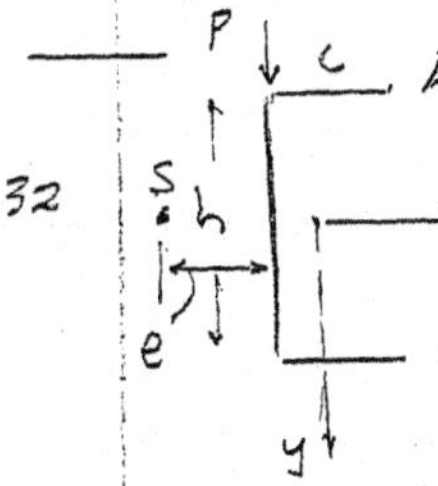

13.32 a) $(\Delta_y)_S = PL^3/3EI, \; (\Delta_z)_S = 0$

$$(\Delta_y)_A = (\Delta_y)_S + \varphi(e+c) = PL^3/3EI_{zz} + \frac{(Pe)L}{GC}(c+e) = PL\left[\frac{L^2}{3EI_{zz}} + \frac{e(c+e)}{GC}\right]$$

$$(13.6.10b) \Rightarrow e = \frac{3c^2}{h+6c}, \quad I_{zz} = th^3/12 + 2ct(h/2)^2 = th^2(h/12 + c/2) = \frac{th^2}{12}(h+6c)$$

$$C = (h+2c)t^3/3$$

$$\therefore (\Delta_y)_A = PL\left\{\frac{4L^2}{Eth^2(h+6c)} + \frac{\left(\frac{3c^2}{h+6c}\right)\left(c + \frac{3c^2}{h+6c}\right)}{G(h+2c)t^3/3}\right\} = \frac{PL}{t(h+6c)}\left\{\frac{4L^2}{Eh^2} + \frac{9c^2(h+3c)}{G(h+2c)(h+6c)}\right\} \downarrow$$

$$(\Delta_z)_A = \varphi h/2 = \frac{PeLh}{2GC} = \frac{9Pc^2hL}{2G(h+2c)(h+6c)t^3} \rightarrow$$

b). $c \to 0 \quad (\Delta_y)_A = \frac{4PL^3}{Eth^3}, \quad (\Delta_z)_A = 0$

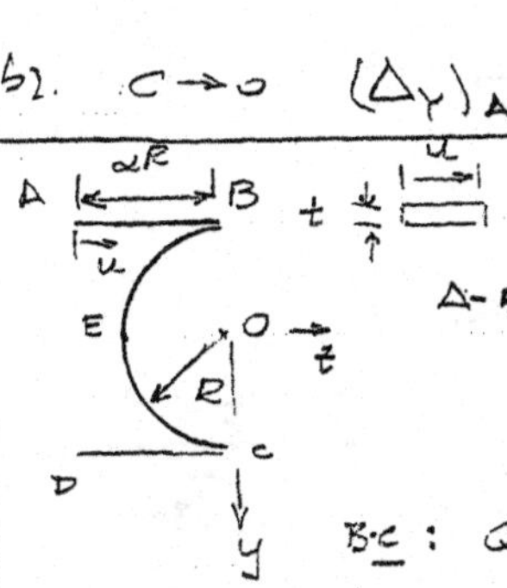

13.33

$$A\text{-}B: \; Q(u) = -tuR \Rightarrow Q_B = Q(u = \alpha R) = -t\alpha R^2$$

$$R_{AB} = \int_0^{\alpha R} \tau(u)t\,du = -\frac{V_y t}{I_{zz}}\frac{\alpha^2 R^3}{2} \rightarrow \qquad \overleftarrow{R_{CD}} = \overrightarrow{R_{AB}}$$

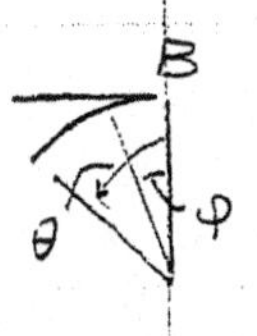

$$B\text{-}C: \; Q_{BC} = Q_B + tR\int_0^\theta R\cos\varphi\,d\varphi = Q_B - tR^2\int_0^\theta \cos\varphi\,d\varphi = -tR^2(\alpha + \sin\theta).$$

$$\sum M_O = 0 \Rightarrow V_y e + R_{AB}(2R) - tR^2\int_\pi \tau(\theta)\,d\theta = 0$$

$$e = 0 \Rightarrow 2R \cdot R_{AB} = tR^4\left[\int_0^\pi(\alpha + \sin\theta)\,d\theta\right]\frac{V_y}{I_{zz}} = \frac{tR^4}{I_{zz}}V_y(\alpha\pi + 2)$$

$$\alpha^2 R^4 = R^4(\alpha\pi + 2) \Rightarrow \alpha^2 - \pi\alpha - 2 = 0$$

$$\alpha = \frac{1}{2}\{\pi + (\pi^2 + 8)^{1/2}\} = 3.684$$

13.34

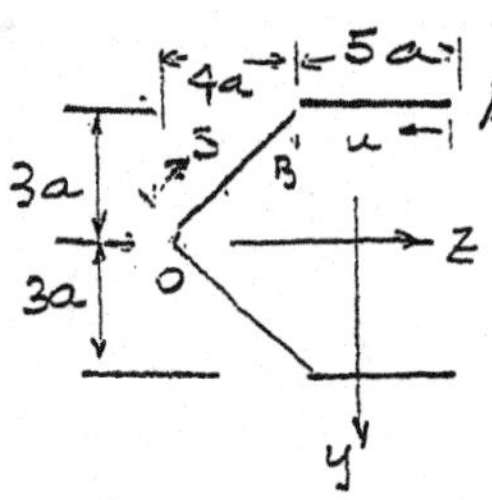

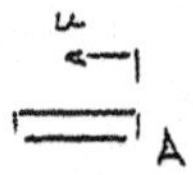

$Q_z(u) = -tu(3a) \quad Q_{zB'} = -15a^2t.$

$R_{A'B'} = \dfrac{75a^3t\,V_y}{2I_{zz}} \leftarrow$

$\sum M_O = 0 \Rightarrow V_y e = 6a R_{A'B'} \Rightarrow e = \dfrac{45a^4t}{I_{zz}}.$

$I_{zz} = 2(5at)(3a)^2 + 2t\int_0^{5a}\left(\frac{3}{5}s\right)^2 ds$

$= 2t\left\{45a^3 + \frac{9}{25}\frac{s^3}{3}\Big|_0^{5a}\right\} = 2t\{45a^3 + 15a^3\} = 120ta^3.$

$e = 15a/8.$

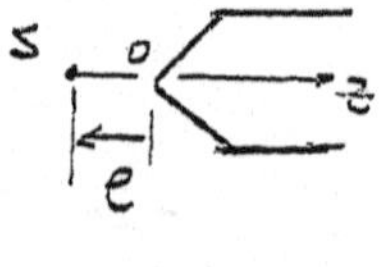

$(\Delta_Y)_O = (\Delta_Y)_{fl} + (\Delta_Y)_T$

From eq (9.7.15c), $(\Delta y)_{fl} = \dfrac{P}{6EI_{zz}}\left\{3(L/3)^2L - (L/3)^3 + 3\left[3(2L/3)^2L - (2L/3)^3\right]\right\}$

$(\Delta y)_{fl} = 46PL^3/81\,EI.$

$\phi_D = \dfrac{T_{AB}L}{3CG} + \dfrac{T_{BC}L}{3CG} = \dfrac{L}{3CG}[4P + 3P]e = \dfrac{7PeL}{3CG} = \dfrac{7PL}{3CG}\cdot\dfrac{15a}{8} = \dfrac{35PaL}{8CG}.$

$C = 20at^3/3 \qquad \phi_D = \dfrac{21PL}{32Gt^3}$

$(\Delta y)_T = \phi e = \left(\dfrac{21PL}{32Gt^3}\right)(15a/8) = \dfrac{315}{256}\dfrac{PLa}{Gt^3}.$

$(\Delta y)_O = \dfrac{46PL^3}{81EI} + \dfrac{315PLa}{256Gt^3}$

13.35

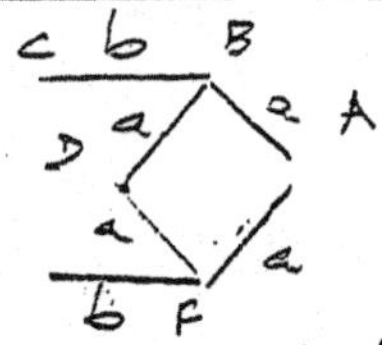

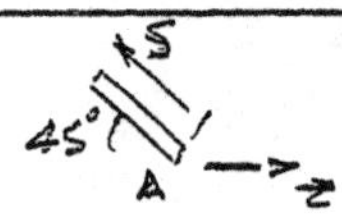

A-B: $Q_z(s) = -ts(s/2\cdot\sqrt{2}/2) = -\sqrt{2}s^2t/4$

$R_{AB} = \int_0^a \tau(s)t\,ds = \dfrac{V_y\sqrt{2}}{4I_{zz}}\int_0^a ts^2 ds = \dfrac{\sqrt{2}V_y ta^3}{12I_{zz}}$

B-C:

$Q_z(u) = -tu(\sqrt{2}a/2) = -\dfrac{\sqrt{2}}{2}tau$

$R_{BC} = -\dfrac{\sqrt{2}ta}{2I_{zz}}\int_0^b u\,du = -\dfrac{\sqrt{2}tV_y}{4I_{zz}}ab^2 \rightarrow$

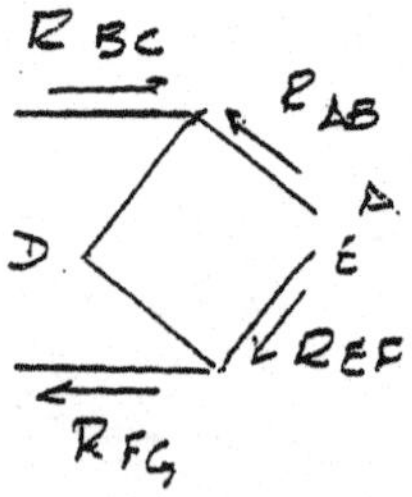

$\sum M_D = 0 \Rightarrow R_{BC}(\sqrt{2}a/2) = R_{AB}a$

$\left(\dfrac{\sqrt{2}a}{2}\right)\left(\dfrac{\sqrt{2}tab^2V_y}{4I_{zz}}\right) = \dfrac{\sqrt{2}V_y ta^3}{12I_{zz}}$

$b^2/4 = \dfrac{\sqrt{2}}{12}a^2 \Rightarrow (b/a)^2 = \sqrt{2}/3$

$b/a = (\sqrt{2}/3)^{1/2}$

13.36

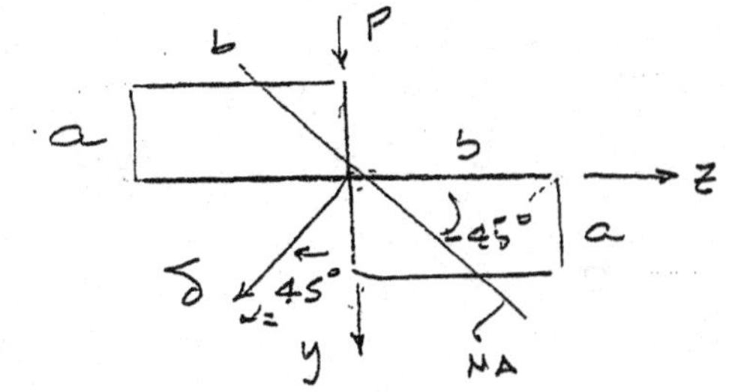

eq. (13.4.19) ⇒

$$\tan\beta = \frac{I_{zz}\sin\theta + I_{yz}\cos\theta}{I_{yy}\cos\theta + I_{yz}\sin\theta}.$$

$\theta = 0, \quad \beta = 45°$

$$\tan\beta = \frac{I_{yz}}{I_{yy}} = 1$$

$$I_{yy} = 2\frac{ab^3}{3}, \quad I_{yz} = 2[(ab)(ab/4)] = a^2b^2/2$$

$$a^2b^2/2 = 2ab^3/3 \Rightarrow b/a = 3/4.$$

13.37

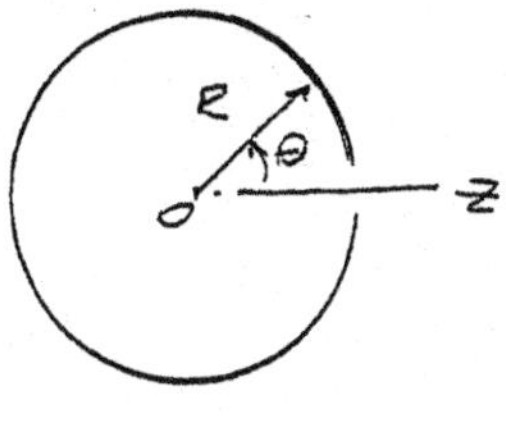

a). $Q(\theta) = -Rt\int_0^\theta R\sin\varphi\, d\varphi = +R^2t(\cos\theta - 1)$

$$\tau(\theta) = \frac{V_Y Q(\theta)}{I_{zz}t} = \frac{V_Y R^2 t(\cos\theta - 1)}{\pi R^3 t^2}$$

$$\Sigma M_O = 0 \Rightarrow V_Y e = R\int_0^{2\pi} \tau(\theta)\, tR\, d\theta = \frac{V_Y R^4}{\pi R^3}\int_0^{2\pi}(1-\cos\theta)\, d\theta = \frac{V_Y R}{\pi}\int_0^{2\pi}(1-\cos\theta)\, d\theta$$

$$e = \frac{R}{\pi}\cdot 2\pi = 2R$$

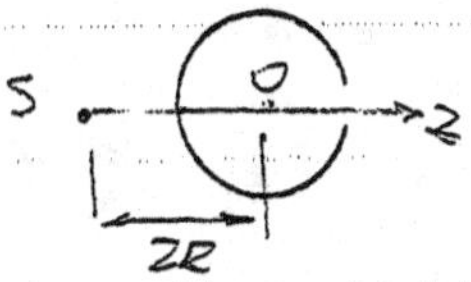

b). $(\Delta_Y)_D = \phi R = \left(\frac{TL}{GC}\right)R =$

$C = 2\pi R t^3/3 \Rightarrow (\Delta_Y)_D = \frac{3TL}{2\pi t^3 G}$ ↑

13.38

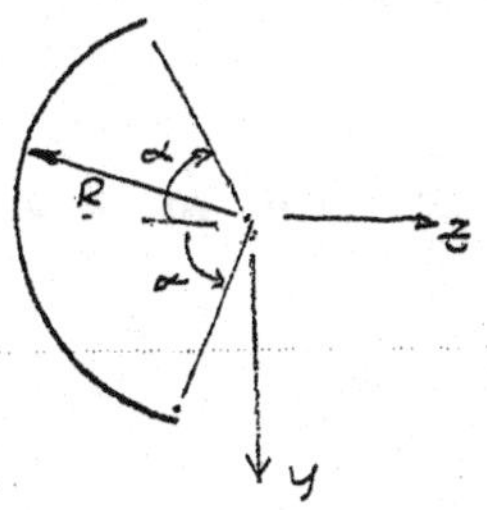

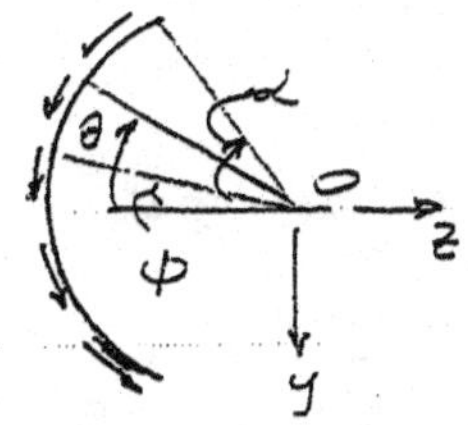

$$Q(\theta) = -\int_\theta^\alpha (R\sin\varphi)(tR\,d\varphi) = -R^2t\int_\theta^\alpha \sin\varphi\, d\varphi$$

$$= R^2t\cos\varphi\Big|_\theta^\alpha = R^2t(\cos\alpha - \cos\theta) \quad < 0$$

$$\tau(\theta) = \frac{V_Y R^2 t(\cos\alpha - \cos\theta)}{I_{zz}t} < 0$$

$$\curvearrowleft + \Sigma M_O = 0 \Rightarrow V_Y e - R\int_{-\alpha}^{\alpha}\tau(\theta)(Rt\,d\theta) = 0 \Rightarrow V_Y e = \frac{V_Y R^4 t}{I_{zz}}\int_{-\alpha}^{\alpha}(\cos\theta - \cos\alpha)\, d\theta$$

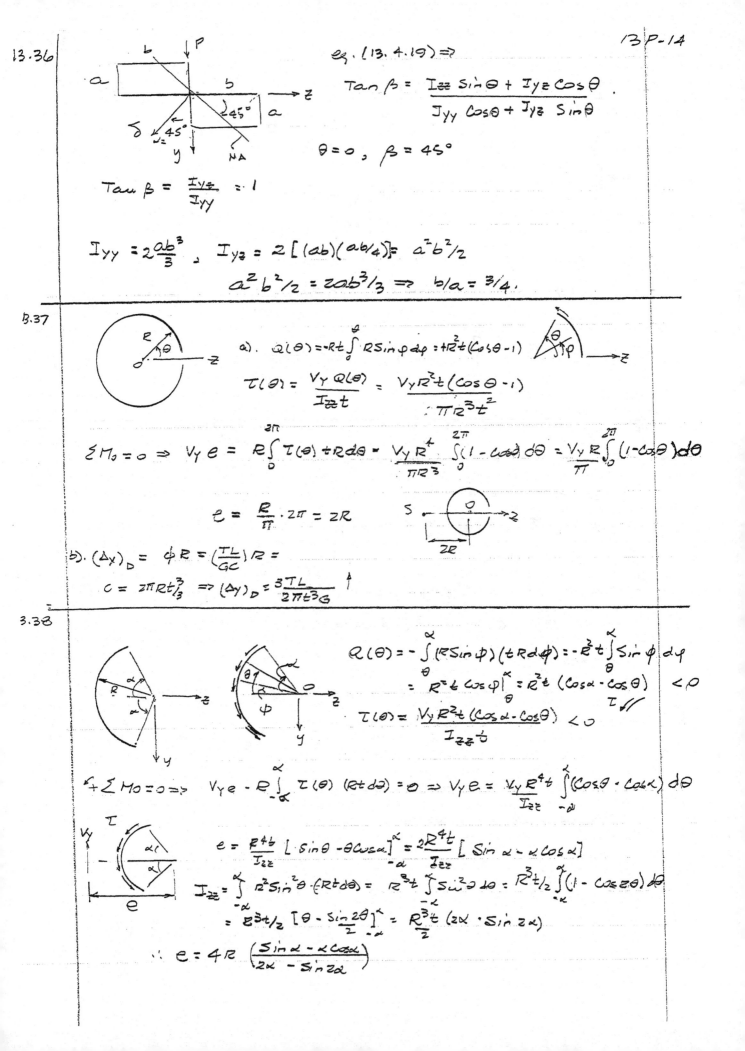

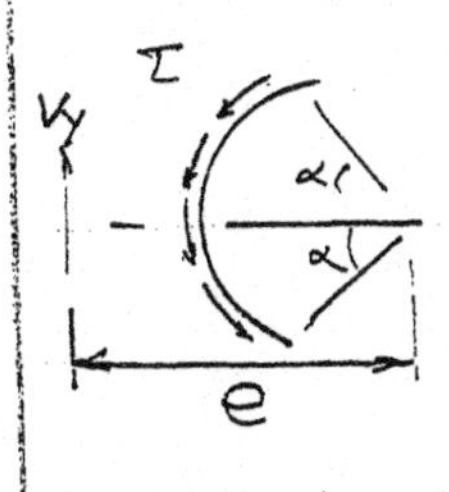

$$e = \frac{R^4t}{I_{zz}}[\sin\theta - \theta\cos\alpha]_{-\alpha}^{\alpha} = \frac{2R^4t}{I_{zz}}[\sin\alpha - \alpha\cos\alpha]$$

$$I_{zz} = \int_{-\alpha}^{\alpha} R^2\sin^2\theta\,(Rt\,d\theta) = R^3t\int_{-\alpha}^{\alpha}\sin^2\theta\, d\theta = R^3t/2\int_{-\alpha}^{\alpha}(1-\cos 2\theta)\, d\theta$$

$$= R^3t/2\left[\theta - \frac{\sin 2\theta}{2}\right]_{-\alpha}^{\alpha} = \frac{R^3t}{2}(2\alpha - \sin 2\alpha)$$

$$\therefore e = 4R\left(\frac{\sin\alpha - \alpha\cos\alpha}{2\alpha - \sin 2\alpha}\right)$$

13.39

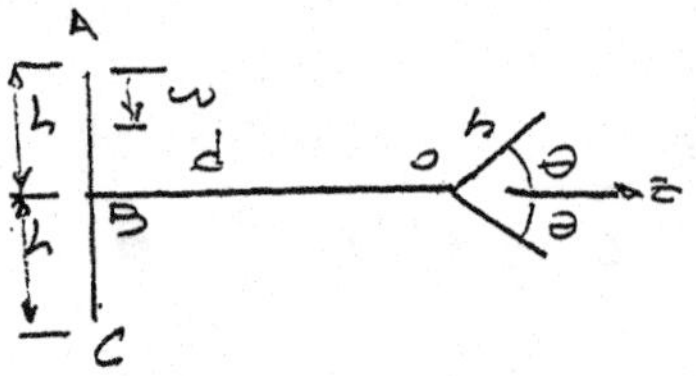

A-B: $Q(w) = -wt(h - w/2) = -\frac{wt}{2}(2h - w)$

$\tau(w) = -\frac{V_y w}{2 I_{zz}}(2h - w)$

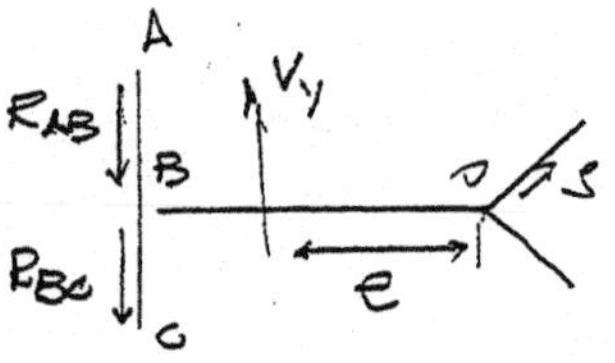

$R_{AB} = \frac{V_y t}{2 I_{zz}} \int_0^h (2hw - w^2)\,dw = \frac{V_y t}{2 I_{zz}} \left[h w^2 - w^3/3 \right]_0^h$

$= V_y h^3 t / 3 I_{zz}$ ↓

By symmetry: $R_{BC} = R_{AB}$ ↓

$\sum M_o = 0 \Rightarrow V_y e = (R_{AB} + R_{BC}) d = \frac{2 V_y h^3 t d}{3 I_{zz}}$

$I_{zz} = t(2h)^3/12 + 2t \int_0^h (s \sin\theta)^2 ds = \frac{2th^3}{3}(1 + \sin^2\theta)$

$\therefore e = \frac{d}{1 + \sin^2\theta}$

13.40 From Prob. 13.20 $\tan 2\theta = \frac{2 I_{yz}}{I_{zz} - I_{yy}}$

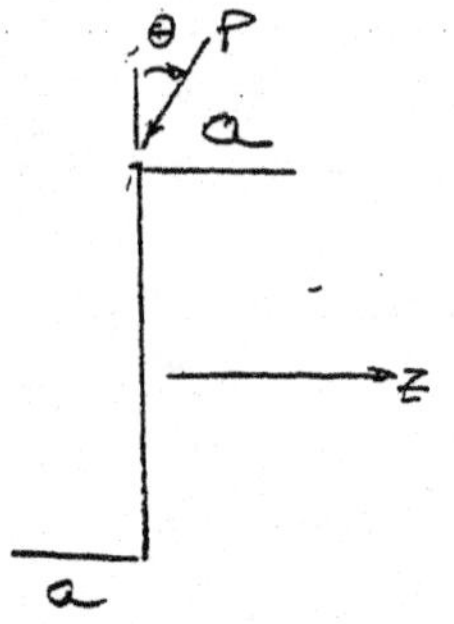

is required condition.

$I_{yy} = 2a^3t/3$, $I_{zz} = \frac{td^2}{12}(d + 6a)$

$I_{yz} = -a^2dt/2$

$\therefore \Rightarrow I_{zz} - I_{yy} = \frac{d^3}{12}(d + 6a) - 2a^3t/3$

$\therefore \left[\frac{d^2}{12}(d + 6a) - 8a^3\right] \tan 2\theta = -a^2 d$

Let $\alpha = d/a$

$\Rightarrow (\alpha^3 + 6\alpha^2 - 8)\tan 2\theta = -12\alpha$ ⇐ Algebraic eq. in α for given θ.

or $\tan 2\theta = -\frac{12\alpha}{\alpha^3 + 6\alpha^2 - 8}$

Note: $\alpha \to \infty \Rightarrow \tan 2\theta \to 0 \Rightarrow \theta = 0, 90°$

$\alpha = 0 \Rightarrow \tan 2\theta \to 0 \Rightarrow \theta = 0, 90°$

14.1

$U_0 = \frac{\tau_{xy}^2}{2G}$ $\quad \tau_{xy} = \frac{VQ_z}{I_{zz} b}$, $Q_z = -bu(d/2 - u/2)$

$Q_z = \frac{bu}{2}(d-u)$

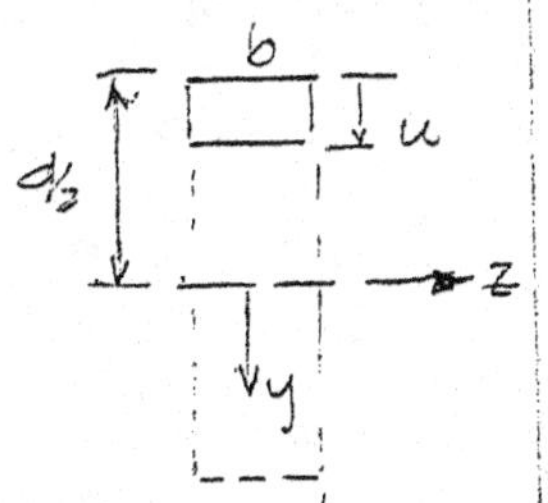

$U_0(u) = \frac{V^2 b^2 u^2 (d-u)^2}{8 G I_{zz}^2 b^2}$

$U = b \int_0^L \frac{V^2}{8 G I_{zz}^2} \left\{ \int_0^d u^2 (d-u)^2 du \right\} dx = \frac{b}{8} \int_0^L \frac{V^2}{G I_{zz}^2} \left[\frac{d^2 u^3}{3} - 2\frac{d u^4}{4} + \frac{u^5}{5} \right]_0^d dx$

$U = \frac{b}{8} \int_0^L \frac{V^2}{G I_{zz}^2} \cdot \frac{d^5}{30} dx = b \int_0^L \frac{V^2(x) d^5}{240 G I_{zz}^2} dx$; $I_{zz} = bd^3/12$

$\therefore U = \int_0^L \frac{6}{10} \frac{V^2(x)}{G b d} dx = \frac{\alpha}{2} \int_0^L \frac{V^2(x)}{AG} dx$, $A = bd$, $\alpha = 1.2$.

14.2

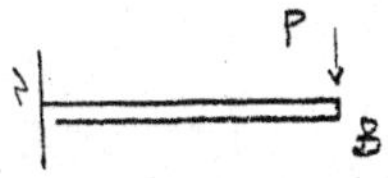

a) $V = P$ $\quad U_s = \frac{\alpha P^2 L}{2AG}$ $\quad U = W = P\Delta_s/2 \Rightarrow \frac{\alpha P^2 L}{AG} = P\Delta_s$

$\Delta_s = \alpha PL/AG$.

b). $\Delta_s/\Delta_f = \frac{\alpha PL/AG}{PL^3/3EI} = 3\alpha (E/G)(I_{zz}/L^2 A) = 3\alpha (E/G)\left(\frac{bd^3}{12 L^2 bd}\right) = \frac{\alpha}{4}(E/G)(d/L)^2$

$\Delta_s/\Delta_f = 0.3 (E/G)(d/L)^2$

c). $\Delta_s/\Delta_f \ll 1$ if $d/L \ll 1$

14.3 a).

$\tau_{xx} = \lambda\Delta + 2\mu\varepsilon_{xx}$, etc.; $\tau_{xy} = 2\mu\varepsilon_{xy}$, etc.

b). Linear elastic, yes. $U = U_0(\varepsilon^4) \Rightarrow \underset{\sim}{\tau} = f(\underset{\sim}{\varepsilon}^3) \Rightarrow$ Non-linear σ-ε relation

c). $U_0 = U_0(\varepsilon^3) \Rightarrow \underset{\sim}{\tau} = f(\varepsilon^2) \Rightarrow$ (1) No one-to-one relation; $\tau(\underset{\sim}{\varepsilon}) = \tau(-\underset{\sim}{\varepsilon})$

(2) $\varepsilon = g(\pm\sqrt{\tau})$

d). $\tau_{xx} = \lambda\Delta + \mu\varepsilon_{xx}$ etc. $\Rightarrow \begin{bmatrix} (\lambda+2\mu) & \lambda & \lambda \\ \lambda & \lambda+2\mu & \lambda \\ \lambda & \lambda & (\lambda+2\mu) \end{bmatrix} \begin{Bmatrix} \varepsilon_{xx} \\ \varepsilon_{yy} \\ \varepsilon_{zz} \end{Bmatrix} = \begin{Bmatrix} \tau_{xx} \\ \tau_{yy} \\ \tau_{zz} \end{Bmatrix}$ Solve for $\varepsilon_{xx}, \varepsilon_{yy}, \varepsilon_{zz}$ ...

$D: \begin{vmatrix} \lambda+2\mu & \lambda & \lambda \\ \lambda & \lambda+2\mu & \lambda \\ \lambda & \lambda & \lambda+2\mu \end{vmatrix} = 4\mu^2(3\lambda + 2\mu)$

$\varepsilon_{xx} = \frac{1}{D} \begin{vmatrix} \tau_{xx} & \lambda & \lambda \\ \tau_{yy} & (\lambda+2\mu) & \lambda \\ \tau_{zz} & \lambda & (\lambda+2\mu) \end{vmatrix} = \frac{\lambda+\mu}{\mu(3\lambda+2\mu)} \tau_{xx} - \frac{\lambda}{2\mu(3\lambda+2\mu)} (\tau_{yy} + \tau_{zz})$

$\varepsilon_{xx} = 1/E[\tau_{xx} - \nu(\tau_{yy} + \tau_{zz})] \Rightarrow E = \frac{\mu(3\lambda+2\mu)}{\lambda+\mu}$, $\nu = \frac{\lambda}{2(\lambda+\mu)}$; $G = \mu$ by inspection

14.4

a) $W = \frac{1}{2}[P_1\Delta_1 + P_2\Delta_2]$

$\Delta_1 = D_{11} + D_{12}$, $\Delta_2 = D_{22} + D_{21}$

$\therefore W = \frac{1}{2}[P_1(D_{11} + D_{12}) + P_2(D_{22} + D_{21})] = \frac{1}{2}[P_1 D_{11} + P_2 D_{22} + P_1 D_{12} + P_2 D_{21}]$

But $P_1 D_{12} = P_2 D_{21}$ (Betti's Law).

$\therefore W = \frac{1}{2}[P_{11} D_{11} + 2P_1 D_{12} + P_2 D_{22}] = P_1 D_{11}/2 + P_1 D_{12} + P_2 D_{22}/2$

b). Using Betti's Law, $W = \frac{1}{2}[P_{11} D_{11} + 2P_2 D_{21} + P_2 D_{22}]$

$= P_{11} D_{11}/2 + P_2 D_{21} + P_2 D_{22}/2$ if P_2 is applied first

14.5

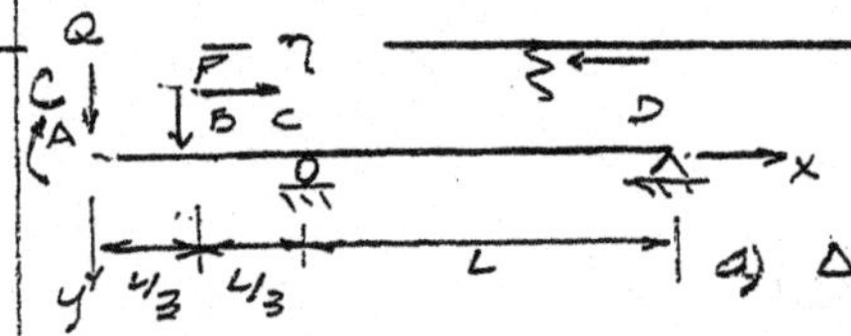

a) Δ_B: $M(\eta) = -P\eta$, $0<\eta<L/3$; $M(\xi) = -P\xi/3$, $0<\xi<L$

$\Delta_B = \frac{1}{EI}\int M\, \partial M/\partial P\, dx$;

$= \frac{1}{EI}\left\{\int_0^{L/3}(-P\eta)(-\eta)\,d\eta + \int_0^L(-P\xi/3)(-\xi/3)\,d\xi\right\}$

$= \frac{P}{EI}\left\{\int_0^{L/3}\eta^2 d\eta + 1/9\int_0^L \xi^2 d\xi\right\} = \frac{P}{EI}\left[\eta^3/3\Big|_0^{L/3} + \frac{1}{27}\xi^3\Big|_0^L\right] = 4PL^3/81EI \downarrow$

b) Δ_A: $M(x) = -Qx$, $0<x<L/3$; $M(x) = -Qx - P(x - L/3)$, $L/3<x<2L/3$

$M(\xi) = -P\xi/3 - 2Q\xi/3$, $0<\xi<L$

$\Delta_A = \frac{1}{EI}\int M\, \partial M/\partial Q\Big|_{Q=0} dx$

$= \frac{1}{EI}\left\{\int_{L/3}^{2L/3} -P(x - L/3)(-x)\,dx + \int_0^L(-P\xi/3)(-2\xi/3)\,d\xi\right\}$

$= \frac{P}{EI}\left\{\int_{L/3}^{2L/3}(x^2 - xL/3)\,dx + \frac{2}{9}\int_0^L \xi^2 d\xi\right\}$

$= \frac{P}{EI}\left\{x^3/3 - x^2L/6\Big]_{L/3}^{2L/3} + \frac{2}{27}\xi^3\Big|_0^L\right\} = \frac{17PL^3}{162EI} \downarrow$

c). θ_A: $M(x) = C$, $0<x<L/3$; $M(x) = C - P(x - L/3)$; $L/3<x<2L/3$

$M(\xi) = C\xi/L - P\xi/3$, $0<\xi<L$

$\theta_A = \frac{1}{EI}\int M\,\frac{\partial M}{\partial C}\Big|_{C=0} dx$

$\theta_A = \frac{1}{EI}\left\{-P\int_{L/3}^{2L/3}(x - L/3)\,dx - P\int_0^L(\xi/3)(\xi/L)\,d\xi\right\}$

$= \frac{P}{EI}\left\{-(x - L/3)^2/2\Big|_{L/3}^{2L/3} - \xi^3/9L\Big|_0^L\right\} = -PL^2/6EI$ ↻

14.6

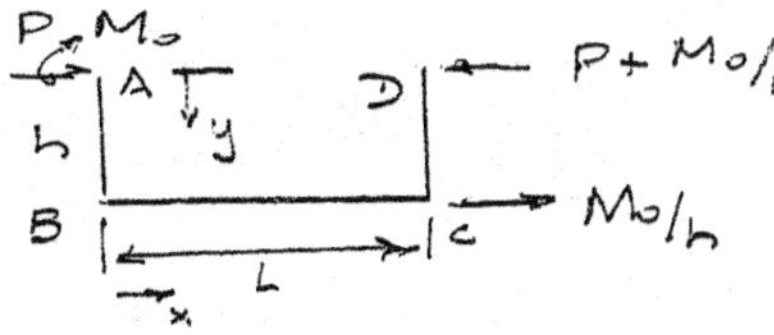

AB: $M(y) = M_o + Py$; $\partial M/\partial P = y$, $\frac{\partial M}{\partial M_o} = 1$

BC: $M = M_o + Ph$; $\partial M/\partial P = h$, $\partial M/\partial M_o = 1$

a). CD: $M(y) = (P + M_o/h)y$; $\partial M/\partial P = y$, $\frac{\partial M}{\partial M_o} = y/h$.

$$\Delta_A = \frac{1}{EI}\int M\,\partial M/\partial P\,ds = \frac{1}{EI}\left\{\int_0^h (M_o + Py)y\,dy + \int_0^L (M_o + Ph)h\,dx + \int_0^h (P + M_o/h)y^2\,dy\right\}$$

$$= \frac{1}{EI}\left\{\left[M_o y^2/2 + Py^3/3\right]_0^h + (M_o + Ph)hL + \left[Py^3/3 + M_o y^3/3h\right]_0^h\right\}$$

$$= \frac{1}{EI}\left\{M_o h^2/2 + Ph^3/3 + (M_o + Ph)hL + Ph^3/3 + M_o h^2/3\right\}$$

$$= \frac{1}{6EI}\left\{3M_o h^2 + 4Ph^3 + 6(M_o + Ph)hL + 2M_o h^2\right\}$$

$$\Delta_A = \frac{1}{6EI}\left\{2h^2(2h + 3L) + M_o h(5h + 6L)\right\}$$

b). Betti's Law: $M_o \theta_P = \Delta_A|_{M_o} P$

$$\theta|_{P=1} = \frac{\Delta_A|_{M_o}}{M_o} = \frac{h(5h + 6L)}{6EI}$$

14.7

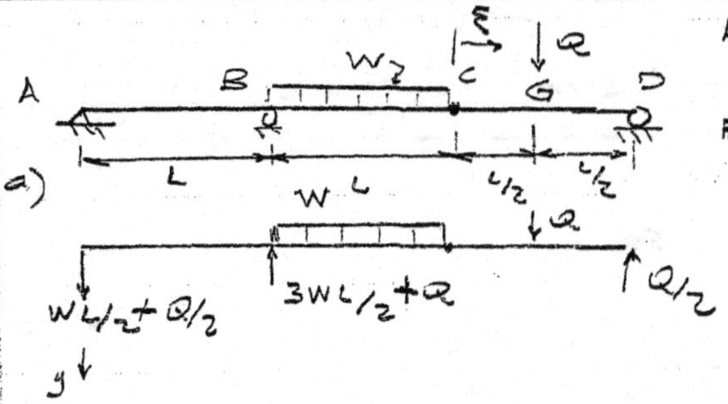

AB:

$0 < x < L$: $M(x) = -WLx/2 - Qx/2$; $\frac{\partial M}{\partial Q} = -x/2$

BC:

$L < x < 2L$: $M(x) = -WLx/2 - Qx/2 + (\frac{3WL}{2} + Q)(x - L) - W(x - L)^2/2$; $\frac{\partial M}{\partial Q} = x/2 - L$

CG: $M(\xi) = Q\xi/2$, $\partial M/\partial Q = \xi/2$;
GC: $M(\xi) = \frac{Q}{2}(L - \xi)$, $\partial M/\partial Q = (L - \xi)/2$ } IRRELEVANT

$$\Delta_G = \frac{1}{EI}\int M \frac{\partial M}{\partial Q}\bigg|_{Q=0} ds$$

$$= \frac{1}{EI}\left\{\int_0^L (-WLx/2)(-x/2)\,dx + \int_L^{2L}\left[-WLx/2 + 3WL/2(x - L) - W(x - L)^2/2\right](x/2 - L)\,dx\right\}$$

$$= \frac{W}{EI}\left\{\frac{L}{4}\int_0^L x^2\,dx + \frac{1}{4}\int_L^{2L}\left[-Lx(x - 2L) + 3L(x - L)(x - 2L) - (x - L)^2(x - 2L)\right]dx\right\}$$

$$= \frac{W}{4EI}\left\{L\int_0^L x^2\,dx + \int_L^{2L}\left[-x^3 + 6Lx^2 - 12L^2x + 8L^3\right]dx\right\}$$

$$= \frac{W}{4EI}\left\{Lx^3/3\Big|_0^L + \left[-x^4/4 + 2Lx^3 - 6L^2x^2 + 8L^3x\right]_L^{2L}\right\} = 7WL^4/48EI \downarrow$$

b)

14.8

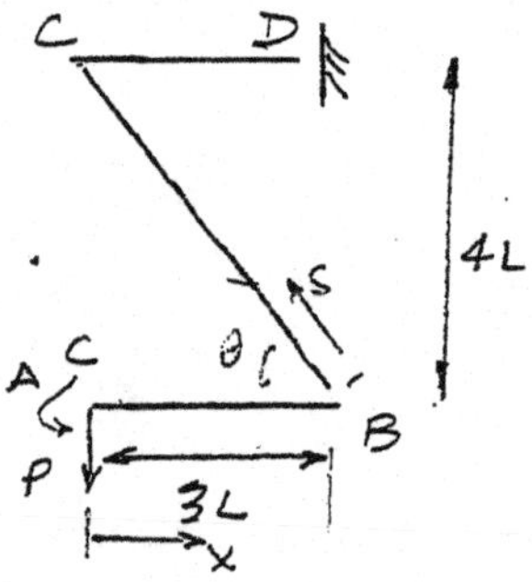

A-B and C-D: $M = -C - Px$; $\partial M/\partial P = -x$, $\frac{\partial M}{\partial C} = -1$

B-C: $M = C + P[3L - s\cos\theta]$

$M = C + P(3L - 3\xi/5)$

$\partial M/\partial P = 3L - 3\xi/5$

$\partial M/\partial C = 1$

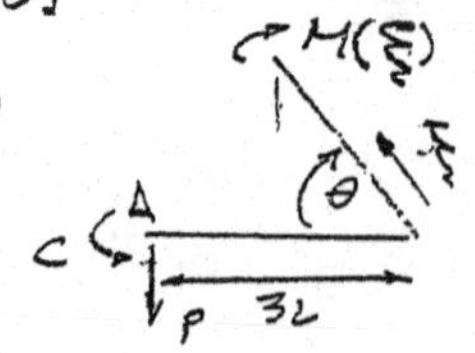

$$\theta_A = \frac{1}{EI}\int M \frac{\partial M}{\partial C}\Big|_{C=0} ds \Rightarrow \theta_A = \frac{1}{EI}\left\{2\int_0^{3L}(-Px)(-1)\,dx + \int_0^{5L} 3P(L-\xi/5)\,d\xi\right\}$$

$$= \frac{P}{EI}\left\{2\int_0^{3L} x\,dx + 3\int_0^{5L}(L-\xi/5)\,d\xi\right\}$$

$$= P/EI\left\{ x^2\Big|_0^{3L} + 3\left[L\xi - \xi^2/10\right]_0^{5L}\right\}$$

$$= P/EI\left\{9L^2 + 3[5L^2 - 25L^2/10]\right\} = 33PL^2/2EI.$$

$$\Delta_A = \frac{1}{EI}\int M\frac{\partial M}{\partial P}\Big|_{C=0} ds \Rightarrow \Delta_A = \frac{1}{EI}\left\{2\int_0^{3L}(-Px)(-x)\,dx + 3P\int_0^{5L}(L-\xi/5)\cdot 3(L-\xi/5)\,d\xi\right\}$$

$$= \frac{1}{EI}\left\{2P\int_0^{3L} x^2\,dx + 9P\int_0^{5L}(L-\xi/5)^2\,d\xi\right\}$$

$$= \frac{P}{EI}\left\{\frac{2}{3}x^3\Big|_0^{3L} + 3(L-\xi/5)^3(-5)\Big|_0^{5L}\right\}$$

$$= \frac{P}{EI}\left\{18L^3 - 15[0 - L^3]\right\} = 33PL^3/EI.$$

14.9

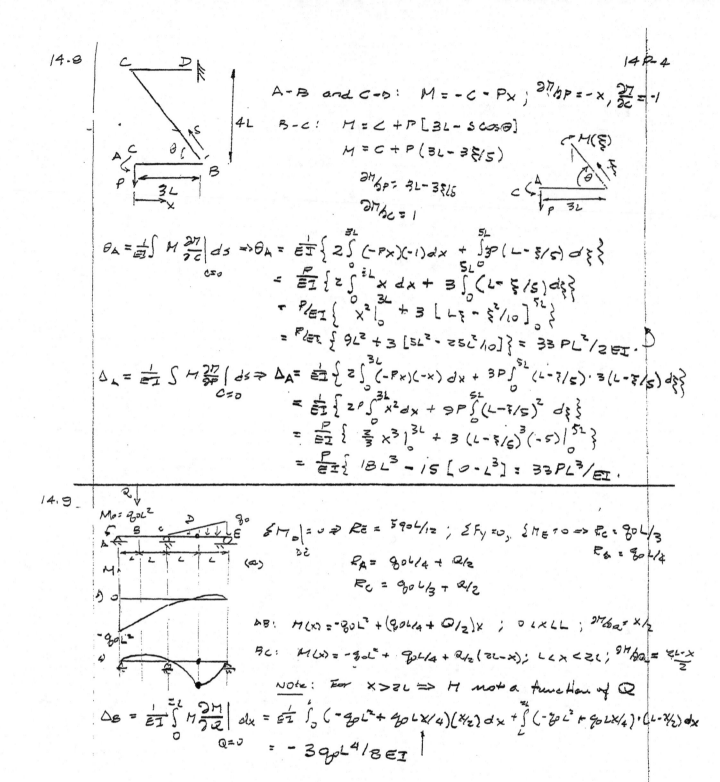

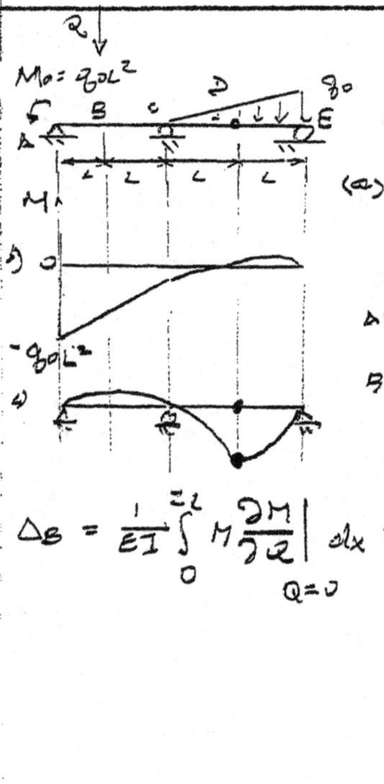

$\sum M_D\big|_{DE} = 0 \Rightarrow R_E = 5q_0L/12$; $\sum F_y = 0$, $\sum M_E = 0 \Rightarrow R_C = q_0L/3$

$R_A = q_0L/4$

$R_A = q_0L/4 + Q/2$

$R_C = q_0L/3 + Q/2$

AB: $M(x) = -q_0L^2 + (q_0L/4 + Q/2)x$; $0 < x < L$; $\partial M/\partial Q = x/2$

BC: $M(x) = -q_0L^2 + q_0L/4 + Q/2(2L - x)$; $L < x < 2L$; $\partial M/\partial Q = \frac{2L-x}{2}$

Note: For $x > 2L \Rightarrow M$ not a function of Q

$$\Delta_B = \frac{1}{EI}\int_0^{2L} M\frac{\partial M}{\partial Q}\Big|_{Q=0} dx = \frac{1}{EI}\int_0^L(-q_0L^2 + q_0Lx/4)(x/2)\,dx + \int_L^{2L}(-q_0L^2 + q_0Lx/4)\cdot(L - x/2)\,dx$$

$$= -3q_0L^4/8EI$$

14.10

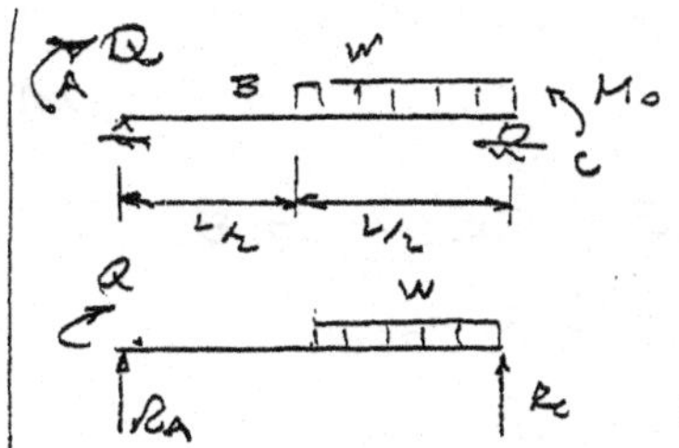

$\sum M_C = 0 \Rightarrow R_A = M_0/L + \frac{WL}{8} - Q/L$

$M(x) = M_0x/L + WLx/8 + Q(1 - x/L)\,;\quad \partial M/\partial Q = (1 - x/L)\,;\quad 0 \le x \le L/2$

$M(x) = M_0x/L + WLx/8 - \frac{W}{2}(x - \frac{L}{2})^2 + Q(1 - x/L)\,;\quad \frac{\partial M}{\partial Q} = 1 - x/L\,;\quad L/2 \le x \le L$

$$\theta_A = \frac{1}{EI}\int M \frac{\partial M}{\partial Q}\Big|_{Q=0} dx = \frac{1}{EI}\Big\{\int_0^{L/2}(M_0x/L + WLx/8)(1 - x/L)dx + \int_{L/2}^{L}\big(M_0x/L + WLx/8 - \frac{W}{8}(2x - L)^2\big)(1 - x/L)\,dx\Big\}$$

$$= \frac{1}{EI}\Big\{\int_0^L (M_0x/L + WLx/8)(1 - x/L)dx - \frac{W}{8}\int_{L/2}^{L}(2x - L)^2(1 - x/L)dx\Big\}$$

$$= \frac{1}{EI}\Big\{(M_0L/6 + WL^3/48) - WL^3/(24\times16)\Big\}$$

$\theta_A = 0 \Rightarrow M_0L/6 + WL^3/48 - WL^3/24\times16 = 0$

$\Rightarrow M_0 = -7WL^2/64$)

14.11 a)

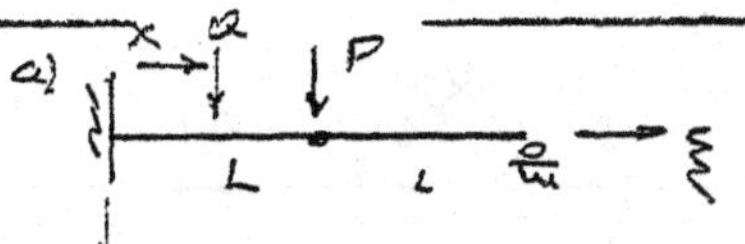

$M(\xi) = -P(L - \xi) - Q(x - \xi)\,;\quad 0 < \xi < L\,;\ 0 < \xi < x,\ 0 < x < L$

$\quad = -P(L - \xi)\,;\quad 0 < \xi < L\,;\ 0 < x < \xi,$

$M(\xi) = -P(L - \xi) + Q(x - 2L) + Q(2 - x/L)\xi\,;\quad L < x < 2L\,;\ 0 < \xi < L$

NOTE: $M(\xi)$ is not a function of P when $L < \xi < 2L$

$$v(x) = \frac{1}{EI}\int M \frac{\partial M}{\partial Q}\Big]_{Q=0} d\xi$$

$\underline{0 < x < L}$

$$v(x) = \frac{1}{EI}\int_0^x + P(L - \xi)(x - \xi)\,d\xi = \frac{P}{EI}\int_0^x (Lx - \xi x - L\xi + \xi^2)d\xi$$

$$= \frac{P}{EI}\Big[Lx\xi - x\xi^2/2 - L\xi^2/2 + \xi^3/3\Big]_0^x = \frac{P}{6EI}(3Lx^2 - x^3)\,,\quad 0 \le x \le L$$

$\underline{L < x < 2L:}$

$$v(x) = \frac{1}{EI}\int_0^L -P(L - \xi)\big[(x - 2L) + (2 - x/L)\xi\big]\,d\xi = \frac{PL^2}{3EI}(2L - x)$$

b). By Betti's Law:

$v_B = \frac{P}{6EI}\big[3La^2 - a^3\big] = \frac{Pa^2}{6EI}(3L - a)\,,\quad 0 < a < L$

$v_B = \frac{PL^2}{3}(2L - a)\,,\quad L < a < 2L$

14.12

a)

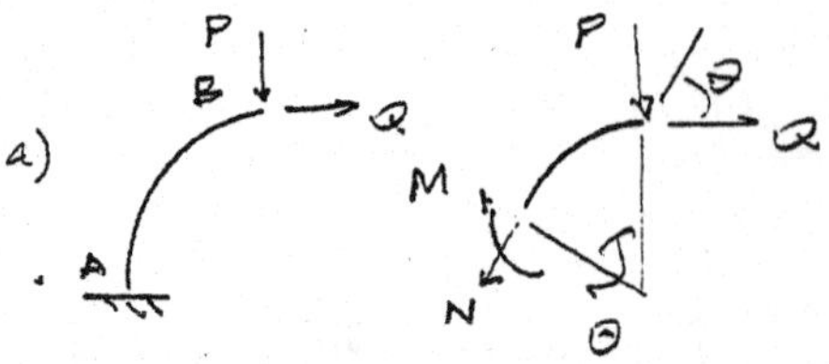

$$M = -PR\sin\theta - QR(1-\cos\theta)$$
$$N = -P\sin\theta + Q\cos\theta$$

$$\Delta_{Bx} = \frac{1}{EI}\int M \frac{\partial M}{\partial Q}\bigg| ds + \frac{1}{AE}\int N \frac{\partial N}{\partial Q}\bigg|_{Q=0} ds$$

$$= \frac{R}{EI}\int_0^{\pi/2}(PR\sin\theta)R(1-\cos\theta)d\theta + \frac{R}{AE}\int_0^{\pi/2} -P\sin\theta\cos\theta\, d\theta$$

$$= P\left\{\frac{R^3}{EI}\int_0^{\pi/2}(\sin\theta - \sin\theta\cos\theta)d\theta - \frac{R}{AE}\int_0^{\pi/2}\sin\theta\cos\theta\, d\theta\right\}$$

$$= P\left\{R^3/EI\int_0^{\pi/2}\left(\sin\theta - \frac{\sin 2\theta}{2}\right)d\theta - \frac{R}{2AE}\int_0^{\pi/2}\sin 2\theta\, d\theta\right\}$$

$$\Delta_{Bx} = P\left\{R^3/EI\left[-\cos\theta + \frac{\cos 2\theta}{4}\right|_0^{\pi/2} + \frac{R}{4AE}\cos 2\theta\Big|_0^{\pi/2} = \frac{PR^3}{2EI} - \frac{PR}{2AE} \quad (\overset{\rightarrow}{+})$$

b). $\frac{(\Delta B)_{ax}}{(\Delta_B)_{flex}} = \frac{I}{AR^2} : (d/R)^2$

c). $d/R \ll 1 \Rightarrow (\Delta)_{ax}/(\Delta)_{fl} \ll 1$

14.13

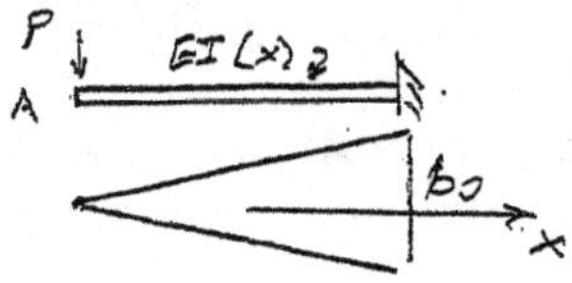

$$\Delta_A = \frac{1}{E}\int \frac{M(x)}{EI(x)}\cdot\frac{\partial M}{\partial P}dx$$

$$M(x) = -Px, \quad \partial M/\partial P = -x$$

$$I(x) = \frac{b_0 x}{L}\cdot d^3/12$$

$$\Delta_A = \frac{12L}{Eb_0d^3}\int_0^L\left(-\frac{Px}{x}\right)(-x)dx = \frac{12PL}{Eb_0d^3}\int_0^L Px\,dx = \frac{6PL^3}{Eb_0d^3}$$

14.14

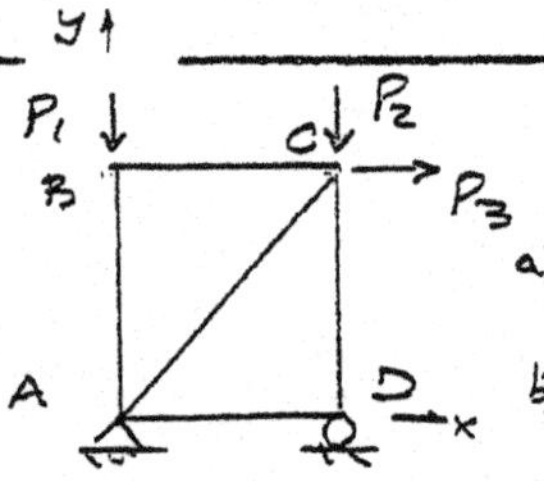

$$F_{AB} = -P_1,\ F_{BC} = 0,\ F_{AC} = \sqrt{2}P_3,\ F_{CD} = -P_2 - P_3$$

a) $\Delta_{YB} = \frac{\partial U}{\partial P_1} = \frac{L}{AE}F_{AB}(-1) = +P_1L/AE \downarrow$

b). $\Delta_{xc} = \sum \frac{L_i}{AE}F_i\frac{\partial F_i}{\partial P_3} = \frac{1}{AE}\{F_{AC}\sqrt{2}(\sqrt{2}L) + F_{CD}(-1)L\}$

$= \frac{L}{AE}\{(2\sqrt{2}+1)P_3 + P_2\} \rightarrow$

c). $\Delta_{Yc} = \frac{L}{AE}\{(F_{CD})(-1)\} = (P_2+P_3)L/AE \downarrow$

14.15

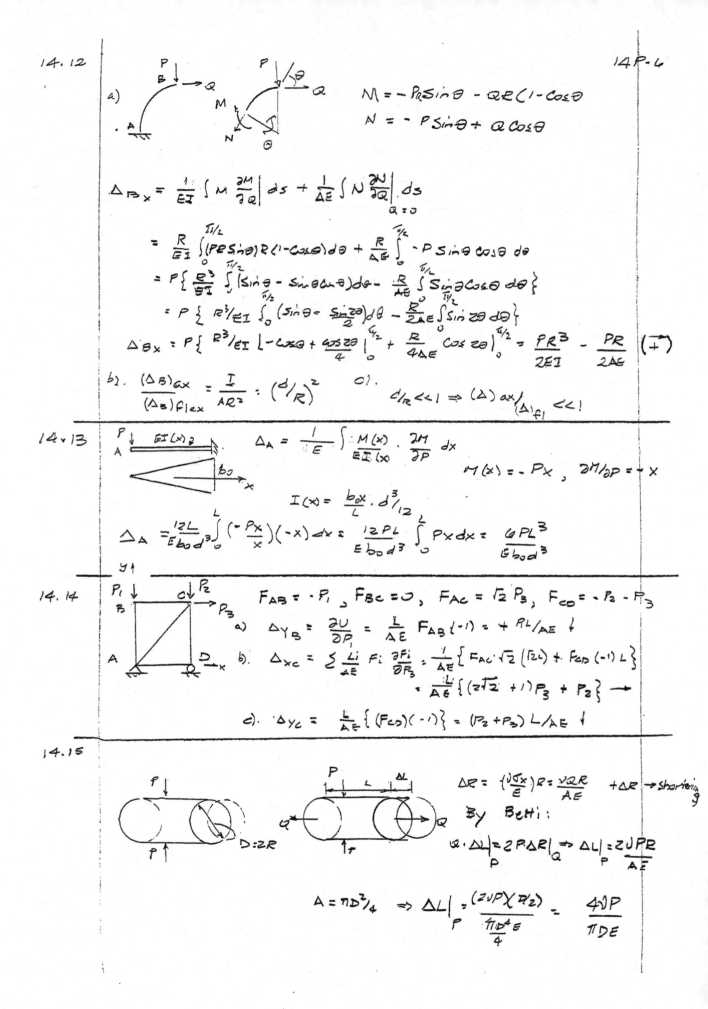

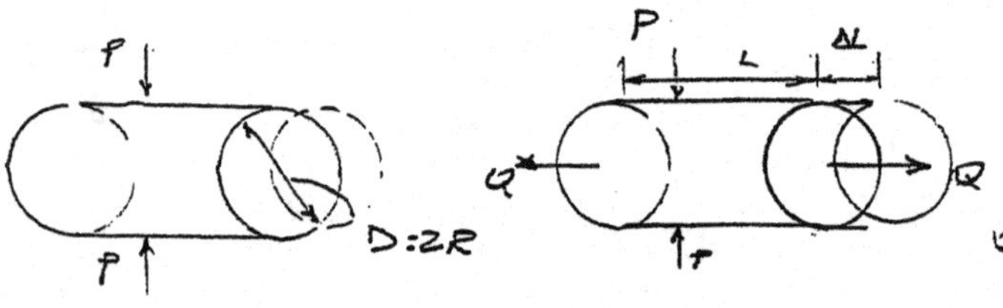

$\Delta R = \left(\frac{\nu\sigma_x}{E}\right)R = \frac{\nu QR}{AE}$ $\quad +\Delta R \rightarrow$ shortening

By Betti:

$$Q\cdot\Delta L\big|_P = 2P\Delta R\big|_Q \Rightarrow \Delta L\big|_P = \frac{2\nu PR}{AE}$$

$$A = \pi D^2/4 \Rightarrow \Delta L\big|_P = \frac{(2\nu P)(D/2)}{\frac{\pi D^2}{4}E} = \frac{4\nu P}{\pi DE}$$

14.16

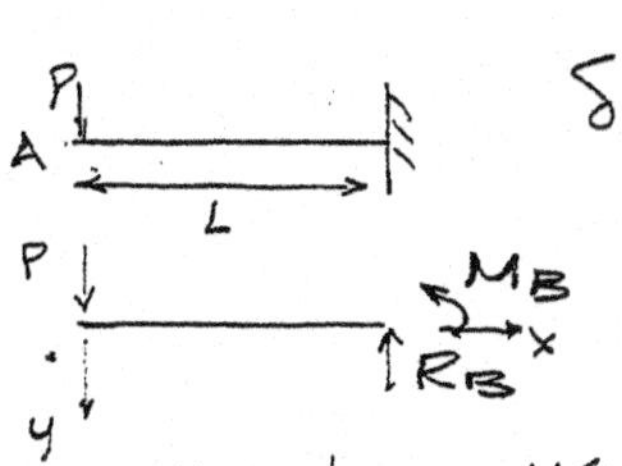

$\delta v(x) = \delta A \sin \frac{n\pi x}{2L}$, $n > 0$ $\Rightarrow$ $\delta v(L) = \delta A \sin n\pi/2$

$\delta v'(x) = \delta A \cdot \frac{n\pi}{2L} \cos \frac{n\pi x}{2L} \Rightarrow \delta v'(L) = + \delta A\, \alpha \cos n\pi/2$

$\delta v''(x) = -\delta A (n\pi/2L)^2 \sin n\pi x/2L \Rightarrow \delta v''(L) = -\delta A \alpha^2 \sin n\pi/2.$

$+v > 0 \downarrow$ $\qquad M(x) = -wx^2/2$ $\qquad \alpha = n\pi/2L$

$$\delta W_{ext} = -R_B \delta v(L) - M_B \delta v'(L) + w\int_0^L \delta v(x)\,dx = -R_B\delta v(L) - M_B \delta v'(L) + w\delta A \int_0^L \sin \alpha x\, dx$$

$$= -R_B \delta v(L) - M_B \delta v'(L) - \frac{w}{\alpha}\delta A \cos \frac{n\pi x}{2L}\Big|_0^L$$

$$= \delta A\left\{ -R_B \sin \frac{n\pi}{2} - M_B \alpha \cos \frac{n\pi}{2} - \frac{2wL}{n\pi}\left(\cos \frac{n\pi}{2} - 1\right)\right\}.$$

$$\delta W_{int} = \int_0^L \delta\kappa\, M(x)\,dx = -\frac{w}{2}\int x^2(-\delta v'')\,dx = -\frac{w\delta A}{2}\alpha^2 \int_0^L x^2 \sin \alpha x\, dx = -\frac{\delta A\, w}{2\alpha}\int_0^{\alpha L} \eta^2 \sin \eta\, d\eta$$

$\eta = \alpha x$

$$\delta W_{int} = -\frac{\delta A w}{2\alpha}\left\{(2-\eta^2)\cos\eta + 2\eta \sin \eta\right\}\Big|_0^{\eta = \alpha L = n\pi/2} = -\frac{\delta w}{2\alpha}\left[(2-\alpha^2L^2)\cos \frac{n\pi}{2} + 2\alpha L \sin\frac{n\pi}{2} - 2\right]$$

$$\delta W_{ext} = \delta W_{int} \Rightarrow -R_B \sin \frac{n\pi}{2} - M_B\alpha \cos\frac{n\pi}{2} - \frac{w}{\alpha}\left(\cos\frac{n\pi}{2} - 1\right) = -\frac{w}{2\alpha}\left[(2-\alpha^2L^2)\cos\frac{n\pi}{2} + 2\alpha L \sin \frac{n\pi}{2} - 2\right]$$

$$R_B(1 - wL)\sin\frac{n\pi}{2} = 0 \Rightarrow R_B = wL$$

$$\left[M_B\alpha + \frac{w}{\alpha} - \frac{w}{\alpha} + \frac{w\alpha L^2}{2}\right]\cos\frac{n\pi}{2} = 0 \Rightarrow M_B = -wL^2/2.$$

14.17

a)

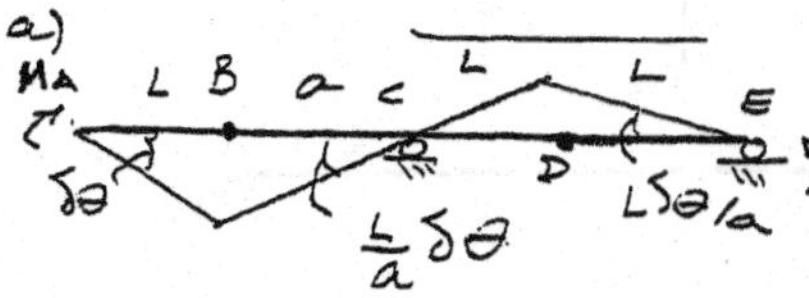

a). $M_A \delta\theta - w\left[\left(\frac{L}{a}\delta\theta\right)(L/2)\cdot 2L\right] - wL^2 (L\delta\theta/a) = 0$

$(M_A - 2wL^3/a)\,\delta\theta = 0 \Rightarrow M_A = +2wL^3/a$

b).

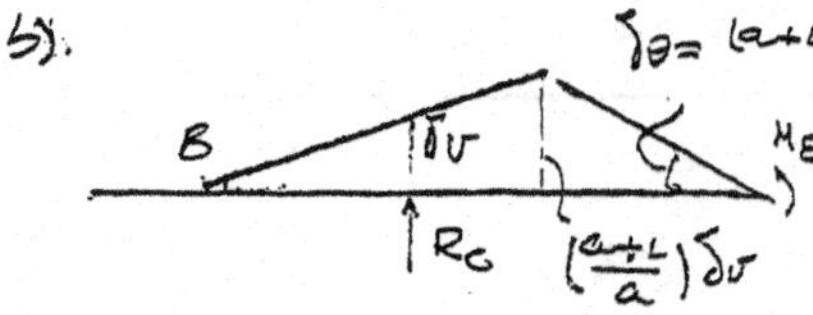

$R_C \delta v - wL\left\{\left[\frac{(1 + L/a) + 1}{2}\right]\delta v + \frac{(1 + L/a)}{2}\delta v\right\} - M_E\left(\frac{a+L}{aL}\right)\delta v = 0$

$\left[R_C - \frac{wL}{2}(3 + 2L/a) - wL\frac{(a+L)}{a}\right]\delta v = 0$

c).

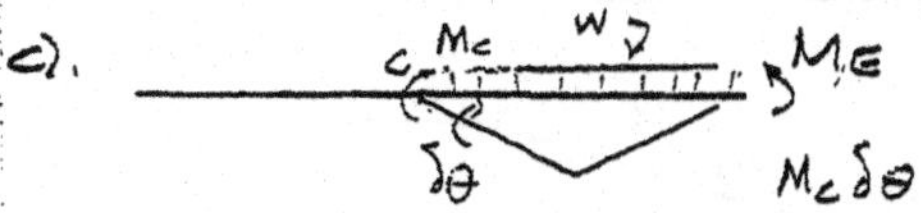

$R_C = (wL/2a)(5a + 4L)$

$M_C \delta\theta + \frac{w}{2}\left[(L\delta\theta)(2L)\right] - wL^2\delta\theta = 0$

$(M_C + 2wL^2)\,\delta\theta = 0 \Rightarrow M_C = -2wL^2.$

14.18

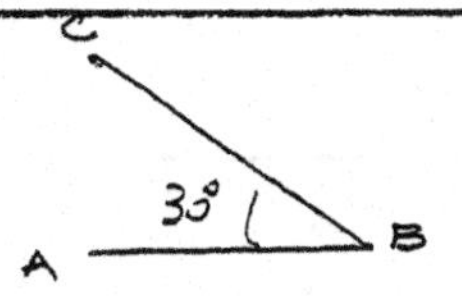

F_{AB}:

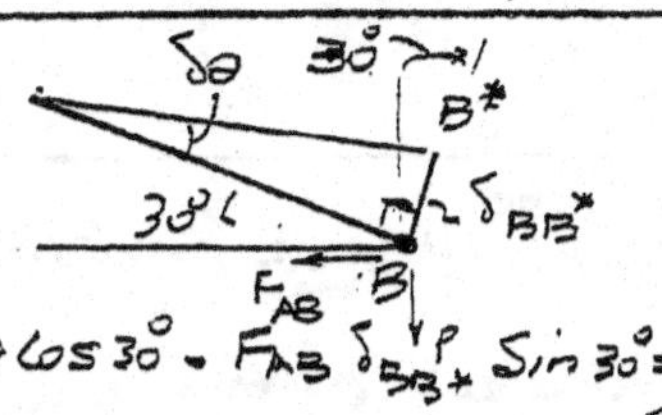

$\delta W_{ext} = -P\delta_{BB^*}\cos 30^\circ - F_{AB}\,\delta_{BB^*}\sin 30^\circ = 0$

$F_{AB} = -P \cot 30^\circ$

F_{BC}:

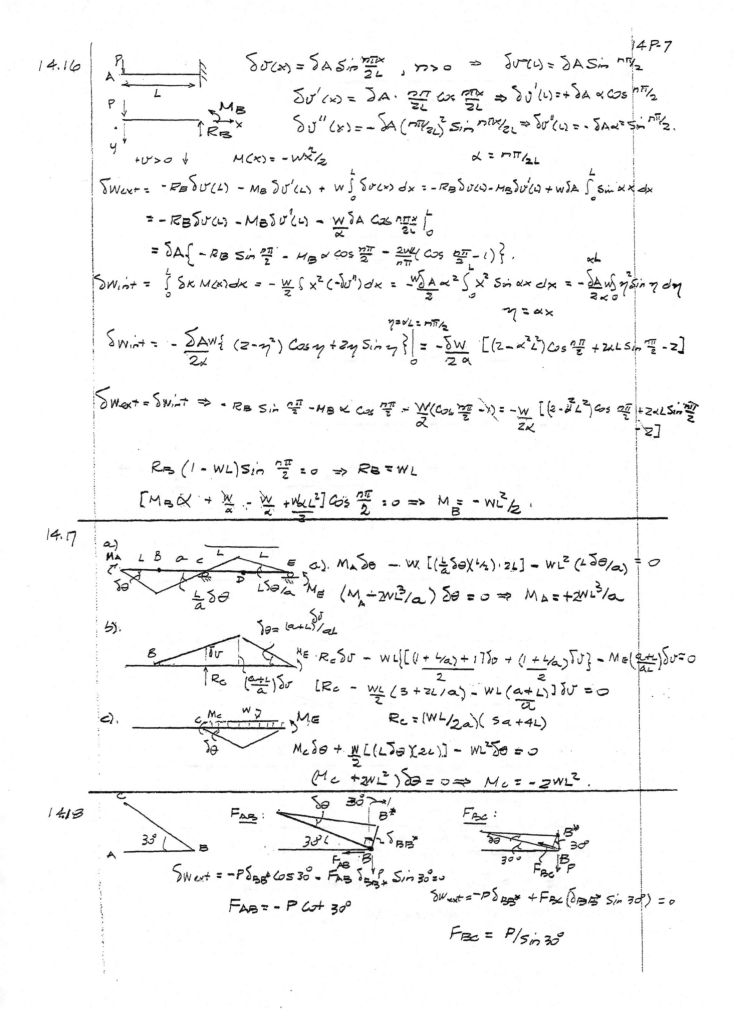

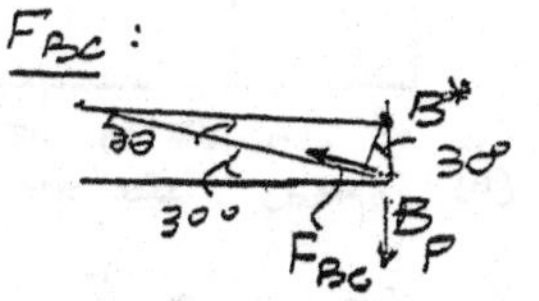

$\delta W_{ext} = -P\delta_{BB^*} + F_{BC}(\delta_{BB^*}\sin 30^\circ) = 0$

$F_{BC} = P/\sin 30^\circ$

14.20

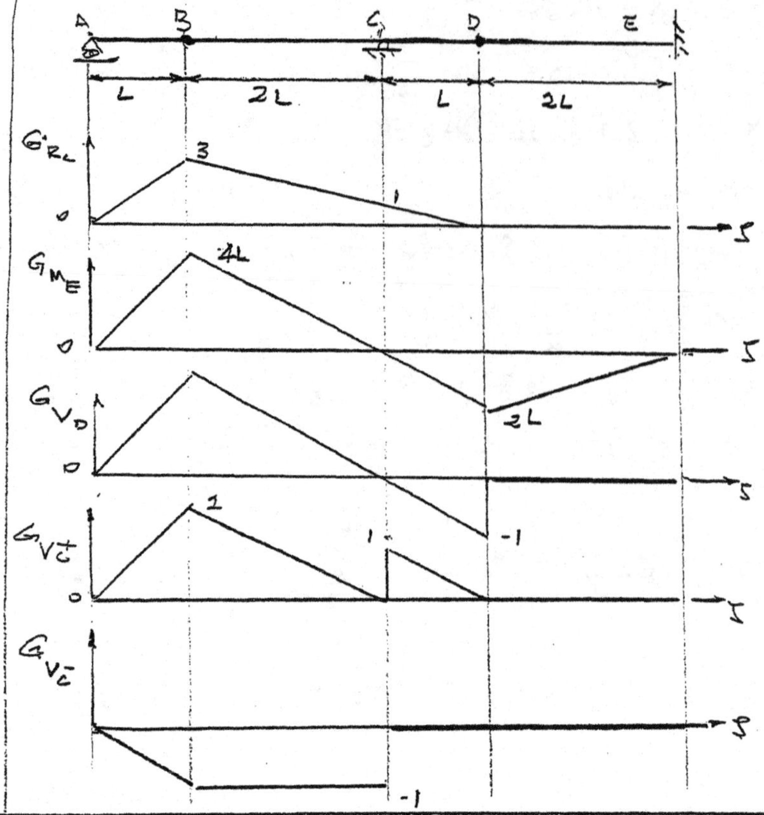

14.21

a)

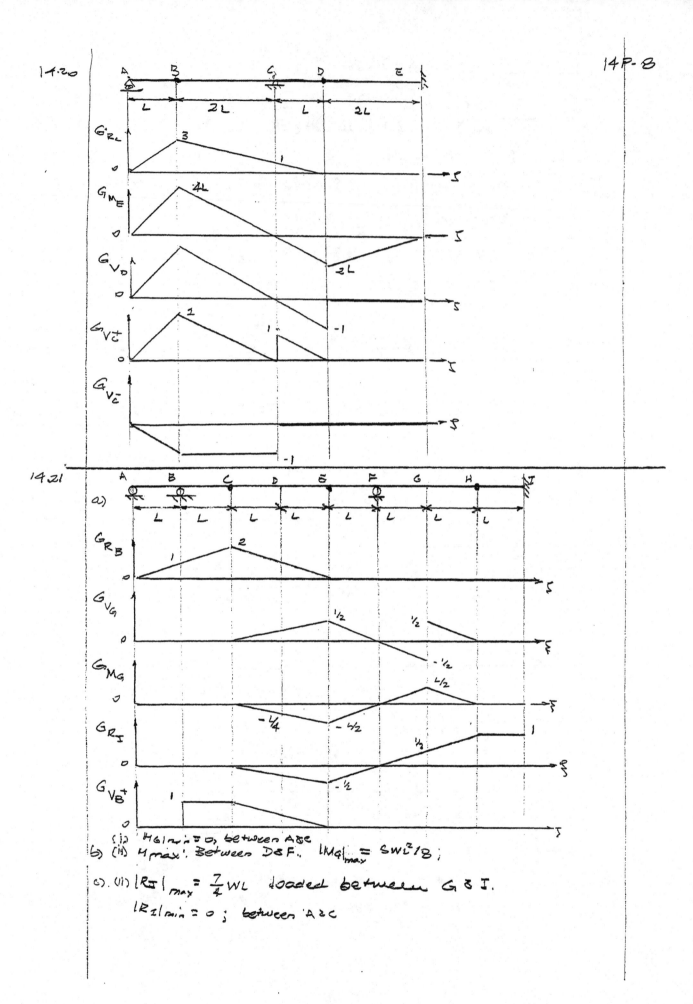

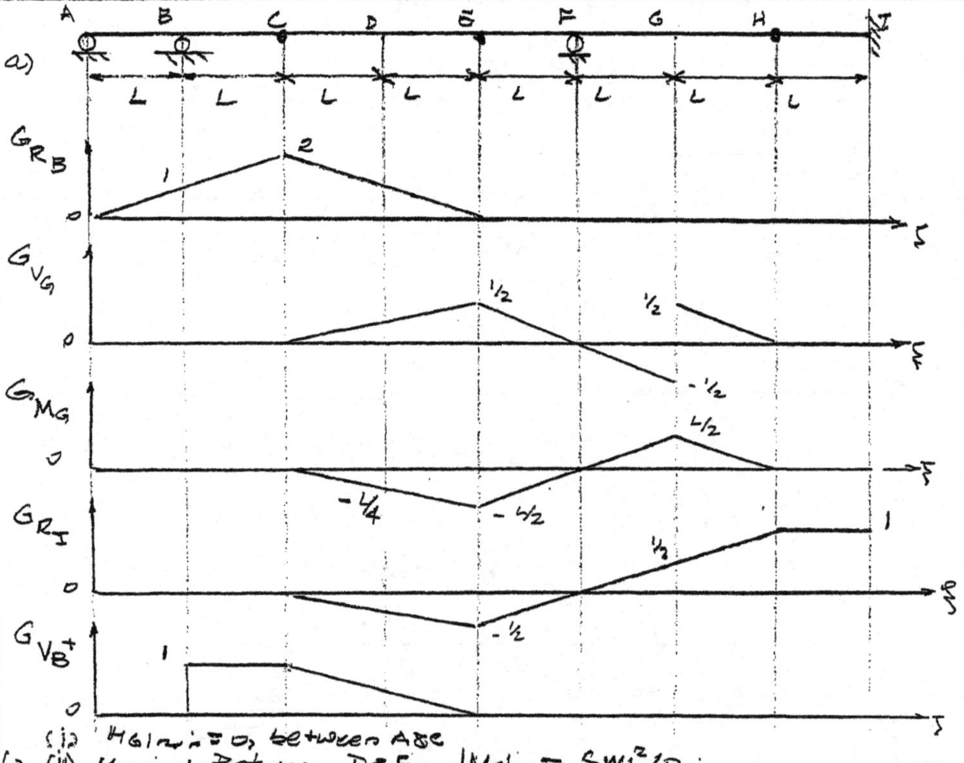

(i) $|M_G|_{min} = 0$, between A&C

b) (ii) M_{max}: Between D&F., $|M_G|_{max} = 5wL^2/8$;

c). (ii) $|R_I|_{max} = \frac{7}{4} wL$ loaded between G & I.

$|R_I|_{min} = 0$; between A&C

14.22

a)

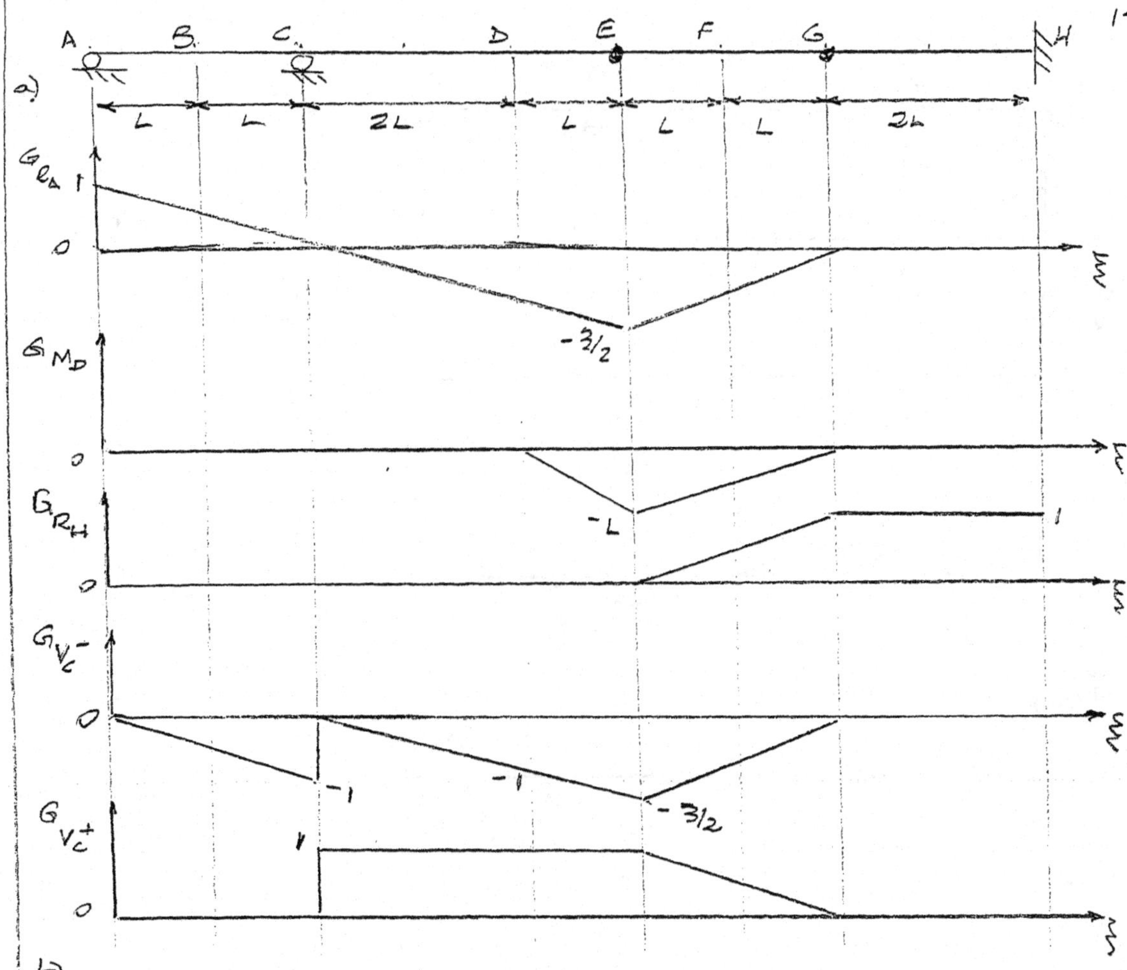

b)

(i) $|M_D|_{max} = 3WL^2/4$ loaded between E & F.

(ii) $|V_{C^-}|_{max} = 5WL/4$ loaded between D & E.

c) (i) $\underline{M_D}$:

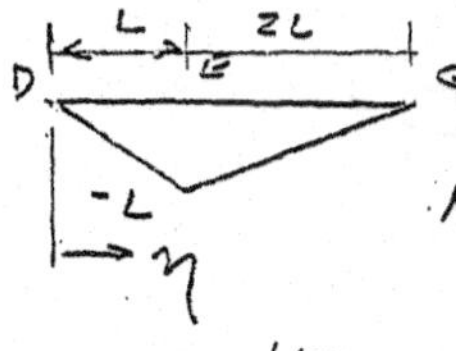

$\eta = \xi - 4L$ $\quad G_{M_D} = \begin{cases} -\eta, & 0<\eta<L \\ \frac{1}{2}(-3L+\eta), & L<\eta<3L \end{cases}$

$M_D = W\int_{\eta_0}^{L+\eta_0} G_{M_D}(\eta)\,d\eta$

$M_D = W\left\{\int_{\eta_0}^{L} -\eta\,d\eta + \frac{1}{2}\int_{L}^{L+\eta_0}(-3L+\eta)\,d\eta\right\} = \frac{W}{2}\left\{-\eta^2\Big|_{\eta_0}^{L} + \left[-3L\eta+\eta^2/2\right]_{L}^{L+\eta_0}\right\}$

$= \frac{W}{2}\left\{-(L^2-\eta_0^2) + \left[-3L(L+\eta_0) + \frac{1}{2}(L+\eta_0)^2 - (-3L^2+L^2/2)\right]\right\} = \frac{W}{2}\left[3\eta_0^2/2 - 2L\eta_0 - L^2\right]$

$dM_D/d\eta_0 = 0 \Rightarrow +3\eta_0 - 2L = 0 \Rightarrow \eta_0 = 2L/3 \Rightarrow |M_D|_{max} = \frac{5}{6}WL^2$; $4.667 < \xi < 5.667L$

(ii) $\underline{V_C^-}$:

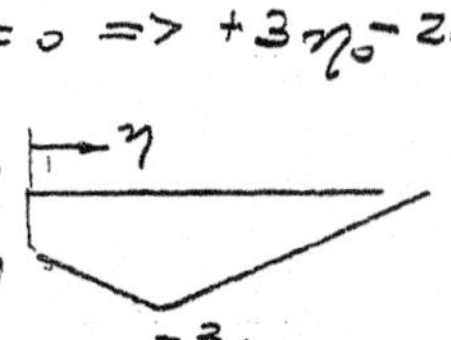

$G_{V_C^-} = \begin{cases} -(1+\eta/2L), & 0<\eta<L \\ 3/4(-3+\eta/L), & L<\eta<3L \end{cases}$

$V_{C^-} = W\left\{-\int_{\eta_0}^{L}(1+\eta/2L)\,d\eta + 3/4\int_{L}^{L+\eta_0}(-3+\eta/L)\,d\eta\right\} = W\left\{-(\eta+\eta^2/4L)\Big|_{\eta_0}^{L} + 3/4(-3\eta+\eta^2/2L)\Big|_{L}^{L+\eta_0}\right\}$

$= \frac{W}{4}\left\{5\eta_0^2/2L - 2\eta_0 - 5L\right\}$

$dV_C/d\eta = 0 \Rightarrow 5\eta/L = 2 \Rightarrow \eta_0 = 0.4 \Rightarrow \xi_0 = 4.4L \qquad 4.4L < \xi < 5.4L$

$|V_{C^-}|_{max} = 27WL/20$

14.23

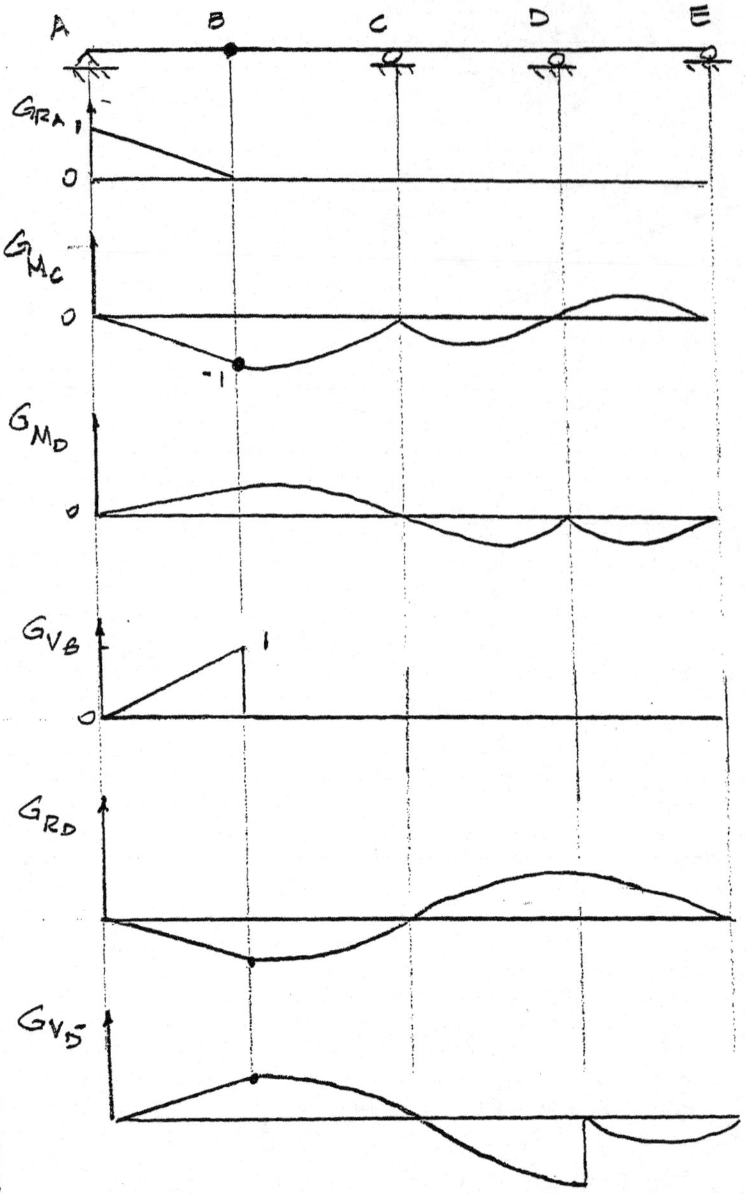

14.24

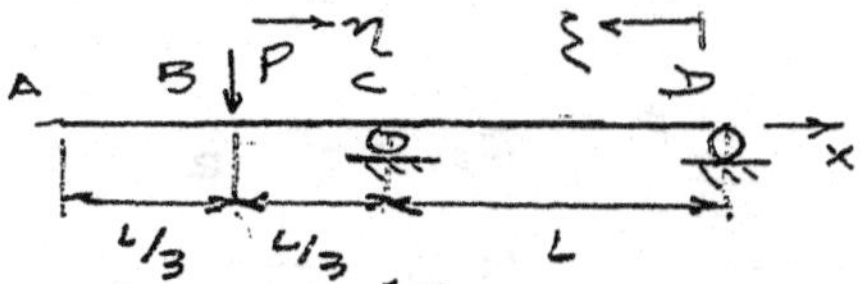

DC

$M = -P\eta;\quad 0<\eta<L/3$

CD:

$M = -P\xi/3,\quad 0<\xi<L$

a). Δ_B:

$\delta M = -\delta P\eta,\quad 0<\eta<L/3;\quad \delta M = -\delta P\xi/3,\quad 0<\xi<L$

$\delta W^*_{ext} = \delta P\cdot\Delta_B;\quad \delta W^*_{int} = \frac{P\delta P}{EI}\left\{\int_0^{L/3}\eta^2 d\eta + 1/9\int_0^L \xi^2 d\xi\right\} = 4P\delta P L^3/81EI$

$\delta W^*_{ext} = \delta W^*_{int} \Rightarrow \Delta_B = 4PL^3/48EI\ \downarrow$

b) Δ_A

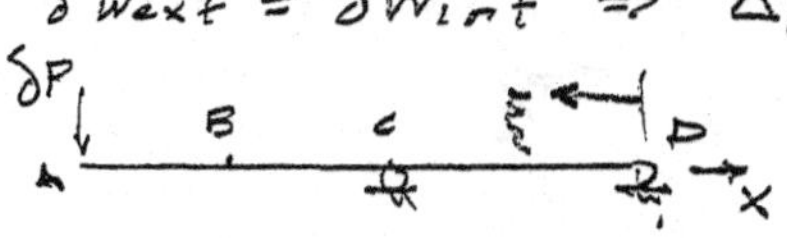

$M = 0,\ 0<x<L/3;\quad M(x) = -P(x-L/3),\ L/3<x<2L/3$

$M = -P\xi/3,\quad 0\le\xi\le L$

$\delta M = -\delta P x,\quad 0<x<2L/3;\quad \delta M = -(2)\delta P/3)\xi,\quad 0\le\xi\le L$

$\Delta_A = \frac{P\delta P}{EI}\left\{\int_{L/3}^{2L/3}(x-L/3)(-x)dx + \int_0^L(-\xi/3)(-2\xi/3)d\xi\right\} = \frac{17P\delta PL^3}{162EI}$

$\delta W^*_{ext} = \delta W^*_{int} \Rightarrow \Delta_A = 17L^3/162EI\ \downarrow$

c). θ_A

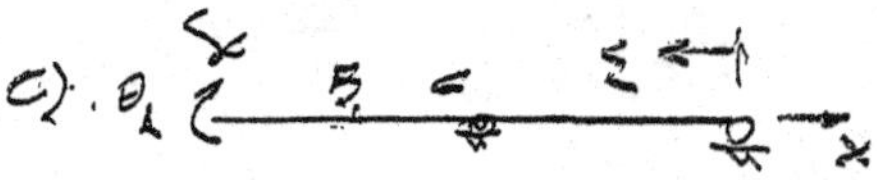

$\delta M = \delta c;\quad 0<x<2L/3;\quad M = (\delta c/L)\xi,\ 0<\xi<L$

$\theta_A = \frac{(\delta c)P}{EI}\left\{\int_{L/3}^{2L/3}-(x-L/3)dx - \int_0^L(\xi/3)(\xi/L)d\xi\right\} = -P\delta c\cdot L^2/6EI$

$\delta W^*_{ext} = \delta W^*_{int} \Rightarrow \Theta_A = -PL^2/6EI$ ↺

14.25 a)

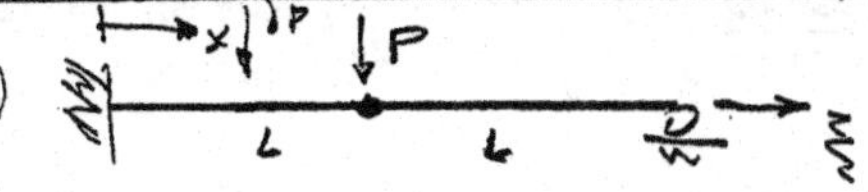

$M(\xi) = -P(L-\xi);\quad 0\le\xi\le L;\quad M(\xi)=0,\ L\le\xi\le 2L$

$\delta M = -\delta P(x-\xi);\quad \xi<x<L$

$\delta M = -\delta P(x-2L) + \delta P(2-x/L)\xi;\quad L<x<2L$

See Prob. 14.11 for detailed integration

$0<x<L$:

$\delta W^*_{int}(x) = \frac{P\cdot\delta P}{EI}\int_0^x (L-x)(x-\xi)d\xi = \frac{P\delta P}{6EI}(3Lx^2 - x^3)$

$\delta W_{ext} = \delta P\cdot v(x) \Rightarrow v(x) = P(3Lx^2 - x^3)/EI$

$L<x<2L$:

$\delta W^*_{int}(x) = \frac{P\cdot\delta P}{EI}\int_0^L -(L-\xi)[(x-2L)+(2-x/L)\xi]d\xi = \frac{P\cdot\delta PL^2}{3EI}(2L-x)$

$\delta W^*_{ext} = \delta P\cdot v(x) \Rightarrow v(x) = PL^2(2L-x)/3EI$

b). By Betti's Law:

$v_B = \frac{P}{6EI}(3La^2 - a^3) = \frac{Pa^2}{6EI}(3L-a),\quad 0<a<L$

$v_B = \frac{PL^2}{3}(2L-a),\quad a\le L\le 2L$

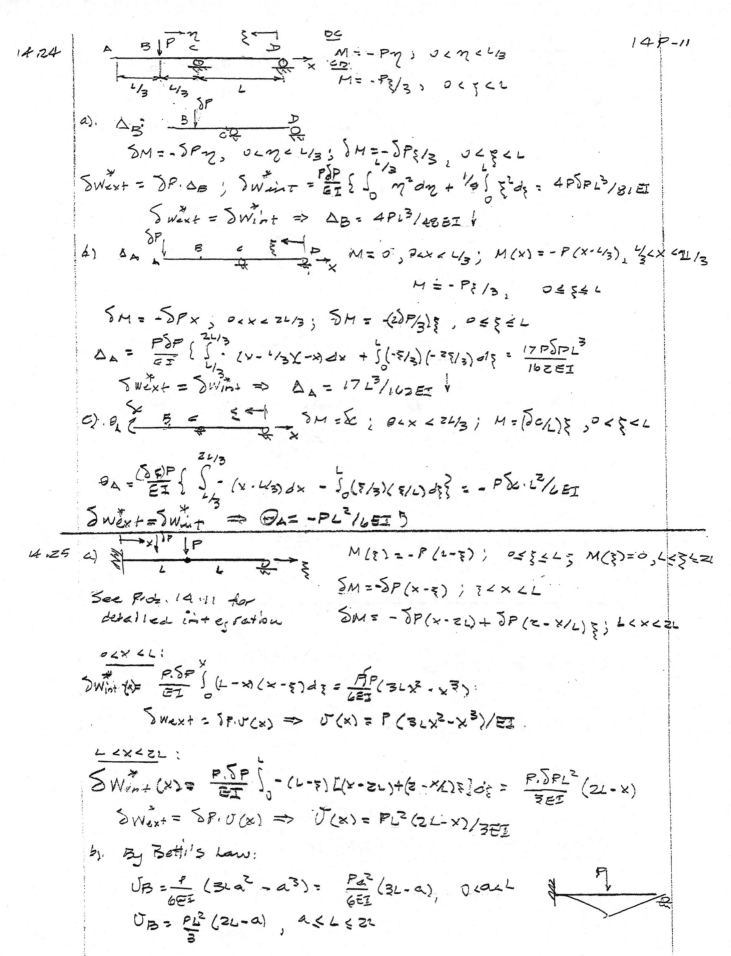

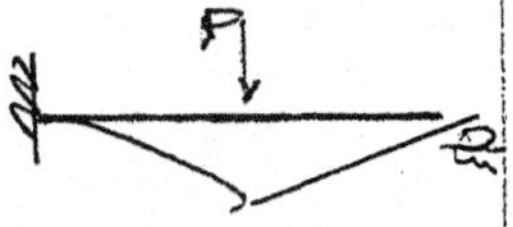

14.26 (a)

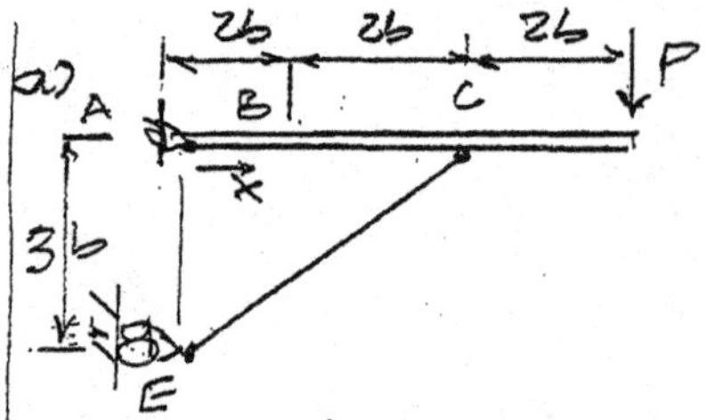

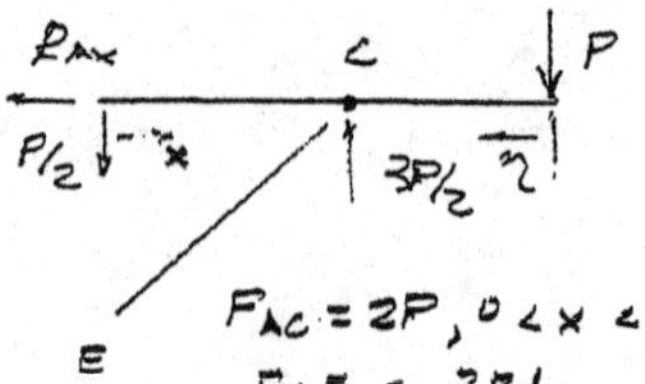

$F_{CE} = -5P/2$ (Comp.)

$R_{Ax} = 2P$

$F_{AC} = 2P,\ 0 < x < 4b;\ M(x) = -Px/2,\ 0 < x < 4b$

$F_{AE} = 3P/2$

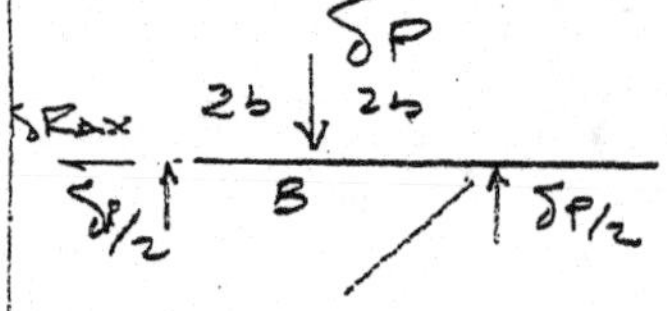

$\delta F_{CE} = -5\delta P/6\ ;\ \delta R_{Ax} = 2\delta P/3\ ;\ \delta F_{AC} = 2\delta P/3$

$\delta M = (\delta P/2)x,\ 0 < x < 2b;\ \delta M = \delta P(2b - x/2)),\ 2b < x < 4b$

$\delta F_{AE} = \delta P/2$

$\Delta_B:\ \delta W^*_{ext} = \delta P \cdot \Delta_B$

$\delta W^*_{int} = \frac{1}{EI}\int M\delta M\, ds + \frac{1}{AE} F \cdot \delta F \cdot l$

$= \frac{1}{E_2 I}\left\{\int_0^{2b}(-Px/2)(\delta Px/2)dx + \int_{2b}^{4b}(-Px/2)\frac{\delta P}{2}(4b-x)dx\right\}$
$+ \frac{1}{6A_0E_2}\left[(2P)(2\delta P/3)\right]^{4b}_{2b} + \frac{1}{2A_0E_1}\left[(-5P/2)(-5\delta P/6)\right](5b) + \frac{1}{A_0E_1}\left[\frac{3P}{2}\cdot\frac{\delta P}{2}\right] \cdot 3b$

$\delta W^*_{int} = P \cdot \delta P \left\{\frac{-1}{4E_2I}\left[\int_0^{2b} x^2 dx + \int_{2b}^{4b} x(4b-x)dx\right] + \frac{b}{A_0E_2}\left[\frac{16}{18}\right] + \frac{179}{24}\frac{b}{A_0E_1}\right\}$

$= P \cdot \delta P\left\{-\frac{2b^3}{E_2I_0} + \frac{8}{9A_0E_2} + \frac{179b}{24A_0E_1}\right\}$

$\delta W^*_{ext} = \delta W^*_{int}$

$$\Delta_B = \left(\frac{179Pb}{24E_1A_0} + \frac{8Pb}{9E_2A_0}\right) - \frac{2Pb^3}{E_2I_0}$$

b). $\Delta_B = (\Delta_B)_{axial} + (\Delta_B)_{flex}$; $(\Delta_B)_{axial} = \frac{(20\times10^3)(2)}{(10^{-4})(200\times10^9)}\left(\frac{179}{24}+\frac{8}{9}\right) = 1.68\times10^{-3}$ m ; $(\Delta_B)_{fl} = 3.2\times10^{-3}$ m

(i) $\Delta_B = 0.15$ cm ↑, (ii) $\Delta_B = -0.3$ cm ↑

14.27

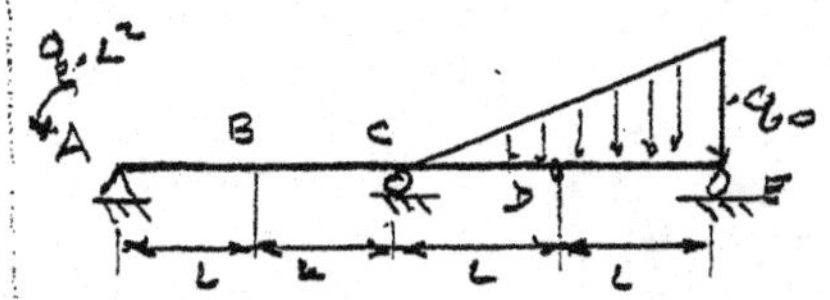

$M(x) = -q_0L^2 + q_0Lx/4\ ,\quad 0 < x < 2L$

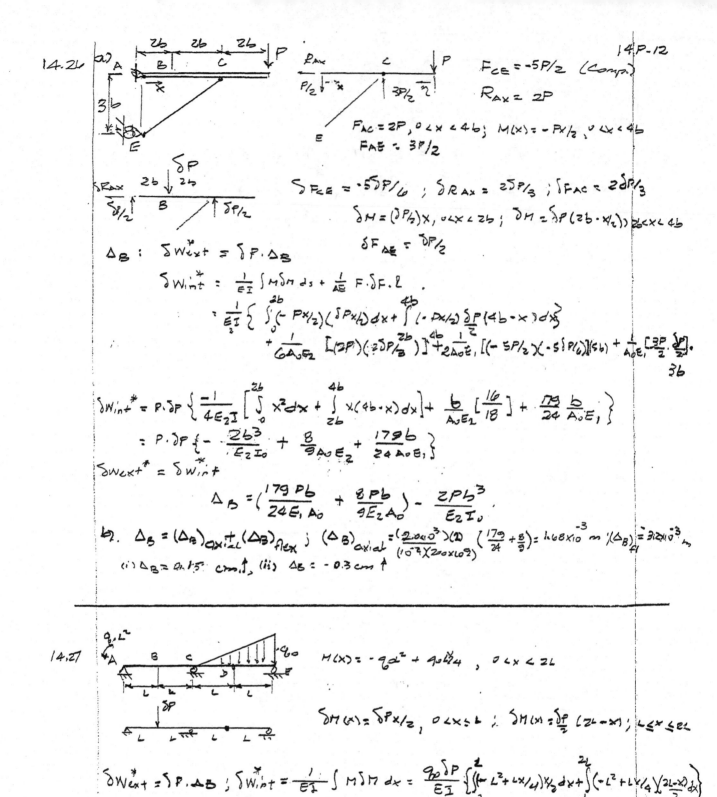

$\delta M(x) = \delta Px/2,\ 0 < x \le L\ ;\ \delta M(x) = \frac{\delta P}{2}(2L - x);\ L \le x \le 2L$

$\delta W^*_{ext} = \delta P \cdot \Delta_B\ ;\ \delta W^*_{int} = \frac{1}{EI}\int M\delta M\, dx = \frac{q_0\delta P}{EI}\left\{\int_0^L(-L^2 + Lx/4)x/2\, dx + \int_L^{2L}(-L^2 + Lx/4)\left(\frac{2L-x}{2}\right)dx\right\}$

$= -3q_0\delta PL^4/8EI$

$\delta W^*_{ext} = \delta W^*_{int} \Rightarrow\quad \Delta_B = -3q_0L^4/8EI$ ↑

14.28

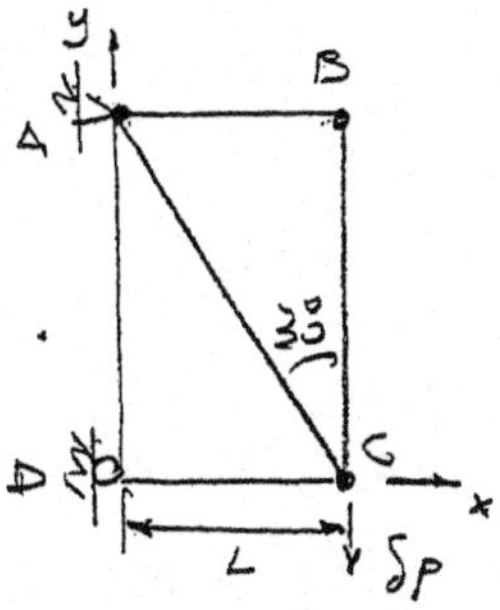

$$\delta W^*_{ext} = \delta P \cdot \Delta_{cy}$$

$$\delta W^*_{int} = \sum \varepsilon_i \cdot \delta F_i \cdot L_i$$

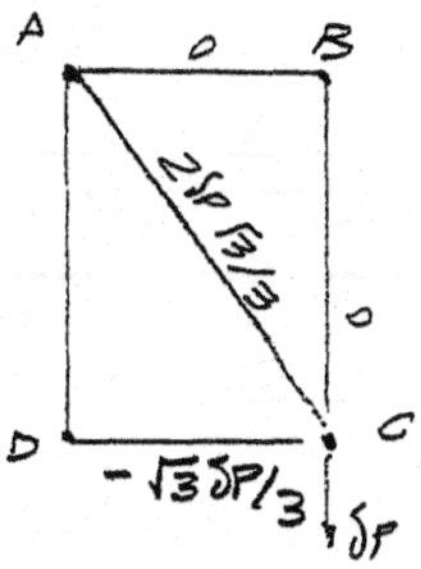

$L_{AC} = 2L$, $L_{CD} = L$

$\varepsilon_0\big|_{AC} = \alpha \Delta T$; $\varepsilon_0\big|_{CD} = 2\alpha \Delta T$

$$\therefore \delta W^*_{int} = [(\alpha \Delta T) \cdot 2\sqrt{3}\,P/3]\,2L + [(2\alpha \Delta T)(-\sqrt{3}\,\delta P/3)]\,L = \frac{2\sqrt{3}\,\alpha L\,\delta P}{3}$$

$$\delta W^*_{ext} = \delta W^*_{int} \Rightarrow \Delta_{cy} = 2\sqrt{3}\,\alpha L/3 \downarrow$$

14.29

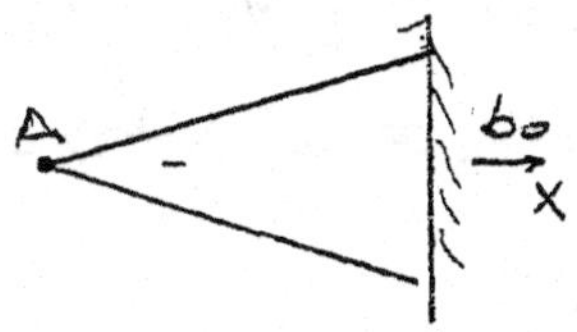

$$\delta W^*_{ext} = \delta P \cdot \Delta_A$$

$$\delta W^*_{int} = \frac{1}{E}\int \frac{M\,\delta M}{I(x)}\,dx \qquad I(x) = b_0 x d^3/12L$$

$M(x) = -Px$, $\delta M(x) = -\delta P x$

$$\delta W^*_{int} = \frac{12L}{E b_0 d^3}\int_0^L (-Px)(-\delta P x)\,dx = 6P\,\delta P L^3/E b_0 d^3.$$

$$\delta W^*_{ext} = \delta W^*_{int} \Rightarrow \Delta_A = 6PL^3/E b_0 d^3.$$

14.30

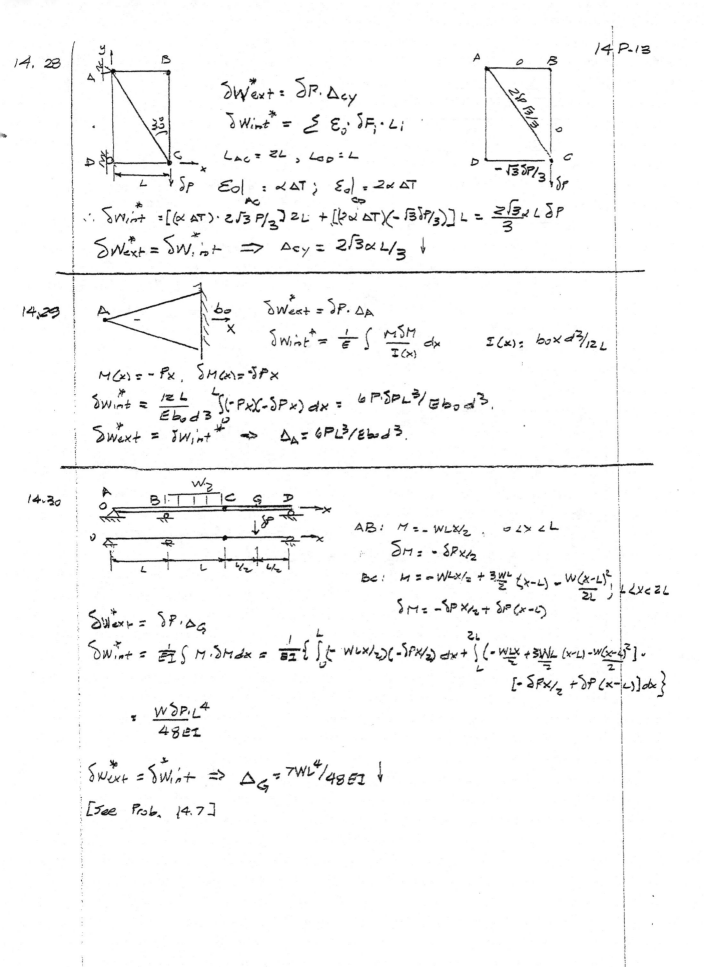

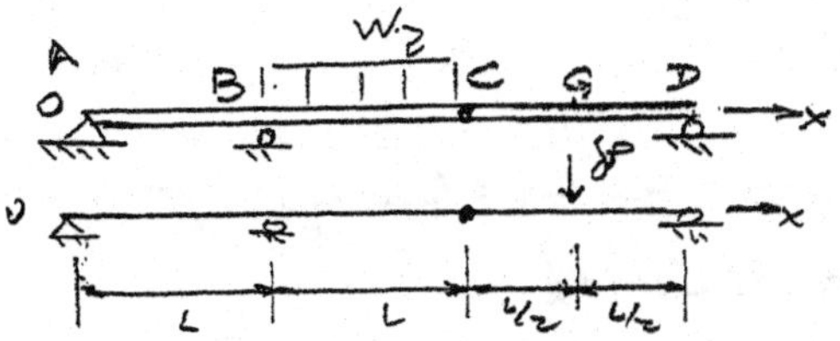

AB: $M = -WLx/2$, $0 < x < L$

$\delta M = -\delta P x/2$

BC: $M = -WLx/2 + \frac{3WL}{2}(x-L) - \frac{W(x-L)^2}{2L}$; $L < x < 2L$

$\delta M = -\delta P x/2 + \delta P (x-L)$

$$\delta W^*_{ext} = \delta P \cdot \Delta_G$$

$$\delta W^*_{int} = \frac{1}{EI}\int M \cdot \delta M\,dx = \frac{1}{EI}\Big\{\int_0^L (-WLx/2)(-\delta P x/2)\,dx + \int_L^{2L}\Big(-\frac{WLx}{2} + \frac{3WL}{2}(x-L) - \frac{W(x-L)^2}{2}\Big)\cdot [-\delta P x/2 + \delta P (x-L)]\,dx\Big\}$$

$$= \frac{W\,\delta P \cdot L^4}{48EI}$$

$$\delta W^*_{ext} = \delta W^*_{int} \Rightarrow \Delta_G = 7WL^4/48EI \downarrow$$

[See Prob. 14.7]

a.

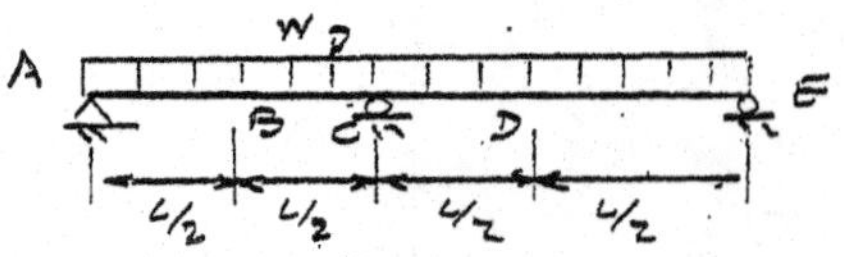

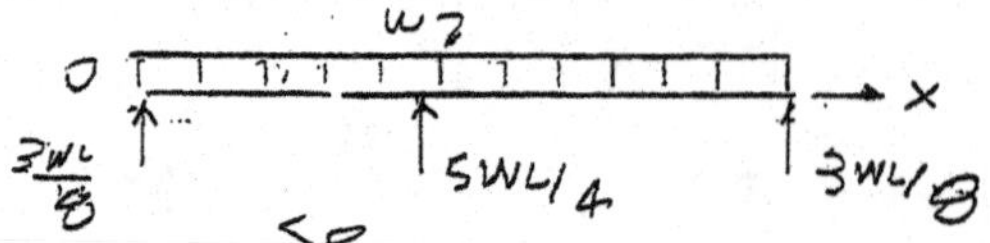

$M(x) = 3wLx/8 - wx^2/2 \,,\quad 0 < x < L$

$M(x) = 3wLx/8 - wx^2/2 + \frac{5wL}{4}(x - L) \,;\quad L \le x \le 2L$

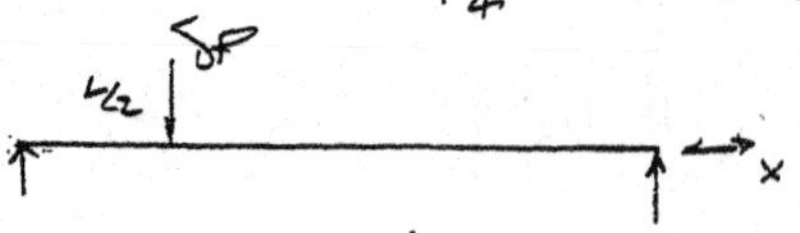

$\delta M(x) = 23\,\delta P x/32 \,;\quad 0 < x < L/2$

$\delta M(x) = 23\,\delta P x/32 - \delta P(x - L/2) \,;\quad L/2 \le x \le L$

$\delta M(x) = 23\,\delta P x/32 - \delta P(x - L/2) + \frac{11\,\delta P}{4}(x - L) \,;\quad L \le x \le 2L$

$$\delta W^*_{int} = \frac{1}{EI}\int M\,\delta M\,dx = \frac{w\,\delta P}{EI}\Big\{\int_0^{L/2}(3Lx/8 - x^2/2)[23x/32]\,dx + \int_{L/2}^{L}(3xL/8 - wx^2/2)\Big[\frac{23x}{32} - (x - L/2)\Big]dx$$
$$+ \int_L^{2L}\Big[3Lx/8 - x^2/2 + \frac{5L}{4}(x - L)\Big]\Big[23x/32 - (x - L/2) + \frac{11}{4}(x - L)\Big]dx\Big\}$$

$$= \frac{w\,\delta P L^4}{192\,EI} \quad \Rightarrow \quad \Delta_B = wL^4/192\,EI$$

b).

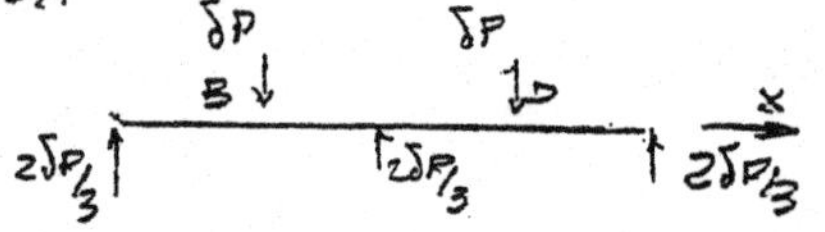

$\delta M = 2\,\delta P x/3 \,,\quad 0 < x < L/2$

$\delta M = -\delta P x/3 + \delta P L/2 \,,\quad L/2 < x < L$

$$\delta W^*_{int} = \frac{2}{EI}\Big[\int_0^{L/2}\frac{2\,\delta P x}{3}(3wLx/8 - wx^2/2)\,dx + \int_{L/2}^{L}(-\delta P x/3 + \delta P L/2)\Big(\frac{3wLx}{8} - wx^2/2\Big)dx\Big]$$

$$= \delta P \cdot wL^4/96\,EI.$$

$\delta W^*_{ext} = \delta P\cdot\Delta_B + \delta P\cdot\Delta_D = 2\,\delta P\cdot\Delta_B$ since $\Delta_B = \Delta_D$ by symmetry.

$\delta W^*_{ext} = \delta W^*_{int} \Rightarrow \Delta_B = \Delta_D = wL^4/192\,EI.$

14.32

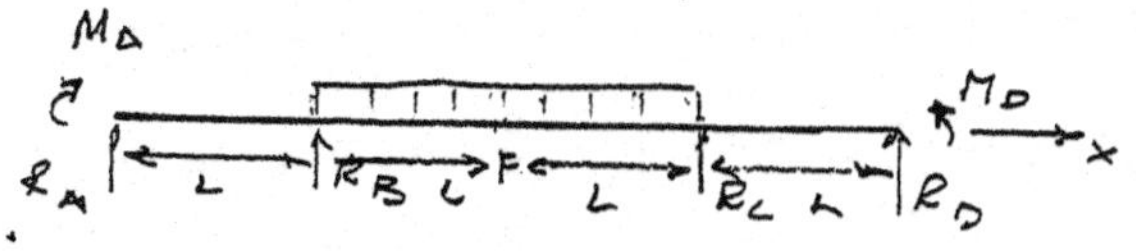

$M(x) = 2wL^2/15 - 2wLx/5, \quad 0 \le x \le L$

$M(x) = 2wL^2/15 - 2wLx/5 + \frac{7wL}{5}(x-L) - \frac{w}{2}(x-L)^2$

$= -wx^2/2 + 2wLx - 53wL^2/30; \quad L \le x \le 2L.$

a). STATICALLY ADMISSIBLE VIRTUAL FORCE SYSTEM

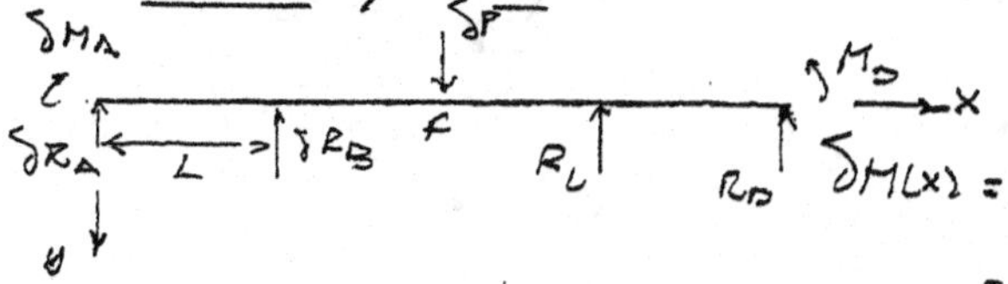

$\delta M(x) = \delta PL/5 - 3\delta Px/5; \quad 0 \le x \le L$

$= \delta PL/5 - \frac{3\delta Px}{5} + \frac{11\delta P}{10}(x-L)$

$= \frac{5\delta Px}{10} - \frac{9\delta PL}{10}, \quad L \le x \le 2L$

BY SYMMETRY

$$\delta W^*_{int} = 2\cdot\frac{1}{EI}\left\{\int_0^L\left(\frac{2wL^2}{15} - \frac{2wLx}{5}\right)(\delta PL/5 - 3\delta Px/5)\,dx + \int_L^{2L}(-wx^2/2 + 2wLx - 53wL^2/30)(5\delta Px/10 - 9\delta PL/10)\,dx\right\}$$

$$= \frac{2w\delta P}{EI}\left\{\frac{2}{25}\int_0^L(L^2/3 - Lx)(L-3x)\,dx + \frac{1}{10}\int_L^{2L}(-x^2/2 + 2Lx - 53/30)(5x - 9L)\,dx\right\}$$

$$= \frac{2w\delta P}{EI}\left\{\frac{2}{25}\int_0^L(3Lx^2 - 2L^2x + L^3/3)\,dx + \frac{1}{10}\int_L^{2L}\left(-5x^3/2 + \frac{29}{2}Lx^2 - \frac{161}{6}L^2x + \frac{159L^3}{10}\right)dx\right\}$$

$$= \frac{2w\delta P}{EI}\left\{\frac{2}{25}\left[Lx^3 - L^2x^2 + L^3x/3\right]_0^L + \frac{1}{10}\left[-5x^4/4 + 29Lx^3/6 - \frac{161}{12}L^2x^2 + \frac{159L^3x}{10}\right]_L^{2L}\right\}$$

$$= \frac{2w\delta P}{EI}\left\{2L^4/75 + \frac{L^4}{10}\left[-10 + \frac{1213}{120}\right]\right\} = \frac{3w\delta PL^4}{40EI}$$

$\delta W^*_{ext} = \delta P\cdot\Delta_F; \quad \delta W_{ext} = \delta W^*_{int} \Rightarrow \Delta_F = 3wL^4/40EI.$

b). STATICALLY INADMISSIBLE VIRTUAL FORCE SYSTEM

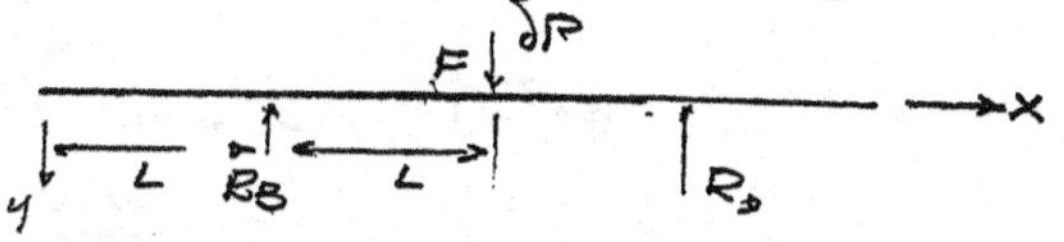

$\delta M(x) = 0, \quad 0 \le x \le L$

$\delta M(x) = \frac{\delta P}{2}(x-L), \quad L \le x \le 2L$

$$\delta W^*_{int} = \frac{2}{EI}\int_L^{2L}(-wx^2/2 + 2wLx - 53wL^2/30)\cdot\frac{\delta P}{2}(x-L)\,dx$$

$$= \frac{\delta P\cdot w}{EI}\int_L^{2L}\left[-x^3/2 + 5Lx^2/2 - 113L^2x/30 + 53L^3/30\right]dx$$

$$= \frac{\delta P\cdot w}{EI}\left\{-x^4/8 + 5Lx^3/6 - 113L^2x^2/60 + 53L^3x/30\right\}\Big|_L^{2L}$$

$$\delta W^*_{int} = \frac{3\delta P\cdot wL^4}{40EI}.$$

$\delta W^*_{ext} = \delta W^*_{int} \Rightarrow \Delta_F = 3wL^4/40EI.$

14.33

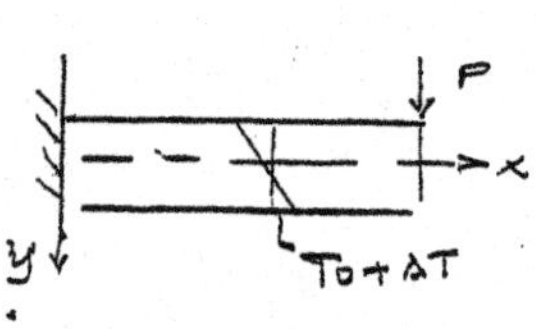

$\delta M(x) = -\delta P(L-x)$, $\varepsilon = \frac{2\alpha y}{d} \cdot \Delta T$

$\delta\sigma = \frac{\delta M y}{I} = -\frac{\delta P(L-x)y}{I}$ satisfies equilibrium.

$$\delta W^*_{int} = \int_0^L \iint_A \delta\sigma \cdot \varepsilon \, dA\, dx = \frac{2\alpha\Delta T}{d} \cdot \frac{\delta P}{I} \int_0^L (L-x) \iint_A y^2 dA \, dx$$

$$= \frac{2\alpha\Delta T \delta P}{d} \int_0^L (L-x)dx = \frac{2\alpha\Delta T\delta P}{d}\left(Lx - x^2/2\right)\Big|_0^L$$

$$= \alpha\Delta T \delta P L^2/d$$

$\delta W^*_{ext} = P\Delta_B \Rightarrow \Delta_B = (\alpha \Delta T L^2)/d.$

14.34

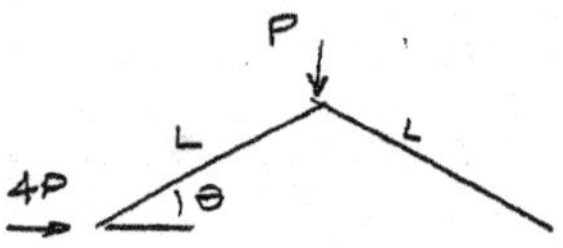

$\Pi = V = -4P(2L - 2L\cos\theta) + PL\sin\theta \quad (U=0)$

$\delta\Pi = \frac{d\Pi(\theta)}{d\theta}\delta\theta = 0 \Rightarrow d\Pi/d\theta = 0$

$PL[-8\sin\theta + \cos\theta] = 0 \Rightarrow \tan\theta = 1/8$

$\theta = 7.125°$

14.35

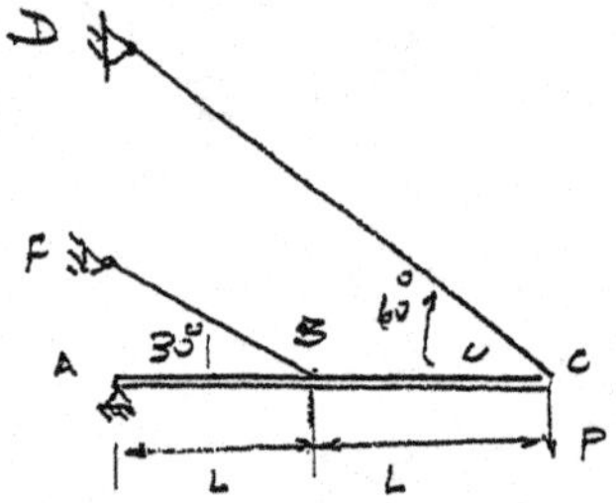

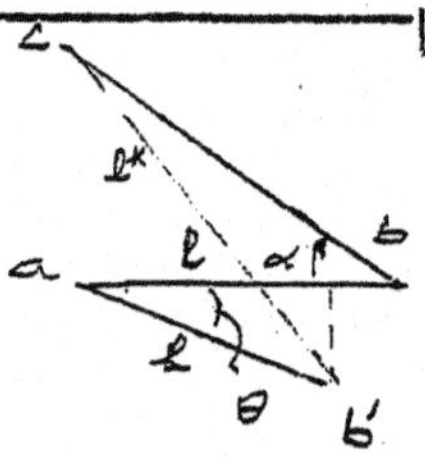

General Case:

$|b'c|^2 = l^{*2} = (l\sin\theta + l\tan\alpha)^2 + l^2\cos^2\theta$

$l^{*2} = l^2[(\sin\theta + \tan\alpha)^2 + \cos^2\theta]$

$l^* = l[1 + \tan^2\alpha + 2\tan\alpha\sin\theta]^{1/2}$

$|\theta| \ll 1 \Rightarrow l^* = l[(1+\tan^2\alpha) + 2\theta\tan\alpha]^{1/2} = l.(1+\tan^2\alpha)^{1/2}\left[1 + \frac{2\theta\tan\alpha}{1+\tan^2\alpha}\right]^{1/2}$

$= l(1+\tan^2\alpha)^{1/2}\left[1 + \frac{\tan\alpha}{1+\tan^2\alpha}\cdot\theta\right] = \frac{l}{\cos\alpha}[1 + \sin\alpha\cos\alpha\cdot\theta]$

Since $1+\tan^2\alpha = 1/\cos^2\alpha$.

$\varepsilon|_{b'c} = \frac{l^*}{l/\cos\alpha} - 1 = [1 + (\sin\alpha\cos\alpha)\theta] - 1 = \frac{\sin 2\alpha}{2}\cdot\theta$

Here $\alpha = 30°$ and $60°$.

$U = \frac{1}{2}\sum EA_i \varepsilon_i^2 L_i = \frac{EA}{2}\left[\frac{\sin^2 60}{4}\frac{L}{\cos 30} + \frac{\sin^2 120°}{4}\cdot\frac{2L}{\cos 60°}\right]\theta^2 = \frac{EA}{16}(6+\sqrt{3})\theta^2.$

$V = -2PL\sin\theta = -2PL\theta$

$\Pi = U + V \Rightarrow \delta\Pi = 0 \Rightarrow d\Pi/d\theta = \frac{EA}{8}(6+\sqrt{3})\theta - 2PL = 0 \Rightarrow \theta = \frac{16P}{33EA}(6-\sqrt{3})$

$\therefore \varepsilon_{BF} = \frac{\sqrt{3}}{4}\cdot\frac{16}{33}(6-\sqrt{3})\frac{P}{EA} = \frac{4\sqrt{3}}{33}(6-\sqrt{3})P/EA. \quad \varepsilon|_{CE} = \text{same.}$

$\Delta_C = \frac{32\sqrt{3}}{33}(6-\sqrt{3})PL/AE$

$F_{BF} = F_{CE} = \varepsilon\cdot EA = \frac{4\sqrt{3}(6-\sqrt{3})P}{33} = 0.8960P$

14.36

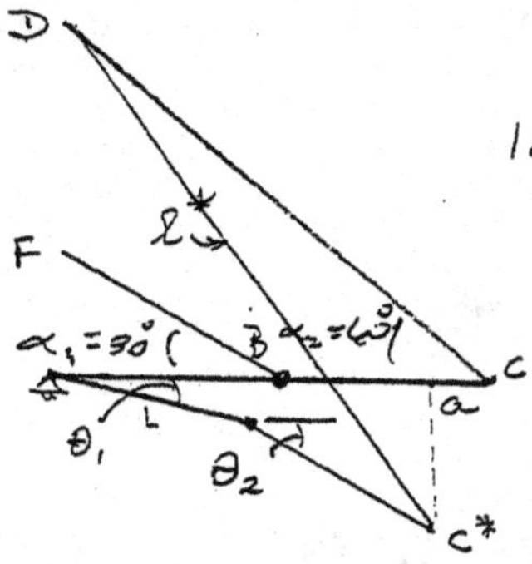

$$|ac^*| = L(\sin\theta_1 + \sin\theta_2)$$

$$l^{*2} = |DC^*|^2 = L^2[\sin\theta_1 + \sin\theta_2 + 2L\tan\alpha_2]^2 + L^2[\cos\theta_1 + \cos\theta_2]^2.$$

$$\Rightarrow l^{*2} = 2L^2[1 + \cos(\theta_2 - \theta_1) + 2\tan^2\alpha_2 + 2\tan\alpha_2\cdot(\sin\theta_1 + \sin\theta_2)]$$

$$|\theta_1|, |\theta_2| << 1 \Rightarrow$$

$$l^{*2} = 4L^2\{1 + \tan^2\alpha_2 + \tan\alpha_2\cdot[(\theta_1+\theta_2) - (\theta_1-\theta_2)^2]\}$$ (the $(\theta_1-\theta_2)^2$ term crossed out → 0)

$$l^* = 2L[1+\tan^2\alpha]^{1/2}\left\{1 + \frac{\tan\alpha_2}{2(1+\tan^2\alpha_2)}\cdot(\theta_1+\theta_2)\right\}.$$

$$= (2L/\cos\alpha_2)\left[1 + \frac{\tan\alpha_2}{2}\cos^2\alpha_2\cdot(\theta_1+\theta_2)\right] = \frac{2L}{\cos\alpha_2}\left\{1 + \frac{\sin 2\alpha_2}{4}\cdot(\theta_1+\theta_2)\right\}.$$

$$\varepsilon_{CD} = \frac{|C^*D|^2}{2L/\cos\alpha_2} - 1 = \frac{\sin 2\alpha_2}{4}(\theta_1+\theta_2); \quad \varepsilon_{BF} = \frac{\sin 2\alpha_1}{2}\cdot\theta_1 \quad \text{[See Prob. 14.35]}$$

$$U = \frac{EA}{2}\sum \varepsilon_i^2 L_i; \quad U = \frac{EA}{2}\left\{\frac{\sin^2 60°}{4}\theta_1^2\cdot\frac{L}{\cos 30°} + \frac{\sin^2 120}{16}(\theta_1+\theta_2)^2\cdot\frac{2L}{\cos 60°}\right\}$$

$$U = \frac{EAL}{32}\{2\sqrt{3}\,\theta_1^2 + 3(\theta_1+\theta_2)^2\}; \quad V = -WL[L\theta_1/2 + (L\theta_1 + \frac{L\theta_2}{2})] = -\frac{WL^2}{2}(3\theta_1+\theta_2)$$

$$\Pi = U + V \Rightarrow \delta\Pi = 0 \Rightarrow \partial\Pi/\partial\theta_1 = 0, \quad \partial\Pi/\partial\theta_2 = 0$$

$$\partial\Pi/\partial\theta_1 = \frac{EAL}{32}[4\sqrt{3}\,\theta_1 + 6(\theta_1+\theta_2)] - 3WL^2/2 = 0 \quad (i)$$

$$\partial\Pi/\partial\theta_2 = \frac{EAL}{32}[6(\theta_1+\theta_2)] - WL^2/2 = 0 \quad (ii) \Rightarrow (\theta_1+\theta_2) = \frac{8}{3}WL/EA.$$

$$(i) \Rightarrow \theta_1 = \frac{8\sqrt{3}}{3}\frac{WL}{AE}, \quad \theta_2 = \frac{8}{3}(1-\sqrt{3})\frac{WL}{AE}$$

$$\Rightarrow \text{i.e. } \varepsilon_{BF} = 2WL/AE, \quad \varepsilon_{CD} = \frac{\sqrt{3}WL}{3AE}$$

$$\Rightarrow F = \varepsilon\cdot AE \Rightarrow F_{BF} = 2WL, \quad F_{CD} = \sqrt{3}WL/3$$

$$\Delta_B = L\cdot\theta_1 = \frac{8\sqrt{3}}{3}WL^2/AE, \quad \Delta_C = 8WL^2/3EA$$

14.37 $v(x) = \sum_{n=\text{odd}} a_n \left[1 - \cos\frac{n\pi x}{2L}\right]$ $\quad U = \frac{EI}{2}\int [v''(x)]^2 dx, \quad V = -w\int_0^L v(x)dx$

$$v''(x) = (\pi/2L)^2 \sum a_n n^2 \cos\frac{n\pi x}{2L}$$

$$U = \frac{EI}{2}(\pi/2L)^4 \sum\sum a_n a_m n^2 m^2 \underbrace{\int_0^L \cos\frac{n\pi x}{2L}\cos\frac{m\pi x}{2L}dx}_{L/2,\ n=m}$$

$\therefore$

$$U = \frac{\pi^4 EI}{64L^3}\sum_{n=1,3,5} a_n^2 n^4. \qquad V = -w\sum a_n \int_0^L \left(1 - \cos\frac{n\pi x}{2L}\right)dx = -w\sum a_n \left(x - \frac{2L}{n\pi}\sin\frac{n\pi x}{2L}\right)\Big|_0^L$$

$$V = -wL\sum a_n \left[1 - (2/n\pi)\sin\frac{n\pi}{2}\right]$$

$$\Pi = U + V \Rightarrow \delta\Pi = \frac{d\Pi}{da_n}da_n = 0$$

$$\Pi = \frac{\pi^4 EI}{64L^3}\sum_{n\ \text{odd}} a_n^2 n^4 - wL\sum_{n\ \text{odd}} a_n \left[1 - \frac{2}{n\pi}(-1)^{\frac{n-1}{2}}\right]$$

$$d\Pi/da_n = 0 \Rightarrow \frac{\pi^4 EI}{32L^3}\sum a_n n^4 - wL\sum\left[1 - \frac{2}{n\pi}(-1)^{\frac{n-1}{2}}\right] = 0$$

$$a_n = \frac{32wL^4}{\pi^4 EI}\cdot\frac{\left[1 - 2(-1)^{\frac{n-1}{2}}/n\pi\right]}{n^4}, \quad n \text{ odd}$$

$$v(x) = \frac{32wL^4}{\pi^4 EI}\sum_{n\ \text{odd}}\left(\frac{1 - 2(-1)^{\frac{n-1}{2}}/n\pi}{n^4}\right)\left(1 - \cos\frac{n\pi x}{2L}\right).$$

$$\text{b).}\ v(L) = \frac{32wL^4}{\pi^4 EI}\Big[\underbrace{(1 - 2/\pi)}_{\substack{c_1\\0.3634}} + \underbrace{\frac{1}{3^4}(1 + 2/3\pi)}_{\substack{c_3\\0.0150}} + \underbrace{\frac{1}{5^4}(1 - 2/5\pi)}_{\substack{c_5\\0.0014}} + \dots\Big] = \frac{0.1248\,wL^4}{EI}$$

Note: $v(L)$ converges as $1/n^4$

Error with first three terms: 0.19%.

$$M(x) = -EIv''(x) = -\frac{8wL^2}{\pi^2}\sum_{n\ \text{odd}}\frac{\left[1 - 2(-1)^{\frac{n-1}{2}}/n\pi\right]}{n^2}\cos\frac{n\pi x}{2L}$$

$$M(0) = -\frac{8wL^2}{\pi^2}\Big[\underbrace{c_1}_{.3634} + \underbrace{c_3/9}_{+0.1347} + \underbrace{c_5/25}_{0.0344} + \dots\Big] = -0.4320\,wL^2$$

Error after 3 first terms: 13.6%. Note: $M(0)$ converges as $1/n^2$

Summing up to $n = 49$: $M(0) = -0.4919\,wL^2$

Error: 1.62%

Note: $M(0) = -\frac{8wL^2}{\pi^2}\left\{\left[1 + \frac{1}{3^2} + \frac{1}{5^2} + \frac{1}{7^2} + \dots\right] - \frac{2}{\pi}\left[1 - \frac{1}{3^3} + \frac{1}{5^3} - \frac{1}{7^3}\dots\right]\right\}$

$$= -8wL^2/\pi^2\left\{(\pi^2/8) - \frac{16}{\pi^2}(\pi^3/32)\right\} = -wL^2/2$$

(↑ Jolley No. 339; ↑ Jolley No. 341)

Ref. Jolley, L.B.W, "Summation of Series", Dover, 1961

14.38

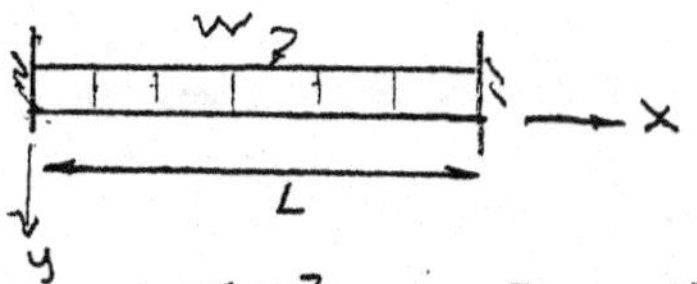

$$v(x) = \sum_{n=2,4,6} a_n [1 - \cos n\pi x/L]$$

$$v''(x) = (\pi/L)^2 \sum a_m n^2 \cos\frac{n\pi x}{L}, \quad U = \frac{EI\pi^4}{2L^4}\sum n^2 m^2 a_n a_m \int_0^L \underbrace{\cos\frac{n\pi x}{L}\cos\frac{m\pi x}{L}}_{L/2,\ n=m} dx$$

$$U = \frac{\pi^4 EI}{4L^3}\sum_{n\ even} a_n^2 n^4 \qquad V = -w\sum a_n \int_0^L (1-\cos\frac{n\pi x}{L})dx = -wL a_n \quad (n\ even)$$

$$\Pi = U + V \Rightarrow \delta\Pi = \frac{\partial\Pi}{\partial a_n}\delta a_n = 0 \Rightarrow \frac{\pi^4 EI}{2L^3} a_i i^4 - wL = 0$$

$$a_n = \frac{2wL^4}{\pi^4 EI}\sum_{n=2,4,6}\frac{1}{n^4}$$

$$v(x) = \frac{2wL^4}{\pi^4 EI}\sum_{n=2,4,6}^{\infty}\frac{1}{n^4}(1-\cos n\pi x/L).$$

b)

$$v(L/2) = \frac{2wL^4}{\pi^4 EI}\sum_{n=2,4,6}^{\infty}\frac{1}{n^4}(1-\cos n\pi/2) = \frac{4wL^4}{\pi^4 EI}\sum_{n=2,6,10,\ldots}^{\infty} 1/n^4$$

$$M(0) = -EIv''(0) = -\frac{2wL^2}{\pi^2}\sum_{n=2,4,6\ldots} 1/n^2$$

Summing to $n=6$: $v(L/2) = 0.002598\ wL^4/EI$ $v_{ex} = wL^4/384 = 0.002604\ wL^4/EI$

% Error = 0.23 %

$M(0) = 0.06895\ wL^2$, $M(0)_{exact} = 0.0833\,wL^2$; % Err = 17.3

Summing to $n=18$: $v(L/2) = 0.002603\ wL^4/EI$; % Error = 0.016 %

$M(0) = 0.07800\ wL^2$; % Err = 6.4 %

Note: 1) $v(L/2)$ converges as $1/n^4$, i.e. very rapidly

$M(0)$ converges as $1/n^2$, i.e. more slowly.

2). $v|_{Rayl-Ritz} < v|_{exact}$. Explanation: stiffer.

NOTE: Ref. Jolley, "Summation of Series," Dover, 1961

$$v(L/2) = \frac{4wL^4}{\pi^4 EI}\left[\left(\frac{1}{2^4}+\frac{1}{4^4}+\frac{1}{6^4}+\frac{1}{8^4}\right) - \left[1/4^4 + 1/8^4 + 1/12^4\right]\right\}$$

Jolley No. 343 − No. 342

$$= \frac{4wL^4}{\pi^4 EI}\{[\text{No.}(343-342)] - 1/4^4[1 + 1/2^4 + 1/3^4]\} = \frac{4wL^4}{\pi^4 EI}\{(J.343 - J.342) - \frac{J.342}{4^4}\}$$

$$= \frac{4wL^4}{\pi^4 EI}\{(1 - 1/4^4)\text{Jolley }343 - \text{Jolley }342\} = \frac{4wL^4}{\pi^4 EI}\left[\left(\frac{255}{256}\right)\pi^4/90 - \pi^4/96\right] = \frac{wL^4}{384EI}\ \text{exact!}$$

$$M(0) = -\frac{2wL^2}{\pi^2}[1/2^2 + 1/4^2 + 1/6^2 \ldots] = -\frac{2wL^2}{\pi^2}[\text{Jolley }(\text{No. }336 - 339)] = -\frac{2wL^2}{\pi^2}(\pi^2/6 - \pi^2/8) = -wL^2/12$$

14.39 $U = W \quad U > 0$ Positive Definite $\Rightarrow \therefore W > 0$ under any load.

But $P \perp \Delta \Rightarrow P \cdot \Delta = 0$; contradiction

---

14.40

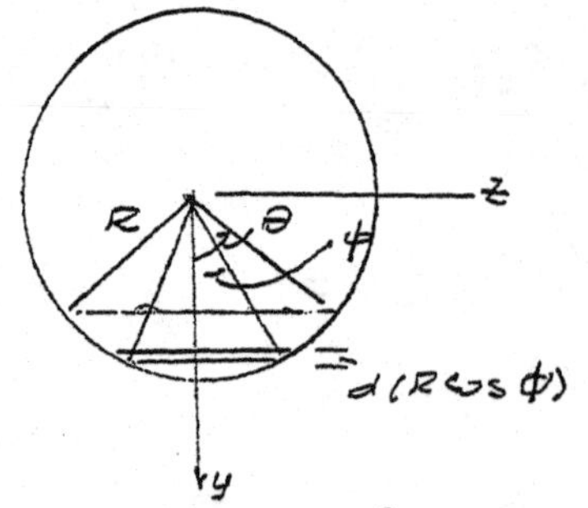

$$\tau = V_y Q_z(\theta)/I_{zz} b \;;\; b = 2R\sin\theta$$

$$Q_z(\theta) = \int_\theta^0 (R\cos\phi)(2R\sin\phi)\, d(R\cos\phi)$$

$$= -2R^3 \int_\theta^0 \sin^2\phi \cos\phi \, d\phi$$

$$Q_z(\theta) = -\frac{2R^3}{3} \sin^3\phi \Big|_\theta^0 = \frac{2R^3 \sin^3\theta}{3}$$

$$\tau(\theta) = \frac{2V_y R^3 \sin^3\theta}{3 I_{zz}(2R\sin\theta)} = \frac{V_y R^2 \sin^2\theta}{3 I_{zz}}$$

$$U_0 = \frac{\tau^2(\theta)}{2G} = \frac{V_y^2 R^4 \sin^4\theta}{18 I_{zz}^2 G}$$

$$U = \int_L \iint_A U_0 \, dA\, dx = \int_L \left\{ \frac{V_y^2 R^4}{18 I_{zz}^2} \int_\pi^0 (\sin^4\theta)(2R\sin\theta)\, d(R\cos\theta) \right\}$$

$$d(R\cos\theta) = -R\sin\theta\, d\theta$$

$$= \int_L \frac{V_y^2 R^6}{9 I_{zz}^2 G} \left(\int_\pi^0 \sin^6\theta \, d\theta \right) dx$$

Note: $\sin^6\theta = \frac{1}{16}\left[5 - 7\cos 2\theta + 3\cos 4\theta - \cos 6\theta - 2\sin^2 2\theta \cos 2\theta\right]$

$$\therefore \int_0^\pi \sin^6\theta\, d\theta = \frac{1}{16}\left[5\theta - \frac{7}{2}\sin 2\theta + \frac{3}{4}\sin 4\theta - \frac{1}{6}\sin 6\theta - \frac{\sin^3 2\theta}{3}\right]_0^\pi = \frac{5\pi}{16}$$

$$\therefore U = \int_L \frac{V_y^2 R^6}{9 I_{zz}^2 G} \cdot \frac{5\pi}{16} \qquad I_{zz} = \pi R^4/4$$

$$\therefore U = \int_L \frac{V_y^2 R^6}{9 \cdot \pi^2 R^8/16\, G} \cdot \frac{5\pi}{16} = \int_L \frac{5V_y^2}{9\pi R^2 G} dx = \int_L \frac{5 V_y^2}{9 AG} dx = \frac{\alpha}{2}\int \frac{V_y^2(x)}{AG} dx$$

$$\alpha = 10/9.$$

14.41

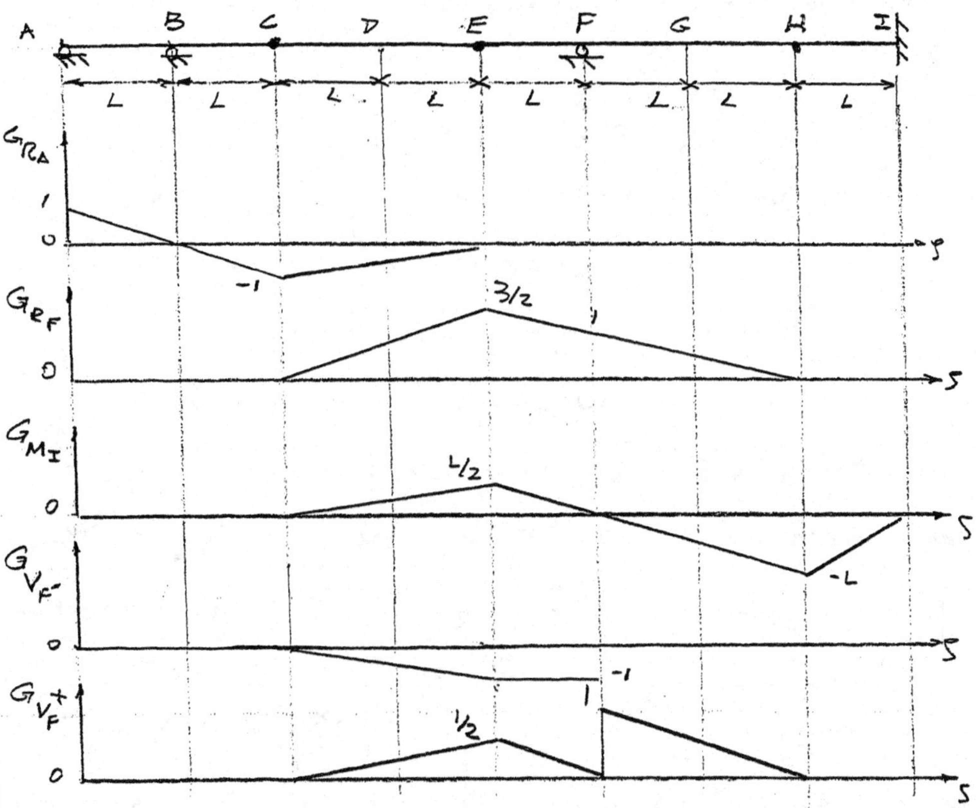

14.42

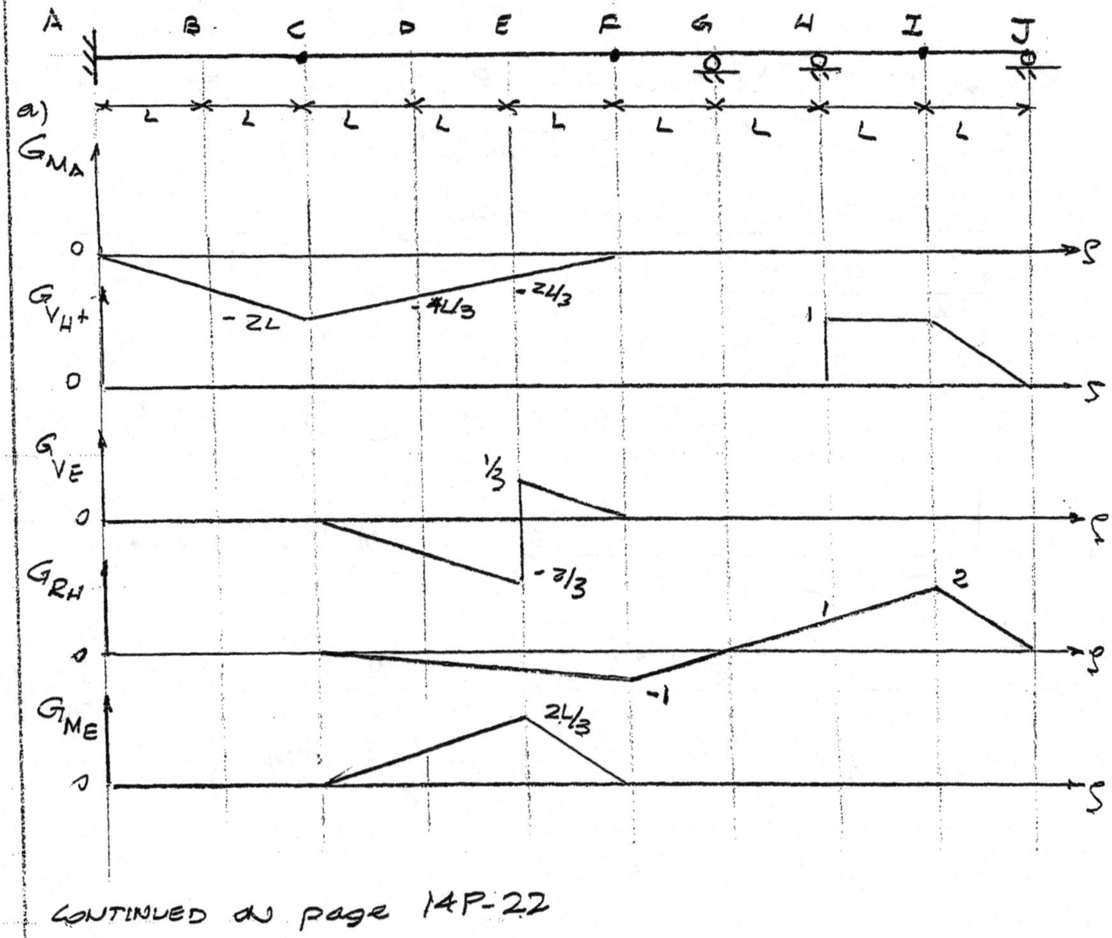

CONTINUED ON page 14P-22

14.42 (CONT'D)

b) (i) Place between B & D ($L \le \xi \le 3L$); $|M_A|_{max} = 19wL^2/6$

(ii)

$$G_{M_A}(\xi) = \begin{cases} -2\xi, & 0 < \xi < 2L \\ -10L/3 + 2\xi/3, & 2L < \xi < 5L \end{cases}$$

$$M_A = w\int_{\xi_0}^{2L+\xi_0} G_{M_A}(\xi)d\xi = w\left\{-2\int_{\xi_0}^{2L}\xi d\xi + \tfrac{1}{3}\int_{2L}^{2L+\xi_0}(-10L+2\xi)d\xi\right.$$

$$= w\left\{-(2L^2 - \xi_0^2/2) + \tfrac{1}{3}\left[-10L\xi + \xi^2\right]_{2L}^{2L+\xi_0}\right.$$

$$M_A = w/6\,(5\xi_0^2 - 12L\xi_0 - 12L^2).$$

$|M_A|_{max} \Rightarrow dM_A/d\xi_0 = 0 \Rightarrow 10\xi_0 - 12L = 0 \Rightarrow \xi_0 = 6L/5$

$$|M_A|_{max} = \frac{wL^2}{6}\left|\left[36/5 - 72/5 - 12\right]\right| = 16wL^2/5 = 3.2wL^2$$

c) (i) Place between H & J ($7L \le \xi \le 9L$); $(R_H)_{max} = 5wL/2$

(ii)

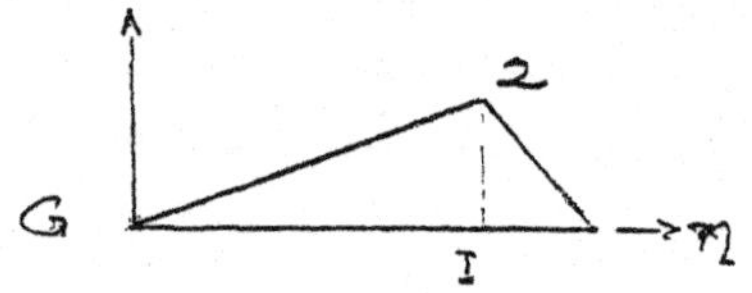

Let $\eta = \xi - 6$ (Transformation)

$$G_{R_H} = \begin{cases} \eta/L, & 0 \le \eta \le 2L \\ 6 - 2\eta/L, & 2L \le \eta \le 3L \end{cases}$$

$$R_H(\eta) = w\int_{\eta_0}^{2L+\eta_0} G_{R_H}(\eta)d\eta = w\left[\int_{\eta_0}^{2L}\eta/L\,d\eta + \int_{2L}^{2L+\eta_0}(6 - 2\eta/L)d\eta\right]$$

$$= w\left[-3\eta_0^2/2L + 2\eta_0 + 2L\right]$$

$$(R_H)_{max} \Rightarrow \frac{dR_H(\eta_0)}{d\eta_0} = 0 \Rightarrow -3\eta_0/2L + 2 = 0 \Rightarrow \eta_0 = 2L/3.$$

$$\Rightarrow (R_H)_{max} = 8wL/3.$$

Place between: $20L/3 \le \xi \le 26L/3$.

14.43

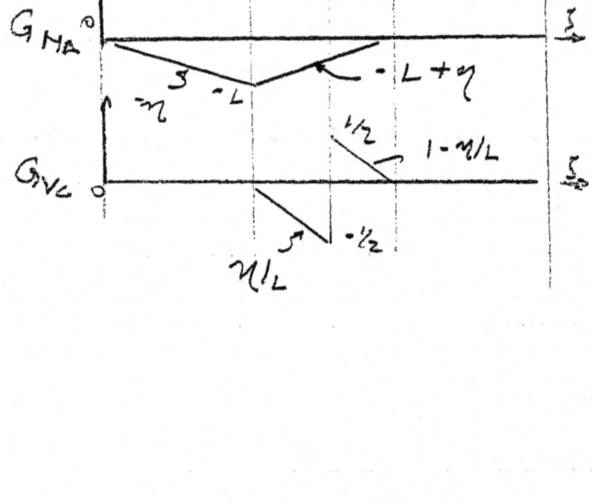

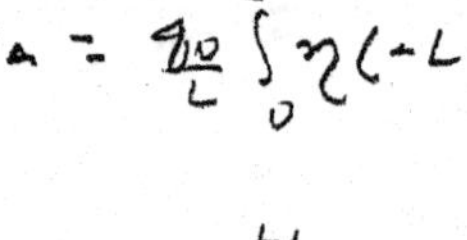

$q(\eta) = \eta q_0/L$

$$R_A = \frac{q_0}{L}\int_0^L \eta(1 - \eta/L)d\eta = q_0L/6$$

$$M_A = \frac{q_0}{L}\int_0^L \eta(-L + \eta)d\eta = -q_0L^2/6$$

$$V_c = \frac{q_0}{L}\left\{\int_0^{L/2} -\eta(\eta/L)d\eta + \int_{L/2}^{L}\eta(1 - \eta/L)d\eta\right.$$

$$= q_0L/24$$

14.44

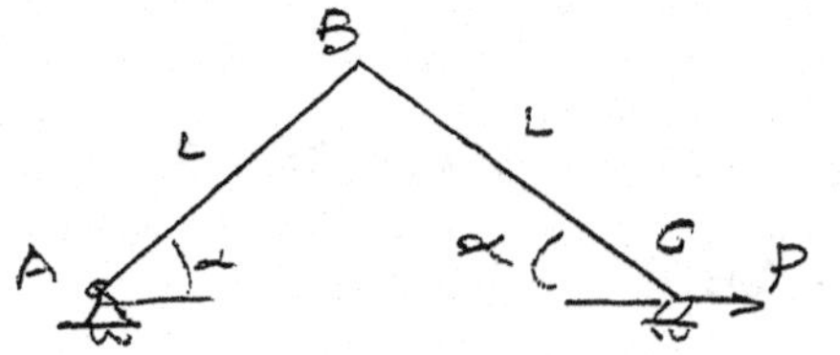

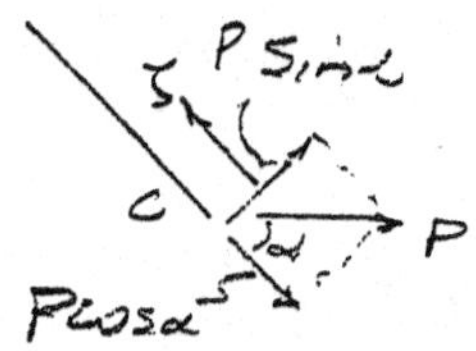

$F_{AB} = F_{BC} = P\cos\alpha$; $M_{AB} = M_{BC} = P\zeta\sin\alpha$

$$\Delta_C = \frac{1}{AE}\int F\, \partial F/\partial P\, d\zeta + \frac{1}{EI}\int M\, \partial M/\partial P\, d\zeta$$

$$= 2\left\{\frac{1}{AE}\int_0^L P\cos^2\alpha\, d\zeta + \frac{1}{EI}\int P^2\sin^2\alpha\, \zeta^2\, d\zeta\right\} = \frac{2PL\cos^2\alpha}{AE} + \frac{2PL^3\sin^2\alpha}{3EI}$$

14.45 Note: May not assume $\sigma = My/I$, since beam is not necessarily elastic.

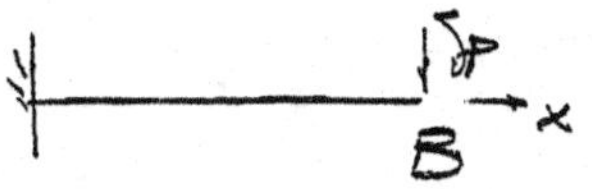

Due to ΔT: $\varepsilon = \frac{2\alpha y}{d} \cdot \Delta T$

$$\delta W_{int} = \iiint \delta\sigma \cdot \varepsilon\, d\Omega = \int_L \iint_A \delta\sigma \cdot \varepsilon\, dA\, dx = \frac{2\alpha\Delta T}{d}\int_L \left(\iint_A \delta\sigma \cdot y\, dA\right) dx$$

But $\iint_A \delta\sigma \cdot y\, dA = \delta M(x)$

$\therefore \delta W_{int} = \frac{2\alpha\Delta T}{d}\int_0^L \delta M(x)\, dx$ For equilibrium: $\delta M = -\delta P(L-x)$

$$\therefore \delta W_{int} = \frac{2\alpha\Delta T}{d}\delta P\left(L^2 - L^2/2\right) = \frac{\alpha\delta P L^2}{d} \cdot \Delta T$$

$\delta W_{ext} = \delta P \cdot \Delta_B$ & $\delta W_{ext} = \delta W_{int} \Rightarrow \Delta_B = \alpha L^2 \Delta T / d$.

14.46

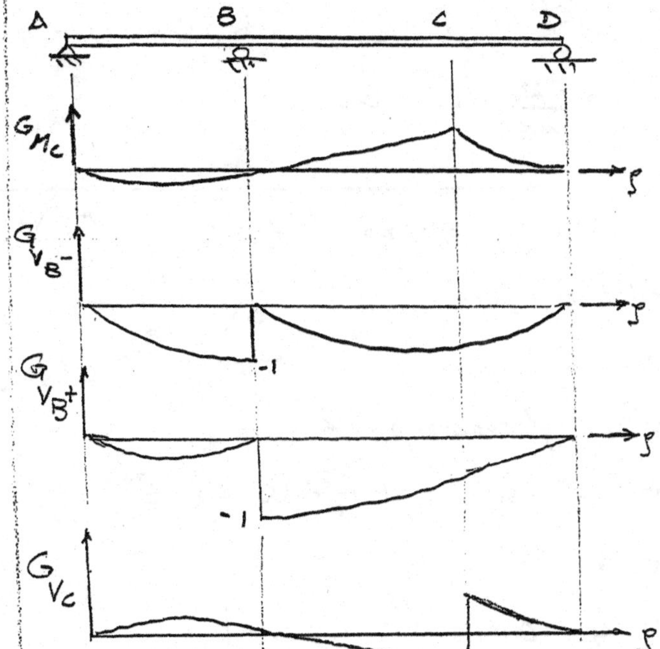

Plane View: (x-y plane).

14.47

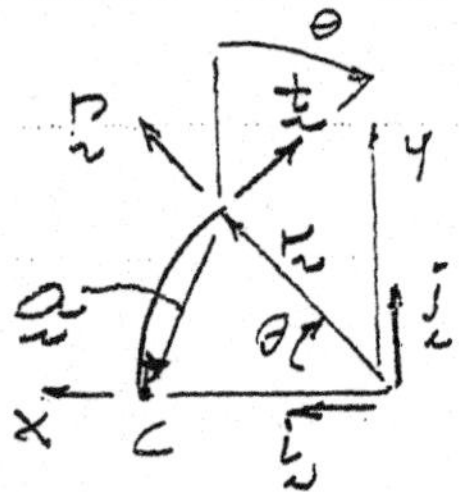

$$\underset{\sim}{a} = R\underset{\sim}{i} - \underset{\sim}{r} = R\underset{\sim}{i} - R(\cos\theta\,\underset{\sim}{i} + \sin\theta\,\underset{\sim}{j})$$
$$= R(1-\cos\theta)\,\underset{\sim}{i} - R\sin\theta\,\underset{\sim}{j}.$$
$$\underset{\sim}{M} = \underset{\sim}{a} \times \underset{\sim}{P} = R[(1-\cos\theta)\,\underset{\sim}{i} - \sin\theta\,\underset{\sim}{j}] \times P\underset{\sim}{k}$$
$$= -PR[\sin\theta\,\underset{\sim}{i} + (1-\cos\theta)\,\underset{\sim}{j}].$$

$M_n = \underset{\sim}{M}\cdot\underset{\sim}{n}$, $\underset{\sim}{n} = \cos\theta\,\underset{\sim}{i} + \sin\theta\,\underset{\sim}{j}$

$$\Rightarrow M_n = -PR[\sin\theta\cos\theta + (1-\cos\theta)\sin\theta] = -PR\sin\theta$$

$M_t = \underset{\sim}{M}\cdot\underset{\sim}{t}$; $\underset{\sim}{t} = -\sin\theta\,\underset{\sim}{i} + \cos\theta\,\underset{\sim}{j}$.

$$T \equiv M_t = -PR[\sin\theta\,\underset{\sim}{i} + (1-\cos\theta)\,\underset{\sim}{j}]\cdot[-\sin\theta\,\underset{\sim}{i} + \cos\theta\,\underset{\sim}{j}]$$
$$= -PR(-\sin^2\theta - \cos^2\theta + \cos\theta) = PR(1-\cos\theta).$$

Flexural deformation:

$$(\Delta_C)_{flex} = \frac{1}{EI}\int_0^{\pi/2} M_n \frac{dM_n}{dP} R\,d\theta = \frac{PR^3}{EI}\int_0^{\pi/2} \sin^2\theta\,d\theta = \frac{PR^3}{2EI}\int_0^{\pi/2}[1-\cos 2\theta]\,d\theta$$
$$(\Delta_C)_{flex} = \pi R^3/4EI.$$

Torsional deformation:

$$(\Delta_C)_{tors} = \frac{1}{GC}\int T\frac{dT}{dP}R\,d\theta = \frac{PR^3}{GC}\int_0^{\pi/2}(1-\cos\theta)^2 d\theta = \frac{PR^3}{GC}\int_0^{\pi/2}[1-2\cos\theta+\cos^2\theta]\,d\theta$$
$$= \frac{PR^3}{GC}\left[\theta - 2\sin\theta\Big|_0^{\pi/2} + \tfrac{1}{2}\int_0^{\pi/2}(1+\cos 2\theta)\,d\theta\right]$$
$$= PR^3/GC\,[\pi/2 - 2 + \tfrac{1}{2}(\pi/2 - 1) + \tfrac{1}{2}]$$
$$= \frac{PR^3}{4GC}[3\pi - 8]$$

$$\Delta_C = \frac{PR^3}{4}\left[\frac{\pi}{EI} + \frac{3\pi-8}{GC}\right]$$

14.48 Consider two systems: (1) given system with P's and (2) system under uniform stress q_0 at surface S.

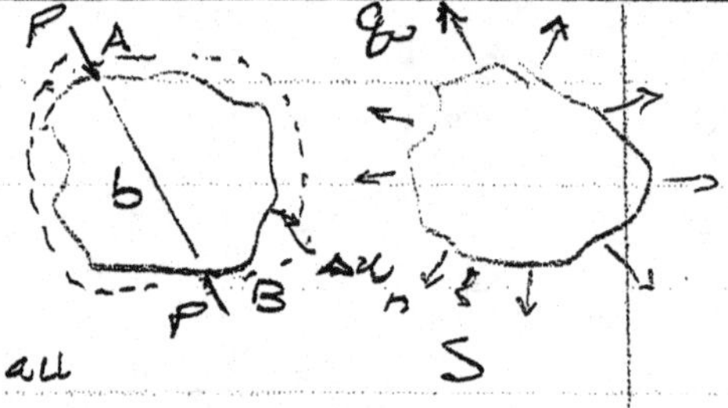

Due q_0, hydrostatic tensile $\sigma = q_0$ (tension) at all points and in all directions $\Rightarrow \epsilon = \left(\frac{1-2\nu}{E}\right)\sigma = \left(\frac{1-2\nu}{E}\right)q_0$

$$\therefore \Delta_{AB} = b\epsilon = \left(\frac{1-2\nu}{E}\right)bq_0.$$

Betti's Law: $-P\cdot\Delta_{AB}\Big|_{q_0} = q_0\cdot\Delta u_n\Big|_{P} = q_0\cdot\Delta V\Big|_P$

$$-\frac{(1-2\nu)}{E}bq_0\cdot P = \Delta V\Big|_P\cdot q_0 \Rightarrow \Delta V = -\frac{(1-2\nu)}{E}bP$$

(contraction)

a). Due to Flexural Deformation:

14.49 a)

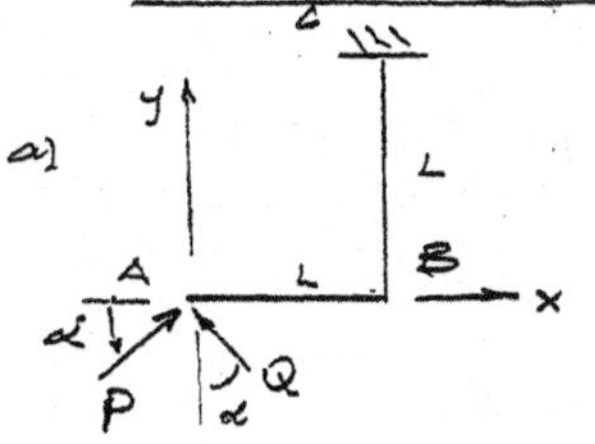

AB: $M(x) = (P\sin\alpha + Q\cos\alpha)x$

BC: $M(y) = (P\sin\alpha + Q\cos\alpha)L + (Q\sin\alpha - P\cos\alpha)y$

$$\Delta_A\Big|_{Q\text{-direct}} = 0 \Rightarrow \frac{1}{EI}\left\{\int_L M\, \partial M/\partial Q\Big|_{Q=0} ds\right\} = 0$$

$$P\int_0^L (x\sin\alpha)(x\cos\alpha)\,dx + P\int_0^L (L\sin\alpha - y\cos\alpha)(L\cos\alpha + y\sin\alpha)\,dy = 0$$

$$\therefore \sin\alpha\cos\alpha\int_0^L x^2 dx + \int_0^L [L^2\sin\alpha\cos\alpha - L(\cos^2\alpha - \sin^2\alpha)y - \sin\alpha\cos\alpha \cdot y^2]\,dy = 0$$

$$L^3\left[\frac{\sin\alpha\cos\alpha}{3} + \sin\alpha\cos\alpha - \frac{\cos 2\alpha}{2} - \frac{\sin\alpha\cos\alpha}{3}\right] = 0$$

$$\sin 2\alpha = \cos 2\alpha \Rightarrow \tan 2\alpha = 1 \Rightarrow 2\alpha = 45^\circ \Rightarrow \alpha = 22.5^\circ, 112.5^\circ$$

b). Due to Axial Deformation:

$$F_{AB} = (-P\cos\alpha + Q\sin\alpha); \quad F_{BC} = -(P\sin\alpha + Q\cos\alpha).$$

$$\Delta_A\Big|_{Q\text{-dir.}} = 0 \Rightarrow \frac{1}{AE}\left\{\int_L F\, dF/dQ\Big|_{Q=0} ds\right\} = 0$$

$$P\left\{\int_0^L [-P\cos\alpha\sin\alpha + P\sin\alpha\cos\alpha]\,dx\right\} = 0 \text{ identically.} \Rightarrow \Delta_A \parallel P \text{ for all } \alpha.$$

14.50 a). Due to flexure:

If $\Delta_A \perp P \Rightarrow \Delta_A\Big|_{P\text{-dir}} = 0 \Rightarrow \int_L M\frac{\partial M}{\partial P} ds = 0$, required condition.

AB: $M(x) = (P\sin\alpha)x$; BC: $M(y) = P(L\sin\alpha - y\cos\alpha)$.

$$P\left\{\int_0^L (\sin^2\alpha)x^2 dx + \int_0^L (L\sin\alpha - y\cos\alpha)^2 dy\right\} = 0$$

$$\sin^2\alpha \cdot L^3/3 + L^2\sin^2\alpha \cdot L - 2L\sin\alpha\cos\alpha(L^2/2) + \cos^2\alpha\, L^3/3 = 0$$

$$L^3[\sin^2\alpha - \cos\alpha\sin\alpha + 1/3] = 0, \text{ quadratic in } \sin\alpha.$$

$$\therefore \sin\alpha = 1/2\left\{\cos\alpha \pm [\cos^2\alpha - 4/3]^{1/2}\right\}$$

$|\cos\alpha| \le 1$

$\Rightarrow [\cos^2\alpha - 4/3]^{1/2}$ has no real roots. $\Rightarrow$ Required condition cannot be satisfied.

b) Due to Axial deformation: $F_{AB} = -P\cos\alpha$, $F_{BC} = -P\sin\alpha$

$$\Delta_A\Big|_{P\text{-dir}} = 0 \Rightarrow \int F\, dF/dP\, ds = 0 \Rightarrow P[\cos^2\alpha + \sin^2\alpha]L = PL > 0;$$ condition not satisfied

14.51

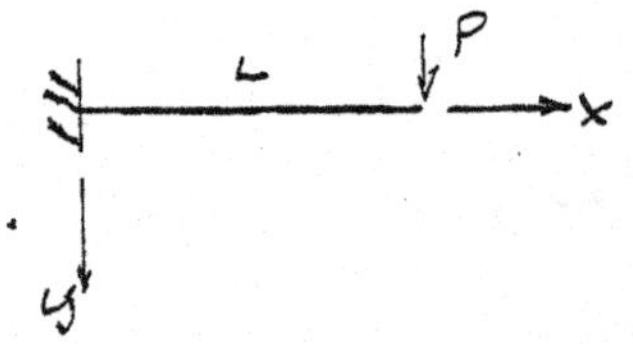

$$v(x) = \sum_{n\,odd} a_n \left[1 - \cos\frac{n\pi x}{2L}\right]$$
$$v'(x) = (\pi/2L) \sum a_n\, n \sin\frac{n\pi x}{2L}$$
$$v''(x) = (\pi/2L)^2 \sum a_n n^2 \cos\frac{n\pi x}{2L}$$

$$U = \frac{EI}{2}\int (v'')^2 dx = \frac{EI}{2}\left(\frac{\pi}{2L}\right)^4 \sum_n \sum_m a_n a_m n^2 m^2 \underbrace{\int_0^L \cos\frac{n\pi x}{2L}\cos\frac{m\pi x}{2L}\,dx}_{L/2,\ n=m}$$

$$U = \frac{\pi^4 EI}{64L^3}\sum a_n^2 n^4 \quad , \quad V = -Pv(L) = -P\sum_{n\,odd} a_n$$

$$\Pi = U + V \;;\; \delta\Pi = \frac{\partial \Pi}{\partial a_s}\delta a_s = 0 \Rightarrow \partial\Pi/\partial a_s = 0$$
$$\Rightarrow \frac{\pi^4 EI}{32L^3} n^4 a_n - P a_n = 0 \Rightarrow a_n = \frac{32PL^3}{\pi^4 EI}\cdot\frac{1}{n^4}.$$

$$v(x) = \frac{32PL^3}{\pi^4 EI}\sum_{n\,odd}\frac{1}{n^4}\left[1-\cos\frac{n\pi x}{2L}\right]$$

$$v(L) = \frac{32PL^3}{\pi^4 EI}\sum_{n\,odd} 1/n^4$$

First 4 terms ($n \le 7$): $v(L) = 0.3332\, PL^3/EI$

% Error : -0.03 %

$$M(x) = -EIv''(x) = -\frac{8PL}{\pi^4}\sum_{n\,odd} 1/n^2 \cos\frac{n\pi x}{2L}$$

$M(0) = -8PL/\pi^2 \sum_{n\,odd} 1/n^2$; First 4 terms ($n \le 7$): $M(0) = -0.9496\, PL$; % Err. = 5 %

First 8 terms: ($n \le 15$), $M(0) = -0.9747\,PL$; % Err. = 2.5 %

$$V(x) = -EIv'''(x) = -\frac{4P}{\pi}\sum_{n\,odd}\frac{1}{n}\sin\frac{n\pi x}{2L}$$

$$V(L) = -\frac{4P}{\pi}\left[1 - \frac{1}{3} + \frac{1}{5} - \frac{1}{7}\right] = 0.9216 \quad ; \ \%\ Err = 8\ \%$$

Up to $n = 15$: $V(L) = 0.9604$; % Err = 4 %.

<u>Note</u>: $v(L)$, M and V converge as $1/n^4$, $1/n^2$, $1/n$ respectively.

Note: Ref. Jolley, "Summation of Series", Dover, 1961

$v(L) = 32PL^3/\pi^4 EI\, [1 + 1/3^4 + 1/5^4 + \cdots] = \frac{32PL^3}{\pi^4 EI}(\pi^4/96) = PL^3/3EI$ by Jolley No 342.

$M(0) = -8PL/\pi^2\, [1 + 1/3^2 + 1/5^2 \cdots] = -8PL/\pi^2 (\pi^2/8) = -PL$ by Jolley No 339

$V(L) = -4P/\pi\, [1 - 1/3 + 1/5 - 1/7 \cdots] = -P$ by Jolley No. 308

15.1 CENTROID OF HEMISPHERE:

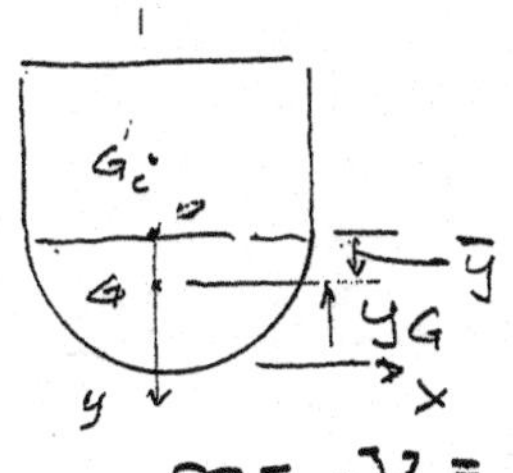

$$\bar{y} = \frac{\int y\,dm}{\int dm}$$

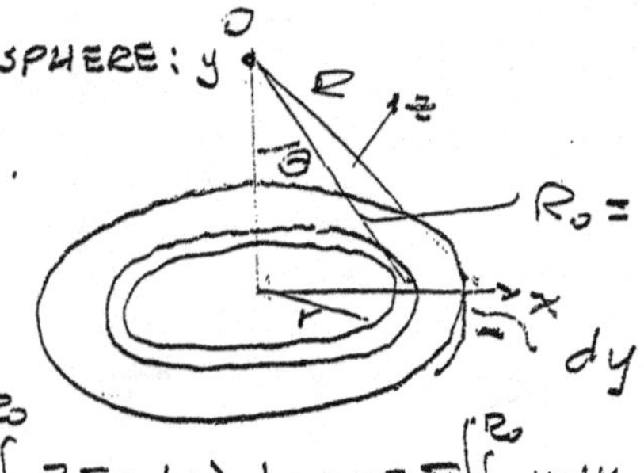

$$m = \rho V = \rho \int dv; \int dv = \int\left(\int_0^{R_0} 2\pi r\,dr\right)dy = -2\pi \int\int_0^{R_0} r\,dr \cdot R\sin\theta\,d\theta$$

$$= -2\pi R \int^{\pi/2}\int_0^{R\sin\theta} r\,dr \cdot \sin\theta\,d\theta = -2\pi \int_0^{\pi/2} \frac{r^2}{2}\Big|_0^{R\sin\theta} R\sin\theta\,d\theta$$

$$\int dm = -\pi R^3 \int_0^{\pi/2} \sin^3\theta\,d\theta = -\pi R^3 \int_0^{\pi/2}(1-\cos^2\theta)\sin\theta\,d\theta$$

$$= -\pi R^3\left\{-\cos\theta + \frac{\cos^3\theta}{3}\right\}\Big|_0^{\pi/2} = +\pi R^3\left\{-1 + 1/3\right\} = 2\pi R^3/3$$

$$m = 2\pi\rho R^3/3.$$

$$\int y\,dm = \rho\int y\,dv = \rho\int\int_0^{R_0}(2\pi r\,dr)(R\cos\theta)\,dy = -2\pi\rho R\int_{\pi/2}\int_0^{R\sin\theta} r\,dr \cdot \cos\theta\,R\sin\theta\,d\theta$$

$$= \pi\rho R^2 \int^{\pi/2}\left[r^2\Big|_0^{R\sin\theta}\right]\cos\theta\,d\theta = \pi\rho R^4\int_0^{\pi/2}\sin^3\theta\cos\theta\,d\theta$$

$$= \pi\rho R^4 \frac{\sin^4\theta}{4}\Big|_0^{\pi/2} = \pi\rho R^4/4$$

CEN

$$\bar{y} = \frac{\pi\rho R^4/4}{2\pi R^3/3} = 3R/8 \;;\quad y_G = R - \bar{y} = 5R/8$$

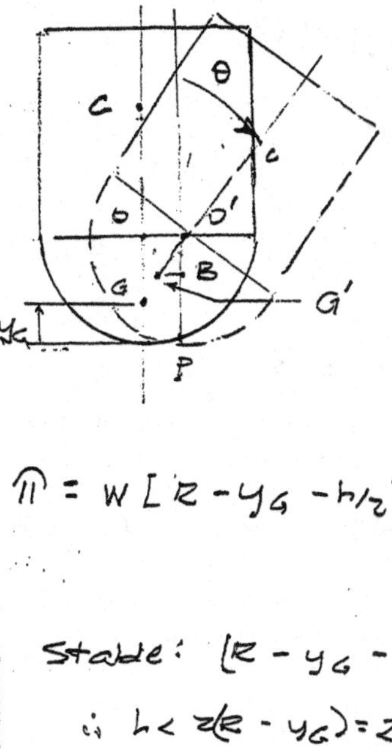

$$PO' = R;\; OG = O'G' = R - \bar{y}_G;\; O'B = O'G'\cos\theta = (R - y_G)\cos\theta$$

$$V_{sph} = W[PB - y_G] = W[(R - O'B) - y_G]$$

$$= W[R - (R - y_G)\cos\theta - y_G] = (R - y_G)(1 - \cos\theta)W$$

$$V_{cyl} = -Wh/2(1 - \cos\theta)$$

$$\Pi = W[(R - y_G)(1 - \cos\theta) - h/2(1 - \cos\theta)]$$

$$\Pi = W[R - y_G - h/2](1 - \cos\theta);\quad \delta\Pi = 0 \Rightarrow d\Pi/d\theta = 0$$

$$W[R - y_G - h/2]\sin\theta = 0 \qquad \theta = 0 \text{ equilib. (Also } \theta = \pi)$$

Stable: $(R - y_G - h/2)\cos\theta\Big|_{\theta=0^\circ} > 0 \Rightarrow R - y_G - h/2 > 0$

$\therefore h < 2(R - y_G) = 2(3R/8) = 3R/4 \Rightarrow$ $h/R < 3/4$ required for stability at $\theta = 0°$

15.2 Mass of Sphere: $M_{sph} = 2\pi\rho R^3/3$; Mass of cone: $M_c = \pi R^2 h\rho/3$.

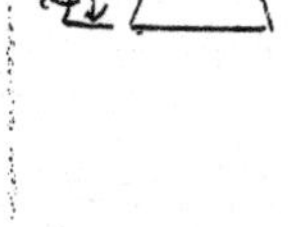

$$\Pi = \left(\frac{2\pi\rho R^3}{3}\right)(3R/8)(1 - \cos\theta) - (\pi\rho R^2 h/3)[h/4(1 - \cos\theta)]$$

$$= (\pi\rho R^2/3)[3R^2/4 - h^2/4](1 - \cos\theta);\quad \delta\Pi = 0 \Rightarrow (3R^2 - h^2)\sin\theta = 0$$

Stable: $d^2\Pi/d\theta^2 = 3R^2 - h^2 > 0 \Rightarrow$ $h < \sqrt{3}R$ for stability at $\theta = 0$.

5.3

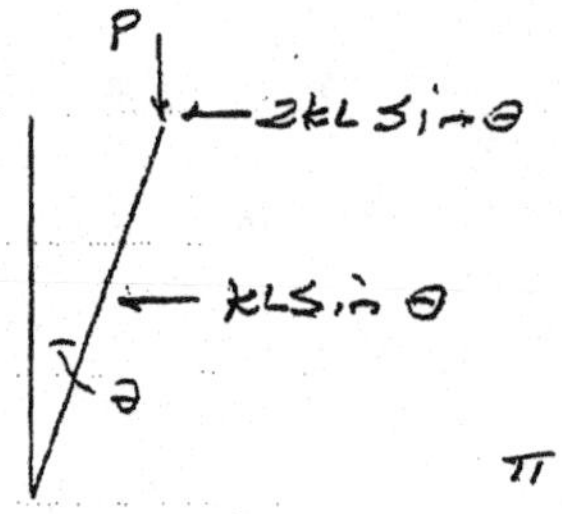

$$U = \frac{1}{2}k(\Delta_1^2 + \Delta_2^2) = \frac{kL^2}{2}[\sin^2\theta + 4\sin^2\theta] = \frac{5kL^2\sin^2\theta}{2}$$

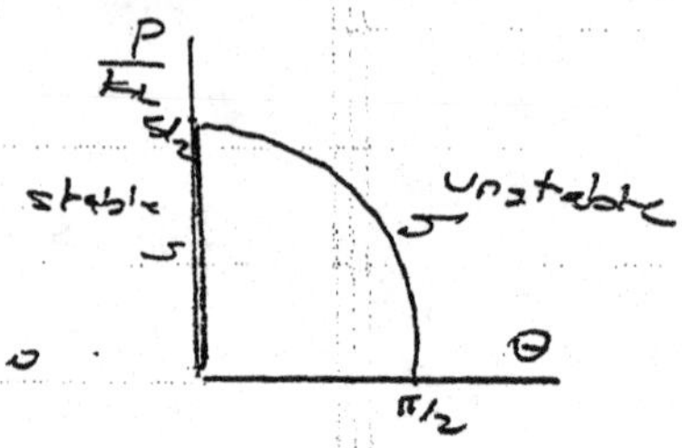

$$V = -2PL(1-\cos\theta)$$

$$\Pi = U + V = \frac{5kL^2\sin^2\theta}{2} - 2PL(1-\cos\theta)$$

$$\delta\Pi = 0 \Rightarrow d\Pi(\theta)/d\theta = 0 \Rightarrow 5kL^2\sin\theta\cos\theta - 2PL\sin\theta = 0$$

$$(5kL^2\cos\theta - 2PL)\sin\theta = 0$$

$$\theta \neq 0 \Rightarrow P = \frac{5}{2}kL\cos\theta \;;\quad \theta = 0 \Rightarrow P_{cr} = 5kL/2$$

For $|\theta| \ll 1$: $d\Pi/d\theta = (5kL^2 - 2PL)\theta$

$$d^2\Pi/d\theta^2 = 5kL^2 - 2PL$$

$\Pi'' > 0 \Rightarrow P < 5kL/2$ Stable; $\Pi'' < 0 \Rightarrow P > kL^2/2$ Unstable, at $\theta =$

$|\theta| \not\ll 1$: $\Pi'' = 5kL^2\cos 2\theta - 2PL\cos\theta$; $\Pi''\big|_{P_{eq}} = -5kL^2\sin^2\theta < 0$ Unstable

---

15.4

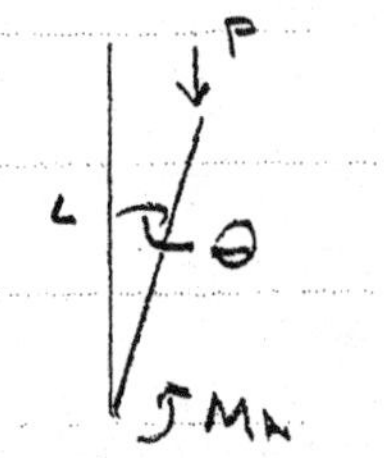

$$U = \int_0^{\theta_f} M_A(\theta)\,d\theta = c\int_0^{\theta^f}\theta^3\,d\theta = \frac{c(\theta^f)^4}{4}$$

$$U = c\theta^4/4 \quad (\theta^f = \theta)$$

$$V = -PL(1-\cos\theta)$$

$$\Pi = U + V = c\theta^4/4 - PL(1-\cos\theta)$$

$$\delta\Pi = 0 \Rightarrow d\Pi/d\theta = c\theta^3 - PL\sin\theta \Rightarrow P = \frac{c}{L}\frac{\theta^3}{\sin\theta}$$

$$d^2\Pi/d\theta^2 = 3c\theta^2 - PL\cos\theta$$

$$\Pi''(\theta)\big|_{\theta=0} = -PL < 0 \Rightarrow \theta = 0 \text{ is always unstable}, (P>0)$$

θ finite: $(P>0)$ $\quad P_{eq} = c\theta^3/L\sin\theta$

$$\therefore \Pi''(\theta)\big|_{P_{eq}} = 3c\theta^2 - (c\theta^3/\sin\theta)\cos\theta = \theta^2 c\left(3 - \frac{\theta}{\tan\theta}\right) > 0$$

$0 < \theta$ position is stable.

Note:
Moment M_A is never sufficient to bring rod back to vertical position no matter how small P or the deflection.

15.5

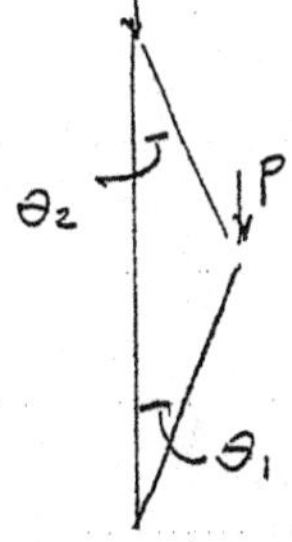

$\theta_1 = \theta_2 = \theta \qquad U = \beta/2\,(\theta_1 + \theta_2)^2 = 2\beta\theta^2$

$V = -3PL(1-\cos\theta)$

$\Pi = U + V = 2\beta\theta^2 - 3PL(1-\cos\theta)$

$\delta\Pi = 0 \Rightarrow d\Pi(\theta)/d\theta = 0$

$\Pi'(\theta) = 4\beta\theta - 3PL\sin\theta \;;\; \Pi'(\theta) = 0 \Rightarrow P = \dfrac{4\beta/3L}{\sin\theta/\theta}$

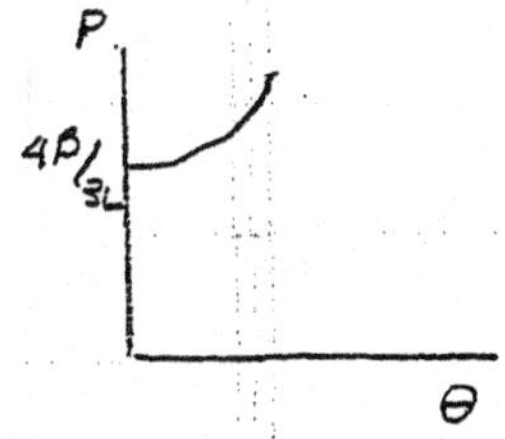

$|\theta| << 1 \Rightarrow \Pi'(\theta) = (4\beta - 3PL)\theta \;;$

$\Pi'\big|_{\theta \neq 0} = 0 \Rightarrow P_{cr} = 4\beta/3L$

15.6

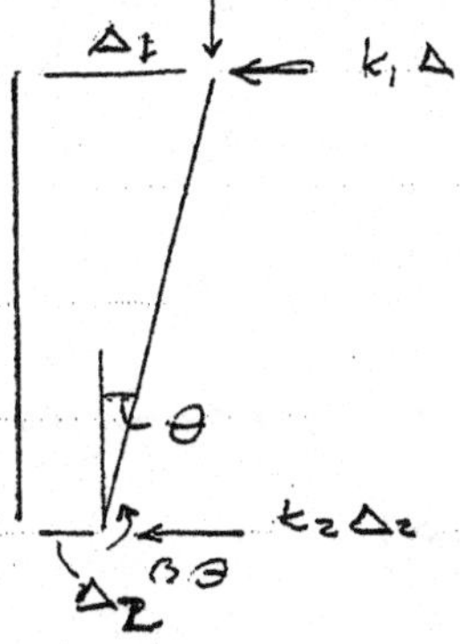

$\Delta_2 = \Delta_1 - L\sin\theta \doteq \Delta_1 - L\theta$

$U = \frac{1}{2}k_1\Delta_1^2 + \frac{1}{2}k_2\Delta_2^2 + \frac{\beta}{2}\theta^2$

$U = \frac{1}{2}\{k_1\Delta_1^2 + k_2(\Delta_1 - L\theta)^2 + \beta\theta^2\}$

$V = -PL(1-\cos\theta) = -PL\theta^2/2$

$\Pi(\Delta_1, \theta) = \frac{1}{2}\{(k_1+k_2)\Delta_1^2 - 2k_2L\cdot\Delta_1\theta + [(\beta + k_2L^2) - PL]\theta^2\}.$

$\delta\Pi = 0 \Rightarrow \dfrac{\partial\Pi}{\partial\Delta}\delta\Delta_1 + \dfrac{\partial\Pi}{\partial\theta}\delta\theta = 0 \Rightarrow \partial\Pi/\partial\Delta_1 = 0,\; \partial\Pi/\partial\theta = 0.$

$\partial\Pi/\partial\Delta_1 = 0 \Rightarrow (k_1+k_2)\Delta - k_2L\theta = 0$

$\partial\Pi/\partial\theta = 0 \Rightarrow -k_2L\Delta_1 + (\beta + k_2L^2 - PL)\theta = 0$

$\Delta_1 \neq 0,\; \theta \neq 0 \Rightarrow |\;\;| = 0 \Rightarrow (k_1+k_2)(\beta + k_2L^2 - PL) - k_2^2L^2 = 0$

$k_1\beta + k_1k_2L^2 + k_2\beta - (k_1+k_2)PL = 0$

a). $\therefore P_{cr} = \left(\dfrac{k_1k_2}{k_1+k_2}\right)L + \beta/L$

b). $P_{cr} = P_{cr}(L) \qquad dP_{cr}(L)/dL = 0 \Rightarrow \dfrac{k_1k_2}{k_1+k_2} - \beta/L^2 = 0$

$L^2 = \beta\dfrac{(k_1+k_2)}{k_1k_2}$

$(P_{cr})_{min} = \left(\dfrac{k_1k_2}{k_1+k_2}\right)\left(\dfrac{k_1+k_2}{k_1k_2}\right)^{1/2}\beta^{1/2} + \beta\left(\dfrac{k_1k_2}{k_1+k_2}\right)^{1/2}\beta^{-1/2} = 2\beta^{1/2}\left(\dfrac{k_1k_2}{k_1+k_2}\right)^{1/2}.$

15.7

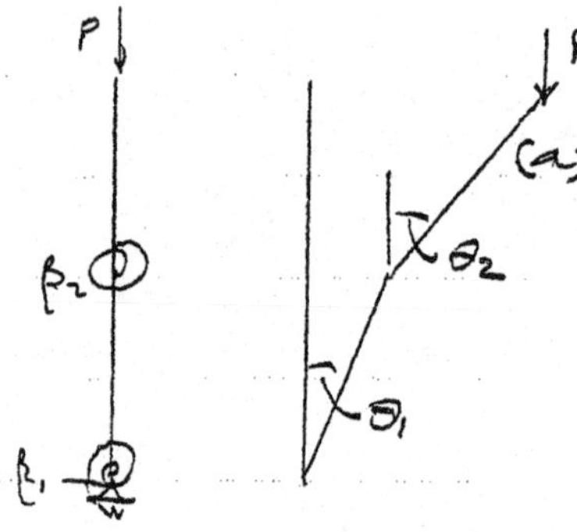

(a) $|\theta_i| \ll 1$:

$$U = \frac{\beta_1}{2}\theta_1^2 + \frac{\beta_2}{2}(\theta_2 - \theta_1)^2$$

$$V = -PL[(1-\cos\theta_1) + (1-\cos\theta_2)] = -\frac{PL}{2}(\theta_1^2 + \theta_2^2)$$

$$\Pi = U + V = \tfrac{1}{2}\{\beta_1\theta_1^2 + \beta_2(\theta_2-\theta_1)^2 - PL(\theta_1^2+\theta_2^2)\}$$

$$\Pi(\theta_1,\theta_2) = \tfrac{1}{2}\{(\beta_1+\beta_2-PL)\theta_1^2 - 2\beta_2\theta_1\theta_2 + (\beta_2 - PL)\theta_2^2\}.$$

$$\delta\Pi = 0 \Rightarrow (\partial\Pi/\partial\theta_1)\delta\theta_1 + (\partial\Pi/\partial\theta_2)\delta\theta_2 = 0 \Rightarrow \partial\Pi/\partial\theta_1 = 0,\ \partial\Pi/\partial\theta_2 = 0$$

$$\partial\Pi/\partial\theta_1 = (\beta_1+\beta_2-PL)\theta_1 - \beta_2\theta_2 = 0 \quad (i)$$
$$\partial\Pi/\partial\theta_2 = -\beta_2\theta_1 + (\beta_2 - PL)\theta_2 = 0 \quad (ii)$$

$\theta_1, \theta_2 \neq 0 \Rightarrow |\quad| = 0$

$$(\beta_1+\beta_2-PL)(\beta_2-PL) - \beta_2^2 = 0$$

$$(PL)^2 - (\beta_1+2\beta_2)(PL) + \beta_1\beta_2 = 0$$

Sol: $P_{cr} = \frac{1}{2L}\{\beta_1 + 2\beta_2 \pm [\beta_1^2 + 4\beta_2^2]^{1/2}\}$ — relevant root.

(b) $\gamma = \beta_2/\beta_1 \gg 1$ Limit case $\gamma \to \infty$ (Fig. b).

$$P = \tfrac{1}{2L}[\beta_1 + 2\beta_1\gamma - (\beta_1 + 4\gamma^2\beta_1^2)^{1/2}]$$
$$= \beta_1/2L\,[1 + 2\gamma - (1+4\gamma^2)^{1/2}] = \frac{\beta_1}{2L}[1+2\gamma - 2\gamma(1+1/4\gamma^2)^{1/2}]$$

$\gamma \gg 1 \Rightarrow P = \beta_1/2L\,[1+2\gamma - 2\gamma(1+1/8\gamma^2)] = \beta_1/2L\,(1 - 1/4\gamma)$.

Limiting case $\gamma \to \infty \Rightarrow P_\infty = \beta_1/2L > P$

15.8

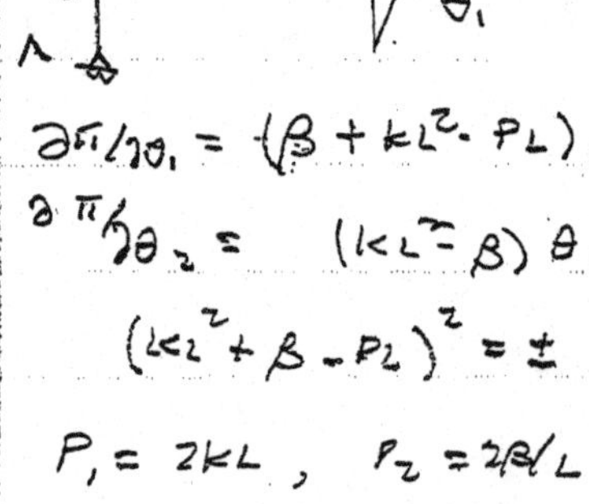

(a) $U = \beta/2\,(\theta_2-\theta_1)^2 + \frac{kL^2}{2}(\theta_1+\theta_2)$; $V = -\frac{PL}{2}(\theta_1^2+\theta_2^2)$

$$\Pi(\theta_1,\theta_2) = \tfrac{1}{2}\{\beta(\theta_2-\theta_1)^2 + kL^2(\theta_1+\theta_2)^2 - PL(\theta_1^2+\theta_2^2)\}$$

$$\delta\Pi = \frac{\partial\Pi}{\partial\theta_1}\delta\theta_1 + \frac{\partial\Pi}{\partial\theta_2}\delta\theta_2 = 0 \Rightarrow \partial\Pi/\partial\theta_1 = \partial\Pi/\partial\theta_2 = 0$$

$$\partial\Pi/\partial\theta_1 = (\beta + kL^2 - PL)\theta_1 + (kL^2-\beta)\theta_2 = 0$$
$$\partial\Pi/\partial\theta_2 = (kL^2-\beta)\theta_1 + (kL^2+\beta-PL)\theta_2 = 0$$

$\theta_1 \neq 0,\ \theta_2 \neq 0 \Rightarrow |\quad| = 0$

$$(kL^2+\beta-PL)^2 = \pm(kL^2-\beta)^2 \Rightarrow P = \tfrac{1}{L}[(kL^2+\beta) \pm (kL^2-\beta)].$$

$P_1 = 2kL$, $P_2 = 2\beta/L$

$P_1 = 2kL \Rightarrow (\beta - kL)(\theta_1-\theta_2) = 0 \Rightarrow \theta_1 = \theta_2$

$P_2 = 2\beta/L \Rightarrow (kL-\beta)(\theta_1+\theta_2) = 0 \Rightarrow \theta_2 = -\theta_2$

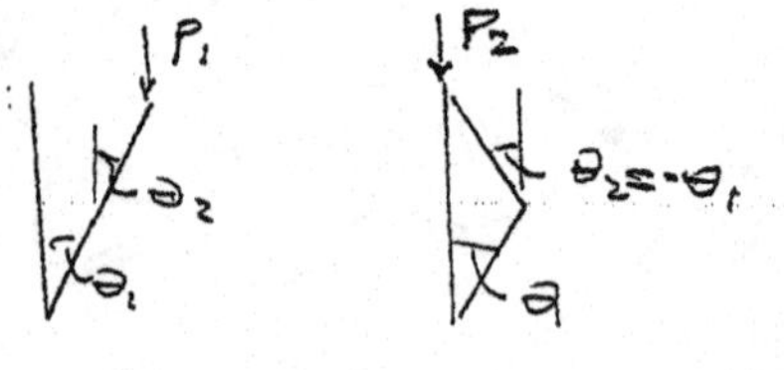

15.8 (cont'd)

b) (i) $P_1 = 2kL = 2(5)(12) = 120\ N$, $P_2 = 2\beta/L = 83.33\ N$ $\quad P_{cr} = P_2 = 83.3\ N$

(ii) $P_1 = 2kL = 80\ N$, $P_2 = 2\beta/L = 125\ N$ $\quad P_{cr} = P_1 = 80\ N$

c).

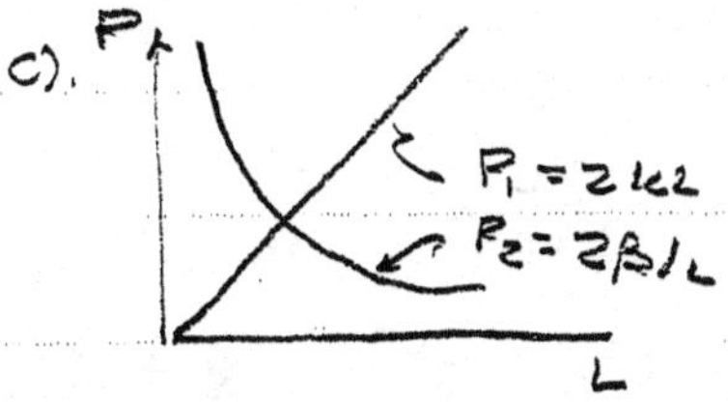

L_{cr} : $P_1 = P_2 \Rightarrow kL = \beta/L \Rightarrow L_{cr} = (\beta/k)^{1/2} = 10\ cm.$

$P_{cr} = 2(k\beta)^{1/2} = 2(2500)^{1/2} = 100\ N$

15.9

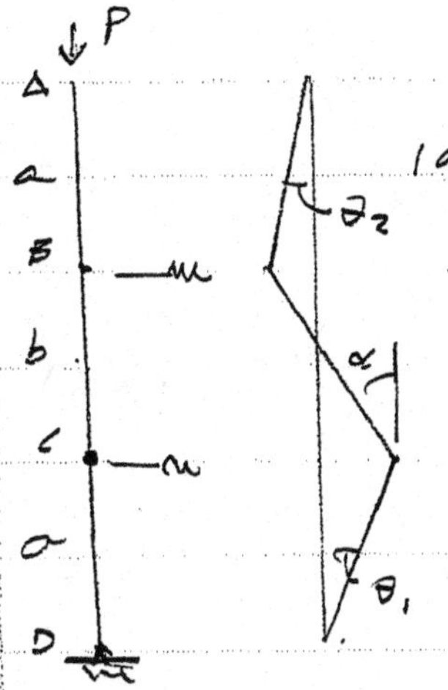

(a) $b\sin\alpha = a(\sin\theta_1 + \theta_2)$; $\alpha, \theta_i \ll 1$

$\Rightarrow b\alpha = a(\theta_1 + \theta_2)$

$U = \frac{k}{2}[(a^2\theta_1^2) + (a^2\theta_2^2)] = \frac{ka^2}{2}(\theta_1^2 + \theta^2).$

$V = -P[a(1-\cos\theta_1) + a(1-\cos\theta_2) + b(1-\cos\alpha)]$

$= -\frac{Pa}{2}[\theta_1^2 + \theta_2^2 + \frac{a}{b}(\theta_1+\theta_2)^2]$

$\Pi(\theta_1, \theta_2) = U + V = \frac{ka^2}{2}(\theta_1^2 + \theta_2^2) - \frac{Pa}{2}[\theta_1^2 + \theta_2^2 + (a/b)(\theta_1+\theta_2)^2]$

$\delta\Pi = (\partial\Pi/\partial\theta_1)\delta\theta + (\partial\Pi/\partial\theta_2)\delta\theta_2 = 0 \Rightarrow \partial\Pi/\partial\theta_1 = \partial\Pi/\partial\theta_2 = 0$

$\partial\Pi/\partial\theta_1 = ka^2\theta_1 - Pa/2[2\theta_1 + 2\,a/b\,(\theta_1+\theta_2)] = 0$

$\partial\Pi/\partial\theta_2 = ka^2\theta_2 - Pa/2[2\theta_2 + 2\,a/b\,(\theta_1+\theta_2)] = 0$

$\theta_1 \neq 0, \theta_2 \neq 0 \Rightarrow |\quad| = 0 \Rightarrow [ka^2 - Pa(1+a/b)] - (Pa^2/b)^2 = 0$

$ka^2 - Pa(1+a/b) = \pm Pa^2/b \Rightarrow Pa(1 \pm 2a/b) = ka^2$

(+ sign) $P_1 = \frac{kab}{b+2a}$, $\theta_1 = \theta_2$

; (− sign) $P_2 = ka$, $\theta_2 = -\theta_1$

$P_1 = ka\left(\frac{1}{1+2a/b}\right)$

<u>Note</u> : $P_1 < P_2$.

15.10

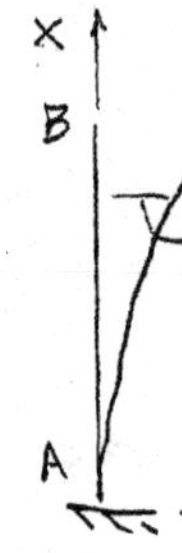

$v(x) = A(x/L)^2(1+x/L)$

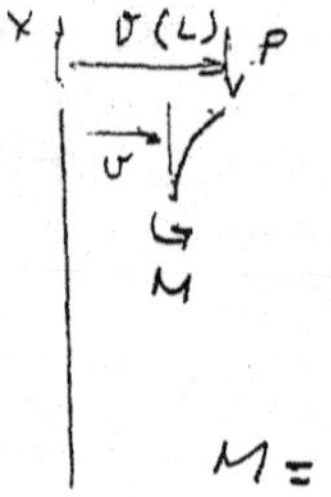

$v_B = v(L) = 2A$

$M = P[v_B - v(x)]$

$M = PA[2-(x/L)^2(1+x/L)]$

$= PA[2-\xi^2(1+\xi)]$, $\xi = x/L$

a).

$U = \frac{1}{2EI}\int_0^L M^2(x)\,dx = \frac{L}{2EI}\int_0^1 M^2(\xi)\,d\xi$

$= \frac{P^2A^2L}{2EI}\int_0^1 [2-\xi^2(1+\xi)]^2 d\xi = \frac{P^2A^2}{2EI}\int_0^1 [4-4\xi^2(1+\xi)+\xi^4(1+\xi)^2]d\xi$

$= \frac{P^2A^2L}{2EI}\{4\xi - 4(\xi^3/3+\xi^4/4)+(\xi^5/5+\xi^6/3+\xi^7/7)\}\Big|_0^1 = 41P^2A^2L/35$

$\lambda = \frac{1}{2}\int_0^L (dv/dx)^2 dx = \frac{1}{L}\int_0^1 (dv(\xi)/d\xi)^2 d\xi$; $dv(\xi)/d\xi = A(3\xi^2+2\xi)$

$\lambda = \frac{A^2}{2L}\int_0^1 (3\xi^2+2\xi)^2 d\xi = \frac{A^2}{2L}\int_0^1 (9\xi^4+12\xi^3+4\xi^2)d\xi = 46A^2/15L$

$U = P\lambda \Rightarrow 41P^2A^2L/35 = (46A^2/15L)P$

b). $\Rightarrow P_B = 2.6179\,EI/L^2 > \pi^2EI/4L^2 = 2.4674\,EI/L^2$

% error = 5.748%

15.11

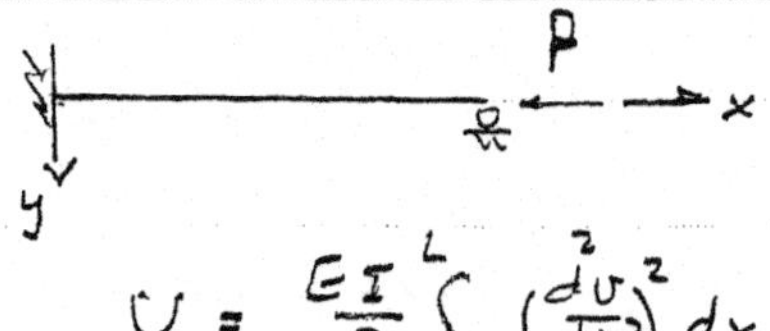

a) $v(x) = A[(x/L)^3 - (x/L)^2] \Rightarrow v(\xi) = A(\xi^3-\xi^2)$, $\xi = x/L$

$U = \frac{EI}{2}\int_0^L \left(\frac{d^2v}{dx^2}\right)^2 dx = \frac{EI}{2L^3}\int_0^1 \left[\frac{d^2v(\xi)}{d\xi^2}\right] d\xi$; $\frac{dv(\xi)}{d\xi} \equiv v'(\xi) = A(3\xi^2-2\xi)$

$v''(\xi) = A(6\xi-2)$

$U = \frac{2EIA^2}{L^3}\int_0^1 (3\xi-1)^2 d\xi = \frac{2EIA^2}{L^3}\int_0^1 (9\xi^2-6\xi+1)d\xi$

$= \frac{2EIA^2}{L^3}[3\xi^3-3\xi^2+1]_0^1 = 2EIA^2/L^3.$

$\lambda = \frac{1}{2}\int_0^L \left[\frac{dv(x)}{dx}\right]^2 dx = \frac{1}{2L}\int_0^1 (dv(\xi)/d\xi)^2 d\xi = \frac{A^2}{2L}\int_0^1 (3\xi^2-2\xi)^2 d\xi$

$= A^2/2L \int_0^1 (9\xi^4-12\xi^3+4\xi^2)d\xi = A^2/2L[9\xi^5/5 - 3\xi^4 + \frac{4}{3}\xi^3]_0^1$

$\lambda = 2A^2/30L$

$P_R = U/\lambda = \frac{2EIA^2/L^3}{2A^2/30} = 30\,EI/L^2$

b) $v(0) = v'(0) = v(L) = 0$ all Geometric B.C satisfied

$v''(L) = 4A \neq 0$ Not satisfied

c). % error = $\frac{30EI/L^2}{\pi^2EI/(0.\ldots)} - 1 = 48.9\%$

.12

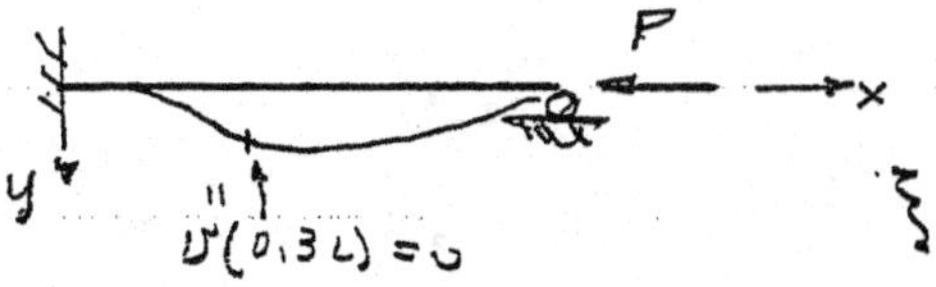

$$v(x) = A\left[\tfrac{2}{3}(x/L)^4 - \tfrac{5}{3}(x/L)^3 + (x/L)^2\right]$$

$$\xi = x/L \Rightarrow v(\xi) = A\left[\tfrac{2}{3}\xi^4 - \tfrac{5}{3}\xi^3 + \xi^2\right].$$

$$v'(\xi) = A\left[8\xi^3/3 - 5\xi^2 + 2\xi\right]$$

$$v''(\xi) = 2A(4\xi^2 - 5\xi + 1).$$

$$U = \frac{EI}{2}\int_0^L\left[\frac{d^2v(x)}{dx^2}\right]^2 dx = \frac{EI}{2L^3}\int_0^1\left[\frac{d^2v(\xi)}{d\xi^2}\right]^2 d\xi = \frac{2EIA^2}{L^3}\int_0^1(4\xi^2 - 5\xi + 1)^2 d\xi$$

$$= \frac{2EIA^2}{L^3}\int_0^1(16\xi^4 - 40\xi^3 + 33\xi^2 - 10\xi + 1)\,d\xi = \frac{2EIA^2}{5L^3}.$$

$$\lambda = \tfrac{1}{2}\int_0^L\left(\frac{dv(x)}{dx}\right)^2 dx = \frac{1}{2L}\int_0^1\left[\frac{dv(\xi)}{d\xi}\right]^2 d\xi = \frac{A^2}{2L}\int_0^1\left(8\xi^3/3 - 5\xi^2 + 2\xi\right)^2 d\xi$$

$$= \frac{A^2}{2L}\int_0^1\left(64\xi^6/9 - 80\xi^5/3 + 107\xi^4/3 - 20\xi^3 + 4\xi^2\right) d\xi$$

$$\lambda = 2A^2/105L$$

$$P_R = U/\lambda = \frac{2EIA^2/5L^3}{2A^2/105L} = 21\,EI/L^2$$

b). $v(0) = v'(0) = v(L) = 0$; Geom. B.C. satisfied; $v''(L) = 0$ Mechanical B.C. satisfied

c). $\%\ \text{Error} = \dfrac{21\cdot EI/L^2}{\pi^2 EI/(.7L)^2} - 1 = 4.3\ \%$

<u>Note</u>: $v''' = 0$ at $x = 0.25$ instead of $x = 0.3$

15.13

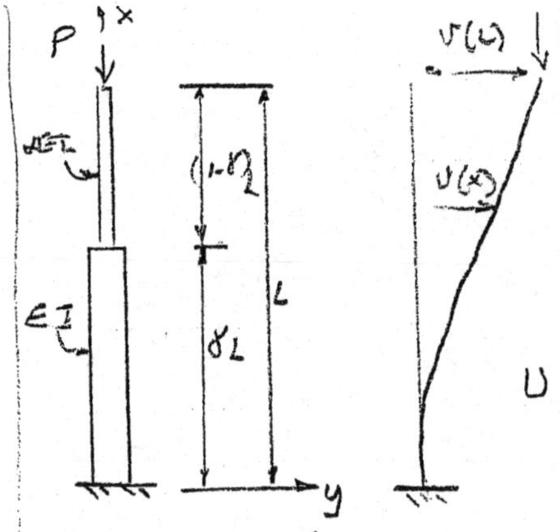

$v(x) = A(1-\cos \pi x/2L)$

a) (i) $M(x) = P[v(L) - v(x)] = PA\cos \pi x/2L$

$$U = \frac{1}{2EI}\int_0^L M^2(x)\,dx = \frac{A^2P^2}{2EI}\left\{\int_0^{\gamma L} \cos^2(\pi x/2L)\,dx + \frac{1}{\alpha}\int_{\gamma L}^{L} \cos^2(\pi x/2L)\,dx\right\}$$

$$U = \frac{A^2P^2}{4EI}\left\{\int_0^{\gamma L}(1+\cos \pi x/L)\,dx + \frac{1}{\alpha}\int_{\gamma L}^{L}(1+\cos \pi x/L)\,dx\right\}$$

$$= \frac{A^2P^2}{4EI}\left\{\gamma L + L(1-\gamma)/\alpha + \int_0^{\gamma L}\cos \pi x/L\,dx + \frac{1}{\alpha}\int_{\gamma L}^{L}\cos \pi x/L\,dx\right\}$$

$$= \frac{A^2P^2}{4EI}\left\{\frac{L}{\alpha}[\alpha\gamma + (1-\gamma)] + \frac{L}{\pi}\left[\sin(\pi\gamma) - \frac{1}{\alpha}\sin(\pi\gamma)\right]\right\}$$

$$U = \frac{A^2P^2L}{4EI\alpha}\left\{\alpha\gamma + (1-\gamma) + \frac{1}{\pi}(\alpha-1)\sin(\pi\gamma)\right\}$$

$$\lambda = \frac{1}{2}\int_0^L (v')^2\,dx = \frac{A^2\pi^2}{8L^2}\int_0^L \sin^2 \pi x/2L\,dx = \frac{A^2\pi^2}{16L^2}\int_0^L(1-\cos \pi x/L)\,dx = \frac{A^2\pi^2}{16L}$$

$$P_R = U/\lambda = \frac{4P^2L^2}{\pi^2 EI\alpha}\left[\alpha\gamma + (1-\gamma) + \frac{1}{\pi}(\alpha-1)\sin(\pi\gamma)\right]$$

$\therefore$

$$P_R = \frac{\pi^2EI\alpha}{4L^2}\left[\alpha\gamma + (1-\gamma) + \frac{1}{\pi}(\alpha-1)\sin(\pi\gamma)\right]^{-1} = \frac{\pi^2EI\alpha}{4L^2}\left\{1 + (\alpha-1)\left[\gamma + \frac{1}{\pi}\sin(\pi\gamma)\right]\right\}^{-1}$$

(ii) $\alpha = 1 \Rightarrow P_R = \pi^2EI/4L^2 = P_{cr}$ or $\gamma = 0$, $\gamma = 1$, $P_R = P_{cr}$

b) (i) $P_R = U/\lambda$; $U = \frac{EI}{2}\int (v'')^2\,dx$; $v''(x) = A(\pi/2L)^2\cos(\pi x/2L)$

$$U = \frac{EIA^2}{2}(\pi/2L)^4\cdot\frac{1}{2}\left\{\int_0^{\gamma L}\left(1+\cos\frac{\pi x}{L}\right)dx + \alpha\int_{\gamma L}^{L}(1+\cos \pi x/L)\,dx\right\}$$

$$= \frac{EIA^2\pi^4}{64L^4}\left\{\gamma L + \frac{L}{\pi}\sin\gamma\pi + \alpha\left[L(1-\gamma) - \frac{L}{\pi}\sin\pi\gamma\right]\right\}$$

$$U = \frac{\pi^4EIA^2}{64L^3}\left\{\gamma + \alpha(1-\gamma) + \frac{1-\alpha}{\pi}\sin\pi\gamma\right\}$$

$\lambda = \pi^2A^2/16L$ from (a) above.

$$P_R = U/\lambda = \frac{\pi^2EI}{4L^2}\left[\gamma + \alpha(1-\gamma) + \frac{1-\alpha}{\pi}\sin\pi\gamma\right]$$

(ii) $P_R = P_{cr}$ if $\alpha = 1$, or if $\gamma = 0$ or $\gamma = 1$.

15.14

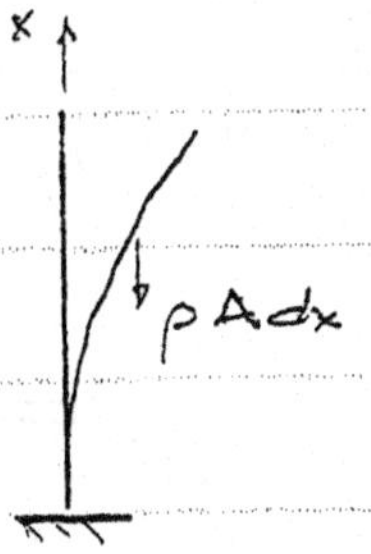

$$v(x) = a(1 - \cos \pi x/2L)$$

$$v'(x) = a\pi/2L \sin \pi x/2L; \quad v''(x) = a(\pi/2L)^2 \cos \pi x/2L$$

$$U = EI/2 \int (v'')^2 dx = \frac{EI}{2} a^2 (\pi/2L)^4 \int_0^L \cos^2(\pi x/2L) dx$$

$$= \frac{\pi^4 EI a^2}{32 L^4} \cdot \frac{1}{2} \int_0^L [1 + \cos \pi x/L] dx$$

$$= \frac{\pi^4 EI a^2}{64 L^4} \cdot L = \pi^4 EI A^2 / 64 L^3.$$

$$V = -\int_0^L \lambda(x) \rho A dx = -\rho A \int_0^L \lambda(x) dx$$

$$\lambda(x) = 1/2 \int_0^x (v')^2 d\xi = \frac{a^2}{2}(\pi/2L)^2 \int_0^x \sin^2 \pi\xi/2L \, d\xi$$

$$= \frac{a^2 \pi^2}{8L^2} \cdot \frac{1}{2} \int_0^x [1 - \cos \pi\xi/L] d\xi = \frac{a^2\pi^2}{16L^2} \left\{ \xi - \frac{L}{\pi} \sin \pi\xi/L \right\}_0^x$$

$$= \frac{a^2\pi^2}{16L^2} [x - L/\pi \sin \pi x/L]$$

$$\int_0^L \lambda(x) dx = \frac{a^2\pi^2}{16L^2} \int_0^L [x - L/\pi \sin \pi x/L] dx$$

$$= \frac{a^2\pi^2}{16L^2} \left\{ x^2/2 + (L/\pi)^2 \cos \pi x/L \right\}_0^L = \frac{a^2\pi^2}{16L^2} [L^2/2 - \frac{L^2}{\pi^2} - (L^2/\pi^2)]$$

$$= \frac{a^2\pi^2}{16L^2} [L^2/2 - 2L^2/\pi^2] = \frac{a^2\pi^2}{32} (1 - 4/\pi^2)$$

$$\Delta \Pi = \Delta U + \Delta V = 0$$

$$\frac{\pi^4 EI A^2}{64 L^3} = \rho A \frac{a^2 \pi^2}{32} [1 - 4/\pi^2]$$

$$W = \rho A L \Rightarrow W = \frac{\pi^2 EI}{2L^2} \cdot \frac{1}{1 - 4/\pi^2} = 0.8407 \, \pi^2 EI/L^2 = 8.298 \, EI/L^2$$

% Error = 5.88 %

5.15

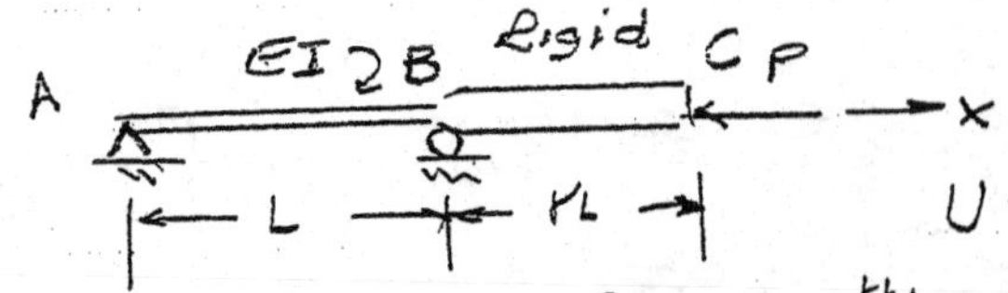

$v(x) = A \sin \pi x/L$

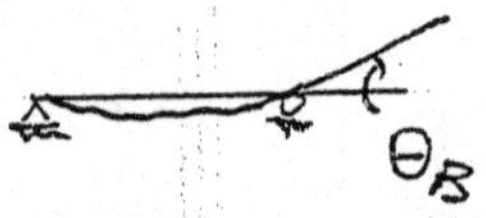

$$U = \frac{EI}{2}\int_0^L [v''(x)]^2 dx$$

$v'(x) = A(\pi/L)\cos \pi x/L$, $v''(x) = -A(\pi/L)^2 \sin \pi x/L$

$$U = (EI/2)A^2(\pi/L)^4 \underbrace{\int_0^L \sin^2 \pi x/L\, dx}_{L/2} = \frac{\pi^4 EI A^2}{4L^3}.$$

$$V = -P\lambda, \quad \lambda_{AB} = \tfrac{1}{2}\int_0^L (v')^2 dx = \frac{A^2}{2}\left(\frac{\pi}{L}\right)^2 \int_0^L \cos^2 \pi x/L\, dx = \pi^2 A^2/4L$$

$$\lambda_{BC} = a(1-\cos\theta_B) \doteq a\frac{\theta_B^2}{2} = a[v'(L)]^2/2 = \frac{a}{2}A^2\pi^2/L^2$$

$$V = -P\lambda = -P[\lambda_{AB}+\lambda_{BC}] = -PA^2\pi^2[1/4L + a/2L^2] = -\frac{PA^2\pi^2}{4L^2}(L+2a)$$

$$\delta(U+V) = 0 \Rightarrow \frac{PA^2\pi^2(L+2a)}{4L^2} = \frac{\pi^4 EI A^2}{4L^3}$$

$$P_R = \frac{\pi^2 EI}{L^2}\left[\frac{1}{1+2a/L}\right] = \frac{\pi^2 EI}{L^2}\left(\frac{1}{1+2\gamma}\right).$$

PROBLEM 15.25

$P_E = \pi^2 EI/L^2$

$\gamma = a/L$	$P_{cr}/(EI/L^2)$	P_{cr}/P_E	P_R/P_E	% ERROR
1/4	6.02997	0.61096	0.6666	9.1 %
1/3	5.23916	0.53084	0.6	13.0 %
1/2	4.11585	0.41702	0.5	21.4 %
2/3	3.371	0.3418	0.4286	25.4 %

15.16

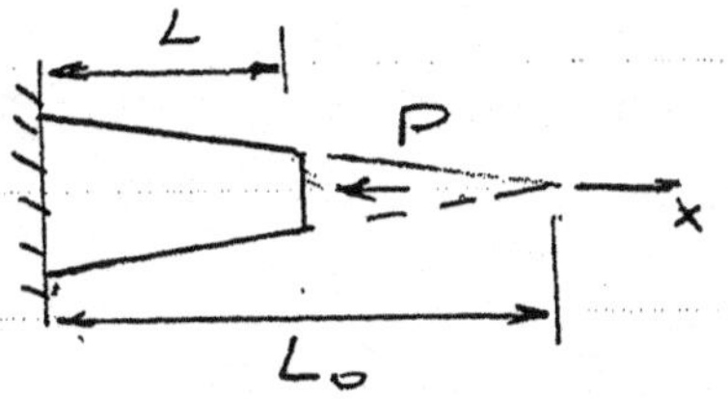

$v(x) = Ax^2/L_0^2$; $\xi = x/L_0$ $\quad x = L \Rightarrow \xi = \gamma$

$v(\xi) = A\xi^2$, $v'(\xi) = 2A\xi$, $v''(\xi) = 2A$

$I(\xi) = I_0(1-\xi)^4$. $\quad \gamma = L/L_0$

$$U = \frac{E}{2}\int_0^L I(x)\left(\frac{d^2v(x)}{dx^2}\right)^2 dx = \frac{EI_0}{2}\int_0^\gamma\left(\frac{2A}{L_0^2}\right)^2(1-\xi)^4 L_0\, d\xi$$

$$U = \frac{4EI_0A^2}{2L_0^3}\int_0^\gamma(1-\xi)^4 d\xi = -\frac{2EI_0A^2}{5L_0^3}(1-\xi)^5\Big|_0^\gamma$$

$$U = -\frac{2EI_0A^2}{5L_0^3}[(1-\gamma)^5 - 1] = \frac{2EI_0A^2}{5L_0^3}[1-(1-\gamma)^5].$$

$$\lambda = \tfrac{1}{2}\int_0^L\left(\frac{dv(x)}{dx}\right)^2 dx = \frac{1}{2}\int_0^\gamma\left(\frac{2A}{L_0}\xi\right)^2 \cdot L_0\, d\xi = \frac{2A^2}{L_0}\int_0^\gamma \xi^2 d\xi = \frac{2A^2\gamma^3}{3L_0}.$$

$V = -P\lambda$

$$\Delta\Pi = \Delta(U+V) = 0 \Rightarrow \frac{2A^2\gamma^3}{3L_0}P_R = \frac{2EI_0A^2}{5L_0^3}[1-(1-\gamma)^5].$$

$$P_R = \frac{3EI_0}{5L_0^2}\,\frac{1-(1-\gamma)^5}{\gamma^3} = \frac{3EI_0}{5L^2}\,\frac{1-(1-\gamma)^5}{\gamma}$$

$\gamma = 1/2$; $P_R = \dfrac{6EI_0}{5L^2}[1 - 1/32] = 1.1625\ EI_0/L^2$

$P_{cr} = 1.026\, EI_0/L^2 \Rightarrow$ % Error $= 13.3$ %

15.17 Note: Refer to figure in Solution of Problem 15.1

Hemisphere: $V_H = \rho(2\pi R^3/3)[(R - y_G)(1-\cos\theta)]$; $y_G = 5R/8$

$= (2\rho\pi R^3/3)(3R/8)(1-\cos\theta) = \rho\pi R^4/4\,(1-\cos\theta)$

Cylinder: $V_C = -\rho\pi R^2 h\,(h/2)(1-\cos\theta)$.

$\Pi = V_H + V_C = \rho\pi R^2\{(R^2/4 - h^2/2)(1-\cos\theta)\}$

$\delta\Pi = \dfrac{d\Pi(\theta)}{d\theta}\delta\theta = 0 \Rightarrow d\Pi/d\theta = 0 \Rightarrow (R^2/4 - h^2/2)\sin\theta = 0$

$d^2\Pi/d\theta^2 = (R^2/4 - h^2/2)\cos\theta > 0$

$h^2 < R^2/2 \Rightarrow h < R/\sqrt{2} = \sqrt{2}R/2$.

5.18 $U = \int_0^\theta M(\psi)\,d\psi = C\int_0^\theta \tanh\psi\, d\psi = C \ln \cosh\psi \Big|_0^\theta = C \ln \cosh\theta - \ln 1$

$U = C \ln(\cosh\theta)$.

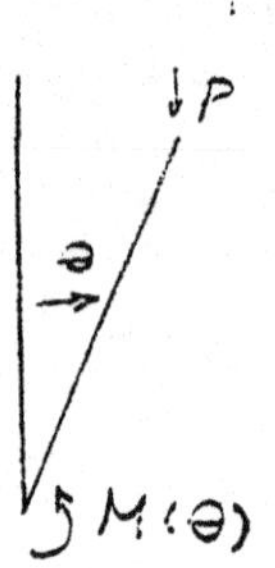

$V = -PL(1-\cos\theta)$

a) $\Pi = U + V = C \ln(\cosh\theta) - PL(1-\cos\theta)$

b) $\delta\Pi = \dfrac{d\Pi(\theta)}{d\theta}\,\delta\theta = 0 \qquad \delta\theta \neq 0 \Rightarrow \dfrac{d\Pi(\theta)}{d\theta} = 0$

$d\Pi/d\theta = \dfrac{C}{\cosh\theta}\cdot\sinh\theta - PL\sin\theta = 0 \Rightarrow P = \dfrac{C}{L}\,\dfrac{\tanh\theta}{\sin\theta}$

c). $|\theta| \ll 1 \Rightarrow P_{cr} = C/L$.

θ	$\frac{P}{C/L}$
0	1
5	0.999
10	0.995
20	0.981
30	0.961
40	0.938
50	0.917
60	0.901
70	0.894
80	0.898
90	0.917
100	0.955
110	1.019
120	1.120
130	1.278
140	1.532
150	1.979
160	2.902
170	5.022
180	∞

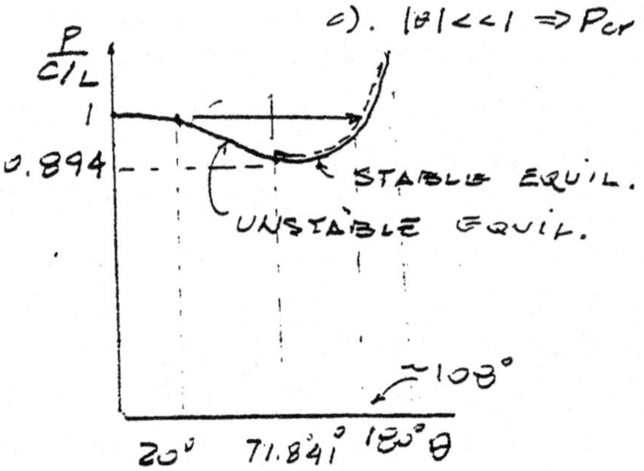

PROBLEM 15.26: Minimum P_{eq}.

$dP/d\theta = C/L\left[\dfrac{1}{\cosh^2\theta}\cdot\dfrac{1}{\sin\theta} - \dfrac{\tanh\theta}{\sin^2\theta}\cdot\cos\theta\right]$

$= \dfrac{C}{L\sin^2\theta}\left[\dfrac{\sin\theta}{\cosh^2\theta} - \tanh\theta\cos\theta\right]$

$dP/d\theta = 0 \Rightarrow \dfrac{\sin\theta}{\cosh\theta} = \sinh\theta\cos\theta$

$\theta_{cr} = 71.841° \Rightarrow P\big|_{\theta_{cr}} = 0.8939\frac{C}{L}$

Note:

$\dfrac{d^2\Pi}{d\theta^2} = C\dfrac{1}{\cosh^2\theta} - PL\cos\theta \Rightarrow \dfrac{d^2\Pi}{d\theta^2}\Big|_{P_{eq}} = C\left(\dfrac{1}{\cosh^2\theta} - \dfrac{\tanh\theta}{\tan\theta}\right)$

$d^2\Pi/d\theta^2\Big|_{P_{eq}} = 0$ at θ_{cr}. $\Rightarrow \dfrac{d^2\Pi}{d\theta^2}$ < 0, $\theta < \theta_{cr}$ unstable; > 0, $\theta > \theta_{cr}$ stable

Note: θ_{cr} (stable/unstable path) occurs at $(P_{eq})_{min}$

15.19

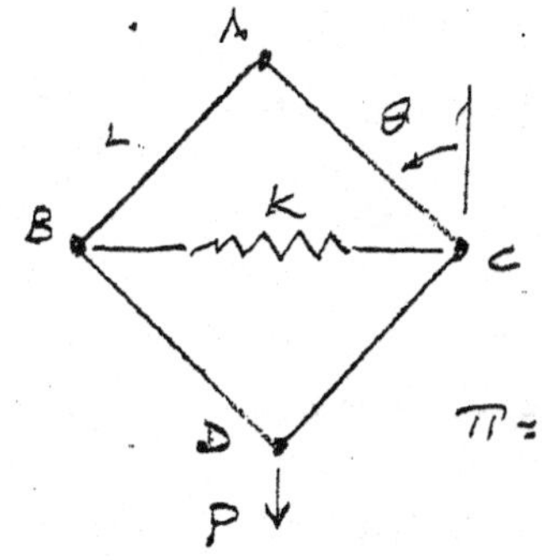

a). $U = \frac{k}{2}[2L\sin\theta - 2L\sin\theta_0]^2$

$= 2kL^2(\sin\theta - \sin\theta_0)^2$.

$V = -P\cdot 2L(\cos\theta - \cos\theta_0)$

$\Pi = U + V = 2kL^2(\sin\theta - \sin\theta_0)^2 - 2PL(\cos\theta - \cos\theta_0)$

$(\partial\Pi(\theta)/\partial\theta)\delta\theta = 0 \Rightarrow [2kL^2(\sin\theta - \sin\theta_0)\cos\theta + 2PL\sin\theta]\delta\theta = 0$

$[kL(\sin\theta - \sin\theta_0)\cos\theta + P\sin\theta]\delta\theta = 0$

$\delta\theta = 0,\ \theta \neq 0$:

$P = kL\left[\frac{\sin\theta_0}{\sin\theta} - 1\right]\cos\theta$

$d^2\Pi/d\theta^2 = kL[\cos^2\theta - \sin^2\theta + \sin\theta_0\sin\theta] + P\cos\theta$

$d^2\Pi/d\theta^2\Big|_{P_{cr}} = kL\left\{\cos^2\theta - \sin^2\theta + \sin\theta_0\sin\theta + \frac{\sin\theta_0\cos^2\theta}{\sin\theta} - \cos^2\theta\right\}$

$= \frac{kL}{\sin\theta}\{\sin\theta_0 - \sin^3\theta\}$

$d^2\Pi/d\theta^2 > 0 \Rightarrow \sin\theta < (\sin\theta_0)^{1/3} \Rightarrow \theta < \sin^{-1}[(\sin\theta_0)^{1/3}]$.

b). $P_{cr} = P(\theta_{cr})$, $\sin\theta_{cr} = (\sin\theta_0)^{1/3}$

$P_{cr} = kL[(\sin\theta_0)^{2/3} - 1][1 - (\sin\theta_0)^{2/3}]^{1/2} = -kL[1 - (\sin\theta_0)^{2/3}]^{3/2}$.

c) $\Pi(\theta,\theta_0) = 2kL^2\left\{(\sin\theta - \sin\theta_0)^2 - \left(\frac{\sin\theta_0}{\sin\theta} - 1\right)(\cos\theta - \cos\theta_0)\cos\theta\right\}$

$= 2kL^2\left\{\sin^2\theta - 2\sin\theta\sin\theta_0 + \sin^2\theta_0 - \frac{\sin\theta_0\cos^2\theta}{\sin\theta} + \cos^2\theta + \frac{\sin\theta_0\cos\theta_0\cos\theta}{\sin\theta} - \cos\theta_0\cos\theta\right\}$

$= \frac{2kL^2}{\sin\theta}\{\sin\theta - 2\sin^2\theta\sin\theta_0 + \sin\theta\sin^2\theta_0 - \sin\theta_0\cos^2\theta + \cos\theta\cos\theta_0(\sin\theta_0 - \sin\theta)\}$

$= \frac{2kL^2}{\sin\theta}\{\sin\theta - \sin\theta_0 - \sin^2\theta\sin\theta_0 + \sin\theta\sin^2\theta_0 + \cos\theta\cos\theta_0(\sin\theta_0 - \sin\theta)\}$

$= \frac{2kL^2}{\sin\theta}\{(\sin\theta - \sin\theta_0)(1 - \cos\theta\cos\theta_0) + \sin\theta\sin\theta_0(\sin\theta_0 - \sin\theta)\}$

$= \frac{2kL^2}{\sin\theta}\{(\sin\theta - \sin\theta_0)(1 - \cos\theta\cos\theta_0 - \sin\theta\sin\theta_0\}$

$\Pi(\theta,\theta_0) = \frac{2kL^2}{\sin\theta}(\sin\theta - \sin\theta_0)[1 - \cos(\theta - \theta_0)]\}$.

15.20

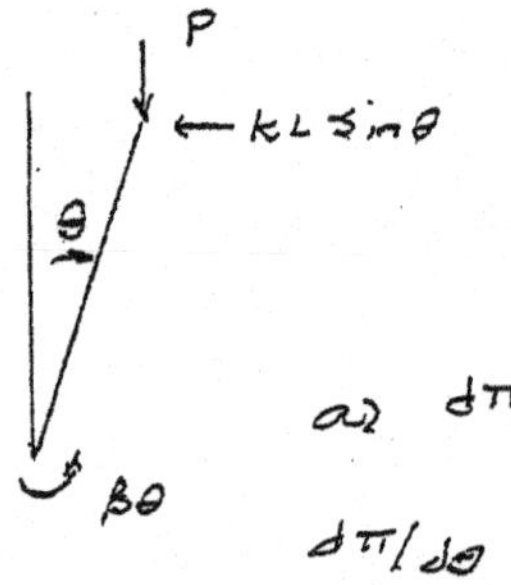

$U = \beta\theta^2/2 + \frac{kL^2}{2}\sin^2\theta$; $V = -PL(1-\cos\theta)$

$\Pi = \beta\theta^2/2 + \frac{kL^2}{2}\sin^2\theta - PL(1-\cos\theta)$

a) $d\Pi/d\theta = \beta\theta + kL^2\sin\theta\cos\theta - PL\sin\theta = \beta\theta + (kL^2\cos\theta - PL)\sin\theta$

$d\Pi/d\theta = 0 \Rightarrow P = \dfrac{\beta\theta}{L\sin\theta} + kL\cos\theta$

b). $0 < \theta << 1$:

$P_{cr} = \beta/L + kL.$

$d^2\Pi/d\theta^2 = \beta + (kL^2\cos\theta - PL)\cos\theta - kL^2\sin^2\theta - PL\cos\theta = \beta + kL^2\cos 2\theta - PL\cos\theta$

$d^2\Pi/d\theta^2\Big|_{P_{eq}} = \beta + kL^2\cos 2\theta - (\beta\theta/\sin\theta)\cos\theta - kL^2\cos^2\theta.$

$= \beta\left[1 - \dfrac{\theta\cos\theta}{\sin\theta}\right] - kL^2\sin^2\theta$

Let $\alpha \equiv \beta/kL^2$

$P = kL[\alpha\theta/\sin\theta + \cos\theta]$

$\Pi'' \equiv d^2\Pi/d\theta^2\Big|_{P_{eq}} = kL^2\left[\alpha\left(1 - \dfrac{\theta\cos\theta}{\sin\theta}\right) - \sin^2\theta\right] \Rightarrow \Pi(\theta_{cr}) = 0$

α	θ_{cr}	$\frac{P_{min}}{kL} = \frac{P(\theta_{cr})}{kL}$	$\Pi >$
0.5	110.992°	0.6792	
1.0	90°	1.5708	
2.0	56.5725°	2.91703	
3.0	0.2272°?	4	

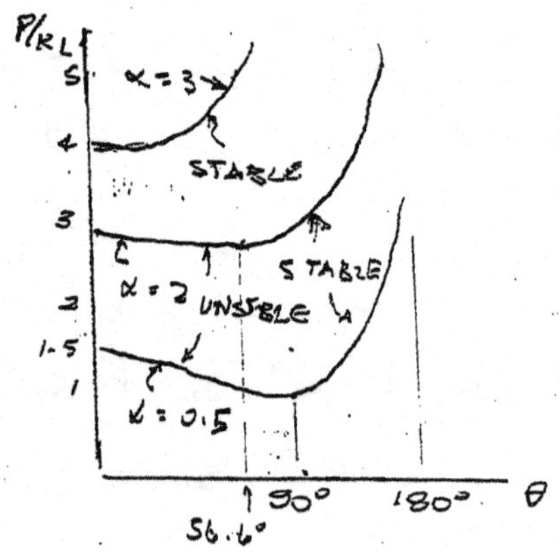

NOTE: $\Pi'' > 0$ for $\theta > \theta_{cr}$

Bifurcation Pt, $\theta = 0$: STABLE/UNSTABLE

$\Pi'' \equiv d^2\Pi/d\theta^2\Big|_{\theta\to 0} = \beta + kL^2 - PL = \beta + kL^2 - (\beta + kL^2) = 0$

$\Pi''' = -2kL^2\sin 2\theta + PL\sin\theta\Big| = 0$

$\Pi^{IV} = -4kL^2\cos 2\theta + PL\cos\theta\Big|_{\theta\to 0} = -4kL^2 + (\beta + kL^2) = \beta - 3kL^2 = (\alpha - 3)kL^2$

$\Pi^{IV} > 0$ if $\alpha > 3 \Rightarrow \Delta\Pi > 0$ if $\alpha > 3$

$\alpha > 3 \Rightarrow$ STABLE BIFURCATION PT.

$\alpha < 3 \Rightarrow$ UNSTABLE " " .

15.21

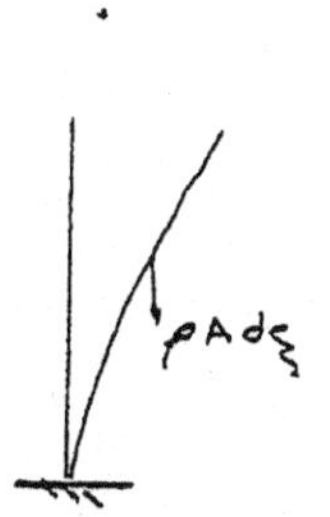

$$v(x) = c_1 (x/L)^2 (x/L - 3) + c_2 (x/L)^2 .$$

$$\Rightarrow v(\xi) = c_1 \xi^2 (\xi - 3) + c_2 \xi^2 \qquad \xi = x/L.$$

$$v'(\xi) = c_1 (3\xi^2 - 6\xi) + 2c_2 \xi, \quad v''(\xi) = c_1 (6\xi - 6) + 2c_2; \quad v'(\xi) \equiv \frac{dv(\xi)}{d\xi}$$

$$U = \frac{EI}{2} \int_0^L (v_{xx})^2 dx = \frac{EI}{L^3} \int_0^1 [v''(\xi)]^2 d\xi = \frac{2EI}{L^3} (3c_1^2 - 3c_1 c_2 + c_2^2)$$

$$V = -\rho A \int_0^L \lambda(x) dx; \quad \lambda(x) = \frac{1}{2} \int_0^x \left(\frac{dv}{dx}\right)^2 dx = \frac{1}{2} \int_0^\xi [3c_1(\xi^2 - 2\xi) + 2c_2 \xi]^2 d\xi$$

$$\lambda(\xi) = \frac{1}{2L} \int_0^\xi [9c_1^2 \xi^2 (\xi^2 - 4\xi + 4) + 4c_2^2 \xi^2 + 12 c_1 c_2 \xi^2 (\xi - 2)] d\xi$$

$$= \frac{1}{2L} \left\{ 9c_1^2 (\xi^5/5 - \xi^4 + 4\xi^3/3) + 4c_2^2 \xi^3/3 + 12 c_1 c_2 (\xi^4/4 - 2\xi^3/3) \right\}_0^\xi$$

$$\lambda(\xi) = \frac{1}{2L} \left\{ 9c_1^2 (\xi^5/5 - \xi^4 + 4\xi^3/3) + 4c_2^2 \xi^3/3 + 12 c_1 c_2 (\xi^4/4 - 2\xi^3/3) \right\}$$

$$V = -\rho A L \int_0^1 \lambda(\xi) d\xi = -\frac{\rho A}{2} \int_0^1 \{ 9c_1^2 (\xi^5/5 - \xi^4 + 4\xi^3/3) + \text{etc.} \} d\xi$$

$$= -\rho A/2 \left\{ 9c_1^2 (\xi^6/30 - \xi^5/5 + \xi^4/3) + c_2^2 \xi^4/3 + 12 c_1 c_2 (\xi^5/20 - \xi^4/6) \right\}_0^1$$

$$= -\rho A/2 \{ 9c_1^2 (1/30 - 1/5 + 1/3) + c_2^2/3 + 12 c_1 c_2 (1/20 - 1/6) \}$$

$$= -\rho A/2 \left[9c_1^2 (1/6) + c_2^2/3 - \frac{7}{5} c_1 c_2 \right]$$

$$V = -\rho A/2 \left[\frac{3}{2} c_1^2 - \frac{7}{5} c_1 c_2 + c_2^2/3 \right] = -\frac{\rho A}{60} [45c_1^2 - 42 c_1 c_2 + 10 c_2^2]$$

$$\Delta U = -\Delta V$$

$$\rho A/60 [45 c_1^2 - 42 c_1 c_2 + 10 c_2^2] = \frac{2EI}{L^3} [3c_1^2 - 3c_1 c_2 + c_2^2]$$

$$W \equiv \rho A L = \frac{120 EI}{L^2} \; \frac{3c_1^2 - 3c_1 c_2 + c_2^2}{45 c_1^2 - 42 c_1 c_2 + 10 c_2^2} = \frac{N}{D}$$

$$\partial N/\partial c_1 = (120EI/L^2)(6c_1 - 3c_2); \quad \partial N/\partial c_2 = \frac{120 EI}{L^2} (-3c_1 + 2c_2)$$

$$\partial D/\partial c_1 = 90 c_1 - 42 c_2, \quad \partial D/\partial c_2 = -42 c_1 + 20 c_2$$

$$3(120EI/L^2)(2c_1 - c_2) - 3W(30c_1 - 14c_2) = 0$$

$$(120 EI/L^2)(-3c_1 + 2c_2) - 2W(-21c_1 + 10c_2) = 0$$

3.21 (cont'd)

Let $\alpha = EI/L^2$

$$(-240\alpha - 30W)c_1 + (-120\alpha + 14W)c_2 = 0$$

$$(-180\alpha + 21W)c_1 + (120\alpha - 10W)c_2 = 0$$

$c_1 \neq 0,\ c_2 \neq 0 \Rightarrow |\ \ | = 0$

$$|\ \ | = 0 \Rightarrow 0.2W^2 - 32\alpha W + 240\alpha^2 = 0$$

$$W = 2.5\alpha(32 \pm \sqrt{832}) = 2.5\alpha(32 \pm 28.84)$$

Relevant root (−): $W = 7.889\alpha = 7.88$

$$W = 7.889\, EI/L^2$$

b) % Error = $(7.889/7.837) - 1 = 0.0066 = 0.66\%$

15.22

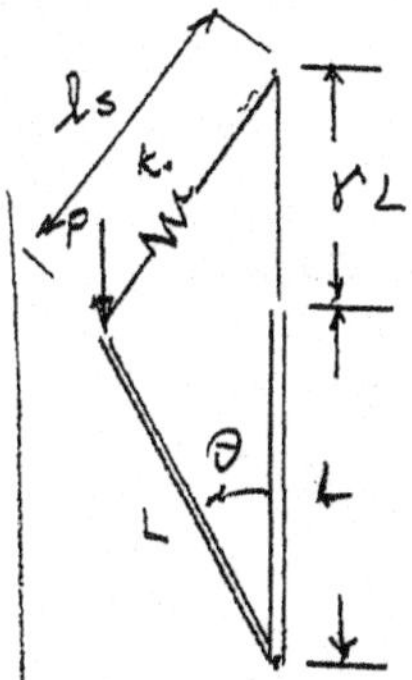

$$\ell_s^2 = (1+\gamma)^2L^2 + L^2 - 2(1+\gamma)L^2\cos\theta$$
$$= [2+2\gamma+\gamma^2 - 2(1+\gamma)\cos\theta]L^2.$$
$$\Delta = \ell_s - \gamma L = L[\gamma^2+2\gamma+2-2(1+\gamma)\cos\theta]^{1/2} - \gamma L$$
$$\Delta^2 = 2L^2\{\gamma^2+\gamma+1-(1+\gamma)\cos\theta - \gamma[\gamma^2+2\gamma+2-2(1+\gamma)\cos\theta]^{1/2}\}$$

$$\Pi = U+V = kL^2\{\gamma^2+\gamma+1-(1+\gamma)\cos\theta - \gamma[\gamma^2+2\gamma+2-2(1+\gamma)\cos\theta]^{1/2}\} - PL(1-\cos\theta).$$

$$d\Pi/d\theta = kL^2\{(1+\gamma)\sin\theta - \frac{\gamma}{2}[\gamma^2+2\gamma+2-2(1+\gamma)\cos\theta]^{-1/2}\cdot 2(1+\gamma)\sin\theta\} - PL\sin\theta$$

$$= \sin\theta\{\{kL^2\{1+\gamma-\gamma[\gamma^2+2\gamma+2-2(1+\gamma)\cos\theta]^{-1/2}(1+\gamma)\} - PL\}\}$$

At $\theta = 0, \pi$: all values of P satisfy equilib.

$\theta \neq 0, \pi$: $P = kL(1+\gamma)\{1-\gamma[\gamma^2+2\gamma+2-2(1+\gamma)\cos\theta]^{-1/2}\}$

STABILITY/INSTABILITY:

$$d^2\Pi/d\theta^2 = \cos\theta\{\{kL^2(1+\gamma)\{1-\gamma[\gamma^2+2\gamma+2-2(1+\gamma)\cos\theta]^{-1/2}\} - PL\}\}$$
$$+ kL^2\gamma(1+\gamma)^2[\gamma^2+2\gamma+2-2(1+\gamma)\cos\theta]^{-3/2}\sin^2\theta$$

At $\theta = 0$: $\left. d^2\Pi/d\theta^2 \right|_{\theta=0} = -PL < 0 \Rightarrow$ Always unstable for $P>0$ (down)
Stable for P up.

At $\theta = \pi$: $\left. d^2\Pi/d\theta^2 \right|_{\theta=\pi} = -\dfrac{2kL^2(1+\gamma)}{\gamma+2} + PL \Rightarrow$ STABLE FOR $P > \dfrac{2kL(1+\gamma)}{2+\gamma}$

$\theta \neq 0,\ \theta \neq \pi$: $\left. d^2\Pi/d\theta^2 \right|_{P_{eq}} = kL^2(1+\gamma)^2[\gamma^2+2\gamma+2-2(1+\gamma)\cos\theta]^{-3/2}\sin^2\theta > 0$ Always stable

PARTIAL RESULTS FOR PROBLEM 15.29

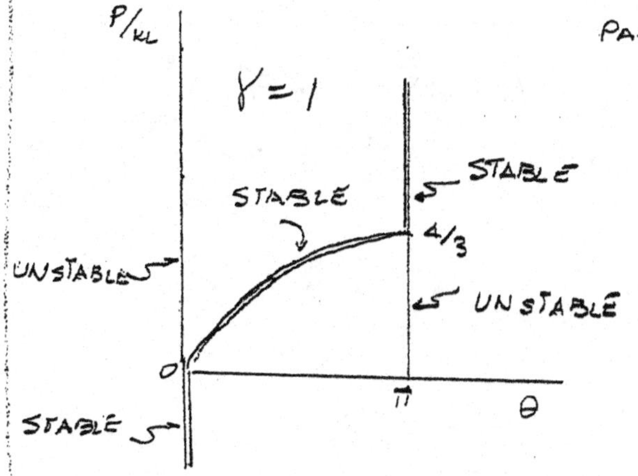

15.23

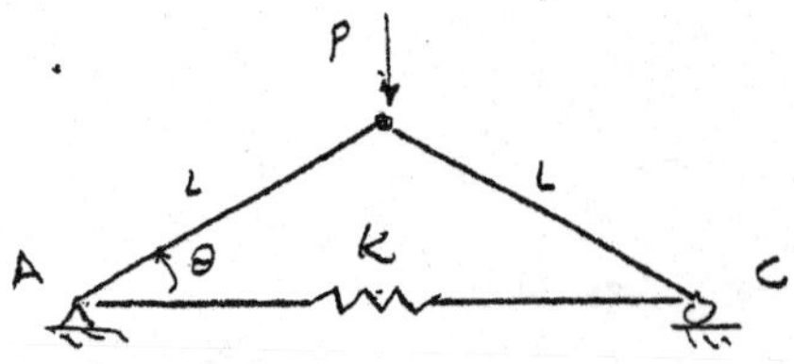

a) $U = \frac{k}{2}\cdot[2L\cos\theta - 2L\cos\theta_0]^2$

$= 2kL^2[\cos\theta - \cos\theta_0]^2$

$V = -PL(\sin\theta_0 - \sin\theta)$

$\Pi(\theta) = U+V = 2kL^2(\cos\theta - \cos\theta_0)^2 - PL(\sin\theta_0 - \sin\theta)$

$\delta\Pi(\theta) = \frac{d\pi}{d\theta}\delta\theta = 0 \; ; \; \delta\theta \neq 0 \Rightarrow d\pi(\theta)/d\theta = 0$

$d\pi/d\theta = -4kL^2(\cos\theta - \cos\theta_0)\sin\theta + PL\cos\theta.$

$P_{eq} = 4kL(\cos\theta - \cos\theta_0)\tan\theta$

b).

$d^2\pi/d\theta^2 = 4kL^2(\sin^2\theta - \cos^2\theta + \cos\theta_0\cos\theta) - PL\sin\theta.$

$d^2\pi/d\theta^2\Big|_{P_{eq}} = 4kL^2\{\sin^2\theta - \cos^2\theta + \cos\theta_0\cos\theta - (\cos\theta - \cos\theta_0)\sin^2\theta/\cos\theta\}$

Simplifying: $d^2\pi/d\theta^2\Big|_{P_{eq}} = \frac{4kL^2}{\cos\theta}(\cos\theta_0 - \cos^3\theta).$

$d^2\pi/d\theta^2 > 0 \Rightarrow \cos\theta_0 > \cos^3\theta \Rightarrow \cos\theta < (\cos\theta_0)^{1/3}$ stable

Note: $d^2\pi/d\theta^2 = 0$ represents limit of stable position.

$\cos\theta_{cr} = (\cos\theta_0)^{1/3}$

$P_{cr} = P_{eq}(\theta_{cr}) = 4kL\left[1 - \frac{\cos\theta_0}{\cos\theta_{cr}}\right]\sin\theta_{cr} = 4kL\left[1 - \frac{\cos\theta_0}{(\cos\theta_0)^{1/3}}\right](1-\cos^2\theta_{cr})^{1/2}$

$= 4kL[1 - (\cos\theta_0)^{2/3}][1 - (\cos\theta_0)^{2/3}]^{1/2}$

$P_{cr} = 4kL[1 - (\cos\theta_0)^{2/3}]^{3/2}.$

$\theta_0 = 30^\circ \quad P_{cr} = 4kL[1 - (3/4)^{1/3}]^{3/2} = 0.1106\,kL$

c) $|\theta_0| \ll 1 \quad \cos\theta_0 \cong 1 - \theta_0^2/2 \; ; \; (\cos\theta)^{2/3} = (1 - \theta_0^2/2)^{2/3} \doteq 1 - \frac{2}{3}(\theta^2/2) = 1 - \theta_0^2/3$

$\therefore P_{cr} = 4kL[1 - (1 - \theta_0^2/3)]^{3/2} = \frac{4kL\theta_0^3}{3\sqrt{3}} = \frac{4\sqrt{3}\,kL\theta_0^3}{9}$